REMEDIAL MATHEMATICS

G.K.RANGANATH
Formerly Professor of Mathematics
AES National College
Gauribidanur – 561 208

Himalaya Publishing House
ISO 9001 : 2015 CERTIFIED

First Edition : 2008
Second Edition : 2010
Third Edition : 2012
Edition : 2020
Edition : 2025

Published by : Mrs. Meena Pandey
for **Himalaya Publishing House Pvt. Ltd.,**
"Ramdoot", Dr. Bhalerao Marg, Girgaon, Mumbai - 400 004.
Phone: 022-23860170, 23863863; **Fax:** 022-23877178
E-mail: himpub@bharatmail.co.in; **Website:** www.himpub.com

Branch Offices :

New Delhi : "Pooja Apartments", 4-B, Murari Lal Street, Ansari Road, Darya Ganj, New Delhi - 110 002. Phone: 011-23270392, 23278631; Fax: 011-23256286

Nagpur : Kundanlal Chandak Industrial Estate, Ghat Road, Nagpur - 440 018. Mobile: 09325409992, 09325908881

Bengaluru : Plot No. 91-33, 2nd Main Road, Seshadripuram, Behind Nataraja Theatre, Bengaluru - 560 020. Phone: 080-41138821; Mobile: 09379847017, 09379847005

Hyderabad : No. 3-4-184, Lingampally, Besides Raghavendra Swamy Matham, Kachiguda, Hyderabad - 500 027. Phone: 040-27560041, 27550139

Chennai : No. 34/44, Motilal Street, T. Nagar, Chennai - 600 017. Mobile: 09380460419

Pune : "Laksha" Apartment, First Floor, No. 527, Mehunpura, Shaniwarpeth (Near Prabhat Theatre), Pune - 411 030. Phone: 020-24496323, 24496333; Mobile: 09370579333

Cuttack : Plot No 5F-755/4, Sector-9, CDA Markat Nagar, Cuttack - 753 014, Odisha. Mobile: 09338746007

Kolkata : 3, S.M. Bose Road, Near Gate No. 5, Agarpara Railway Station, North 24 Parganas, West Bengal - 700109. Mobile: 09674536325

DTP by : Page Designers, Bengaluru

Printed at : Geetanjali Press Pvt. Ltd., Nagpur. On behalf of HPH (P).

Preface to the First Edition

The present book, A Text Book of Remedial Mathematics, is designed to serve as a handbook for the students of I year B. Pharmacy of Jawaharlal Nehru Technological University, Hyderabad.

Atmost care has been taken to treate the subject matter in a lucid manner. Throughout the book numerous worked examples of different nature have been incorporated to help the students to understand various concepts practically. The book also contains exercises with problems at the end of each section, for the students to practice.

I hope that the students and teachers will find the book useful.

I express my sincere thanks to Sri Niraj Pandey, Director of Himalaya Publishing House Pvt. Ltd., for his efforts in publishing this book.

Thanks to Sri B.S. Madhu and Ms. Divya of Page Designers for excellent type setting of the text.

Suggestions for improvement of the book are welcome.

May 2008 **G.K. Ranganath**

Preface to the Third Edition

The present book, A Text Book of Remedial Mathematics, is designed to serve as a handbook for the students of I year B. Pharmacy of Jawaharlal Nehru Technological University, Hyderabad.

In this edition the chapter on Laplace Transformation is included.

Suggestions for improvement of the book are welcome.

G.K. Ranganath

Preface to the Second Edition

The present book, A Text Book of Remedial Mathematics, is designed to serve as a handbook for the students of I year B. Pharmacy of Jawaharlal Nehru Technological University, Hyderabad and Jawaharlal Nehru Technological University, Anantapur.

Atmost care has been taken to treate the subject matter in a lucid manner. Throughout the book numerous worked examples of different nature have been incorporated to help the students to understand various concepts practically. The book also contains exercises with problems at the end of each section, for the students to practice.

I hope that the students and teachers will find the book useful.

I express my sincere thanks to Sri Niraj Pandey, Director of Himalaya Publishing House Pvt. Ltd., for his efforts in publishing this book.

Thanks to Sri B.S. Madhu and Ms. Divya of Page Designers for excellent type setting of the text.

Suggestions for improvement of the book are welcome.

May 2010 **G.K. Ranganath**

Syllabus

JAWAHARLAL NEHRU TECHNOLOGICAL UNIVERSITY, HYDERABAD

I Year B. Pharmacy

	T	P	C
	4 + 1*	**0**	**8**

Unit I

Algebra

Arithmetic Progression-Geometric Progression : Permutations and Combinations – Binomial theorem – Partial Fractions – Matrices – Determinants – Applications of Determinants to Solve Simultaneous Equations (Cramer's Rule).

Unit II

Trigonometry : Trigonometric ratios and the relations between them sin $(A + B)$, cos $(A + B)$, tan $(A + B)$ formulae only. Trigonometric ratios of multiple angles – Heights and distances (simple 000 problems there on).

Unit III

Co-ordinate Geometry : Distances between points – Area of a triangle, Co-ordinates of a point dividing a given segment in a given ratio-locus-equation to a straight line in different forms – Angle between straight lines-point of intersection.

Unit IV

Differential Calculus : Continuity and Limit : Differentiation, derivability and derivative, R.H. derivatives and L.H. derivatives, Differentiation, General theorems of derivation.

Unit V

Derivatives of trigonometric functions (excluding inverse trigonometric and hyperbolic functions). Logarithmic differentiation. Partial differentiation, maxima and minima (elementary).

Unit VI

Integral Calculus : Integration as on inverse process of differentiation, Definite integrals, Integration by substitution, Integration by parts, Integration of algebraic function of e^x evolution of area in simple cases.

Unit VII

Differential Equations : Formation of a differential equation, Order and degree, Solution of first order differential equations.

Syllabus

JAWAHARLAL NEHRU TECHNOLOGICAL UNIVERSITY, ANANTAPUR

I Year B. Pharmacy	**Th**	**Pu**	**C**
	3	**1**	**6**

Unit I

Algebra

Arithmetic Progression-Geometric Progression : Binomial theorem – Partial fractions – Permutations and Combinations – Matrices – Basic matrix operations – Determinants – Applications of determinants to solve simultaneous equations (Cramer's rule and Cali-Hamilton's theorem).

Unit II

Trigonometry : Trigonometric ratios and the relations between them sin $((A + B)$, cos $(A + B)$, tan $(A + B)$ formulae only. Trigonometric ratios of multiple and sub-multiple angles – Heights and distances (simple problems) – Complex numbers and Demoivre's theorem.

Unit III

Co-ordinate Geometry : Distances between points – Area of a triangle, Co-ordinates of a point dividing a given segment in a given ratio – Locus equation to a straight line in different forms – Angle between straight lines-point of intersection – Circles – Conic sections.

Unit IV

Differential Calculus : Continuity and Limit : Differentiation, derivability and derivative – R.H. derivatives and L.H. derivatives – Differentiation – General theorems of derivation.

Unit V

Derivatives of trigonometric functions (excluding inverse trigonometric and hyperbolic functions) – Logarithmic differentiation – Partial differentiation – Maxima and minima (elementary) – Successive differentiation up to second order.

Unit VI

Integral Calculus : Integration as an inverse process of differentiation – Definite integrals – Integration by substitution – Integration by parts – Integration of algebraic function of e^x, Evolution of area and volumes in simple cases.

Unit VII

Differential Equations : Formation of a differential equation – Order and degree – Solution of first order differential equations.

Unit VIII

Applications of first order and first degree differential equation – Law of natural growth and decay – Newton's law of coding – Definition of linear differential equations for homogeneous – Non homogeneous – Second and higher order equation.

Contents

UNIT II – TRIGONOMETRY 83 - 157

UNIT III – ANALYTICAL GEOMETRY 159 - 257

UNIT IV & V – DIFFERENTIAL CALCULUS 259 - 313

UNIT I

Algebra

- **Chapter 1** – Arithmetic and Geometric Progression
- **Chapter 2** – Permutations and Combinations
- **Chapter 3** – Binomial Theorem
- **Chapter 4** – Partial Fractions
- **Chapter 5** – Matrices
- **Chapter 6** – Determinants

Chapter 1

Arithmetic and Geometric Progression

1.1 Introduction

In this chapter, we shall see the concepts of Arithmetic progression, geometric progression and corresponding results with illustrations.

1.2 Arithmetic Progression

Defintion : A sequence of numbers in which different elements are written by increasing (or decreasing) its previous element by the same quantity, is called an Arithmetic Progression (A.P.).

The different elements of an arithmetic progression can be written as

$$a, \quad a+d, \quad a+2d, \quad a+3d, \; \ldots\ldots\ldots$$

Here *'a'* is called the **first element** and *'d'* the fixed number by which the elements are increased or decreased, is called the **common difference**.

Now first element can be written as $a + 0 \cdot d$

second element is $a + 1 \cdot d$

third element is $a + 2 \cdot d$

fourth element is $a + 3 \cdot d$

....

....

tenth element is $a + 9 \cdot d$

....

We observe that the coefficient of d in any element is one less than the number of the element.

Thus the nth element or the general element of the Arithmetic progression with *'a'* as first element and *'d'* as common difference is $a + (n-1)d$.

The nth term of an arithmetic progression is denoted by T_n.

The following are few examples of A.P.

1. $$1, 3, 5, 7, \ldots\ldots$$

First element = $a = 1$, common difference = $d = 2$. $T_n = 2n - 1$.

2. $$15, 12, 9, 6, \ldots\ldots$$

Here first element = $a = 15$, common difference = $d = -3$.

$$T_n = 15 + (n-1)(-3) = 18 - 3n$$

3. $$2, \frac{7}{2}, 5, \frac{13}{2}, 8, \ldots\ldots$$

Here first element = $a = 2$, common difference = $d = \frac{3}{2}$.

$$T_n = 2 + (n-1)\frac{3}{2} = \frac{3n+1}{2}$$

Note : The common difference of an arithmetic progression can be obtained by taking the difference between any element and its previous element.

Example 1. Find the nth term and the 13th term of the A.P. 7, 10, 13, ...

Solution : In the given A.P., $a = 7$, $d = 10 - 7 = 13 - 10 = 3$

Now, n^{th} term $= T_n = a + (n-1)d = 7 + (n-1)3 = 3n + 4$

Now, $T_{13} = 3(13) + 4 = \mathbf{43}$

Example 2. Find the 10th term of the A.P., 2, 0, –2, –4, ...

Solution : Here, $a = 2$, $d = 0 - 2 = -2$

$\therefore$ $T_{10} = a + (10-1)d = 2 + 9(-2) = \mathbf{-16}$

Example 3. The first term of an A.P. is 6 and the common difference is 2, find the 15th term.

Solution : Here, $a = 6$, $d = 2$

$\therefore$ $T_{15} = a + (15-1)d = 6 + 14(2) = \mathbf{34}$

Example 4. Find the common difference of A.P. whose first term is 5 and 11th term is 25.

Solution : Let d be the common difference. By data $a = 5$ and 11th term = $T_{11} = 25$.

Now, $T_{11} = a + (11-1)d \Rightarrow 25 = 5 + 10d \Rightarrow \mathbf{d = 2}$

$\therefore$ the common difference is 2.

Example 5. Which term of A.P., 4, $5\frac{1}{3} - 4$, $6\frac{2}{3}$, ... is 104?

Solution : In the given A.P., $a = 4$, $d = 5\frac{1}{3} - 4 = \frac{16}{3} - 4 = \frac{4}{3}$

Let 104 be the nth term.

$\Rightarrow$ $T_n = 104 \Rightarrow a + (n-1)d = 104$

$$\Rightarrow 4 + (n-1)\frac{4}{3} = 104 \Rightarrow 4n - 8 = 312 \Rightarrow 4n = 304 \Rightarrow n = 76$$

$\therefore$ The 76th term is **104**.

Example 6. The 8th term and 20th term of an A.P are 22 and 46 respectively. Find the A.P and hence 18th term.

Solution : By data $T_8 = 22$ and $T_{20} = 46$.

$\Rightarrow$ $22 = a + 7d$ .... (1) and $46 = a + 19d$ .... (2)

Consider (2) – (1), we get $12d = 24 \Rightarrow d = 2$

Now, $a + 7d = 22 \Rightarrow a + 14 = 22 \Rightarrow a = 8$

Thus, the A.P is, 8, 10, 12, 14, ...

Now the 18th term is given by, $T_{18} = a + (18 - 1)d \Rightarrow T_{18} = 8 + 17\ (2) = \mathbf{42}$

Example 7. The 12th term of an A.P exceeds the 3rd term by 36. If the 16th term is 64. Find the A.P.

Solution : We have $T_{12} = a + 11d$ and $T_3 = 22 = a + 2d$

By data $T_{12} = T_3 + 36 \Rightarrow a + 11d = a + 2d + 36 \Rightarrow 9d = 36 \Rightarrow d = 4$

Now again by data $T_{16} = 64 \Rightarrow a + 15d = 64 \Rightarrow a + 60 = 64 \Rightarrow a = 4$

Thus, the A.P. is **4, 8, 12, 16, ...**

Example 8. Find the three numbers which are in A.P whose sum is 12 and the sum of their cubes is 408.

Solution : Let the three numbers be, $a - d,\ a,\ a + d$

These numbers are in A.P. whose common difference is d.

By data, $a - d + a + a + d = 12 \Rightarrow 3a = 12 \Rightarrow a = 4$

Again by data, $(a - d)^3 + a^3 + (a + d)^3 = 408 \Rightarrow 3a^3 + 6ad^2 = 408$

$\Rightarrow 192 + 24d^2 = 408 \quad (\because a = 4)$

$\Rightarrow 24d^2 = 216 \Rightarrow d^2 = 9 \Rightarrow d = \pm 3$

If $a = 4$, $d = 3$, the required numbers are **1, 4, 7**

If $a = 4$, $d = -3$, the required numbers are **7, 4, 1**

Observe that we get same three numbers (but different in order) for $d = 3$ and for $d = -3$.

Example 9. The sum of four numbers which are in A.P is 32 and the product of whose extremes is 55. Find the numbers.

Solution : Let the three numbers be, $a - 3d,\ a - d,\ a + d,\ a + 3d$

These four numbers are in A.P. whose common difference $2d$.

By data, $a - 3d + a - d + a + d + a + 3d = 32 \Rightarrow 4a = 32 \Rightarrow a = 8$

Again by data, $(a - 3d)(a + 3d) = 55 \Rightarrow a^2 - 9d^2 = 55$

$\Rightarrow 9d^2 = a^2 - 55$

$\Rightarrow 9d^2 = 64 - 55 \quad (\because a = 8)$

$\Rightarrow d^2 = 1 \Rightarrow d = \pm 1$

If $d = 1$, $a = 8$, the required numbers are **5, 7, 9, 11**

If $d = -1$, $a = 8$, the required numbers are **11, 9, 7, 5.**

Exercise

I.

1. Find the
   (a) 20th term of the A.P 2, 6, 10, ... **(b)** 10th term of the A.P 7, 10, 13, ...
   (c) 13th term of the A.P 2, 0, –2, –4, ... **(d)** 25th term of the A.P 0.3, 1, 1.7, ...
2. Is –300 a term of the A.P 10, 7, 4, ...

3. Which term of the A.P is5, 13, 21, ... is 181.
4. Find 6th, 8th and 17th term of the A.P whose n^{th} term is $4n - 3$.
5. Which term of the A.P 3, 7, 11, ... is 43.
6. Which term of the A.P $\frac{1}{6}, \frac{1}{2}, \frac{5}{6}$, ... is $31\frac{1}{2}$.
7. Find the common difference of an A.P whose first term is 5 and 11th term is 125.
8. Determine k, so that $\frac{2}{3}, k, \frac{5}{8}k$ are the three consecutive elements of A.P.

II.

1. The first term of A.P is 6 and the common difference is 2 find the 15th term.
2. If the first term of an A.P is 2, the 20th term is 59 find 32nd term.
3. The 10th term of an A.P is 2 and 16th term is –10 find the 11th term.
4. The third term of an A.P is 25 and the tenth term is –3. Find the first term and the common difference.
5. Find the 13th term of A.P in which the 6th term is –2 and the 9th term is –5.
6. Determine the 2nd term and the rth term of the A.P, whose 6th term is 12 and 8th term is 22.
7. If 7 times 7th term of an A.P is equal to 11 times its 11th term show that 18th term of A.P is zero.
8. If m times the mth term of an A.P is equal to n times its nth term, show that $(m + n)^{th}$ term of A.P is zero.
9. If p^{th} term of an A.P is q and the qth term is p show that r^{th} term is $p + q - r$.

III.

1. The sum of three numbers in A.P is 54 and the product of two extremes is 275, find them.
2. The sum of three numbers in A.P is 24 and their product is 440, find them.
3. The sum of three numbers in A.P is 9 and the sum of their squares is 77, find them.
4. The sum of the first four terms of an A.P is 16 and the sum of their squares is 84. Find them.
5. Each of the following A.P's

   3, 5, 7, ...

   and 4, 7, 10, ...

   is written upto 200 terms. Find how many terms are identical.
6. The fourth term of an A.P is equal to 3 times the first term and the seventh term exceeds twice the third term by 1. Find the A.P.
7. For what value of n, the nth term of the A.P 3, 10, 17, ... and 63, 65, 67, ... are equal.
8. If p^{th}, q^{th} and r^{th} terms of an A.P are x, y and z respectively show that
$$x (q - r) + y (r - p) + z (p - q) = 0$$
9. If a, b, c are in A.P show that $b + c, c + a, a + b$ are also in A.P.
10. If t_n denoted the nth term of an A.P and if $t_2 : t_4 = 3 : 7$, find the value of $t_6 : t_{11}$.
11. The ratio of 7th to the 3rd term of an A.P is 12 : 5. Find the ratio of the 13th to the 4th term.

Answers

I.	**1. (a)** 78	**(b)** 34	**(c)** –18	**(d)** 17.1			
	2. No	**3.** 23	**4.** 23, 29, 85	**5.** 11	**6.** 95	**7.** 12	**8.** $\frac{16}{33}$
II.	**1.** 34	**2.** 95	**3.** 0	**4.** 33, –4	**5.** –9	**6.** –8, 5*r* – 18	
III.	**1.** 11, 18, 25	**2.** 5, 8, 11	**3.** –2, 3, 8	**4.** 7, 3, 5, 3, 1			
	5. 66	**6.** 3, 5, 7, 9, ...	**7.** 13	**10.** 11 : 21		**11.** 10 : 3	

1.3 Sum to *n* terms of an Arithmetic Progression

In this section we shall find a formula for the sum of n terms of an arithmetic progression.

Consider the arithmetic progression 'd'. That is

$$a, \quad a + d, \quad a + 2d, \; ....$$

Let the nth term $a + (n - 1)d$ be denoted by 'l'. That is $l = a + (n - 1)d$.

Let S_n be the sum of the first n elements of the A.P.

Thus, $\qquad S_n = a + (a + d) + (a + 2d) + ... + (l - d) + l$

Also, $\qquad S_n = l + (l - d) + ... + (a + d) + a$

Adding the two expressions term by term, we get

$$2S_n = (a + l) + (a + l) + ... + (a + l) + (a + l)$$

Since there are n terms in the right hand side, we get

$$2S_n = n\,(a + l) \qquad \Rightarrow \qquad S_n = \frac{n}{2}\,(a + l)$$

But $\qquad l = T_n = a + (n - 1)d$

$\therefore \qquad S_n = \frac{n}{2}\,[2a + (n - 1)d]$

Thus, the sum to n terms of an arithmetic progression

$$a, a + d, a + 2d, + ... + a\,(n - 1)d = l, ...$$

is given by

$$\boldsymbol{S_n = \frac{n}{2}\,[2a + (n - 1)d] \qquad \text{or} \qquad S_n = \frac{n}{2}\,[a + l]}$$

Note : The first formula is useful, to find the sum of n terms of an A.P., if the first element and the common difference are known. Whereas the second formula is useful to find the sum to n terms, if the first element and the nth elements are known.

Example 1. Find the sum of 30 elements of the A.P. 1, 3, 5, 7, ...

Solution : Here $a = 1.\ d = 2$ and $n = 30$.

Now we have, $\qquad S_n = \frac{n}{2}\,[2a + (n - 1)d] = \frac{30}{2}\,[2 + (30 - 1)2] = 15 \times 60 = \mathbf{900}$

Example 2. Evaluate $\frac{1}{2}+\frac{1}{3}+\frac{1}{6}+$ ..... to 10 terms.

Solution : The terms of the given expression form an A.P. First element, $a=\frac{1}{2}$.

common difference $= d = \frac{1}{3}-\frac{1}{2}=-\frac{1}{6}$. By data $n = 10$.

$$S_n=\frac{n}{2}[2a+(n-1)d]=\frac{10}{2}\left[2\left(\frac{1}{2}\right)+(10-1)\left(-\frac{1}{6}\right)\right]=5\left[1-\frac{9}{6}\right]=-\mathbf{\frac{5}{2}}$$

Example 3. Evaluate 2 + 5 + 8 + ... + 152.

Solution : The terms of the given expression form an A.P, with first element 2 and common difference as 3. That is $a = 2$, $d = 3$. The last term $l = 152$.

Let $\quad l = T_n \Rightarrow 152 = a + (n-1)d$

$\Rightarrow 152 = 2 + (n-1)3 \Rightarrow 150 = 3n - 3 \Rightarrow 153 = 3n \Rightarrow n = 51$

Now, $\quad S_n = \frac{n}{2}(a+l) \Rightarrow S_{51} = \frac{51}{2}(2+152) = 51 \times 77 = \mathbf{3927}$

Example 4. How many terms of the A.P 15, 10, 5, ... should be added to make the sum –75.

Solution : For the A.P., $a = 15$, $d = -5$ and $S_n = -75$.

Now we have

Now, $\quad S_n = -75 \Rightarrow \frac{n}{2}[2a+(n-1)d] = -75 \Rightarrow n[30+(n-1)(-5)] = -150$

$\Rightarrow 5n^2 - 35n - 150 = 0$

$\Rightarrow n^2 - 7n - 30 = 0$

$\Rightarrow (n-10)(n+3) = 0$

$\Rightarrow n = 10$ or $n = -3$

Rejecting the negative value, we have $\boldsymbol{n} = \mathbf{10}$. Thus 10 terms must be added to make the sum –75.

Example 5. The first term of an A.P is 42 and the sum of the first five terms is 178.5. Find the fifth term.

Solution : By data $a = 4.2$ and $S_5 = 178.5$.

Now, $\quad S_5 = 178.5 \Rightarrow \frac{5}{2}[2a+(5-1)d] = 178.5 \Rightarrow \frac{5}{2}[8.4+4d] = 178.5$

$\Rightarrow 42 + 20d = 357$

$\Rightarrow 20d = 315 \Rightarrow d = 15.75$

Now, $\quad T_5 = a + 4d \Rightarrow T_5 = 4.2 + 4(15.75) \Rightarrow T_5 = 4.2 + 63 = 67.2$

Thus, the fifth term is **67.2**.

Second method : By data $a = 4.2$, $n = 5$ and $S_5 = 178.5$.

We have, $\quad S_5 = \frac{5}{2}(a+l) \quad (l = 5^{th}$ term)

$$\Rightarrow \quad 178.5 = \frac{5}{2}(4.2 + l) \quad \Rightarrow \quad 357 = 21 + 5l \quad \Rightarrow \quad l = 67.2$$

That is the 5th term = **67.2.**

Example 6. The sum of *n* elements of A.P. 25, 22, 19, 16, ... is 116. Find the number of terms and the last term.

Solution : The given A.P is 25, 22, 19, 16, ...

Here, $a = 25$ and $d = -3$

By data, $S_n = 116 \Rightarrow \frac{n}{2}[2a + (n-1)d] = 116 \Rightarrow n[50 + (n-1)(-3)] = 232$

$$\Rightarrow 3n^2 - 53n + 232 = 0$$

$$\Rightarrow (n-8)(3n-28) = 0$$

$$\Rightarrow \mathbf{n = 8} \text{ or } n = \frac{28}{3}$$

We reject the value $n = \frac{28}{3}$ as the number of terms is a positive whole number. Thus, $\mathbf{n = 8}$.

Now last term is given by

$$l = a + (n-1)d \quad \Rightarrow \quad l = 30 + 7(-3) \quad \Rightarrow \quad \mathbf{l = 4}$$

Example 7. Find the sum of all numbers between 100 and 1000 which are divisible by 13.

Solution : The first integer greater than 100 and divisible by 13 is 104. The last integer which is less than 1000 and divisible by 13 is 988. Thus, the required numbers are

104, 117, 130, ..., 988

These numbers form an A.P with $a = 104$ and $d = 13$.

Now, let $T_n = 988 \Rightarrow 104 + (n-1)13 = 988 \Rightarrow 13n + 91 = 988$

$$\Rightarrow 13n = 897$$

$$\Rightarrow n = 69$$

Now, $S_{69} = \frac{69}{2}[a + l] = \frac{69}{2}[104 + 988] = 69 \times 546 = \mathbf{37{,}674}$

Example 8. A person buys every year Bank's cash certificate of value exceeding the last year's purchase by 250. After 20 years, he finds that the total value of the certificates purchased by him is Rs. 72,500. Find the value of the certificates purchased by him (a) in the first year (b) in the 13th year.

Solution : (a) Let the value of the certificate purchased by the person in the first year be Rs. *a*. Let d = Rs. 250. By data $n = 20$ and $S_{20} = 72{,}500$

$$\therefore S_{20} = \frac{n}{2}[2a + (n-1)d] \quad \Rightarrow \quad 72{,}500 = \frac{20}{2}[2a + (20-1)\,250]$$

$$\Rightarrow 72{,}500 = 10[2a + 4750]$$

$$\Rightarrow 7{,}250 = 2a + 4750$$

$$\Rightarrow \quad 2a = 7250 - 4750$$

$$\Rightarrow \quad 2a = 2500 \quad \Rightarrow \quad a = 1250$$

$\therefore$ the value of the certificate purchased in the first year = **Rs. 1250**

(b) Here $a = 1250$, $d = 250$, $n = 13$

$\therefore$ the value of the certificate purchased in the 13th year

$= a + (n - 1)\, d = 1250 + 12\,(250) = 1250 + 3000 =$ **Rs. 4250.**

Example 9. A farmer buys a used tractor for Rs. 12000. He pays Rs. 6,000 cash and agrees to pay the balance in annual installments of Rs. 500 plus 12% interest on the unpaid amount. How much will the tractor cost him.

Solution : Total cost = Rs. 12000

Initial Payment = Rs. 6000

$\therefore$ The unpaid amount = Rs. 6000

Now, interest amount on Rs. 6000 for one year = Rs. $\dfrac{6000 \times 12}{100}$ = Rs. 720

Let the instalment paid at the end of nth year be I_n

Now, $$I_1 = \text{Rs.}\left[500 + \frac{6000 \times 12}{100}\right] = \textbf{Rs. 1220}$$

$$I_2 = \text{Rs.}\left[500 + \frac{5500 \times 12}{100}\right] = \text{Rs. } 1160$$

$$I_3 = \text{Rs.}\left[500 + \frac{5000 \times 12}{100}\right] = \text{Rs. } 1100$$

and so on.

$$C = \text{Last instalment} = \text{Rs.}\left[500 + \frac{500 \times 12}{100}\right] = \text{Rs. } 560$$

$\therefore$ cost of the tractor = 6000 + [1220 + 1160 + 1100 + ... + 560]

The terms within the brackets, form an A.P. with first element $a = 1220$ and the common difference $d = -60$, the number of terms is $n = 12$. Thus

$$C = 6000 + \frac{12}{2}[2(1220) + (12 - 1)(-60)]$$

$$= 6000 + 6[2440 - 660] = 6000 + 10{,}680 = \textbf{Rs. 16,680}$$

Exercise

I.

1. Find the sum of the following A.P.

(a) 1, 3, 5, 7, ... to 30th term

(b) 2, 6, 10, ... to 50th term

(c) −8, −3, 2, ... to 20th term

(d) 5.3, 3.9, 2.5, ... to 15th term

2. How many terms of the A.P.

   (a) 2, 8, 14, ... amount to 352? **(b)** 2, 5, 8, ... amount to 610?

   (c) 1, 4, 7, ... amount to 715?

II.

1. Find the sum of all natural numbers between 100 and 1000, which are multiples of 5.
2. Find the sum of all even integers from 92 to 768.
3. Find the sum of all integers between 200 and 800, which are divisible by 9.
4. Find the sum of all integers between 50 and 500, which are divisible by 7.
5. Find the sum of the first hundred even natural numbers divisible by 5.
6. How may terms are there in A.P. whose first and fifth terms are −14 and 2 respectively and the sum of the terms is 40?
7. Find the sum of n terms of an A.P. whose 7th term is 30 and 13th term is 54.
8. The sum of p terms of A.P. is $3P^2 + 4p$ find the nth term.

III.

1. In the following problems, three out of a, d, n, T_n and S_n (with usual notations) are given find the remaining.

   (a) $a = 3$, $d = 5$ $T_n = 63$ **(b)** $a = -5$, $d = 3$ $S_n = 470$

   (c) $d = 3$, $n = 5$ $T_n = 14$ **(d)** $a = 1$, $n = 12$ $S_n = 342$

   (e) $a = -\frac{3}{2}$, $n = \frac{39}{2}$ $S_n = 72$ **(f)** $d = \frac{3}{2}$, $d = 12$ $S_n = 141$

2. Show that if unity be added to the sum of any number of terms of the A.P. 8, 16, 24, ... the result will be the square of an odd number.
3. If 4 be added to the sum of any number of terms of the A.P, 21, 39, 57, ... show that the result will be the square of an even number.
4. An A.P. consists of 15 terms. The middle term is 20. Find the sum of all the terms.
5. In an A.P. the sum of p terms is q and the sum of q terms is p. Find the sum of $(p + q)$ terms.
6. If S_1, S_2, S_3 are the sums of n, $2n$, $3n$ terms respectively of an A.P. show that $S_3 = 3(S_2 - S_1)$.
7. Divide 20 into 4 parts which are in A.P. and such that the product of the first and fourth is to the product of the second and third in the ratio 2 : 1.
8. A manufacturer of radio sets produced 600 units in the third year and 700 units in the seventh year. Assuming the production uniformly increased by a fixed number every year, find **(i)** the production in the first year, **(ii)** total production in 7 years, **(iii)** the production in 10th year.
9. A man borrows Rs. 1000 and agrees to repay with a total of Rs. 140 in 12 installments, each instalment being less than the immediately preceding one by Rs. 10. What should be his first instalment.
10. A person pays Rs. 975 in monthly installments each instalment being less than the former by Rs. 5. The amount of the first instalment is Rs. 100. In what time will the entire amount be paid?

Answers

I. 1. **(a)** 900 **(b)** 5000 **(c)** 340 **(d)** −6.75

2. **(a)** 11 **(b)** 20 **(c)** 22

II. **1.** 98450 **2.** 145770 **3.** 32967 **4.** 17696

5. 505000 **6.** 10 **7.** $n(2n+2)$ **8.** $6n+1$

III. **1. (a)** $n = 13, S_n = 429$ **(b)** $n = 20, T_n = 52$ **(c)** $a = 2, S_n = 40$

(d) $d = 5, T_n = 56$ **(e)** $n = 8, T_n = 52$ **(e)** $a = \frac{7}{2}, T_n = 20$

4. 300 **5.** $-(p+q)$ **7.** 8, 6, 4, 2 **8.** 550, 4375, 775

9. 150 **10.** 15 months.

1.4 Geometric Progression

Definition : The geometric progression is defined as a sequence of numbers in which successive elements are written by multiplying (or dividing) its previous element by the same quantity.

The different elements of the sequence are $a, ar, ar^2, ar^3, \ldots.\ ar^{n-1}, \ldots$

Here ***a*** is called the **first element** and r is called the **common ratio** of the Geometric progression. (G.P.)

We can observe that the ratio of each element to its previous element is always same and equal to r. That

$$\frac{ar}{r} = \frac{ar^2}{ar} = \frac{ar^3}{ar^2} = \ldots.. = \frac{ar^{n-1}}{ar^{n-2}} = \ldots.. = r$$

Thus the Geometr ic progression can also be defined as below.

A Geometric progression is a sequence of numbers in which the ratio of every element to its previous element is a fixed constant. This fixed constant is called common ratio.

The element ar^{n-1} is called the nth element or the general term of the G.P. and it is denoted by T_n.

The following are examples of few Geometric progression.

1. 2, 4, 8, 16, ... $a = 2,\ r = 2,\ T_n = 2^n$

2. $6, -2, \frac{2}{3}, \frac{-2}{3}, \ldots..$ $a = 6,\ r = \frac{-1}{3},\ T_n = 6 \cdot \left(-\frac{1}{3}\right)^{n-1}$

Example 1. Find the 7th and 9th elements of G.P. $4, 1, \frac{1}{4}, \frac{1}{16}, \ldots.$

Solution : In the given G.P. $a = 4,\ r = \frac{1}{4}$

Now, $T_7 = a \cdot r^{7-1} = 4 \cdot \left(\frac{1}{4}\right)^6 = \frac{1}{4^5}$; $T_9 = a \cdot r^{9-1} = 4 \cdot \left(\frac{1}{4}\right)^8 = \frac{1}{4^7}$

Example 2. The 3rd and 6th element of a G.P. are 3 and 81 respectively, find the first element and common ratio of the G.P.

Solution : By data, $T_3 = 3$ and $T_6 = 81$ $\Rightarrow$ $a \cdot r^2 = 3$ and $a \cdot r^5 = 81$

Now, $\frac{a \cdot r^5}{a \cdot r^2} = \frac{81}{3}$ $\Rightarrow$ $r^3 = 27$ $\Rightarrow$ $r = 3$

Now, $a \cdot r^2 = 3 \Rightarrow 9a = 3$ $(\because r = 3)$

$\Rightarrow a = \frac{1}{3}$

Thus the G.P is $\mathbf{\frac{1}{3}, 1, 3, 9, \ldots\ldots}$

Example 3. Find the 9th element of the G.P. whose 4th term is 1 and 7th is $\frac{1}{9}$.

Solution : By data, $T_4 = 1$ and $T_7 = \frac{1}{8} \Rightarrow a \cdot r^3 = 1$ and $a \cdot r^6 = \frac{1}{8}$

Consider $\frac{a \cdot r^6}{a \cdot r^3} = \frac{1}{8} \Rightarrow r^3 = \frac{1}{8} \Rightarrow r = \frac{1}{2}$

Now, $a \cdot r^3 = 1 \Rightarrow \frac{a}{8} = 1 \Rightarrow a = 8$

Now, $T_9 = ar^8 \Rightarrow T_9 = 8\left(\frac{1}{2}\right)^8 \Rightarrow T_9 = \frac{1}{32}$

Thus the 9th element is $\mathbf{\frac{1}{32}}$.

Example 4. The third element of a G.P., is the square of the first and fifth element is 64, find the G.P.

Solution : By data, $T_3 = (T_1)^2$ and $T_5 = 64$

$\Rightarrow a \cdot r^2 = a^2$ and $a \cdot r^4 = 64$

$\Rightarrow r^2 = a$ and $a \cdot (r^2)^2 = 64$

Now $a \cdot (r^2)^2 = 64 \Rightarrow a \cdot a^2 = 64$ $(\because r^2 = a)$

$\Rightarrow a^3 = 64 \Rightarrow a = 4$

Now, $r^2 = a \Rightarrow r^2 = 4 \Rightarrow r = \pm 2$

Thus the G.P is 4, 8, 16, 32, ... or **4, –8, 16, –32, ...**

Example 5. If the three numbers $x + 9$, $x - 6$, 4 are the first three elements of a G.P., find x.

Solution : By data $a = x + 9$, $ar = x - 6$, $ar^2 = 4$

$\therefore \frac{ar}{a} = \frac{ar^2}{ar} \Rightarrow \frac{x-6}{x+9} = \frac{4}{x-6}$

$\Rightarrow (x-6)^2 = 4(x-6) \Rightarrow x^2 - 16x = 0 \Rightarrow x(x-16) = 0 \Rightarrow \mathbf{x = 0}$ or $\mathbf{x = 16}$

Example 6. The second, third and sixth elements of an A.P. are consecutive terms of a G.P. Find the common ratio of the G.P.

Solution : Let a be the first element and d be the common difference of the A.P. Then by data

$$a + d, \quad a + 2d, \quad a + 5d$$

are the second, third and sixth elements of G.P.

Thus, Common ratio $= \frac{a+2d}{a+d} = \frac{a+5d}{a+2d} = \frac{(a+2d)-(a+5d)}{(a+d)-(a+2d)} = \frac{3d}{d} = \mathbf{3}$

Example 7. If *a*, *b*, *c*, are in G.P. and $a^x = b^y = c^z$ prove that $\frac{1}{x} + \frac{1}{z} = \frac{2}{y}$.

Solution : By data *a, b, c* are in G.P.

$$\Rightarrow \quad \frac{b}{a} = \frac{c}{b} \quad \Rightarrow \quad b^2 = ac$$

Again by data $a^x = b^y = c^z \quad \Rightarrow \quad a = b^{y/x}, \; c = b^{y/z}$

Now, $b^2 = ac \quad \Rightarrow \quad b^2 = b^{y/x} \cdot b^{y/z}$

$$\Rightarrow \quad b^2 = b^{y/x + y/z} \quad \Rightarrow \quad 2 = \frac{y}{x} + \frac{y}{z} \quad \Rightarrow \quad \frac{1}{x} + \frac{1}{z} = \frac{2}{y}$$

Example 8. The sum of three numbers which are in G.P, is $\frac{42}{5}$ and their product is −8. Find the numbers.

Solution : Let the number be $\frac{a}{r}$, a, ar

These three numbers are in G.P.

By data, $\frac{a}{r} + a + ar = \frac{42}{5}$ and $\frac{a}{r} \cdot a \cdot ar = -8 \quad \Rightarrow \quad a^3 = -8 \quad \Rightarrow \quad a = -2$

$$\therefore \quad \frac{a}{r} + a + ar = \frac{42}{5} \quad \Rightarrow \quad \frac{-2}{r} - 2 - 2r = \frac{42}{5}$$

$$\Rightarrow \quad 10r^2 + 52r + 10 = 0$$

$$\Rightarrow \quad 5r^2 + 26r + 5 = 0$$

$$\Rightarrow \quad (r + 5)(5r + 1) = 0 \quad \Rightarrow \quad r = -5 \quad \text{or} \quad r = \frac{-1}{5}$$

Thus the required G.P is

$\frac{2}{5}$, **−2, 10** when $a = -2, \; r = -5$; **10, −2,** $\frac{2}{5}$ when $a = -2, \; r = \frac{-1}{5}$

Example 9. The three numbers whose sum is 18 are in A.P. If 2, 4 and 11 are added to them respectively, the resulting numbers are in G.P. Find the numbers.

Solution : Let the three numbers be $(a - d)$, a, $a + d$

By data $a - d + a + a + d = 18 \quad \Rightarrow \quad 3a = 18 \quad \Rightarrow \quad a = 6$

Thus the numbers are $6 - d, \; 6, \; 6 + d$

Again by data $6 - d + 2, \; 6 + 4, \; 6 + d + 11$

are in G.P.

i.e., $8 - d, \; 10, \; 17 + d$ are in G.P.

$$\Rightarrow \quad \frac{10}{8 - d} = \frac{17 + d}{10} \quad \Rightarrow \quad (17 + d)(8 - d) = 100$$

$$\Rightarrow \quad d^2 + 9d - 36 = 0$$

$$\Rightarrow \quad (d - 3)(d + 12) = 0 \quad \Rightarrow \quad d = 3 \quad \text{or} \quad d = -12$$

$\therefore$ the required numbers are **3, 6, 9** or **18, 6, −12**

Exercise

I. 1. Write

(a) the fifth element of G.P. 4, 12, 36, ... **(b)** the ninth element of G.P. 1, 4, 16, 64, ...

(c) the ninth element of G.P. $1, \frac{1}{2}, \frac{1}{4}, \frac{1}{8}$ ......

(d) the tenth element of G.P. $3, -1, \frac{1}{3}, \frac{-1}{9}, \frac{1}{27}$ ......

2. The second and fifth elements are respectively 3 and $\frac{81}{8}$ write down the common ratio and first element.

3. The 4th element of a G.P. is 9 and the 7th element is 72. Find the first element and common ratio.

4. The fourth element of a G.P. is 27 and 7th element is 729 find the G.P.

5. The first element of G.P. is 54 and the common ratio is $\frac{2}{3}$. Find its 10th element.

6. If $x - 2, x, x + 3$ are in G.P. find the value of x.

II. 1. The seventh element of a G.P. is 8 times the fourth element. Find the G.P. when its 5th element is 48.

2. The fifth element of G P 4 times the 3rd and the sum of the first two element is -4. Find the G.P.

3. The third element of a G.P. is the square of the first and fifth element is 64. Find the G.P.

4. The ratio of 9th element of a G.P. to the 6th element is -8 and 5th element is 16. Find the G.P.

5. The 10th element of a G.P. is double the 12th element. If the third element is 6, find the 5th element.

6. Which element of the G.P. $1, -2, 4, -8,$ ...... is 256

7. Which element of the G.P. $1, -2, 4, -8,$ ...... is 1024

8. Which element of the G.P. $2, 1, \frac{1}{2}, \frac{1}{4},$ ...... is $\frac{1}{128}$.

III. 1. Find the three numbers which are in G.P. if their sum is 28 and their product is 572.

2. The sum of the three numbers in G.P. is 31 and their product is 125. Find the numbers.

3. The sum of three numbers in G.P. is 70, if the two extremes be multiplied each by 4 and the middle by 5, the product are in A.P. Find the numbers.

4. The product of three numbers in G.P. is 27 and the sum of their products in pairs is 39, find the numbers.

5. Three numbers whose sum is 15 are in A.P. If 1, 4 and 19 are added to them respectively, results are in G.P. Determine the numbers.

6. If a, b, c are in A.P. and $a, b\ d$ are in G.P. then show that $a, \ a - b, d - c$ are in G.P.

7. If 4th, 7th and 10th elements of a G.P. are a, b, c show that $b^2 = ac$.

8. If $\frac{1}{x+y}, \frac{1}{2y}, \frac{1}{y+z}$ are three consecutive elements of A.P., show that x, y, z are three consecutive element of a G.P.

9. If a, b, c are in G.P. show that $(a + b), (b + c), (c + d)$ are also in G.P.

Answers

I. **1. (a)** 324 **(b)** 65536 **(c)** 256 **(d)** $\frac{-1}{6561}$

2. $a = \frac{9}{2}, r = \frac{2}{3}$ **3.** $a = 72, r = \frac{1}{2}$ **4.** $a = 729, r = \frac{1}{3}$ **5.** $54\left(\frac{2}{3}\right)^9$ **6.** $x = 6$

II. **1.** 1, 3, 6, 12, ... **2.** $-\frac{4}{3}, -\frac{8}{3}, -\frac{16}{3}$ ...... and 4, –8, 16, ... **3.** 4, 8, 16, ...

4. 1, –2, –4, –8, ... **5.** 3 **6.** 9 **7.** 11 **8.** 9

III. **1.** 4, 8, 16 **2.** 1, 5, 25 **3.** 10, 20, 40 **4.** 1, 3, 9 **5.** 2, 5, 8 or 26, 5, –16

1.5 Sum to *n* elements of a Geometric Progression

Consider the Geometric Progression

$$a, ar, ar^2, ... , a \cdot r^{n-1}, ...$$

Let S_n be the sum of the elements of the G.P.

i.e., $S_n = a + a \cdot r + a \cdot r^2 + ... + a \cdot r^{n-1}$

Now $r \cdot S_n = a \cdot r + a \cdot r^2 + ... + a \cdot r^{n-1} + a \cdot r^n$

Consider, $S_n - r \cdot S_n = a - a \cdot r^n \Rightarrow S_n(1 - r) = a(1 - r^n) \Rightarrow S_n = \frac{a(1 - r^n)}{1 - r}$

Thus, if *a* is the first element and *r* is the common ratio of the G.P., then the sum of *n* elements is

$$\boldsymbol{S_n = \frac{a(1 - r^n)}{1 - r}}$$

Example 1. Find the sum of *n* elements of the G.P 1, 4, 6, ...

Solution : Here $a = 1$, $r = 4$.

Now, $S_n = \frac{a(1 - r^n)}{1 - r} \Rightarrow S_n = \frac{1(1 - 4^n)}{1 - 4} = \frac{1}{3}(4^n - 1)$

Example 2. Find the sum of how many elements of the G.P 1, 3, 9, ... will be 9841?

Solution : Let the sum of n elements be 9841.

In the given G.P., $a = 1$, $r = 3$.

Now, $S_n = 9841 \Rightarrow \frac{a(1 - r^n)}{1 - r} = 9841$

$$\Rightarrow \frac{1 - 3^n}{1 - 3} = 9841$$

$$\Rightarrow \frac{3^n - 1}{2} = 9841 \Rightarrow 3^n - 1 = 19682 \Rightarrow 3^n = 19683 \Rightarrow \boldsymbol{n = 9}$$

Example 3. The sum of the first eight elements of G.P is five times the sum of the first four terms. Find the common ratio.

Solution : By data $S_8 = 5 \cdot S_4 \Rightarrow \dfrac{a(1-r^8)}{1-r} = 5 \cdot \dfrac{a(1-r^4)}{1-r}$

$$\Rightarrow \quad 1 - r^8 = 5(1 - r^4)$$
$$\Rightarrow \quad r^8 - 5r^4 + 4 = 0$$
$$\Rightarrow \quad (r^4 - 4)(r^4 - 1) = 0$$
$$\Rightarrow \quad r^2 = 2 \quad \text{or} \quad r^2 = 1$$
$$\Rightarrow \quad \boldsymbol{r = \pm\sqrt{2}} \quad \text{or} \quad \boldsymbol{r = \pm 1}$$

Example 4. The first and the last elements of a G.P, are respectively 3 and 768 and the sum is 1533. Find the common ratio and the number of terms.

Solution : Let T_n be the last element. Now by data $a = 3$, $T_n = ar^{n-1} = 768$ and $S_n = 1533$

Now, $T_n = ar^{n-1} = 768 \quad \Rightarrow \quad 3\,r^{n-1} = 768 \qquad (\because a = 3)$

$$\Rightarrow \quad r^{n-1} = 256$$

Now, $S_n = 1533 \quad \Rightarrow \quad \dfrac{a(1-r^n)}{1-r} = 1533$

$$\Rightarrow \quad \frac{3\left[1 - r \cdot r^{n-1}\right]}{1-r} = 1533$$

$$\Rightarrow \quad \frac{1 - 256r}{1-r} = 511 \qquad (\because r^{n-1} = 256)$$

$$\Rightarrow \quad 1 - 256\,r = 511\,r$$
$$\Rightarrow \quad 255\,r = 510 \quad \Rightarrow \quad r = 2$$

Now, $r^{n-1} = 256 \quad \Rightarrow \quad 2^{n-1} = 2^8 \quad \Rightarrow \quad n - 1 = 8 \quad \Rightarrow \quad \boldsymbol{n = 9}$

$\therefore$ The common ratio is **2** and the number of terms is **9**.

Example 5. Find the sum of *n* terms of 5 + 55 + 555 + ...

Solution : Let $S = 5 + 55 + 555 + \ldots\ldots$ to n terms

$\Rightarrow \quad S = 5\,[1 + 11 + 111 + \ldots\ldots$ to n terms$]$

$$= \frac{5}{9}\,[9 + 99 + 999 + \ldots\ldots \text{ to } n \text{ terms}]$$

$$= \frac{5}{9}\,[(10 - 1) + (100 - 1) + (1000 - 1) + \ldots\ldots \text{ to } n \text{ terms}]$$

$$= \frac{5}{9}\,[10 + 10^2 + \ldots. \text{ to } n \text{ terms}] + (1 + 1 + \ldots. + n \text{ terms})$$

$$= \frac{5}{9}\left[\frac{10\left(1-10^n\right)}{1-10} - n\right] \qquad (\because a = 10, r = 10)$$

$$= \frac{5}{9}\left[\frac{10\left(10^n - 1\right)}{9} - n\right]$$

Exercise

I. 1. Find the sum of the G.P

(a) $\sqrt{2}, \frac{1}{\sqrt{2}}, \frac{1}{2\sqrt{2}}$, ... to 8 element **(b)** $\frac{2}{9}, \frac{-1}{3}, \frac{1}{2}$,........, to 5 element

(c) 27, 9, 3, 1,..... to 8 element **(d)** $\sqrt{3}, 3, 3\sqrt{3}$, ... to 6 element

(e) $1, \frac{1}{2}, \frac{1}{4}$, ... to n terms.

II. 1. How many elements of the G.P. $\frac{2}{9}, -\frac{1}{3}, \frac{1}{2}$ must be taken amount to $\frac{55}{72}$.

2. How many elements of the G.P 1, 2, 4, 8, ... must be taken amount to 255?

3. The sum of first ten elements of a G.P is equal to 244 times the sum of the five elements. Find the common ratio.

4. Determine the third element of the G.P, whose common ratio is 3 and the sum of first 7 element is 2186.

5. Determine n, if $a = 3$, $T_n = 96$ and $S_n = 189$.

6. The sum of the first two element of G.P is 36 and the product of the first and the third element is 9 times the second element. Find the sum of first 8 elements.

7. Find the sum to n terms of

(a) 1 + 11 + 111 + 1111 + ... **(b)** 7 + 77 + 777 + 7777 + ...

(c) 2 + 22 + 222 + ... **(d)** 0.3 + 0.33 + 0.333 + 0.3333 + ...

(e) 0.5 + 0.55 + 0.555 + 0.5555 + ...

8. Find the value of $\sqrt{3\sqrt{3\sqrt{3\sqrt{3.........\text{to } \infty}}}}$.

9. In the following problems 3 elements out of a, r, n, T_n and S_n of G.P are given. Find the remaining two.

(a) $a = 2$, $n = 4$, $T_n = 16$ **(b)** $r = 3$, $n = 5$, $T_n = 162$

(c) $a = 12$, $r = \frac{2}{3}$, $S_n = \frac{76}{3}$

10. Find the G.P., which is such that the sum of the first two elements is 2 and the sum of first four elements is 20.

11. Four numbers are in G.P. If the product of the extremes is 243 and the sum of the middle two is 36 find the numbers.

12. If S_1, S_2, S_3 are the sums of n, $2n$, $3n$ elements respectively of a G.P show that,

$$S_1(S_3 - S_2) = (S_2 - S_1)^2.$$

13. The sum of n elements of a G.P. is $3 - \frac{3^{n+1}}{4^{2n}}$. Find the first term and common ratio.

Answers

I. **1. (a)** $\frac{255\sqrt{2}}{128}$ **(b)** $\frac{55}{72}$ **(c)** $\frac{3280}{82}$ **(d)** $39 + 13\sqrt{3}$ **(e)** $2\left(1 - \frac{1}{2n}\right)$

II. **1.** 5 **2.** 8 **3.** 3 **4.** 18 **5.** 6 **6.** $\frac{3280}{82}$

7. (a) $\frac{10\left(10^{n-1}\right)}{81}$ **(b)** $\frac{7}{9}\left[\frac{10}{9}\left(10^n - 1\right) - n\right]$ **(c)** $\frac{20}{81}\left(10^n - 1\right) - \frac{2n}{9}$

(d) $\frac{n}{3} - \frac{1}{27}\left(1 - (0.6)^n\right)$ **(e)** $\frac{5}{9}\left[n - \frac{1}{9}\left(1 - \frac{1}{10n}\right)\right]$

8. 3 **9. (a)** $r = 2,\ S_n = 30$ **(b)** $a = 2,\ S_n = 242$ **(c)** $n = 3,\ T_n = \frac{16}{3}$

10. $\frac{1}{2}, \frac{3}{2}, \frac{9}{2}$ **11.** 3, 9, 27, 81 **13.** $\frac{39}{16}, \frac{3}{16}$

Chapter 2

Permutations and Combinations

2.1 Introduction

Most of the times in our daily life as well in our study of mathematics, we come across the problem of selection of things and arrangement of the selected things, from the given set of things. The process of selecting things is called combination and that of arranging the selected things is called permutation. In this chapter we shall discuss the problems relating to combinations and permutations.

2.2 Factorial Notation

If n is a natural number, then we denote

$$1 \cdot 2 \cdot 3 \cdot 4 \ldots\ldots n \text{ by } n! \text{ or } \lfloor n,$$

and read as n factorial. We define $\mathbf{0\,! = 1}$.

That is product of n consecutive positive integer is denoted by n !. For example,

$$5! = 1 \cdot 2 \cdot 3 \cdot 4 \cdot 5 = 120, \qquad 7! = 1 \cdot 2 \cdot 3 \cdot 4 \cdot 5 \cdot 6 \cdot 7 = 5040 \text{ etc.}$$

Following results are useful.

1. $\boldsymbol{n! = n \cdot (n-1)!}$

Proof : $n! = 1 \cdot 2 \cdot 3 \cdot 4 \ldots\ldots (n-1) \cdot n$

$\Rightarrow$ $n! = [1 \cdot 2 \cdot 3 \cdot 4 \ldots\ldots (n-1)] \cdot n \quad \Rightarrow \quad \boldsymbol{n! = (n-1)! \cdot n}$

2. $\boldsymbol{n \cdot (n-1) \cdot (n-2) \ldots\ldots (n-r+1) = \dfrac{n!}{(n-r)!}}$.

Proof : Consider,

$$n \cdot (n-1)(n-2) \ldots\ldots (n-r+1) = n \cdot (n-1) \cdot (n-2) \ldots\ldots (n-(r-1))$$

$$= \frac{n \cdot (n-1) \cdot (n-2) \ldots\ldots \big(n-(r-1)\big)(n-r)\big(n-(r+1)\big) \ldots\ldots 3 \cdot 2 \cdot 1}{(n-r) \cdot \big(n-(r+1)\big) \ldots\ldots 3 \cdot 2 \cdot 1}$$

$$= \frac{\boldsymbol{n!}}{\boldsymbol{(n-r)!}} = \text{R.H.S.}$$

Example : $17 \times 16 \times 15 \times 14 \times \ldots\ldots \times 10 = \dfrac{17 \times 16 \times 15 \times 14 \times \ldots\ldots \times 10 \times 9 \times \ldots\ldots \times 1}{9 \times 8 \times 7 \times \ldots\ldots \times 1} = \dfrac{\mathbf{17!}}{\mathbf{9!}}$

2.3 Fundamental Principle

The principle that plays a very important role in discussing the problem relating to permutation and combination is, the fundamental principle. This we shall understand by the following illustration.

Let us suppose that there are 3 independent entrance gates and 4 independent exist gates for a football stadium. Now the question is how many ways can a person enter and leave the stadium.

Let the entrance gates be denoted by *A*, *B*, *C* and the exit gates by 1, 2, 3 and 4.

Now a person can enter the stadium in 3 ways - i.e., either by *A* or by *B* or by *C*. Let us suppose he enters by the gate *A*. Now to come out from the stadium he has 4 ways - i.e., he can come out either by gate 1 or 2 or 3 or 4. Thus corresponding to his "entry" by gate *A* there are 4 ways of "exit". Since there are 3 different entry gates, the number of ways that a person can enter and leave the stadium in $3 \times 4 = 12$ ways. This we can state it as,

"If there are 3 ways of doing one act (entry) and 4 ways of doing another act (exit), independent of the first, then both the acts (entry and exist) can be done together in 3×4 ways".

This is the essence of the fundamental principle. We shall state this principle in the generalised form.

A fundamental principle : If one event *A* can occur in *m* different ways and corresponding to each way of occurrence of event *A*, another event *B* (independent of *A*) can occur in *n* ways, then both the events *A* and *B* can occur together in $m \times n$ ways.

The above principle can be extended to the case in which the different operations can be performed in *m*, *n*, *p* ..... ways. In this case the number of ways of performing all the operations together would be $m \cdot n \cdot p$ .....

Example 1. There are three mathematics teachers in a college in which there are six classes, in how many different ways can they choose the classes, provided one teacher teaches one class only.

Solution : Any one of the teacher can select one class out of six classes. This can be done 6 ways. After this we are left with 2 teachers and 5 classes. Now any one of the two teachers can select one class out of 5 classes This can be done in 5 ways. Thus the first two teachers can select the classes together 6×5 ways. After this one teacher is left to select any one of the class out of 4 classes left. This can be done in 4 ways. Thus, the number of ways of choosing the classes by teachers in $6 \times 5 \times 4 = 120$ ways.

Example 2. A tennis club consists of 8 boys and 5 girls. In how many ways can a mixed doubles team be chosen?

Solution : To each boy one girl can be associated in 5 ways. But 8 boys can be selected in 8 ways. Hence a team of a boy and a girl can be selected in $8 \times 5 = 40$ ways.

2.4 Permutations and Combinations

Consider two letters *a* and *b*. Supposing we are asked to select two things out of the two letters. This can be done in only one way by selecting both *a* and *b* and putting it together as *ab*. If we are asked to arrange these two letters by taking both at a time, this we can do it by two ways by writing *ab* or *ba*. Thus, in the case of arrangement the order of appearance of *a* and *b* is important.

If we are given three letters say *a, b* and *c* and asked to select two letters at a time, this we can do it by writing

$$ab, \quad ac, \quad bc$$

That is 2 letters can be selected out of three letters in 3 ways. Now each of these selection made above, give rise to two arrangements - i.e., *ab* can be arranged in two ways as *ab* and *ba.* Thus the total number of arrangements of 3 letters taking two at a time, is 6, which are

$$ab,\ ba,\ ac,\ ca,\ bc,\ cb$$

Thus we have two types of classification of things **(i)** selection and **(ii)** arrangements.

Selection of things is called **combination** and the arrangements of things is called **permutation**.

We shall study these two concepts independently. First we shall see permutation then the combination.

The formal definition of permutation and combination are given below.

Definition (Permutation) : Each of the different arrangements which can be made by taking some or all of a number of things at a time is called a permutation.

The number of permutations of n different things, taken r at a time is denoted by nP_r.

Definition (Combination) : A group or selection that can be made by taking some or all of a number of things is called combination.

The number of combinations of n different things taken r at a time is denoted by nC_r.

2.5 Permutations

In this section we shall find the value of nP_r and few problems connected to permutations.

Theorem : The number of permutations of n things taken r at a time is

$$nP_r = n\,(n-1)\,(n-2)\,.......[n-(r-1)].$$

Proof : The number of permutations of n things taken r at a time is same as the number of ways of filling r blank spaces with n given things.

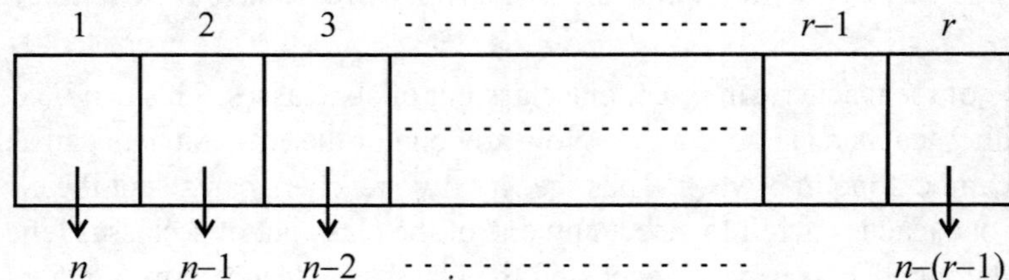

Now the first place can be filled by any one of the n things. Thus there are n ways of filling up the first place.

After filling up the first place by any one of the n things, we are left with $(n-1)$ things. Now the second place can be filled up in $(n-1)$ ways. Thus the first two places can be filled up together by $n \cdot (n-1)$ ways (by fundamental principle).

Now after filling up the first two places in any one of the above ways, we are left with $(n-2)$ things. Thus the third place can be filled up in $(n-2)$ ways.

Again by fundamental principle, the first three places can be filled up by $n\,(n-1)\,(n-2)$ ways.

Proceeding like this, the rth place can be filled up by $n-(r-1)$ ways.

Thus the number of ways of filling up r places by n things, together is given by

$$n \cdot (n-1) \cdot (n-2)\ .....\ [(n-(r-1)]$$

Thus the number of permutation of *n* things taken *r* at a time is given by

$$nP_r = n \cdot (n-1)\,(n-2)\,.....\,[n-(r-1)]$$

We know that the product of first n natural numbers is denoted by $n!$ or $\lfloor n$ and is read as "factorial n".

That is, $n! = 1 \cdot 2 \cdot 3 \,.....\, n$

$$4! = 1 \cdot 2 \cdot 3 \cdot 4 = 24\,, \qquad 6! = 1 \cdot 2 \cdot 3 \cdot 4 \cdot 5 \cdot 6 = 720$$

Further we have,

$$n! = n \cdot (n-1)! \quad \text{and} \quad (n-r+1) = (n-r+1) \cdot (n-r)!$$

Note : We define, $0! = 1$

Now we shall write the value of nP_r in factorial notation.

We have,

$$nP_r = n \cdot (n-1)\,(n-2)\,....\,[n-(r-1)]$$

$$= \frac{n \cdot (n-1)\,(n-2) .... [n-(r-1)] \cdot (n-r)\,[n-(r+1)] \,....\, 3 \cdot 2 \cdot 1}{(n-r)\,[n-(r+1)] \,.....\, 3 \cdot 2 \cdot 1} \qquad \text{(Note this step)}$$

$$= \frac{n \cdot (n-1) \cdot (n-2) .....3 \cdot 2 \cdot 1}{(n-r)\,[n-(r+1) .....3 \cdot 2 \cdot 1} = \frac{n!}{(n-r)!}$$

Thus the number of permutations of *n* things taken *r* at a time is given by

$$nP_r = \frac{n!}{(n-r)!}$$

We shall see more particular cases.

Case (i). Let $r = 0$. Then we have, $nP_0 = \dfrac{n!}{n!} = 1$

Case (ii). Let $r = n$. Then we have, $nP_n = \dfrac{n!}{0!} = n!$ $\quad (\because 0! = 1)$

Example 1. Find the values of $9P_4$, $7P_3$ and $5P_2$.

Solution : We have, $nP_r = \dfrac{n!}{(n-r)!} = n \cdot (n-1)\,(n-2)\,....\,[n-(r-1)]$

If $n = 9$ and $r = 4$, we have,

$$9P_4 = \frac{9!}{(9-4)!} = \frac{\cancel{1 \cdot 2 \cdot 3 \cdot 4 \cdot 5} \cdot 6 \cdot 7 \cdot 8 \cdot 9}{\cancel{1 \cdot 2 \cdot 3 \cdot 4 \cdot 5}} = 6 \cdot 7 \cdot 8 \cdot 9 = \mathbf{3024}$$

If $n = 7$ and $r = 3$, we have,

$$7P_3 = 7 \cdot 6 \cdot 5 = 210 \text{ or } 7P_3 = \frac{7!}{(7-3)!} = 5 \cdot 6 \cdot 7 = \mathbf{210}$$

If $n = 5$ and $r = 2$, we have,

$$5P_3 = 5 \cdot 4 = 20 \quad \text{or} \quad 5P_2 = \frac{5!}{(5-2)!} = 4 \cdot 5 = \mathbf{20}$$

Example 2. Find the value of n if (i) $nP_3 = 24$ and (ii) $nP_4 = 10 \cdot nP_3$.

Solution : (i) By data $nP_2 = 12$

$$\Rightarrow \quad n(n-1) = 12 \quad \Rightarrow \quad n^2 - n - 12 = 0 \quad \Rightarrow \quad (n-4)(n+3) = 0 \quad \Rightarrow \quad n = 4 \text{ or } n = 3$$

As n cannot be negative, we have $\boldsymbol{n = 4}$

(ii) By data, $nP_4 = 10 \cdot nP_3$.

$$\Rightarrow \quad n(n-1)(n-2)(n-3) = 10 \cdot n(n-1)(n-2)$$

We are considering nP_4 and nP_3, thus n cannot be 0, 1 or 2. Thus we can cancel, $n(n-1)(n-2)$ both sides.

$\therefore$ We have, $\quad n - 3 = 10 \quad \Rightarrow \quad \boldsymbol{n = 13}$

Example 3. Show that $10P_3 = 9P_3 + 3 \cdot 9P_2$.

Solution : We have, $\quad \text{L.H.S} = 10P_3 = 10 \cdot 9 \cdot 8 = 720$

Now, $\quad \text{R.H.S} = 9P_3 + 3 \cdot 9P_2 = 9 \cdot 8 \cdot 7 + 3 \cdot 9 \cdot 8 = 504 + 216 = \mathbf{720}$

Example 4. Find r if $10P_{r+1} : 11P_r = 30 : 11$.

Solution : By data we have,

$$\frac{10\,P_{r+1}}{11P_r} = \frac{30}{11} \quad \Rightarrow \quad \frac{10\,!}{[10-(r+1)]\,!} \cdot \frac{(11-r)\,!}{11\,!} = \frac{30}{11}$$

$$\Rightarrow \quad \frac{(11-r)\,!}{11 \cdot (9-r)!} = \frac{30}{11} \qquad \left(\because \frac{10\,!}{11\,!} = 11\right)$$

$$\Rightarrow \quad \frac{(11-r)\cdot(10-r)\cdot(9-r)\,!}{11\cdot(9-r)\,!} = \frac{30}{11} \qquad (\because n\,! = n \cdot (n-1)!)$$

$$\Rightarrow \quad (11-r)(10-r) = 30$$

$$\Rightarrow \quad r^2 - 21r + 80 = 0$$

$$\Rightarrow \quad (r-16)(r-5) = 0 \quad \Rightarrow \quad \boldsymbol{r = 16} \text{ or } \boldsymbol{r = 5}$$

As $r \le n$, $r = 16$ is rejected. Hence, $\boldsymbol{r = 5}$.

Example 5. If $5P_r = 2 \cdot 6\,P_{(r-1)}$ find r.

Solution : By data we have,

$$5P_r = 2 \cdot 6\,P_{(r-1)} \quad \Rightarrow \quad \frac{5\,!}{(5-r)\,!} = 2 \cdot \frac{6\,!}{[6-(r-1)]\,!}$$

$$\Rightarrow \quad \frac{(7-r)\,!}{(5-r)\,!} = \frac{2 \cdot 6\,!}{5\,!}$$

$$\Rightarrow \quad \frac{(7-r)\cdot(6-r)\cdot(5-r)\,!}{(5-r)\,!} = 12 \qquad \left(\because \frac{6\,!}{5\,!} = 6\right)$$

$$\Rightarrow \quad (7-r)(6-r) = 12 \quad \Rightarrow \quad r^2 - 13r + 30 = 0$$

$$\Rightarrow \quad (r-10)(r-3) = 0 \quad \Rightarrow \quad r = 10 \text{ or } r = 3$$

Since $n \ge r$, we reject $r = 10$, Hence, $\boldsymbol{r = 3}$.

Example 6. How many 4 digit numbers can be formed by using the digits 1, 2, 3, 7, 8, 9 repetitions not allowed.

(i) How many of these are less than 6000. **(ii) How many of these are even.**

(iii) How many of these are end with 7.

Solution : We have been given 6 different digits. The number of 4 digit number is the number permutations of 6 things taken 4 at a time. That is $6P_4$.

$\therefore$ the required number $= 6P_4 = 6 \cdot 5 \cdot 4 \cdot 3 = 360$

(i) By data the number to be formed should be less than 6000. Thus, the digit that occupies 1000th place must be less than 6. Thus, the 1000th place can be filled up by any of 1, 2 and 4. This can be done in 3 ways.

Now after filling up the 1000th place, we are left with 5 digits to fill up the remaining 3 places. The remaining 3 places can be filled up by $5P_3$ ways.

$\therefore$ Number of numbers less than 6000 $= 3 \times 5P_3 = 3 \cdot 5 \cdot 4 \cdot 3 = \mathbf{180}$.

(ii) Since the required number is an even, the unit place can take 2, 4 or 8. Thus the unit place can be filled up in 3 ways.

Now after filling up the unit place the remaining 5 digits by $5P_3$ ways.

Thus, the number of numbers which are even is given by, $3 \times 5P_3 = \mathbf{180}$.

(iii) In this case the unit place is fixed by 7. The remaining 3 places can be filled up by the remaining 5 digits in $5P_3$ ways.

$\therefore$ the required number $= 5P_3 = 5 \cdot 4 \cdot 3 = \mathbf{60}$.

Example 7. In how many ways 6 examination question papers, out of which two are of mathematics, can be arranged, so that the two mathematics papers never come together.

Solution : First let us regard the two mathematics papers together as one paper.

Now we have (6 – 1) papers which can be arranged in $5P_3$ ways - i.e., 5! ways.

Now these two mathematics papers wherever they are can be arranged in 2! ways.

$\therefore$ the number of ways the two papers are always together is $2! \times 6!$.

Now the number of ways of arranging of 6 papers, without any restrictions is $6P_6 = 6!$

$\therefore$ The number of ways that the two mathematics papers never together

= Total number of permutation – number of ways the two papers are always together

$= 61 - (2! \times 5!) = 720 - 240 = \mathbf{480}$

Example 8. In how many ways 3 boys and 5 girls can be arranged in a row so that

(i) no two boy are together **(ii) all the girls are together.**

Solution :

(i) Let us denote the boys by B_1, B_2 and B_3 and the girls are denoted by G_1, G_2, G_3, G_4 and G_5.

Since no two boys are to be together, first we shall arrange the girls in a row. This can be done in 5! ways.

Now the two boys can only be placed in the places marked $\times$ in the following arrangement.

$$\times G_1 \times G_2 \times G_3 \times G_4 \times$$

That is the three boys can be placed in any one of 6 places which are marked ×. This can be done in $6P_3$ ways.

Hence the total number of arrangements is given by

$$5! \times 6P_3 = 120 \times 6 \times 5 \times 4 = \mathbf{14400}.$$

(ii) We require all the 5 girls to be together. We shall consider 5 girls as one unit.

Now 3 boys and 1 unit of 5 girls can be arranged in 4! ways.

Again the 5 girls wherever they are can be arranged in 5! ways.

Thus, the required total number of ways = 4! × 5! = 24 × 120 = **2880**.

Example 9. How many five digit numbers can be formed with the digits 2, 3, 5, 7, 9 which lies between 30,000 and 90,000.

Solution : Since we require numbers which are greater than 30,000 and less than 90,000 the extreme left digit of the required number can take either of the digits 3, 5, 7. This place can be filled up in 3 ways.

The remaining four places are to be filled up with the remaining four digits. This can be done in 4! ways.

Hence, the number of numbers lying between, 30,000 and 90,000 = 3 × 24 = **72.**

Example 10. How many different ways can be letters of the word VOWEL be arranged?

(i) How many of these arrangements begin with *W*?

(ii) How many of these arrangements, in which the letters *O*, *E* occupy even places?

(iii) How many of these arrangements which begin with *V* and end with *L*?

(iv) In how many arrangements *A* and *E* are together?

Solution : The word VOWEL, has 5 different letters. These letters can be arranged in 5! ways.

∴ 5 different letters can be arranged in 5! = 120 ways.

(i) We shall fix the first letter by *W*. The remaining 4 places can be filled up with remaining 4 letters in 4! ways. Thus the required number is 4! = **24.**

(ii) There are two even places - i.e., 2nd and the 4th places. These two places can be filled by *O* and *E*, in 2 way.

The remaining 3 places can be filled up by remaining 3 letters by 3! ways.

∴ the required number = 2 × 3! = 2 × 6 = **12.**

(iii) The first and the last places are fixed with *V* and *L*. Now the remaining three places can be filled with the remaining 3 letters by 3! ways. Thus, the required number of arrangements is **6.**

(iv) Let us regard *A* and *E* together as one letter. The 4 remaining letters together with one unit of *A* and *E*, can be arranged in 4! ways. The letters *A* and *E* wherever they are can be arranged in 2 ways. Thus, the required number of ways is 4! × 2 = 24 × 2 = **48.**

Exercise

I.

1. Show that, **(a)** $nP_3 = 2 \cdot nP_{3-2}$ **(b)** $nP_n = nP_{n-1}$

2. Find the values of n, if

(a) $nP_4 = 360$ **(b)** $nP_5 = 42 \cdot nP_3$ $(n \geq 5)$ **(c)** $nP_3 = 8 \cdot nP_2$ **(d)** $2nP_3 = 100 \times nP_2$

3. Find the value of r if, **(a)** $11p_r = 12p_{r-1}$ **(b)** $10p_r = 2 \cdot 9P_r$ **(c)** $4 \cdot 6P_r = 6P_{r+1}$
4. In how many ways five girls be seated on a bench.
5. Ten students are participating in a race. In how many ways can the first three prizes be won?
6. How many words with or without meanings can be formed using all the letters of the word EQUATION, using each letter exactly once.
7. Four books, one each in chemistry, physics, biology and mathematics are to be arranged in a shelf. In how many ways this be done?
8. In how man ways 5 women draw water from 5 taps, if no tap remains unused.

II.

1. If there are six periods in each working day of a school in how many ways can one arrange 5 subjects such that each subject is allowed atleast one period?
2. How many different 5 letter words can be formed out of the letters of the word DELHI? How many of these will begin with *D* and end with *I*.
3. How many words can be formed from the letters of the word DAUGHTER so that **(i)** the vowels always come together **(ii)** the vowels are never together
4. How many different 6 digit number can be formed from the digits 4, 2, 5, 0, 6 and 7 if no digit be repeated? How many of these will have the digit 0 in the tenth place?
5. How many numbers greater than 100 and less than 10,000 be formed with the digit 1, 2, 3, 4, 5, 6 when there are no repetitions of the same digit in any number?
6. How many five digit numbers can be formed with 0, 1, 2, 3, 5 which are divisible by 5?
7. If the arrangements of the letters of the word NOVELTY, taken all at a time how many do not end with TY.

III.

1. How many even numbers greater than 10,000 can be formed with the digits 0, 1, 2, 3, 4 if no digit is repeated in the same number.
2. Find the sum of the five digit numbers that can be formed with the digits 1, 3, 5, 7, 9 no digit being repeated.
3. In how many ways 4 Hindi, 7 Kannada and 5 English books be arranged in a row so that
   (i) Kannada books are together
   (ii) the Hindi books are together and Kannada books are together
   (iii) no two English books are together.
4. Show that, **(i)** $nP_r = (n-1)P_r + r \cdot (n-1)P_r$ **(ii)** $nP_r = (n-r+1) \cdot nP_{r-1}$

Answers

I. **2.** **(a)** 6	**(b)** 10	**(c)** 12	**(d)** 13
3. **(a)** 9	**(b)** 5	**(c)** 2	**4.** 120
5. 720	**6.** 40320	**7.** 24	**8.** 120
II. **1.** 3600	**2.** 120, 6	**3.** **(i)** 4320	**(ii)** 3600
4. 720, 120	**5.** 480	**6.** 42	**7.** 4320
III. **1.** 60	**2.** 66, 66, 600		
3. **(i)** 10! · 7!	**(ii)** 7! · 7! · 4!	**(iii)** $12P_5 \times 11!$	

2.6 Combinations

A group or selection that can be made by taking some or all of a number of things is called a combination.

The number of combination of n different things taken r at a time is denoted by nC_r.

Theorem : The number of combinations of n different things taken r at a time is given by

$$nC_r = \frac{n!}{(n-r)! \cdot r!}$$

Proof : Now the required number of combinations is nC_r. Consider one such combination. This combination consists of r different things. These r different things can be arranged among themselves in $r!$ ways. Thus, one combination give rise to $r!$ permutations. But there are nC_r combinations. Thus,

the total number of permutations $= nC_r \times r!$

But the total number of permutations of n things taken r at a time is nP_r. Thus we have

$$nP_r = nC_r \times r! \quad \Rightarrow \quad nC_r = \frac{nP_r}{r!} \quad \Rightarrow \quad nC_r = \frac{n!}{(n-r)! \cdot r!} \qquad \left(\because nP_r = \frac{n!}{(n-r)!} \right)$$

Thus, the number of combinations of n things taken r at a time is

$$nC_r = \frac{n!}{(n-r)!\, r!}$$

To find the number of Combinations of n different things taken r at a time, from the first principles

Let the n different things be $a_1, a_2, \ldots\ldots a_n$.

The number of selections of r letters out of n letters in nC_r.

But each selection contain r letters. Thus the total number of letters in nC_r selection is $r \times nC_r$.

Now we shall find the total number of letters in nC_r in a different way.

We shall select the letter a_1. The number of ways of selecting $(r-1)$ things in the remaining $(n-1)$ things is $n-1\ C_{r-1}$.

Thus the letter a_1 appears $(n-1)\ C_{r-1}$ selections.

That is each letter appear in $n-1\ C_{r-1}$ combinations.

Thus the total number of letters in all letters in nC_r combinations is $n \times n-1\ C_{r-1}$.

$\therefore$ We have, $\quad r \times nC_r = n \times n-1\ C_{r-1} \Rightarrow nC_r = \frac{n}{r} \times n-1\ C_{r-1}$

Changing n to $n-1, n-2, n-3 \ldots.$ and r to $r-1, r-2, r-3 \ldots.$ successively, we get

$$nC_r = \frac{r}{n} \cdot \frac{n-1}{r-1} \cdot \frac{n-2}{r-2} \ldots\ldots n-r+1C_1$$

But, $\quad n-r+1\ C_1 = n-r+1$

$$\therefore \quad nC_r = \frac{r}{n} \cdot \frac{n-1}{r-1} \cdot \frac{n-2}{r-2} \ldots\ldots \frac{n-r+1}{1} = \frac{n(n-1)(n-2)\ldots\ldots(n-r+1)}{r \cdot (r-1) \cdot (r-2) \ldots\ldots}$$

$$\therefore \quad nC_r = \frac{n!}{r!\,(n-r)!}$$

Properties of nC_r

1. $nC_n = 1$ and $nC_0 = 1$.

Proof : We have, $$nC_r = \frac{n!}{(n-r)!\cdot r!}$$

Putting, $r = n$, we get, $$nC_n = \frac{n!}{0!\cdot n!} = \frac{n!}{n!} = 1 \qquad (\because 0! = 1)$$

Again putting $r = 0$, we get, $$nC_0 = \frac{n!}{n!\cdot 0!} = \frac{n!}{n!} = 1$$

2. $nC_r = nC_{n-r}$

Proof : We changing r to $(n-r)$ in nC_r, we get

$$nC_{n-r} = \frac{n!}{[n-(n-r)]!\,(n-r)!} = \frac{n!}{r!\cdot(n-r)!} = nC_r$$

Thus, $$nC_r = nC_{n-r}$$

3. If $nC_r = nC_k$ then either $r = k$ or $r + k = n$.

Proof : We have from the property (2),

$$nC_r = nC_k = nC_{n-k}$$

Now, $$nC_r = nC_k \Rightarrow r = k$$

and $$nC_r = nC_{n-k} \Rightarrow r = n-k \Rightarrow r+k = n$$

Thus either $$r = k \quad \text{or} \quad r+k = n$$

4. $nC_r + nC_{r-1} = n + 1C_r$

Proof : Consider,

$$nC_r + nC_{r-1} = \frac{n!}{r!\cdot(n-r)!} + \frac{n!}{(r-1)!\cdot(n-r+1)!}$$

$$= \frac{n!}{r\cdot(r-1)!\,(n-r)!} + \frac{n!}{(r-1)!\,(n-r+1)\cdot(n-r)!} \qquad [\because k! = k\cdot(k-1)!]$$

$$= \frac{[(n-r+1)+r]\cdot n!}{r\cdot(r-1)!\,(n-r+1)\cdot(n-r)!}$$

$$= \frac{(n+1)\cdot n!}{r!\,(n-r+1)!} = \frac{(n+1)!}{r!\,[(n+1)-r]} = (n+1)\,C_r$$

Hence, $$nC_r + nC_{r-1} = (n+1)\,C_r$$

Analytic proofs for (a) $nC_n = nC_{n-r}$ (b) $nC_{r-1} + nC_r = (n+1)\,C_r$.

(a) To prove $nC_r = nC_{n-r}$.

Proof : Supposing, r things are selected from n things. Then we are left with $(n-r)$ things.

$\Rightarrow$ the number of ways in which r things can be selected from n things is same as the number of ways in which $(n-r)$ things can be selected from n things.

$\Rightarrow$ $$nC_r = nC_{n-r}.$$

(b) To prove $nC_{r-1} + nC_r = (n + 1)\, C_r$

Proof : We shall consider $(n + 1)\, C_r$ in two ways.

(i) The number of combinations of $(n + 1)$ things taken r at a time in which one particular thing never occur.

This can be done in nC_r ways, by keeping aside the particular thing which never occur.

(ii) The number of combinations of $(n + 1)$ things taken r at a time in which one particular thing always occur.

This can be done nC_{r-1} ways, by first selecting $(r - 1)$ things from n things (keeping particular element aside) then including that particular thing which should always occur.

Hence, $$nC_r + nC_{r-1} = (n + 1)\, C_r.$$

Example 1. If $2nC_3 : nC_2 = 44 : 3$ find the value of n.

Solution : We have, $$2nC_3 : nC_2 = 44 : 3$$

$$\Rightarrow \quad \frac{(2n)!}{3!\,(2n-3)!} \cdot \frac{2!\,(n-2)!}{n!} = \frac{44}{3}$$

$$\Rightarrow \quad \frac{2n\,(2n-1)\,(2n-2)}{1 \cdot 2 \cdot 3} = \frac{2}{n\,(n-1)} = \frac{44}{3} \quad \Rightarrow \quad \frac{4\,(2n-1)}{3} = \frac{44}{3} \quad \Rightarrow \quad 2n - 1 = 11 \quad \Rightarrow \quad \boldsymbol{n = 6}$$

Example 2. Verify $2 \cdot 7C_4 = 8C_4$.

Solution : $$\text{L.H.S} = 2 \cdot \frac{7 \cdot 6 \cdot 5 \cdot 4}{1 \cdot 2 \cdot 3 \cdot 4} = 70$$

$$\text{R.H.S.} = \frac{8 \cdot 7 \cdot 6 \cdot 5}{1 \cdot 2 \cdot 3 \cdot 4} = 70 \qquad \therefore \quad \text{LHS} = \text{RHS}$$

Example 3. If $nC_{10} = nC_{12}$ determine n and hence nC_5.

Solution : By data, $$nC_{10} = nC_{12} \quad \Rightarrow \quad nC_{10} = nC_{n-12} \qquad (\because nC_r = nC_{n-r})$$

$$\Rightarrow \quad n - 12 = 10 \quad \Rightarrow \quad n = 22$$

Now, $$nC_5 = 22C_5 = \frac{22 \times 21 \times 20 \times 19 \times 18}{1 \cdot 2 \cdot 3 \cdot 4 \cdot 5} = \mathbf{26334}$$

Example 4. There are 6 boys and 3 girls in a class. An entertainment committee of 5 is to be selected such that there are 3 boys and 2 girls in the committee. In how many ways can the committee be selected? What is the number ways, if there is atleast one girl in the committee.

Solution : The number of ways of selecting 3 boys out of 5 is $6C_3$ and the number of ways of selecting 2 girls out of 3 is $3C_2$.

$$\text{Thus the required number} = 6C_3 \times 3C_2 = \frac{6 \cdot 5 \cdot 4}{1 \cdot 2 \cdot 3} \times \frac{3 \cdot 2}{1 \cdot 2} = 60$$

As there should be atleast one girl, the committee may consist of as follows.

(a)	4 boys	1 girl
(b)	3 boys	2 girls
(c)	2 boys	3 girls

Number of selection for (a) $= 6C_4 \times 3C_1 = \dfrac{6 \cdot 5 \cdot 4 \cdot 3}{1 \cdot 2 \cdot 3 \cdot 4} \times 3 = 45$

Number of selection for (b) $= 6C_3 \times 3C_2 = \dfrac{6 \cdot 5 \cdot 4}{1 \cdot 2 \cdot 3} \cdot \dfrac{3 \times 2}{1 \cdot 2} = 60$

Number of selection for (c) $= 6C_2 \times 3C_3 = \dfrac{6 \cdot 5}{1 \cdot 2} \times 1 = 15$

$\therefore$ total number of selections $= 45 + 60 + 15 =$ **120**

Example 5. In how many ways can a foot ball team of 11 players selected from 15 players. How many of these will

(i) include one particular player (ii) exclude one particular player.

Solution : The number of was of selecting 11 players out of 15 players is

$$15C_{11} = 15C_{15-11} = 15C_4 = \frac{15 \cdot 14 \cdot 13 \cdot 12}{1 \cdot 2 \cdot 3 \cdot 4} = 1365$$

(i) If one player is always included, then we have to select 10 players out of remaining 14 players. This can be done in $14C_{10}$ ways.

$$\therefore \quad 14C_{10} = 14C_{14-10} = 14C_4 = \frac{14 \times 13 \times 12 \times 11}{1 \times 2 \times 3 \times 4} = \mathbf{1001}$$

(ii) If one player is always excluded, we have to select 11 players out of out 14 players. This can be done in $14C_{11}$ ways.

$$\therefore \quad 14C_{11} = 14C_{14-11} = 14C_3 = \frac{14 \times 13 \times 12}{3 \times 2 \times 1} = \mathbf{364}$$

Example 6. There are 15 points in a plane of which 5 are Collinear. Find the number of (a) straight lines (b) triangles which can be formed from these points.

Solution : (a) Since we are given 5 collinear points, these 5 points give only one line.

Now the number of line with 15 points is given by $15C_2$.

$$\therefore \quad \text{Number of lines} = 15C_2 - 5C_2 + 1 = \frac{15 \times 14}{1 \cdot 2} - \frac{5 \times 4}{1 \cdot 2} + 1 = 105 - 10 + 1 = \mathbf{95}$$

(b) We need three points to form a triangle. Thus the total number of triangles would be $15C_3$, provided none of the 3 points are collinear.

But 5 points are collinear, we would not get $5C_3$ triangles from these points.

$$\therefore \quad \text{Number of triangles} = 15C_3 - 5C_3 = \frac{15 \times 14 \times 13}{1 \times 2 \times 3} - \frac{5 \times 4 \times 3}{1 \times 2 \times 3} = \mathbf{445}$$

Exercise

I. **1.** If $nC_8 = nC_{12}$ then find nC_{18}.

2. If $nC_2 = 105$ find n.

3. If $nC_{20} = nC_8$ find nC_{26}.

4. If $nC_3 : (2n - 1)C_2 = 8 : 15$ find n.

II. **1.** In how many ways can one select 4 numbers from the set of numbers 1, 5, 7, 3, 9, 8.

2. If the number combination of n things taken atleast one at a time is 225 find n.

3. In how many ways can 5 red and 4 white balls be drawn from a bag containing 10 red and 8 white balls?

4. In how many different selections of 4 books can be made from 10 different books if two particulars books are always selected?

5. Find the number of diagonals of a polygon of n sides.

6. If a convex polygon has 170 diagonals what is the number of sides of the polygon.

7. In how many ways can 5 members forming a committee out of 10 be selected so that

(i) two particular members must be included

(ii) two particular members must be excluded

III. **1.** An examination paper consists of 12 questions divided into parts A and B. Part A contains 7 questions and part B contains 5 questions. A candidate is required to attempt 8 questions selecting at least 3 from each part. In how many ways can the candidate select the questions.

2. How many different cricket teams of 11 players can be selected from 14 players of which only two can play as wicket keeper? Given each team must have exactly one wicket keeper.

3. A team of eleven is to chosen out of 16 cricket players of whom 4 are bowlers and 2 others are wicket keepers. In how many ways can the team be chosen so that

(i) there are exactly 3 bowlers and one wicket keeper

(ii) there are atleast 3 bowlers and atleast one wicket keeper.

4. A Committee of 6 has to be formed out of 7 English men and 4 Americans. In how many ways can this be done if the committee contains atleast two Americans?

5. From a class of 12 boys and 10 girls, 10 students are to be chosen for a competiton, at least including 4 boys and 4 girls. The 2 girls who won th eprizes last year should be included. In how many ways can be the selections be made.

6. A boy has 3 library tickets and 8 books of his interest in this library. Of these 8 books, he does not want to borrow chemistry part II, unless Chemistry Part I is also borrowed. In how many ways can he choose the three books to be borrowed.

Answers

I. **1.** 190 **2.** 15 **3.** 378 **4.** 8

II. **1.** 15 **2.** 8 **3.** $10C_5 \times 8C_4$ **4.** 28 **5.** $\dfrac{n(n-3)}{2}$ **6.** 20 **7.** 56, 56

III. **1.** 420 **2.** 132 **3.** 960, 2472 **4.** 371 **5.** 104874 **6.** 35

Chapter 3

Binomial Theorem

3.1 Introduction

In this chapter we shall see a very useful and important theorem called Binomial theorem, which gives us a formula for the nth power of $(x + a)$. An expression consisting of two terms is called binomial.

This theorem was first established by **Sir Issac Newton.**

3.2 Binomial Theorem

We mention the statement of binomial theorem when the index is a positive integer, without proof.

Theorem (Binomial theorem for positive integral power)

If n is a positive integer, then,

$$(x + a)^n = nC_0 x^n + nC_1 x^{n-1} \cdot a + nC_2 x^{n-2} \cdot a^2 + ... + nC_n a^n$$

We observe the following common features in the binomial expansion.

1. The power of x in each term goes on decreases by one and where as the power of a goes on increases by one.
2. The number of terms in the expansion is one greater than the index of $(x + a)$. That is $n + 1$.
3. The power of x in any term is equal to difference of upper and lower suffixes of C. That is in third term the power of x is $(n - 2)$ which is the difference of n and 2 of nC_2.
4. The power of a in any term is same as lower suffix of C. For example the power of a in third term is 2, and suffix of nC_2 is 2.
5. The sum of the powers of x and a in each term is equal to the index of left side i.e., n.

Observing these common properties of different terms of the binomial expansion we shall write the expression for $(r + 1)^{th}$ term of the expansion, denoted by T_{r+1}.

$$T_{r+1} = nC_r \cdot x^{n-r} \cdot a^r$$

This term is called the general term of the binomial expansion.

The general term (i.e., $(r + 1)^{th}$ term) in the expansion of $(x + a)^n$ is given by

$$T_{r+1} = nC_r \cdot x^{n-r} \cdot a^r$$

By putting $r = 0, 1, 2, 3, \ldots\ldots, n$, we get different terms of the binomial expansion.

Note : Since $nC_r = nC_{n-r}$, the coefficients of the terms in the expansion, which are equidistant from the beginning and the end are equal.

Example 1. Write the Expansion of $(3 - 2x)^5$

Solution : Here $n = 5$, $x = 3$, $a = -2x$.

$$\therefore\ [3 + (-2x)]^5 = 3^5 + 5C_1 \cdot 3^4 \cdot (-2x) + 5C_2 \cdot 3^3 \cdot (-2x)^2 + 5C_3 \cdot 3^2 \cdot (-2x)^3 + 5C_4 \cdot 3 \cdot (-2x)^4 + 5C_5 \cdot (-2x)^5$$

Now, $5C_o = 1;\ 5C_1 = 5;\quad 5C_2 = \dfrac{5 \cdot 4}{1 \cdot 2} = 10;\quad 5C_3 = \dfrac{5 \cdot 4 \cdot 3}{1 \cdot 2 \cdot 3} = 10$

$$5C_4 = 5C_{5-4} = 5C_1 = 5\ ;\ 5C_5 = 1$$

Thus, $(3 - 2x)^5 = 243 - 5 \cdot 81 \cdot 2x + 10 \cdot 27 \cdot 4x^2 - 10 \cdot 9 \cdot 8x^3 + 5 \cdot 3 \cdot 16x^4 - 32x^5$

$\mathbf{(3 - 2x)^5 = 243 - 810x + 1080x^2 - 720x^3 + 240x^4 - 32x^5}$

Example 2. Find the eleventh term in the expansion of $(2x - y)^{11}$

Solution : We have $'x' = 2x$, $a = -y$, $n = 11$.

We have the general term, $T_{r+1} = nC_r \cdot x^{n-r} \cdot a^r$

Now, putting $r = 10$, we get the 11th term

i.e., $T_{11} = 11C_{10} \cdot (2x)^{11-10} \cdot (-y)^{10} \Rightarrow T_{11} = 11C_{11-10}\ 2x \cdot y^{10}$

$\Rightarrow T_{11} = 11C_1 \cdot 2x \cdot y^{10}$

$\Rightarrow T_{11} = 11 \cdot 2x \cdot y^{10} = \mathbf{22xy^{10}}$

Example 3. Write the tenth term in the expansion of $\left(2x^2 + \dfrac{1}{x}\right)^{12}$.

Solution : We have $n = 12$, $'x' = 2x^2$, $a = \dfrac{1}{x}$.

We have the general term, $T_{r+1} = nC_r \cdot x^{n-r} \cdot a^r$

Now, putting $r = 9$, we get the 10th term T_{10}

i.e., $T_{10} = 12C_9 \cdot (2x^2)^{12-9} \cdot \left(\dfrac{1}{x}\right)^9 \Rightarrow T_{10} = 12C_{12-9}\ (2x^2)^3 \cdot \dfrac{1}{x^9}$

$\Rightarrow T_{10} = 12C_3 \cdot 8x^6 \cdot \dfrac{1}{x^9}$

$\Rightarrow T_{10} = \dfrac{12 \cdot 11 \cdot 10}{1 \cdot 2 \cdot 3} \cdot 8 \cdot \dfrac{1}{x^3} = \mathbf{1760 \cdot \dfrac{1}{x^3}}$

Example 4. Find the coefficient of x^7 in the expansion of $\left(x^2 + \dfrac{2}{x}\right)^{11}$.

Solution : We have $n = 11$, $'x' = x^2$, $a = \dfrac{2}{x}$.

Let us assume that x^7 occurs in $(r + 1)^{th}$ term

we have, $T_{r+1} = nC_r \cdot x^{n-r} \cdot a^r \Rightarrow T_{r+1} = 11C_r \cdot (x^2)^{11-r} \cdot \left(\frac{2}{x}\right)^r$

$$\Rightarrow T_{r+1} = 11C_r \cdot x^{22-2r} \cdot \frac{2^r}{x^r} \Rightarrow T_{r+1} = 11C_r \cdot 2^r \cdot x^{(22-3r)}$$

By data the power of x must be 7. Thus we must have

$$22 - 3r = 7 \Rightarrow -3r = -15 \Rightarrow \boldsymbol{r = 5}$$

Thus T_6 contains x^7 and therefore the coefficient of x^7 can be obtained by putting $r = 5$ in $11C_r \cdot 2^r$.

i.e., coefficient of $x^7 = 11C_5 \cdot 2^5 = 32 \cdot \left(\frac{11 \cdot 10 \cdot 9 \cdot 8 \cdot 7}{1 \cdot 2 \cdot 3 \cdot 4 \cdot 5}\right) = \mathbf{14784}.$

Example 5. Find the coefficient of $a^6 \cdot b^3$ in the expansion of $\left(2a - \frac{b}{3}\right)^9$.

Solution : Here $n = 9$, $x = 2a$, $'a' = -\frac{b}{3}$.

Let the $(r + 1)^{\text{th}}$ term contains $a^6 \cdot b^3$. Now,

$$T_{r+1} = nC_r \cdot x^{n-r} \cdot a^r \Rightarrow T_{r+1} = 9C_r \cdot (2a)^{9-r} \cdot \left(-\frac{b}{3}\right)^r$$

$$\Rightarrow T_{r+1} = 9C_r \cdot 2^{9-r} \cdot \left(-\frac{1}{3}\right)^r \cdot a^{9-r} \cdot b^r$$

By data the power of a must be 6 and power of b must be 3.

$\Rightarrow$ $(9 - r = 6 \text{ and } r = 3) \Rightarrow r = 3$

Thus $(3 + 1) = 4^{\text{th}}$ term contains $a^6 \cdot b^3$. The coefficient of $a^6 \cdot b^3$ is given by

$$9C_3 \cdot 2^{9-3} \cdot \left(-\frac{1}{3}\right)^3 = 9C_3 \cdot 2^6 \cdot (-1)^3 \cdot \frac{1}{27} = -\left(\frac{9 \cdot 8 \cdot 7}{1 \cdot 2 \cdot 3}\right)\frac{64}{27} = -\left(\mathbf{\frac{1792}{9}}\right)$$

Example 6. Find the term containing x^3 if any in $\left(3x - \frac{1}{2x}\right)^8$.

Solution : Here $n = 8$, $'x' = 3x$, $a = -\frac{1}{2x}$.

Let the $(r + 1)^{\text{th}}$ term contain x^3. Now,

$$T_{r+1} = nC_r \cdot x^{n-r} \cdot a^r \Rightarrow T_{r+1} = 8C_r \cdot (3x)^{8-r} \cdot \left(-\frac{1}{2x}\right)^r$$

$$\Rightarrow T_{r+1} = 8C_r \cdot 3^{8-r} \cdot (-1)^r \cdot x^{8-r} \cdot \left(\frac{1}{2}\right)^r \cdot \left(\frac{1}{x}\right)^r$$

$$\Rightarrow T_{r+1} = 8C_r \cdot (-1)^r \cdot 3^{8-r} \cdot \left(\frac{1}{2}\right)^r \cdot x^{8-2r}$$

By data the power of x must be 3.

$$\Rightarrow \quad 8 - 2r = 3 \quad \Rightarrow \quad -2r = -5 \quad \Rightarrow \quad r = \frac{5}{2} \neq \text{integer}$$

Thus **there exists no term containing x^3 in the expansion.**

Example 7. Find the middle term in the expansion of $\left(2x - \dfrac{1}{y}\right)^8$.

Solution : We know that in the expansion of $(x + a)^n$, there are $(n + 1)$ terms.

Thus in the expansion of $\left(2x - \dfrac{1}{y}\right)^8$ there are 9 terms, therefore the middle term is 5th term.

[Middle term is $\left(\dfrac{n}{2}+1\right)^{th}$ term].

Here $n = 8$, $'x' = 2x$, $a = -\dfrac{1}{y}$. Now $(r + 1)^{th}$ term is given by,

$$T_{r+1} = nC_r \cdot x^{n-r} \cdot a^r$$

Putting $r = 4$ we get 5th term,

$$\Rightarrow \quad T_5 = 8C_4 \cdot (2x)^{8-4} \cdot \left(-\frac{1}{y}\right)^4 \quad \Rightarrow \quad T_5 = 8C_4 \cdot 2^4 \cdot x^4 \cdot (-1)^4 \cdot \frac{1}{y^4}$$

$$\Rightarrow \quad T_5 = 2^4 \cdot \frac{8 \cdot 7 \cdot 6 \cdot 5}{1 \cdot 2 \cdot 3 \cdot 4} \cdot \left(\frac{x}{y}\right)^4 = \mathbf{1120\left(\frac{x}{y}\right)^4}$$

Example 8. Find the middle terms in the expansion of $\left(\dfrac{2}{3}x^2 - \dfrac{3}{2x}\right)^{11}$

Solution : Here $n = 11$. Thus the expansion will have 12 terms. Therefore there are two middle terms, which are 6th and 7th terms. [middle terms are $\left(\dfrac{n+1}{2}\right)^{th}$, $\left(\dfrac{n+1}{2}+1\right)^{th}$ terms]

Now $(r + 1)^{th}$ term is given by, $T_{r+1} = nC_r \cdot x^{n-r} \cdot a^r$

We have, $n = 11$, $'x' = \dfrac{2}{3}x^2$, $a = -\dfrac{3}{2x}$

Putting $r = 5$ we get 6th term

$$\Rightarrow \quad T_6 = 11C_5 \cdot \left(\frac{2}{3}x^2\right)^{11-5} \cdot \left(-\frac{3}{2x}\right)^5 \quad \Rightarrow \quad T_6 = 11C_5 \cdot \left(\frac{2}{3}\right)^6 \cdot (x^2)^6 \cdot \left(-\frac{3}{2}\right)^5 \cdot \frac{1}{x^5}$$

$$\Rightarrow \quad T_6 = \frac{2^6}{3^6} \cdot \left(-\frac{3^5}{2^5}\right) \cdot \frac{x^{12}}{x^5} \cdot \frac{11 \cdot 10 \cdot 9 \cdot 8 \cdot 7}{1 \cdot 2 \cdot 3 \cdot 4 \cdot 5}$$

$$\Rightarrow \quad T_6 = -\frac{2}{3} \cdot 462 \cdot x^7 = \mathbf{-308\, x^7}$$

Putting $r = 6$ we get 7th term

$$\Rightarrow \quad T_7 = 11C_6 \cdot \left(\frac{2}{3}x^2\right)^{11-6} \cdot \left(-\frac{3}{2x}\right)^6 \quad \Rightarrow \quad T_7 = \frac{2^5}{3^5} \cdot \frac{3^6}{2^6} \cdot x^{10} \cdot \frac{1}{x^6} \cdot 11C_6$$

$$\Rightarrow \quad T_7 = \frac{3}{2} \cdot 462 \cdot x^4 = \mathbf{693\ x^4}$$

Following results are useful to write middle term or terms directly.

1. If n is even, then there will be only one middle term in the expansion of $(x + a)^n$. The middle term will be $\left(\frac{n}{2}+1\right)^{th}$ term. It is given by, $T_{\left(\frac{n}{2}+1\right)} = nC_{\frac{n}{2}} \cdot x^{\frac{n}{2}} \cdot a^{\frac{n}{2}}$

2. If n is odd then there will be two middle terms in the expansion of $(x + a)^n$. The middle terms are $\left(\frac{n+1}{2}\right)^{th}$ and $\left(\frac{n+3}{2}\right)^{th}$ terms. These are given by

$$nC_{\frac{n-1}{2}} \cdot x^{\frac{n+1}{2}} \cdot a^{\frac{n-1}{2}} \quad \text{and} \quad nC_{\frac{n+1}{2}} \cdot x^{\frac{n-1}{2}} \cdot a^{\frac{n+1}{2}}$$

Example 9. Find the middle term of $\left(2x - \frac{1}{y}\right)^8$.

Solution : Here $n = 8$, $'x' = 2x$, $a = -\frac{1}{y}$

Since $n = 8$ is even, the middle term is $\frac{n}{2} + 1 = 5^{th}$ term. The middle term is

$$T_5 = 8C_4 \cdot (2x)^4 \cdot \left(-\frac{1}{y}\right)^4 = 2^4 \cdot 8C_4 \cdot \left(\frac{x}{y}\right)^4 = \mathbf{1120}\left(\frac{x^4}{y^4}\right) \qquad \left(\because \frac{n}{2} = 4\right)$$

Example 10. Find the middle terms of $\left(3x - \frac{x^3}{6}\right)^7$.

Solution : Here $n = 7$, $'x' = 3x$, $a = -\frac{x^3}{6}$

Since $n = 7$ is odd, the middle terms are $\frac{n+1}{2} = 4^{th}$ term and $\frac{n+3}{2} = 5^{th}$ term

These are, $T_4 = 7C_3 \cdot (3x)^4 \cdot \left(-\frac{x^3}{6}\right)^3 = 7C_3 \cdot 3^4 \cdot \left(-\frac{1}{6}\right)^3 \cdot x^{13} = -\mathbf{\frac{108}{3}x^{13}}$

and $T_5 = 7C_4 \cdot (3x)^3 \cdot \left(-\frac{x^3}{6}\right)^4 = 7C_4 \cdot 3^3 \cdot \left(-\frac{1}{6}\right)^4 \cdot x^{15} = \mathbf{\frac{35}{48}x^{15}}$

Example 10. If the coefficient of x^2 and x^3 in the expansion of $(3 + ax)^9$ are equal. Find the value of a.

Solution : Consider the expansion,

$$(3 + ax)^9 = 3^9 + 9C_1 \cdot 3^8 \cdot (ax) + 9C_2 \cdot 3^7 \cdot (ax)^2 + 9C_3 \cdot 3^6 \cdot (ax)^3 + ... + (ax)^9$$

Clearly x^2 and x^3 occurs in 3rd and 4th terms

Thus the coefficient of $x^2 = 9C_2 \cdot 3^7 \cdot a^2$ and the coefficient of $x^3 = 9C_3 \cdot 3^6 \cdot a^3$

By data, $9C_2 \cdot 3^7 \cdot a^2 = 9C_3 \cdot 3^6 \cdot a^3$

$$\Rightarrow \quad 3 \cdot 9C_2 = 9C_3 \cdot a \Rightarrow 3 \cdot \frac{9 \cdot 8}{1 \cdot 2} = \frac{9 \cdot 8 \cdot 7}{1 \cdot 2 \cdot 3} \cdot a \Rightarrow 3 = \frac{7}{3}a \Rightarrow \boldsymbol{a} = \frac{\mathbf{9}}{\mathbf{7}}$$

Example 11. If 21st and 22nd terms in the expansion of $(1 + x)^{44}$ are equal find the value of x $(x \neq 0)$.

Solution : Here, $n = 44$, 'x' $= 1$, 'a' $= x$

The $(r + 1)^{th}$ term of the expansion is given by is, $T_{r+1} = nC_r \cdot x^{n-r} \cdot a^r$

Putting $r = 20$, we get 21th term

i.e $\quad T_{21} = 44C_{20} \cdot (1)^{44-20} \cdot x^{20} \Rightarrow T_{21} = 44C_{20} \cdot x^{20}$

Again by putting $r = 21$, we get 22nd term

i.e $\quad T_{22} = 44C_{21} \cdot (1)^{44-21} \cdot x^{21} \Rightarrow T_{22} = 44C_{21} \cdot x^{21}$

by data, $\quad 44C_{20} \cdot x^{20} = 44C_{21} \cdot x^{21} \Rightarrow x = \dfrac{44!}{20! \cdot 24!} \cdot \dfrac{21! \cdot 23!}{44!} \qquad \left(\because nC_r = \dfrac{n!}{(n-r)! \cdot r!}\right)$

$$\Rightarrow \quad x = \frac{21}{24} = \frac{\mathbf{7}}{\mathbf{8}}$$

Example 12. Find the term independent of x (i.e., absolute term) in the expansion of $\left(\dfrac{3x^2}{2} - \dfrac{1}{3x}\right)^9$.

Solution : Here $\quad n = 9$, 'x' $= \dfrac{3x^2}{2}$, 'a' $= -\dfrac{1}{3x}$

Let $(r + 1)^{th}$ term be term independent of x. We have,

$$T_{r+1} = nC_r \cdot x^{n-r} \cdot a^r \quad \Rightarrow \quad T_{r+1} = 9C_r \cdot \left(\frac{3x^2}{2}\right)^{9-r} \cdot \left(-\frac{1}{3x}\right)^r$$

$$\Rightarrow \quad T_{r+1} = 9C_r \cdot \left(\frac{3}{2}\right)^{9-r} \cdot \left(-\frac{1}{3}\right)^r \cdot x^{18-2r} \cdot \frac{1}{x^r}$$

$$\Rightarrow \quad T_{r+1} = 9C_r \cdot \left(\frac{3}{2}\right)^{9-r} \cdot \left(-\frac{1}{3}\right)^r \cdot x^{18-3r}$$

By data the power of x in $(r + 1)^{th}$ term must be zero (i.e., the $(r + 1)^{th}$ term is independent of x.

$\Rightarrow \quad 18 - 3r = 0 \quad \Rightarrow \quad 3r = 18 \quad \Rightarrow \quad \mathbf{r = 6}$

The 7th term is independent of x

i.e., $T_7 = 9C_6 \cdot \left(\frac{3}{2}\right)^{9-6} \cdot \left(-\frac{1}{3}\right)^6 \Rightarrow T_7 = \left(\frac{3}{2}\right)^3 \cdot \left(\frac{1}{3^6}\right) \cdot 9C_3$ $\left(\because nC_r = nC_{n-r}\right)$

$\Rightarrow T_7 = \left(\frac{3}{2}\right)^3 \cdot \left(\frac{1}{3^6}\right) \cdot 9C_3$

$\Rightarrow T_7 = \frac{1}{216} \cdot \frac{9 \cdot 8 \cdot 7}{1 \cdot 2 \cdot 3} = \mathbf{\frac{7}{18}}$

Example 13. Using binomial theorem find the values of (a) $(10.1)^5$ (b) $(0.99)^5$ (c) $(101)^7$

Solution : (a) We have, $10.1 = 10 + 0.1$

$\therefore (10.1)^5 = (10 + 0.1)^5$

$= 5C_0 \cdot 10^5 + 5C_1 \cdot 10^4 \cdot (0.1) + 5C_2 \cdot 10^3 \cdot (0.1)^2 + 5C_3 \cdot 10^2 \cdot (0.1)^3 + 5C_4 \cdot 10^1 \cdot (0.1)^4 + 5C_5 \cdot (0.1)^5$

$= (10)^5 + 5 \cdot (10)^4 \cdot (0.1) + \frac{5 \cdot 4}{1 \cdot 2} \cdot (10)^3 \cdot (0.1)^2 + \frac{5 \cdot 4}{1 \cdot 2} \cdot (10)^2 \cdot (0.1)^3 + 5 \cdot 10 \cdot (0.1)^4 + (0.1)^5$

$= (10)^5 + 5 \cdot \frac{10^4}{10} + 10 \cdot \frac{10^3}{100} + 10 \cdot \frac{10^2}{1000} + 5 \cdot \frac{10}{10000} + \frac{1}{10000}$

$= 100000 + 5000 + 100 + 1 + 5\,(.001) + .00001$

$= 105101 + 0.005 + 0.00001 = \mathbf{105101.00501}$

(b) We have, $0.99 = 1 - 0.01$

$\therefore (0.99)^5 = (1 - 0.01)^5$

$= 5C_0 - 5C_1 \cdot (0.01) + 5C_2 \cdot (0.01)^2 - 5C_3 \cdot (0.01)^3 + 5C_4 \cdot (0.01)^4 - 5C_5 \cdot (0.01)^5$

$= 1 - 5\,(0.01) + \frac{5 \cdot 4}{1 \cdot 2}\,(0.01)^2 - \frac{5 \cdot 4}{1 \cdot 2}\,(0.01)^3 + 5\,(0.01)^4 - (0.01)^5$

$= 1 - 0.05 + 10\,(0.0001) - 10\,(0.000001) + 5\,(0.00000001) - (0.0000000001)$

$= 1 - 0.05 + 0.001 - 0.00001 + 0.00000005 - 0.0000000001 = 0.950990049$

(c) We have, $101 = 100 + 1$

$\therefore (101)^7 = (100 + 1)^7$

$= 100^7 + 7C_1 \cdot 100^6 + 7C_2 \cdot 100^5 + 7C_3 \cdot 100^4 + 7C_4 \cdot 100^3 + 7C_5 \cdot 100^2 + 7C_6 \cdot 100 + 7C_7$

$= 100^7 + 7 \cdot 100^6 + \frac{7 \cdot 6}{1 \cdot 2} \cdot 100^5 + \frac{7 \cdot 6 \cdot 5}{1 \cdot 2 \cdot 3} \cdot 100^4 + \frac{7 \cdot 6 \cdot 5}{1 \cdot 2 \cdot 3} \cdot 100^3 + \frac{7 \cdot 6}{1 \cdot 2} \cdot 100^2 + 7 \cdot 100 + 1$

$= 100^7 + 7 \cdot 100^6 + 21 \cdot 100^5 + 35 \cdot 100^4 + 35 \cdot 100^3 + 21 \cdot 100^2 + 700 + 1$

$$= 100000000000000 + 7000000000000 + 210000000000 + 3500000000 + 35000000 + 210000 + 700 + 1$$

$$= 107213535210701$$

Following results are useful.

1. In the expansion of $(x + a)^n$, $\dfrac{T_{r+1}}{T_r} = \dfrac{n - r + 1}{r} \cdot \left(\dfrac{a}{x}\right)$

2. The greatest coefficient in the expansion of $(x + a)^n$ is given by

(i) $nC_{n/2}$ if n is even (ii) $nC_{(n-1)/2}$ and $nC_{(n+1)/2}$ if n is odd (both being equal)

For example, the greatest coefficient in the expansion of $(2x^2 - x)^{10}$ is $10C_5$, i.e., 252

The greatest coefficient in the expansion of $\left(x + \dfrac{1}{x^2}\right)$ are $17C_8$ and $17C_9$ (which are equal), is 24,310.

Exercise

I. 1. Using binomial theorem expand the following

(i) $\left(x + \dfrac{1}{x}\right)^6$ **(ii)** $(3x + 2y)^4$ **(iii)** $\left(\sqrt{x} + \sqrt{y}\right)^{10}$

(iv) $\left(x - \dfrac{1}{x}\right)^6$ **(v)** $\left(\dfrac{2x}{3} - \dfrac{3}{2x}\right)^6$ **(vi)** $\left(4x + \dfrac{1}{2x}\right)^6$

2. Find

(i) the 7th term in $\left(\dfrac{4x}{5} + \dfrac{5}{2x}\right)^8$ **(ii)** the 16th term in $\left(\sqrt{x} - \sqrt{y}\right)^{17}$

(iii) the 6th term $\left(x^3 - \dfrac{1}{x^2}\right)^{10}$ **(iv)** the 10th term in $\left(\dfrac{a}{b} - \dfrac{2b}{a^2}\right)^{12}$

(v) Find the $(m + 1)^{th}$ term in $\left(x + \dfrac{1}{x}\right)^{2m}$ **(vi)** the 4th term from the end in $\left(\dfrac{x^3}{2} + \dfrac{2}{x^2}\right)^9$

(vii) the 3rd term from the end in $\left(\dfrac{a}{x} + bx\right)^{12}$ **(vii)** the tenth term in $\left(2x^2 + \dfrac{1}{x}\right)^{12}$

3. Find the term independent of x in the expansion of

(a) $\left(3x-\frac{2}{x^2}\right)^{15}$ **(b)** $\left(\frac{2}{x^2}-\sqrt{x}\right)^{10}$ **(c)** $\left(x^2+\frac{1}{x}\right)^9$ **(d)** $\left(\frac{3}{2}x^2-\frac{1}{3}x\right)^9$

(e) $\left(2x-\frac{1}{x}\right)^{10}$ **(f)** $\left(\sqrt{x}+\frac{1}{3x^2}\right)^{10}$ **(g)** $\left(x-\frac{1}{x^2}\right)^{3x}$ **(h)** $\left(x^2-\frac{2}{x^3}\right)^5$

4. Find the coefficient of

(i) x^5 in $(x+3)^6$ **(ii)** x^{10} in $(x^2-2)^{11}$ **(iii)** x^{16} in $(2x^2-x)^{10}$

5. Find the middle term in the expansion of

(a) $(3+x)^6$ **(b)** $\left(x^2-\frac{2}{x}\right)^{10}$ **(c)** $\left(3x-\frac{1}{6}x^3\right)^8$ **(d)** $\left(1-\frac{x^2}{2}\right)^{14}$

II.

1. Find the coefficient of

(i) x^{13} in $\left(3x-\frac{1}{6}x^3\right)^7$ **(ii)** x^{18} in $\left(x^2+\frac{3a}{x}\right)^{15}$ **(iii)** x^5 in $\left(x+\frac{1}{x^2}\right)^{17}$

(iv) x^{-2} in $\left(x+\frac{1}{x^2}\right)^{17}$ **(v)** x^6 in $\left(3x^2-\frac{1}{3x}\right)^9$ **(vi)** x^8 in $\left(x^2-\frac{1}{x}\right)^{10}$

(vii) x^{17} in $\left(3x-\frac{1}{6}x^3\right)^9$

2. Find the middle terms in the expansion of

(a) $\left(x+\frac{1}{x^2}\right)^{17}$ **(b)** $\left(3x-\frac{2}{x^2}\right)^{15}$ **(c)** $\left(3x-\frac{x^3}{6}\right)^7$

(d) $\left(x-\frac{2}{x}\right)^{11}$ **(e)** $\left(x^4-\frac{1}{x^3}\right)^{11}$ **(f)** $\left(3a-\frac{a^3}{6}\right)^9$

3. Show that there is no term independent of x in the expansion of $\left(2\sqrt{x}-\frac{1}{x\sqrt{x}}\right)^{35}$.

4. In the binomial expansion $(1+a)^{m+n}$ show that coefficients of a^m and a^n are equal

5. Simplify

(i) $\left(1+\sqrt{5}\right)^5+\left(1-\sqrt{5}\right)^5$ **(ii)** $\left(3+\sqrt{2}\right)^4+\left(3-\sqrt{2}\right)^4$

(iii) $\left(2+\sqrt{5}\right)^4+\left(2-\sqrt{5}\right)^4$ **(iv)** $\left(\sqrt{3}+1\right)^5+\left(\sqrt{3}-1\right)^5$

III.

1. Use binomial theorem to evaluate

(a) $(1.02)^6$ upto five places of decimals **(b)** $(102)^6$

(c) $(0.998)^6$ upto six places of decimals **(d)** $(0.98)^4$ upto 4 decimal places

2. Show that the coefficient of the middle term in $(1+x)^{2n}$ is equal to sum of the coefficients of two middle terms of $(1+x)^{2n-1}$.

3. If first, third and fifth terms in $(x+y)^n$ are respectively 32, 240 and 90, find x, y and n.

4. If p is real and the fourth term in the expansion of $\left(px+\frac{1}{x}\right)^n$ is $\frac{5}{2}$ find n and p.

5. If three successive coefficients in the expansion of $(1+x)^n$ are 462, 330 and 165 find n.

6. If the coefficients of y^2 and y^3 in the expansion of $\left(3+\frac{9y}{7}\right)^n$ are equal show that $n=9$

Answers

I. 2. (i) $\frac{4375}{x^4}$ **(ii)** $-136\,xy^{15/2}$ **(iii)** $-252\,x^5$ **(iv)** $-12\,C_3\cdot 2^9\cdot\frac{b^6}{a^{15}}$

(v) $(2m)C_m$ **(vi)** $\frac{672}{x^3}$ **(vii)** $1760\,\frac{1}{x^3}$

3. (a) $-2^5\cdot 3^{10}\cdot 15C_5$ **(b)** 180 **(c)** 84 **(d)** $\frac{7}{18}$ **(e)** -8064 **(f)** 5 **(g)** $(-1)^n(3n)\,C_n$ **(h)** 40

4. (i) 18 **(ii)** 29568 **(iii)** 13440

5. (i) $540\,x^3$ **(ii)** $-8064\,x^5$ **(iii)** $\frac{35}{8}x^{16}$ **(iv)** $-\frac{429}{16}x^{14}$

II.1. (i) $-\frac{105}{8}$ **(ii)** $110565a^4$ **(iii)** 2380 **(iv)** 0 **(v)** 378 **(vi)** 210 **(vii)** $\frac{189}{8}$

2. (a) $17C_8\,x^{-7}$, $17C_9\,x^{-10}$ **(b)** $-15C_7\,3^8\cdot 2^7\frac{1}{x^6}$, $15C_7\,3^7\cdot 2^8\frac{1}{x^9}$ **(c)** $-\frac{105}{8}x^{13}$, $\frac{35}{24}x^{15}$

(d) $11C_5\,2^5\cdot x$, $11C_5\,2^5\cdot\frac{1}{x}$ **(e)** $-462\,x^9$, $462\,x^2$ **(f)** $\frac{189}{8}a^{17}$, $-\frac{21}{16}a^{19}$

5. (i) 352 **(ii)** $168\sqrt{2}$ **(iii)** 222 **(iv)** 152

III. 1. (a) 1.12616 **(b)** 1126162419264 **(c)** 0.984112 **(d)** 0.9224

3. 2, $\sqrt{3}$, 5 **4.** $n=6$, $p=\frac{1}{2}$ **5.** 11

Chapter 4

Partial Fractions

4.1 Introduction

Consider the expression $\dfrac{2}{x+1} + \dfrac{3}{x-2}$

This can be written as a single expression

i.e., $$\frac{2}{x+1} + \frac{3}{x-2} = \frac{2x-4+3x+3}{(x+1)(x-2)} \quad \text{or} \quad \frac{2}{x+1} + \frac{3}{x-2} = \frac{5x-1}{(x+1)(x-2)}$$

Some times it is required to perform the reverse operation. That is given an expression of the form in the right side, we have to write it in the left hand side form.

In the above illustration the expressions $\dfrac{2}{x+1}$ and $\dfrac{3}{x-2}$ are called **partial fractions** of the right side expression.

The process of writing the given expression of the form $\dfrac{5x-1}{(x+1)(x-2)}$ as a sum of several partial fractions is known as **"resolving into partial fractions"**.

In this chapter we shall see the methods of resolving into partial fractions.

To this end we shall make few definitions.

Definition 1.

An expression of the form $a_0 x^n + a_1 x^{n-1} + a_2 x^{n-2} + \ldots. + a_n$

where $a_0, a_1, a_2, \ldots.. a_n$, with $a_0 \neq 0$, are constants, is called a polynomial of degree n in the variable x.

Examples

1. $3x^3 - 4x^2 + 7x - 2$ is a polynomial of degree 3

2. $3x^2 - 4x + 2$ is a polynomial of degree 2

3. $4x - 2$ is a polynomial of degree 1.

Note: A non zero constant k is considered as a polynomial of degree 0 – for k can be written as $k \cdot x^0$.

Definition 2. Rational function (Polynomial fraction)

An expression of the form $\frac{P}{Q}$, where P and Q are polynomials in x, is called a **rational function or a polynomial fraction.**

Examples

(i) $\frac{3x-2}{4x^2+2x-1}$ **(ii)** $\frac{2}{3x^2-4x+2}$ **(iii)** $\frac{2x^2-4x+1}{4x^2-2x+3}$ **(iv)** $\frac{3x^2-4x+5}{3x-1}$

Definition 3. Proper rational ~function

A rational function $\frac{P}{Q}$ is said to be a **proper rational function,** if the degree of the polynomial P is less than the degree of the polynomial Q.

Consider the rational function $\frac{3x-2}{4x^2+2x-1}$.

Here the degree of the numerator polynomial $3x - 2$ is 1 and that of denominator polynomial $4x^2 + 2x - 1$ is 2. Thus it is a proper rational function.

Definition 4. Improper rational function

A rational function $\frac{P}{Q}$ is said to be an **improper rational function,** if the degree of the polynomial P is greater than or equal to that of the denominator polynomial Q.

For example, the following rational functions are improper rational functions.

$$\frac{3x^2-4x+2}{4x^2+2x-1}, \quad \frac{3x^3+2x+1}{2x^2-4x+1}, \quad \frac{x^3-1}{x^2+1}$$

Given an improper rational function $\frac{P}{Q}$, we can express it as a sum of a polynomial and a proper rational function. This can be done by dividing the numerator polynomial P by the denominator polynomial.

Consider the improper rational function $\frac{4x^2-3x+7}{x+2}$.

We shall divide the polynomial $4x^2-3x+7$ by $x + 2$.

$$\begin{array}{r|l} x+2) & 4x^2 - 3x + 7 \ (4x - 11 \\ & \underline{4x^2 + 8x } \\ & -11x + 7 \\ & \underline{-11x - 22} \\ & 29 \end{array}$$

Thus, we have,

$$\frac{4x^2-3x+7}{x+2} = (4x - 11) + \frac{29}{x+2} = \text{a polynomial + proper rational function}$$

Again consider the improper rational function $\dfrac{2x^2 - 14x + 7}{x^2 + 3x - 2}$.

Consider,

$$\begin{array}{r|l} x^2 + 3x - 2) & 2x^2 - 14x + 7 \ (2 \\ & \underline{2x^2 + 6x - 4} \\ & -20x + 11 \end{array}$$

$$\therefore \qquad \frac{2x^2 - 14x + 7}{x^2 + 3x - 2} = 2 + \frac{11 - 20x}{x^2 + 3x - 2}$$

In what follows, given a rational function, we always express it as a proper rational function (if it is not already).

Exercise

1. Express the following as a sum of a polynomial and a proper rational function.

(a) $\dfrac{4x^2 - 4x - 1}{2x - 1}$ **(b)** $\dfrac{x^3 - 4}{x^2 + 3}$ **(c)** $\dfrac{4x^3 - 2x^2 + 3x + 1}{2x^2 + 4x - 1}$

(d) $\dfrac{3x^2 - 4x + 7}{x + 7}$ **(e)** $\dfrac{x^3 - 1}{x^2 + 1}$ **(f)** $\dfrac{x^3 + 7}{x^2 - 2x + 1}$

Answers

(a) $(2x - 1) - \dfrac{2}{2x - 1}$ **(b)** $x - \dfrac{3x + 4}{x^2 + 3}$ **(c)** $(2x - 5) + \dfrac{25x - 4}{2x^2 + 4x - 1}$

(d) $(3x - 25) + \dfrac{182}{x + 7}$ **(e)** $x - \dfrac{x + 1}{x^2 + 1}$ **(f)** $(x + 2) + \dfrac{3x + 5}{x^2 - 2x + 1}$

4.2 Partial fractions

Consider a proper rational function $\dfrac{P}{Q}$. The process of writing this into sum of two or more proper rational functions is called **"resolving into partial fractions"**.

Working procedure for resolving a rational function into partial fractions.

Step 1. If the given rational function is an improper rational function, express it as a sum of a polynomial and a proper rational function. Let the corresponding proper function be $\dfrac{P}{Q}$.

Step 2. Factorise the denominator polynomial Q in to the products of several factors.

Step 3. Use the following rules to resolve the proper rational function $\dfrac{P}{Q}$ into partial fractions.

(a) Corresponding to each non repeated linear factor $(ax + b)$ of the polynomial Q, write a partial fraction $\dfrac{A}{ax + b}$, where A is a constant to be determined.

For example, $$\frac{x-3}{(x+5)(x-2)} = \frac{A}{(x+5)} + \frac{B}{(x-2)}$$

where A and B are constants to be determined.

(b) If the denominator Q has a linear factor $(ax + b)$ repeated r times (i.e., if $(ax + b)^r$ is a factor of Q), then corresponding to this factor write r partial fractions,

$$\frac{A_1}{ax+b}, \frac{A_2}{(ax+b)^2}, \frac{A_3}{(ax+b)^3}, \ldots\ldots\ldots\ldots, \frac{A_r}{(ax+b)^r}$$

where A_1, A_2, A_3, ... A_r are r constants to be determined.

For example, we write, $\frac{x-3}{(x-5)(x-2)^2} = \frac{A}{x-5} + \frac{B}{x-2} + \frac{C}{(x-2)^2}$ where A, B and C are constants to be determined.

Observe that corresponding to the factor $(x - 5)$, we have written one partial fraction $\frac{A}{x-5}$ (by rule (a)) and corresponding to the factor $(x - 2)^2$, we have written two partial fractions $\frac{B}{x-2}$ and $\frac{C}{(x-2)^2}$ (by rule (b)).

(c) If the denominator Q has a non factorizable quadratic factor $ax^2 + bx + c$, corresponding to this factor write a partial fraction of the form $\frac{Ax+B}{ax^2+bx+c}$, where A and B are constants to be determined.

For example, we write $\frac{x-1}{(x-2)(3x^2+1)} = \frac{A}{x-2} + \frac{Bx+C}{3x^2+1}$ where A, B and C are constants to be determined.

Method of determining the constants A, B, ... etc., are explained in the following examples.

Example 1. Resolve $\frac{5x-4}{(x+1)(x-2)}$ into partial fractions.

Solution : The given rational function is proper and the denominator has two distinct non repeated factors. Thus we write

$$\frac{5x-4}{(x+1)(x-2)} = \frac{A}{x+1} + \frac{B}{x-2} \quad \ldots. (1)$$

Where A and B are constants to be determined.

To find A and B we shall follow the following method.

Multiply both sides by $(x + 1)$ $(x - 2)$, we get

$$5x - 4 = A (x - 2) + B (x + 1) \quad \ldots. (2)$$

This is an identity (i.e., the equality is true for all values of x).

To find the value of A, we shall put $\boldsymbol{x = -1}$ in (2) (this makes the second term zero).

$$\Rightarrow \qquad -5-4 = A(-1-2) + B\cdot 0 \;\Rightarrow\; -9 = -3A \;\Rightarrow\; \mathbf{A = 3}$$

Again we shall put $\mathbf{x = 2}$ in (2) (this makes the first term zero).

$$\Rightarrow \qquad 10-4 = A\cdot 0 + B(2+1) \;\Rightarrow\; 6 = 3B \;\Rightarrow\; \mathbf{B = 2}$$

Putting these values A and B, in (1) we get

$$\frac{5x-4}{(x+1)(x-2)} = \frac{3}{x+1} + \frac{2}{x-2}$$

Example 2. Resolve into partial fractions $\dfrac{x-1}{x(x+2)(x+6)}$.

Solution : The given rational function is proper rational function. Let

$$\frac{x-1}{x(x+2)(x+6)} = \frac{A}{x} + \frac{B}{x+2} + \frac{C}{x+6} \qquad \text{.... (1)}$$

Multiply both sides by $x(x+2)(x+6)$, we get

$$x-1 = A(x+2)(x+6) + B\cdot x(x+6) + C\cdot x(x+2) \qquad \text{.... (2)}$$

Put $x = 0$ in (2) (this makes second and third term simultaneously zero), we get,

$$-1 = A(2)(6) \;\Rightarrow\; \mathbf{A = -\frac{1}{12}}$$

Put $x = -2$ in (2) (this makes first and third term simultaneously zero), we get,

$$-2-1 = B(-2)(-2+6) \;\Rightarrow\; -3 = -8B \;\Rightarrow\; \mathbf{B = \frac{3}{8}}$$

Again putting $x = -6$ in (2), we get

$$-6-1 = C(-6)(-6+2) \;\Rightarrow\; -7 = 24C \;\Rightarrow\; \mathbf{C = -\frac{7}{24}}$$

Putting these values A, B and C in (1) we get

$$\frac{x-1}{x(x+2)(x+6)} = \frac{-1}{12x} + \frac{3}{8(x+2)} - \frac{7}{24(x+6)}$$

Example 3. Resolve $\dfrac{x^2-2}{x^2+x-12}$ **into partial fractions.**

Solution : The given rational function is an improper rational function (because the degree of the numerator is equal to that of the denominator). Thus, we shall express it as a sum of a polynomial and a proper rational function, by dividing x^2-2 by x^2+x-12.

$$\begin{array}{r} x^2+x-12)\; x^2-2 \quad (1 \\ x^2+x-12 \\ \hline -x+10 \end{array}$$

$$\therefore \qquad \frac{x^2-2}{x^2+x-12} = 1 + \frac{10-x}{x^2+x-12} \qquad \text{.... (1)}$$

Further, $\qquad x^2+x-12 = (x+4)(x-3)$

Let $$\frac{10-x}{(x+4)(x-3)} = \frac{A}{x+4} + \frac{B}{x-3} \qquad \text{.... (2)}$$

Multiply both sides by $(x + 4)(x - 3)$, we get

$$10 - x = A(x - 3) + B(x + 4) \qquad \text{.... (3)}$$

Putting $x = -4$, in (3) we get,

$$\Rightarrow \quad 10 + 4 = A(-4 - 3) \Rightarrow 14 = -7A \Rightarrow \mathbf{A = -2}$$

Again putting $x = 3$, in (3) we get,

$$\Rightarrow \quad 10 - 3 = B(3 + 4) \Rightarrow 7 = 7B \Rightarrow \mathbf{B = 1}$$

Putting these values of A and B in (2) we get

$$\frac{10-x}{(x+4)(x-3)} = \frac{-2}{x+4} + \frac{1}{x-3}$$

$\therefore$ (1) becomes $$\frac{\mathbf{x^2-2}}{\mathbf{x^2+x-12}} = \mathbf{1 - \frac{2}{x+4} + \frac{1}{x-3}}$$

Example 4. Resolve $\dfrac{6x^2-8x-29}{(2x+3)(x-2)}$ into partial fractions.

Solution : Consider, $$\frac{6x^2-8x-29}{(2x+3)(x-2)} = \frac{6x^2-8x-29}{2x^2-x-6}$$

This is an improper rational function.

$$\begin{array}{r} 2x^2 - x - 6)\ 6x^2 - 8x - 29\ (3 \\ \underline{6x^2 - 3x - 18} \\ -5x - 11 \end{array}$$

$$\therefore \quad \frac{6x^2-8x-29}{(2x+3)(x-2)} = 3 - \frac{5x+11}{(2x+3)(x-2)} \qquad \text{.... (1)}$$

Let $$\frac{5x+11}{(2x+3)(x-2)} = \frac{A}{2x+3} + \frac{B}{x-2} \qquad \text{.... (2)}$$

Multiply both sides by $(2x + 3)(x - 2)$ we get

$$5x + 11 = A(x - 2) + B(2x + 3) \qquad \text{.... (3)}$$

Putting $2x = -3$, i.e., $x = -3/2$ in (3) we get

$$\Rightarrow \quad -\frac{15}{2} + 11 = A\left(-\frac{3}{2} - 2\right) \Rightarrow \frac{7}{2} = -\frac{7}{2}A \Rightarrow \mathbf{A = -1}$$

Again putting $x = 2$, in (3) we get

$$\Rightarrow \quad 10 + 11 = B(4 + 3) \Rightarrow 21 = 7B \Rightarrow \mathbf{B = 3}$$

Putting these values A and B in (2) we get

$$\frac{5x+11}{(2x+3)(x-2)} = \frac{-1}{2x+3} + \frac{3}{x-2}$$

$\therefore$ (1) becomes

$$\frac{6x^2-8x-29}{(2x+3)(x-2)} = 3 - \left[\frac{-1}{2x+3} + \frac{3}{x-2}\right] \Rightarrow \frac{\mathbf{6x^2-8x-29}}{\mathbf{(2x+3)(x-2)}} = \mathbf{3} + \frac{\mathbf{1}}{\mathbf{2x+3}} - \frac{\mathbf{3}}{\mathbf{x-2}}$$

Example 5. Resolve $\frac{3x+5}{(x+2)^2(x-3)}$ into partial fractions.

Solution : The denominator has two linear factors and one of them (i.e., $x + 2$) is repeated twice. Thus, we will have two partial fractions corresponding to this factor.

Let $$\frac{3x+5}{(x+2)^2(x-3)} = \frac{A}{x+2} + \frac{B}{(x+2)^2} + \frac{C}{x-3} \quad \text{.... (1)}$$

Multiply both sides by $(x + 2)^2 (x - 3)$ we get

$$3x + 5 = A(x + 2)(x - 3) + B(x - 3) + C(x + 2)^2 \quad \text{.... (2)}$$

Observe that we can not find the value of A by our previous method (because we can not make the second and third term of R.H.S. simultaneously zero for one value of x).

Now we shall put $x = -2$ in (2), we get

$$\Rightarrow \quad -6 + 5 = B(-2-3) \quad \Rightarrow -1 = -5B \Rightarrow \mathbf{B} = \frac{\mathbf{1}}{\mathbf{5}}$$

Again putting $x = 3$ in (2), we get

$$\Rightarrow \quad 9 + 5 = C(3+2)^2 \quad \Rightarrow 14 = 25C \Rightarrow \mathbf{C} = \frac{\mathbf{14}}{\mathbf{25}}$$

To find the value A, we shall put some value for x in (2) (other than–2 and 3) say $x = 0$, we get

$$5 = A(2)(-3) + B(-3) + C(2)^2$$

$$\Rightarrow \quad 5 = -6A - 3\left(\frac{1}{5}\right) + 4\left(\frac{14}{25}\right) \quad \left(\because B = \frac{1}{5},\ C = \frac{14}{25}\right)$$

$$\Rightarrow \quad 6A = -5 - \frac{3}{5} + \frac{56}{25} \Rightarrow 6A = \frac{-125-15+56}{25} \Rightarrow 6A = -\frac{84}{25} \Rightarrow \mathbf{A} = -\frac{\mathbf{14}}{\mathbf{25}}$$

Putting these values of A, B and C in (1) we have,

$$\frac{\mathbf{3x+5}}{\mathbf{(x+2)^2(x-3)}} = -\frac{\mathbf{14}}{\mathbf{25(x+2)}} + \frac{\mathbf{1}}{\mathbf{5(x+2)^2}} + \frac{\mathbf{14}}{\mathbf{25(x-3)}}$$

Example 6. Resolve $\frac{x^3-2}{x(x+1)^2}$ into partial fractions.

Solution : Consider $$\frac{x^3-2}{x(x+1)^2} = \frac{x^3-2}{x^3+2x^2+x}$$

This is an improper rational function.

$$\begin{array}{r|l} x^3 + 2x^2 + x) & x^3 - 2 \quad (1 \\ & \underline{x^3 + 2x^2 + x} \\ & -2x^2 - x - 2 \end{array}$$

$$\therefore \quad \frac{x^3-2}{x(x+1)^2} = 1 - \frac{2x^2+x+2}{x(x+1)^2}$$

Let $$\frac{2x^2+x+2}{x(x+1)^2} = \frac{A}{x} + \frac{B}{x+1} + \frac{C}{(x+1)^2} \quad \text{.... (1)}$$

Multiply both sides $x\ (x + 1)^2$, we get

$$2x^2+x+2 = A\ (x+1)^2 + B\ x\ (x+1) + C\ x \quad \text{.... (2)}$$

Put $x = 0$ in (2), we get, $2 = A\ (1)^2 \Rightarrow \boldsymbol{A = 2}$

Put $x = -1$ in (2), we get, $2 - 1 + 2 = C\ (-1) \Rightarrow -C = 3 \Rightarrow \boldsymbol{C = -3}$

To find the value of C, we shall put $x = 1$ (i.e., other than 0 and -1) in (2), we get

$$2 + 1 + 2 = A\ (2)^2 + B \cdot 1\ (1+1) + C \cdot 1$$

$$\Rightarrow \quad 5 = 4\ A + 2\ B + C$$

$$\Rightarrow \quad 5 = 8 + 2\ B - 3 \qquad (\because\ A = 2,\ C = -3)$$

$$\Rightarrow \quad \boldsymbol{B = 0}$$

Putting these values of A, B and C in (1) we have

$$\frac{2x^2+x+2}{x(x+1)^2} = \frac{2}{x} - \frac{3}{(x+1)^2}$$

$$\therefore \quad \boldsymbol{\frac{x^3-2}{x(x+1)^2} = 1 - \frac{2}{x} + \frac{3}{(x+1)^2}}$$

Example 7. Resolve $\boldsymbol{\frac{4x^2+5x+8}{(x^2+5)(x+2)}}$ into partial fractions.

Solution : The denominator has a non factorisable quadratic factor. Thus we shall write

$$\frac{4x^2+5x+8}{(x^2+5)(x+2)} = \frac{Ax+B}{x^2+5} + \frac{C}{x+2} \quad \text{.... (1)}$$

Multiply both sides by $(x^2 + 5)\ (x + 2)$, we get

$$4x^2+5x+8 = (A\ x + B)\ (x+2) + C\ (x^2+5) \quad \text{.... (2)}$$

Put $x = -2$ in (2), we get

$$16 - 10 + 8 = C\ (4+5) \Rightarrow 14 = 9\ C \Rightarrow \boldsymbol{C = \frac{14}{9}}$$

To find the other constants A and B, we shall rewrite (2).

$$4x^2 + 5x + 8 = A\,x^2 + B\,x + 2\,A\,x + 2\,B + C\,x^2 + 5\,C$$

i.e., $$4x^2 + 5x + 8 = (A + C)\,x^2 + (B + 2\,A)\,x + (2\,B + 5\,C)$$

[Two polynomials are equal if their corresponding coefficients are equal]. Equating Coefficients of x^2: $A + C = 4 \Rightarrow A = 4 - C$

$$\Rightarrow \quad A = 4 - \frac{14}{9} \Rightarrow \boldsymbol{A = \frac{22}{9}} \qquad \left(\because\ C = \frac{14}{9}\right)$$

Coefficients of x: $B + 2A = 5 \Rightarrow B = 5 - 2A$

$$\Rightarrow B = 5 - \frac{44}{9} \Rightarrow \mathbf{B = \frac{1}{9}} \qquad \left(\because A = \frac{22}{9}\right)$$

Putting the values of A, B and C in (1) we get,

$$\mathbf{\frac{4x^2 + 5x + 8}{(x^2+5)(x+2)} = \frac{22x+1}{9(x^2+5)} + \frac{14}{9(x+2)}}$$

Example 8. Resolve $\frac{x^4+1}{x^3+x}$ into partial fractions.

Solution : The given rational function is an improper rational function

$$\begin{array}{r} x^3 + x)\ x^4 + 1\ (x \\ \underline{x^4 + x^2} \\ -x^2 + 1 \end{array}$$

$$\therefore \quad \frac{x^4+1}{x^3+x} = x + \frac{1-x^2}{x(x^2+1)}$$

Let
$$\frac{1-x^2}{x(x^2+1)} = \frac{A}{x} + \frac{Bx+C}{x^2+1} \qquad \text{.... (1)}$$

Multiplying both sides by $x(x^2+1)$, we get,

$$1 - x^2 = A(x^2+1) + (Bx + C)x \qquad \text{.... (2)}$$

Put $x = 0$ in (2), we get, $1 = A(0+1) \Rightarrow \mathbf{A = 1}$

(2) can be written as, $1 - x^2 = Ax^2 + A + Bx^2 + Cx$

$\Rightarrow$ $1 - x^2 = (A+B)x^2 + Cx + A$

Equating,

Coefficients of x^2: $A + B = -1 \Rightarrow B = -1 - A \Rightarrow \mathbf{B = -2}$ $(\because A = 1)$

Coefficients of x: $\mathbf{C = 0}$

Putting the values of A, B and C in (1) we get,

$$\frac{1-x^2}{x(x^2+1)} = \frac{1}{x} - \frac{2x}{x^2+1}$$

$$\therefore \quad \mathbf{\frac{x^4+1}{x^3+x} = x + \frac{1}{x} - \frac{2x}{x^2+1}}$$

Shortcut methods in certain cases

We mention below the methods of writing the partial fractions directly in certain cases. These methods are useful, when one has to resolve the rational fraction into partial fractions, when it is a part of a main problem.

I. If $\dfrac{f(x)}{(x-a)(x-b)}$ **is a proper rational function and**

$$\frac{f(x)}{(x-a)(x-b)} = \frac{A}{x-a} + \frac{B}{x-b}$$

then $\quad A = \dfrac{f(a)}{a-b}, \quad B = \dfrac{f(b)}{b-a}$

That is, to get the value of A, put $x = a$ in L.H.S. (except in $x - a$) and to get the value of B, put $x = b$ in L.H.S. (except in $x - b$).

This procedure can be applied, if the denominator of L.H.S. has more than two non repeated linear factors.

Examples

1. If $\quad \dfrac{4x-2}{(x-2)(x+3)} = \dfrac{A}{x-2} + \dfrac{B}{x+3}$

then $\quad A = \dfrac{4(2)-2}{2+3} = \dfrac{6}{5}, \quad B = \dfrac{4(-3)-2}{-3-2} = \dfrac{14}{5}$

$$\therefore \quad \frac{4x-2}{(x-2)(x+3)} = \frac{6}{5(x-2)} + \frac{14}{5(x+3)}$$

2. If $\quad \dfrac{2x+3}{(x+1)(x-2)(2x-1)} = \dfrac{A}{x+1} + \dfrac{B}{x-2} + \dfrac{C}{2x-1}$

then $\quad A = \dfrac{2(-1)+3}{(-1-2)(-2-1)} = \dfrac{1}{9}, \quad B = \dfrac{2(2)+3}{(2+1)(4-1)} = \dfrac{7}{9}, \quad C = \dfrac{1+3}{\left(\dfrac{1}{2}+1\right)\left(\dfrac{1}{2}-2\right)} = -\dfrac{16}{9}$

$$\therefore \quad \frac{2x+3}{(x+1)(x-2)(2x-1)} = \frac{1}{9(x+1)} + \frac{7}{9(x-2)} - \frac{16}{9(2x-1)}$$

Note: In particular we have,

$$\frac{1}{(x-a)(x-b)} = \frac{1}{a-b}\left[\frac{1}{x-a} - \frac{1}{x-b}\right]$$

II. If $\dfrac{f(x)}{(x-a)(x-b)^2}$ **is a proper rational function and**

$$\frac{f(x)}{(x-a)(x-b)^2} = \frac{A}{x-a} + \frac{B}{x-b} + \frac{C}{(x-b)^2}$$

then $\quad A = \dfrac{f(a)}{(a-b)^2}, \quad C = \dfrac{f(b)}{b-a}$

The value of B can be obtained by putting some value for x (other than a and b).

In particular if $f(x)$ is of the form $Px + Q$ or a constant Q then

$$B = -\frac{f(a)}{(a-b)^2} \quad \text{i.e.,} \quad B = -A$$

Again if $f(x)$ is of the form $Px^2 + Qx + R$ and $P \neq 0$ then

$$B = P - A$$

Examples

1. If $$\frac{4}{(x-3)(x+1)^2} = \frac{A}{x-3} + \frac{B}{x+1} + \frac{C}{(x+1)^2}$$

then $$A = \frac{4}{(3+1)^2} = \frac{1}{4}, \quad C = \frac{4}{(-1-3)} = -1$$

To find B, we shall put $x = 0$, we get,

$$\frac{4}{(-3)(1)} = \frac{A}{-3} + \frac{B}{1} + \frac{C}{1} \Rightarrow -\frac{4}{3} = -\frac{1}{12} + B - 1$$

$$\Rightarrow B = \frac{1}{12} + 1 - \frac{4}{3} = \frac{1+12-16}{12} = -\frac{1}{4}$$

$$\therefore \quad \frac{4}{(x-3)(x+1)^2} = \frac{1}{4(x-3)} - \frac{1}{4(x+1)} - \frac{1}{(x+1)^2}$$

2. If $$\frac{3x+2}{(x-2)(x+3)^2} = \frac{A}{x-2} + \frac{B}{x+3} + \frac{C}{(x+3)^2}$$

then $$A = \frac{6+2}{(2+3)^2} = \frac{8}{25}, \quad B = -A = -\frac{8}{25}, \quad C = \frac{-9+2}{(-3-2)} = \frac{7}{5}$$

$$\therefore \quad \frac{3x+2}{(x-2)(x+3)^2} = \frac{8}{25(x-2)} - \frac{8}{25(x+3)} + \frac{7}{5(x+3)^2}$$

3. If $$\frac{2x^2-4x+1}{(x-2)(x-3)^2} = \frac{A}{x-2} + \frac{B}{x-3} + \frac{C}{(x-3)^2}$$

then $$A = \frac{8-8+1}{(2-3)^2} = 1, \quad B = 2 - 1 = 1 \text{ (here } P = 2\text{)}, \quad C = \frac{18-12+1}{(3-2)} = 7$$

$$\therefore \quad \frac{2x^2-4x+1}{(x-2)(x-3)^2} = \frac{1}{x-2} + \frac{1}{x-3} + \frac{7}{(x-3)^2}$$

III. $$\frac{1}{(x^2+a^2)(x^2+b^2)} = \frac{1}{b^2-a^2}\left(\frac{1}{x^2+a^2} - \frac{1}{x^2+b^2}\right)$$

Example.

$$\frac{1}{(x^2+4)(x^2-9)} = \frac{1}{-9-4}\left(\frac{1}{x^2+4} - \frac{1}{x^2-9}\right) = -\frac{1}{13}\left(\frac{1}{x^2+4} - \frac{1}{x^2-9}\right)$$

Exercises

I. Resolve the following rational functions into partial fractions.

1. $\dfrac{x}{(x+1)(x-4)}$
2. $\dfrac{x^2}{(x+1)(x+2)(x+3)}$
3. $\dfrac{3x+1}{x^2-6x+8}$
4. $\dfrac{x-1}{2x^2-5x-3}$
5. $\dfrac{5x+1}{(x+2)(x-1)}$
6. $\dfrac{3x+20}{x^2+4x}$
7. $\dfrac{x+3}{(x-1)(x^2-4)}$
8. $\dfrac{5x^2+19x-18}{x(x+3)(x-2)}$
9. $\dfrac{7x-1}{(1-2x)(1-3x)}$
10. $\dfrac{1+3x+2x^2}{(1-2x)(1-x^2)}$
11. $\dfrac{2x^3+x^2-x-3}{x(x-1)(2x+3)}$

II. Resolve the following rational functions into partial fractions.

1. $\dfrac{2x^2+16x+29}{(x+3)^2(x+4)}$
2. $\dfrac{x^2}{(x+1)^2(x-5)}$
3. $\dfrac{4x^2+-11x+12}{x^2(x-4)}$
4. $\dfrac{9}{(x-1)(x+2)^2}$
5. $\dfrac{x^4-3x^3-3x^2+10}{(x-1)^2(x-3)}$
6. $\dfrac{4+7x}{(2+3x)(1+x)^2}$

III. Resolve the following rational functions into partial fractions.

1. $\dfrac{x^2}{(2x+3)(x^2+9)}$
2. $\dfrac{2x+1}{(x-1)(x^2+1)}$
3. $\dfrac{2x^2-14x+8}{(x^2+3x-2)(x-3)}$
4. $\dfrac{3x-1}{x^3-1}$
5. $\dfrac{1}{x^3+1}$
6. $\dfrac{3x-1}{(x+2)(1-x+x^2)}$

Answers

I.

1. $\dfrac{1}{5(x+1)}+\dfrac{4}{5(x-4)}$
2. $\dfrac{1}{2(x+1)}-\dfrac{4}{x+2}+\dfrac{9}{2(x+3)}$
3. $\dfrac{13}{2(x-4)}-\dfrac{7}{2(x-2)}$
4. $\dfrac{2}{7(x-3)}+\dfrac{3}{7(2x+1)}$
5. $\dfrac{3}{x+2}+\dfrac{2}{x-1}$
6. $\dfrac{5}{x}-\dfrac{2}{x+4}$
7. $\dfrac{-4}{3(x-1)}+\dfrac{5}{4(x-2)}+\dfrac{1}{12(x+2)}$
8. $\dfrac{3}{x}-\dfrac{2}{x+3}+\dfrac{4}{x-2}$
9. $\dfrac{4}{1-3x}-\dfrac{5}{1-2x}$
10. $\dfrac{4}{1-2x}-\dfrac{3}{1-x}$
11. $1+\dfrac{1}{x}-\dfrac{1}{5(x-1)}-\dfrac{8}{5(2x+3)}$

II.

1. $\dfrac{5}{x+3}-\dfrac{1}{(x+3)^2}-\dfrac{3}{x+4}$
2. $\dfrac{11}{36(x+1)}-\dfrac{1}{6(x+1)^2}+\dfrac{25}{36(x-5)}$
3. $\dfrac{2}{x}-\dfrac{3}{x^2}+\dfrac{2}{x-4}$
4. $\dfrac{1}{x-1}-\dfrac{1}{x+2}-\dfrac{3}{(x+2)^2}$

5. $(x-2)+\frac{17}{16(x+1)}-\frac{11}{4(x+1)^2}-\frac{17}{16(x-3)}$

6. $\frac{2}{1+x}+\frac{3}{(1+x)^2}-\frac{6}{2+3x}$

III. 1. $\frac{1}{5(2x+3)}+\frac{2x-3}{5(x^2+9)}$

2. $\frac{3}{2(x-1)}-\frac{3(x-1)}{2(x^2+1)}$

3. $\frac{3x-2}{x^2+3x-2}-\frac{1}{x-3}$

4. $\frac{2}{3(x-1)}+\frac{5-2x}{3(x^2+x+1)}$

5. $\frac{1}{3(x+1)}+\frac{2-x}{3(x^2-x+1)}$

6. $\frac{-1}{x+2}+\frac{x}{1-x+x^2}$

Chapter 5

Matrices

5.1 Introduction

In this chapter we shall introduce an important mathematical tool called matrix. Further we shall develope the algebra of matrices. Theory of matrices has wide application in different branch of mathematics. One of the application is in connection with solving the system of linear equations.

5.2 Matrices

Definition : An arrangement of numbers (real or complex) in the form of rows and columns within brackets is called a Matrix.

The numbers that form a matrix are called **elements** of the matrix.

The matrices are denoted by capital letter A, B, C, ...

If a matrix A has m rows and n columns, then A is called the matrix of **order** $\boldsymbol{m \times n}$ (read as m cross n or m by n *matrix*).

Examples :

$\begin{pmatrix} 2 & -1 \\ 3 & 2 \\ 4 & 2 \end{pmatrix}_{3 \times 2}$ is a matrix of order 3×2. $\begin{pmatrix} 3 & 2 \\ 4 & -1 \end{pmatrix}_{2 \times 2}$ is a matrix of order 2×2

$\begin{pmatrix} 2 \\ -1 \\ 3 \end{pmatrix}_{3 \times 1}$ is a matrix of order 3×1 $(i \quad 2i \quad 0)_{1 \times 3}$ is a matrix of order 1×3

5.3 Types of Matrices

1. Null matrix or Zero matrix

A matrix in which each element is a zero, is called a null matrix or a zero matrix and denoted by *O*.

$\begin{pmatrix} 0 & 0 \\ 0 & 0 \end{pmatrix}$ is a zero matrix of order 2×2 $(0 \quad 0 \quad 0)$ is a zero matrix of order 1×3

2. Row matrix

A matrix having only one row is called a row matrix. That is a matrix of $1 \times n$ is a row matrix.

For example, $(1 \quad 3)_{1 \times 2}$, $(3 \quad -2 \quad 4)_{1 \times 3}$ are row matrices.

3. Column matrix

A matrix having only one column is called a column matrix. That is a matrix of $m \times 1$ is a column matrix.

For example, $\begin{pmatrix} 1 \\ 3 \end{pmatrix}_{2 \times 1}$, $\begin{pmatrix} 3 \\ -2 \\ 4 \end{pmatrix}_{3 \times 1}$ are column matrices.

4. Square matrix

A matrix in which the number of rows equal to number of columns is called a square matrix.

A square matrix having n rows and n columns is called a **square matrix of order n.**

For example, $\begin{pmatrix} 2 & 1 \\ -1 & 3 \end{pmatrix}$, $\begin{pmatrix} 2 & -1 & -1 \\ 3 & 1 & 2 \\ 1 & 1 & 0 \end{pmatrix}$ are square matrices of order 2 and 3 respectively.

In a square matrix the elements along the diagonal which runs from left top to right bottom are called **diagonal elements**. In the above examples 2, 3 are the **diagonal elements** of the first square matrix, whereas the elements 2, 1, 0 are the **diagonal elements** of the second square matrix.

5. Rectangular matrix

A matrix in which the number of rows is not equal to the number of columns is called a rectangular matrix.

The matrices $\begin{pmatrix} 2 & -1 & 3 \\ 1 & 2 & 4 \end{pmatrix}_{2 \times 3}$; $\begin{pmatrix} 2 & 1 \\ 4 & 1 \\ 0 & 3 \end{pmatrix}_{3 \times 2}$ are rectangular matrices.

6. Diagonal matrix

A square matrix in which all the elements except the diagonal elements are zero, is called a diagonal matrix.

The following are diagonal matrices $\begin{pmatrix} 2 & 0 \\ 0 & -1 \end{pmatrix}_{2 \times 2}$; $\begin{pmatrix} -1 & 0 & 0 \\ 0 & 2 & 0 \\ 0 & 0 & 4 \end{pmatrix}_{3 \times 3}$

7. Scalar matrix

A diagonal matrix in which all the diagonal elements are equal, is called a scalar matrix.

For example, $\begin{pmatrix} 2 & 0 \\ 0 & 2 \end{pmatrix}_{2 \times 2}$; $\begin{pmatrix} -3 & 0 & 0 \\ 0 & -3 & 0 \\ 0 & 0 & -3 \end{pmatrix}_{3 \times 3}$ are scalar matrices.

8. Unit matrix or Identity matrix

A diagonal matrix in which each diagonal entry is unity, is called unit matrix or identity matrix.

The unit matrix of order *n* is denoted by I_n.

For example, $I_2 = \begin{pmatrix} 1 & 0 \\ 0 & 1 \end{pmatrix}_{2 \times 2}$ and $I_3 = \begin{pmatrix} 1 & 0 & 0 \\ 0 & 1 & 0 \\ 0 & 0 & 1 \end{pmatrix}_{3 \times 3}$

5.4 Algebra of Matrices

1. Equality of two matrices

Two matrices are said to be equal, if their orders are same and the corresponding elements are equal.

Example 1. Find the values of *a* and *b* if the following matrices are equal.

$$A = \begin{pmatrix} a+b & b^2+2 \\ 0 & -6 \end{pmatrix}, \quad B = \begin{pmatrix} 2a+1 & 3b \\ 0 & b^2-5b \end{pmatrix}$$

Solution : By data $A = B$. Thus, by definition of equality of two matrices we have

$$a + b = 2a + 1, \quad b^2 + 2 = 3b \quad \text{and} \quad -6 = b^2 - 5b$$

Now, $b^2 + 2 = 3b \Rightarrow b^2 - 3b + 2 = 0 \Rightarrow (b-2)(b-1) = 0 \Rightarrow b = 2$ or $b = 1$

Now, $b = 1$ will not satisfy the equation $-6 = b^2 - 5b \quad \therefore \quad \mathbf{b = 2.}$

Now, $a + b = 2a + 1 \Rightarrow a = b - 1 \Rightarrow \mathbf{a = 1}$

2. Addition of two matrices

Let A and B be two matrices of order $m \times n$. The sum $A + B$ of the matrices A and B is a matrix of order $m \times n$ and whose elements are the sum of the corresponding elements of A and B.

Let, $A = \begin{pmatrix} 2 & 1 \\ 3 & -1 \\ 4 & 2 \end{pmatrix}_{3 \times 2}$ and $B = \begin{pmatrix} -1 & 4 \\ 2 & 3 \\ 1 & -1 \end{pmatrix}_{3 \times 2}$ then

$$A + B = \begin{pmatrix} 2 & 1 \\ 3 & -1 \\ 4 & 2 \end{pmatrix} + \begin{pmatrix} -1 & 4 \\ 2 & 3 \\ 1 & -1 \end{pmatrix} = \begin{pmatrix} 2-1 & 1+4 \\ 3+2 & -1+3 \\ 4+1 & 2-1 \end{pmatrix} = \begin{pmatrix} 1 & 5 \\ 5 & 2 \\ 5 & 1 \end{pmatrix}_{3 \times 2}$$

Subtraction of two matrices is similarly defined. For example,

$$A - B = \begin{pmatrix} 2 & 1 \\ 3 & -1 \\ 4 & 2 \end{pmatrix} - \begin{pmatrix} -1 & 4 \\ 2 & 3 \\ 1 & -1 \end{pmatrix} = \begin{pmatrix} 3 & -3 \\ 1 & -4 \\ 3 & 3 \end{pmatrix}$$

3. Scalar multiple of a matrix

Let A be a matrix of order $m \times n$ and k be any scalar. The scalar multiple $k \cdot A$ of the matrix A with scalar k is a matrix of order $m \times n$ and its elements are obtained by multiplying each element of A by k.

Let, $A = \begin{pmatrix} 2 & -1 & 3 \\ 4 & 1 & 2 \end{pmatrix}$ then $4A = \begin{pmatrix} 8 & -4 & 12 \\ 16 & 4 & 8 \end{pmatrix}$

The addition of matrices and scalar multiple of a matrix satisfy the following properties.

If A, B and C are matrices of same order, say $m \times n$, then

(a) $A + B = B + A$ **Commutative law**

(b) $A + (B + C) = (A + B) + C$ **Associative law**

(c) $k \cdot (A + B) = k \cdot A + k \cdot B$ where k is a scalar.

(d) $A + O = O + A = A$ where O is the zero matrix of order $m \times n$.

(e) $A + (-A) = -A + A = O$ where $-A$ is the matrix obtained from A by multiplying each element of A by -1. The matrix $-A$ is called the negative of the matrix A.

Example 2. If $A = \begin{pmatrix} 3 & -1 & 2 \\ 3 & 1 & 2 \end{pmatrix}$ and $B = \begin{pmatrix} 1 & 4 & 6 \\ 1 & 3 & -1 \end{pmatrix}$, find $2A - 3B$.

Solution : Consider

$$2A - 3B = \begin{pmatrix} 6 & -2 & 4 \\ 6 & 2 & 4 \end{pmatrix} - \begin{pmatrix} 3 & 12 & 18 \\ 3 & 9 & -3 \end{pmatrix} = \begin{pmatrix} 3 & -14 & -14 \\ 3 & -7 & 7 \end{pmatrix}$$

Example 3. If $A = \begin{pmatrix} 2 & 3 & 4 \\ -3 & 0 & 2 \end{pmatrix}$, $B = \begin{pmatrix} 3 & -4 & -5 \\ 1 & 2 & 1 \end{pmatrix}$ and $C = \begin{pmatrix} 5 & -1 & 2 \\ 7 & 0 & 3 \end{pmatrix}$ find the matrix X such that $2A + 3B - X = C$.

Solution : We have,

$$2A + 3B - X = C \Rightarrow X = 2A + 3B - C$$

Consider, $$2A + 3B = \begin{pmatrix} 4 & 6 & 8 \\ -6 & 0 & 4 \end{pmatrix} + \begin{pmatrix} 9 & -12 & -15 \\ 3 & 6 & 3 \end{pmatrix} = \begin{pmatrix} 13 & -6 & -7 \\ -3 & 6 & 7 \end{pmatrix}$$

Now, $$2A + 3B - C = \begin{pmatrix} 13 & -6 & -7 \\ -3 & 6 & 7 \end{pmatrix} - \begin{pmatrix} 5 & -1 & 2 \\ 7 & 0 & 3 \end{pmatrix} \Rightarrow X = \begin{pmatrix} 8 & -5 & -9 \\ -10 & 6 & 4 \end{pmatrix}$$

Example 4. If $2A + B = \begin{pmatrix} 3 & -1 \\ 2 & 4 \end{pmatrix}$ and $A - 2B = \begin{pmatrix} 4 & 2 \\ -1 & 5 \end{pmatrix}$, find A and B.

Solution : The given equations are

$$2A + B = \begin{pmatrix} 3 & -1 \\ 2 & 4 \end{pmatrix} \quad \text{.... (1)}$$

$$A - 2B = \begin{pmatrix} 4 & 2 \\ -1 & 5 \end{pmatrix} \quad \text{.... (2)}$$

Consider $2 \times (1) + (2)$, we get

$$5A = \begin{pmatrix} 6 & -2 \\ 4 & 8 \end{pmatrix} + \begin{pmatrix} 4 & 2 \\ -1 & 5 \end{pmatrix} = \begin{pmatrix} 10 & 0 \\ 3 & 13 \end{pmatrix} \Rightarrow A = \begin{pmatrix} 2 & 0 \\ 3/5 & 13/5 \end{pmatrix}$$

Again consider $(1) - 2 \times (2)$, we get

$$5B = \begin{pmatrix} 3 & -1 \\ 2 & 4 \end{pmatrix} - \begin{pmatrix} 8 & 4 \\ -2 & 10 \end{pmatrix} = \begin{pmatrix} -5 & -5 \\ 4 & -6 \end{pmatrix} \Rightarrow \boldsymbol{B} = \begin{pmatrix} \mathbf{-1} & \mathbf{-1} \\ \mathbf{4/5} & \mathbf{-6/5} \end{pmatrix}$$

Example 5. Solve for x : $\begin{pmatrix} \mathbf{x^2} & \mathbf{1} \\ \mathbf{2} & \mathbf{3} \end{pmatrix} + \begin{pmatrix} \mathbf{2x} & \mathbf{3} \\ \mathbf{1} & \mathbf{4} \end{pmatrix} = \begin{pmatrix} \mathbf{3} & \mathbf{4} \\ \mathbf{3} & \mathbf{7} \end{pmatrix}$.

Solution : We have from the given equation

$$\begin{pmatrix} x^2 + 2x & 4 \\ 3 & 7 \end{pmatrix} = \begin{pmatrix} 3 & 4 \\ 3 & 7 \end{pmatrix}$$

$$\Rightarrow \quad x^2 + 2x = 3 \quad \Rightarrow \quad x^2 + 2x - 3 = 0 \quad \Rightarrow \quad (x+3)(x-1) = 0 \quad \Rightarrow \quad \boldsymbol{x = -3} \text{ or } \boldsymbol{x = 1}$$

4. Multiplication of matrices

Consider two matrices A and B where A is of order $m \times n$ and B is of order $n \times p$. Here, the number of columns of A is equal to the number of rows of B. Two such matrices are said to be **compatible**.

Further the number of elements in any row of A is equal to the number of elements is any column of B.

The product AB of the matrices A and B is the matrix of order $m \times p$ and its ith row, j^{th} column element is given by the sum of the product of ith row elements of A with the corresponding element of jth column elements of B.

Consider the matrices $A = \begin{pmatrix} 3 & 1 & -2 \\ 1 & 2 & 1 \end{pmatrix}$ and $B = \begin{pmatrix} 1 & 1 \\ -1 & 2 \\ 2 & 1 \end{pmatrix}$

Here the number of columns of A is equal to the number of rows of B.

Observe that the number of elements in each row of A is equal to the number of elements in any column of B.

Now the product matrix AB is of order 2×2. To obtain different elements of AB, we apply the following procedure.

To get the first row first column element of AB, we shall multiply the elements of the first row of A with the corresponding elements of the first column of B and add. That is

$$3 \cdot 1 + 1 \cdot (-1) + (-2) \cdot 2 = 3 - 1 - 4 = -2$$

That is -2 is the first row first column element of AB.

Supposing we require the second row second column element of AB, we multiply the second row elements of A with the corresponding elements of second column elements of B and add. That is

$$1 \cdot 1 + 2 \cdot 2 + 1 \cdot 1 = 1 + 4 + 1 = 6$$

The following steps shows the complete multiplication.

1. $\begin{pmatrix} 3 & 1 & -2 \\ 1 & 2 & 1 \end{pmatrix} \begin{pmatrix} 1 & 1 \\ -1 & 2 \\ 2 & 1 \end{pmatrix} = \begin{pmatrix} 3-1-4 & * \\ * & * \end{pmatrix}$

2. $\begin{pmatrix} 3 & 1 & -2 \\ 1 & 2 & 1 \end{pmatrix} \begin{pmatrix} 1 & 1 \\ -1 & 2 \\ 2 & 1 \end{pmatrix} = \begin{pmatrix} * & 3+2-2 \\ * & * \end{pmatrix}$

3. $\begin{pmatrix} 3 & 1 & -2 \\ 1 & 2 & 1 \end{pmatrix} \begin{pmatrix} 1 & 1 \\ -1 & 2 \\ 2 & 1 \end{pmatrix} = \begin{pmatrix} * & * \\ 1-2+2 & * \end{pmatrix}$

4. $\begin{pmatrix} 3 & 1 & -2 \\ 1 & 2 & 1 \end{pmatrix} \begin{pmatrix} 1 & 1 \\ -1 & 2 \\ 2 & 1 \end{pmatrix} = \begin{pmatrix} * & * \\ * & 1+4+1 \end{pmatrix}$

Thus, we have, $\begin{pmatrix} 3 & 1 & -2 \\ 1 & 2 & 1 \end{pmatrix} \begin{pmatrix} 1 & 1 \\ -1 & 2 \\ 2 & 1 \end{pmatrix} = \begin{pmatrix} 3-1-4 & 3+2-2 \\ 1-2+2 & 1+4+1 \end{pmatrix} = \begin{pmatrix} \mathbf{-2} & \mathbf{3} \\ \mathbf{1} & \mathbf{6} \end{pmatrix}$

Students are advised to practice to write all the different steps illustrated in one step.

The matrix multiplication satisfies the following properties.

(a) If A, B and C are three matrices whose orders respectively are $m \times n$, $n \times p$ and $p \times n$ then

$A \cdot (B \cdot C) = (A \cdot B) \cdot C$ **Associative law**

(b) If A is a matrix of order $m \times n$ and B, C are matrices of order $n \times p$, then

$A \cdot (B + C) = AB + AC$ **Distributive law**

(c) If A is a $m \times n$ matrix, $A \cdot I_n = A$ **and** $I_m \cdot A = A$

Matrix multiplication is **not commutative**. That is if A and B are two matrices, then AB need not be equal to BA, even if both the products are defined.

For example, consider the matrices

$$A = \begin{pmatrix} 2 & -1 \\ 3 & 2 \end{pmatrix}_{2\times2} \quad \text{and} \quad B = \begin{pmatrix} 3 & 1 \\ 2 & 1 \end{pmatrix}_{2\times2}$$

Now, $A \cdot B = \begin{pmatrix} 2 & -1 \\ 3 & 2 \end{pmatrix} \begin{pmatrix} 3 & 1 \\ 2 & 1 \end{pmatrix} = \begin{pmatrix} 6-2 & 2-1 \\ 9+4 & 3+2 \end{pmatrix} = \begin{pmatrix} \mathbf{4} & \mathbf{1} \\ \mathbf{13} & \mathbf{5} \end{pmatrix}$

Now, $B \cdot A = \begin{pmatrix} 3 & 1 \\ 2 & 1 \end{pmatrix} \begin{pmatrix} 2 & -1 \\ 3 & 2 \end{pmatrix} = \begin{pmatrix} 6+3 & -3+2 \\ 4+3 & -2+2 \end{pmatrix} = \begin{pmatrix} \mathbf{9} & \mathbf{-1} \\ \mathbf{7} & \mathbf{0} \end{pmatrix}$

Observe that, $AB \neq BA$

5. Transpose of a matrix

Let A be a matrix of order $m \times n$. The matrix obtained from A by interchanging the rows into columns is called the **transpose** of the matrix A and it is denoted by A' or A^T.

Clearly, if A is of order of $m \times n$ then the order of A' is $n \times m$.

If, $A = \begin{pmatrix} 2 & -1 & 3 \\ 1 & 2 & 4 \end{pmatrix}_{2\times3}$ then $A' = \begin{pmatrix} 2 & 1 \\ -1 & 2 \\ 3 & 4 \end{pmatrix}_{3\times2}$

The following properties are true for the transpose of a matrix.

Let A and B be any two matrices of same order, then

(i) $(A + B)' = A' + B'$ **(ii)** $(A')' = A$

(iii) If A and B are two matrices such that $A \cdot B$ is defined then $(A \cdot B)' = B' \cdot A'$

Example 6. Find AB and BA, if $A = \begin{pmatrix} \mathbf{1} & \mathbf{2} & \mathbf{-3} \\ \mathbf{5} & \mathbf{0} & \mathbf{2} \\ \mathbf{1} & \mathbf{-1} & \mathbf{1} \end{pmatrix}$ **and** $B = \begin{pmatrix} \mathbf{3} & \mathbf{-1} & \mathbf{2} \\ \mathbf{4} & \mathbf{2} & \mathbf{5} \\ \mathbf{2} & \mathbf{0} & \mathbf{3} \end{pmatrix}$.

Solution :

$$AB = \begin{pmatrix} 1 & 2 & -3 \\ 5 & 0 & 2 \\ 1 & -1 & 1 \end{pmatrix} \begin{pmatrix} 3 & -1 & 2 \\ 4 & 2 & 5 \\ 2 & 0 & 3 \end{pmatrix}$$

$$= \begin{pmatrix} 3+8-6 & -1+4+0 & 2+10-9 \\ 15+0+4 & -5+0+0 & 10+0+6 \\ 3-4+2 & -1-2+0 & 2-5+3 \end{pmatrix} = \begin{pmatrix} \mathbf{5} & \mathbf{3} & \mathbf{3} \\ \mathbf{19} & \mathbf{-5} & \mathbf{16} \\ \mathbf{1} & \mathbf{-3} & \mathbf{0} \end{pmatrix}$$

Now, $$BA = \begin{pmatrix} 3 & -1 & 2 \\ 4 & 2 & 5 \\ 2 & 0 & 3 \end{pmatrix}\begin{pmatrix} 1 & 2 & -3 \\ 5 & 0 & 2 \\ 1 & -1 & 1 \end{pmatrix}$$

$$= \begin{pmatrix} 3-5+2 & 6+0-2 & -9-2+2 \\ 4+10+5 & 8+0-5 & -12+4+5 \\ 2+0+3 & 4+0-3 & -6+0+3 \end{pmatrix} = \begin{pmatrix} \mathbf{0} & \mathbf{4} & \mathbf{-9} \\ \mathbf{19} & \mathbf{3} & \mathbf{-3} \\ \mathbf{5} & \mathbf{1} & \mathbf{-3} \end{pmatrix}$$

Note that , $\boldsymbol{AB \neq BA}$

Example 7. If $A = \begin{pmatrix} -2 & 1 \\ 0 & 3 \end{pmatrix}$ find $2A^2 - 3A - 7I$.

Solution : Consider, $$A^2 = A \cdot A = \begin{pmatrix} -2 & 1 \\ 0 & 3 \end{pmatrix}\begin{pmatrix} -2 & 1 \\ 0 & 3 \end{pmatrix} = \begin{pmatrix} 4+0 & -2+3 \\ 0+0 & 0+9 \end{pmatrix} = \begin{pmatrix} 4 & 1 \\ 0 & 9 \end{pmatrix}$$

Now, $$2A^2 - 3A = \begin{pmatrix} 8 & 2 \\ 0 & 18 \end{pmatrix} - \begin{pmatrix} -6 & 3 \\ 0 & 9 \end{pmatrix} = \begin{pmatrix} 14 & -1 \\ 0 & 9 \end{pmatrix}$$

Now, $$2A^2 - 3A - 7I = \begin{pmatrix} 14 & -1 \\ 0 & 9 \end{pmatrix} - \begin{pmatrix} 7 & 0 \\ 0 & 7 \end{pmatrix} = \begin{pmatrix} \mathbf{7} & \mathbf{-1} \\ \mathbf{0} & \mathbf{2} \end{pmatrix}$$

Example 8. If $A = \begin{pmatrix} \cos\theta & \sin\theta \\ \sin\theta & \cos\theta \end{pmatrix}$, $B = \begin{pmatrix} \cos\varphi & \sin\varphi \\ \sin\varphi & \cos\varphi \end{pmatrix}$. Show that $AB = BA$.

Solution : $$AB = \begin{pmatrix} \cos\theta & \sin\theta \\ \sin\theta & \cos\theta \end{pmatrix}\begin{pmatrix} \cos\varphi & \sin\varphi \\ \sin\varphi & \cos\varphi \end{pmatrix}$$

$$= \begin{pmatrix} \cos\theta\cos\varphi + \sin\theta\sin\varphi & \cos\theta\sin\varphi + \sin\theta\cos\varphi \\ \sin\theta\cos\varphi + \cos\theta\sin\varphi & \sin\theta\sin\varphi + \cos\theta\cos\varphi \end{pmatrix}$$

$$= \begin{pmatrix} \cos(\theta-\varphi) & \sin(\theta+\varphi) \\ \sin(\theta+\varphi) & \cos(\theta-\varphi) \end{pmatrix}$$

Consider, $$BA = \begin{pmatrix} \cos\varphi & \sin\varphi \\ \sin\varphi & \cos\varphi \end{pmatrix}\begin{pmatrix} \cos\theta & \sin\theta \\ \sin\theta & \cos\theta \end{pmatrix}$$

$$= \begin{pmatrix} \cos\varphi\cos\theta + \sin\varphi\sin\theta & \sin\theta\cos\varphi + \cos\theta\sin\varphi \\ \sin\varphi\cos\theta + \cos\varphi\sin\theta & \sin\varphi\sin\theta + \cos\varphi\cos\theta \end{pmatrix} = \begin{pmatrix} \cos(\theta-\varphi) & \sin(\theta+\varphi) \\ \sin(\theta+\varphi) & \cos(\theta-\varphi) \end{pmatrix}$$

Thus, $\boldsymbol{AB = BA}$

Example 9. If $A = \begin{pmatrix} \cos\theta & -\sin\theta \\ \sin\theta & \cos\theta \end{pmatrix}$, find AA'.

Solution : $$A = \begin{pmatrix} \cos\theta & -\sin\theta \\ \sin\theta & \cos\theta \end{pmatrix},\quad A' = \begin{pmatrix} \cos\theta & \sin\theta \\ -\sin\theta & \cos\theta \end{pmatrix}$$

Now, $$AA' = \begin{pmatrix} \cos\theta & -\sin\theta \\ \sin\theta & \cos\theta \end{pmatrix} \begin{pmatrix} \cos\theta & \sin\theta \\ -\sin\theta & \cos\theta \end{pmatrix}$$

$$= \begin{pmatrix} \cos^2\theta + \sin^2\theta & \cos\theta\sin\theta - \sin\theta\cos\theta \\ \sin\theta\cos\theta - \cos\theta\sin\theta & \sin^2\theta + \cos^2\theta \end{pmatrix} = \begin{pmatrix} \mathbf{1} & \mathbf{0} \\ \mathbf{0} & \mathbf{1} \end{pmatrix}$$

Example 10. If $A = \begin{pmatrix} 2 & -1 \\ 3 & 2 \end{pmatrix}$, $B = \begin{pmatrix} 3 & 1 \\ -1 & 2 \end{pmatrix}$. Verify $(AB)' = B' \cdot A'$.

Solution : $$AB = \begin{pmatrix} 2 & -1 \\ 3 & 2 \end{pmatrix} \begin{pmatrix} 3 & 1 \\ -1 & 2 \end{pmatrix} = \begin{pmatrix} 6+1 & 2-2 \\ 9-2 & 3+4 \end{pmatrix} = \begin{pmatrix} 7 & 0 \\ 7 & 7 \end{pmatrix} \Rightarrow (AB)' = \begin{pmatrix} 7 & 7 \\ 0 & 7 \end{pmatrix}$$

Now, $$B'A' = \begin{pmatrix} 3 & -1 \\ 1 & 2 \end{pmatrix} \begin{pmatrix} 2 & 3 \\ -1 & 2 \end{pmatrix} = \begin{pmatrix} 6+1 & 9-2 \\ 2-2 & 3+4 \end{pmatrix} = \begin{pmatrix} 7 & 7 \\ 0 & 7 \end{pmatrix}$$

$\therefore$ $$\boldsymbol{(AB)' = B'A'}$$

Example 11. $E(x) = \begin{pmatrix} \cos x & \sin x \\ -\sin x & \cos x \end{pmatrix}$ show that $E(\alpha) \cdot E(\beta) = E(\alpha + \beta)$.

Solution : $$E(\alpha) = \begin{pmatrix} \cos\alpha & \sin\alpha \\ -\sin\alpha & \cos\alpha \end{pmatrix} \text{ and } E(\beta) = \begin{pmatrix} \cos\beta & \sin\beta \\ -\sin\beta & \cos\beta \end{pmatrix}$$

$$E(\alpha)\cdot E(\beta) = \begin{pmatrix} \cos\alpha & \sin\alpha \\ -\sin\alpha & \cos\alpha \end{pmatrix} \begin{pmatrix} \cos\beta & \sin\beta \\ -\sin\beta & \cos\beta \end{pmatrix}$$

$$= \begin{pmatrix} \cos\alpha\cos\beta - \sin\alpha\sin\beta & \cos\alpha\sin\beta + \sin\alpha\cos\beta \\ -\sin\alpha\cos\beta - \cos\alpha\sin\beta & -\sin\alpha\sin\beta + \cos\alpha\cos\beta \end{pmatrix}$$

$$= \begin{pmatrix} \cos(\alpha+\beta) & \sin(\alpha+\beta) \\ -\sin(\alpha+\beta) & \cos(\alpha+\beta) \end{pmatrix} = E(\alpha+\beta)$$

$\therefore$ $$\boldsymbol{E(\alpha)\cdot E(\beta) = E(\alpha+\beta)}$$

Example 12. If $A = \begin{pmatrix} 1 \\ 3 \end{pmatrix}$, $B = (1\ \ 6\ \ 7)$, find AB.

Solution : A is of order 2×1 and B is of order 1×3. Thus AB is defined.

Now, $$AB = \begin{pmatrix} 1 \\ 3 \end{pmatrix} (1\ \ 6\ \ 7) = \begin{pmatrix} \mathbf{1} & \mathbf{6} & \mathbf{7} \\ \mathbf{3} & \mathbf{18} & \mathbf{21} \end{pmatrix}_{\mathbf{2\times 3}}$$

Example 13. If $A = \begin{pmatrix} 4 & 3 \\ 1 & 2 \end{pmatrix}$ and $B = \begin{pmatrix} -4 \\ 3 \end{pmatrix}$, find AB.

Solution : A is of order 2×2 and B is of order 2×1. Thus, AB is defined and of order 2×1.

$$AB = \begin{pmatrix} 4 & 3 \\ 1 & 2 \end{pmatrix} \begin{pmatrix} -4 \\ 3 \end{pmatrix} = \begin{pmatrix} -16+9 \\ -4+6 \end{pmatrix} = \begin{pmatrix} \mathbf{-7} \\ \mathbf{2} \end{pmatrix}_{\mathbf{2\times 1}}$$

Example 14. If $A = \begin{pmatrix} 3 & 2 \\ 1 & 0 \end{pmatrix}$, $B = \begin{pmatrix} 4 & 5 & 6 \\ 0 & 1 & 2 \end{pmatrix}$ and $C = \begin{pmatrix} 1 & -4 & -1 \\ -2 & 5 & -3 \\ 3 & 6 & 5 \end{pmatrix}$ **Verify $(AB)\,C = A\,(BC)$.**

Solution : The matrix A is of order 2×2, the matrix B is of order 2×3 and C is of order 3×3. Thus, the product AB, BC are defined. Now, AB will be of order 2×3. Thus, $(AB)\,C$ is defined. Again the matrix BC is of order 2×3 and therefore $A\,(BC)$ is also defined.

Now, $$AB = \begin{pmatrix} 3 & 2 \\ 1 & 0 \end{pmatrix}\begin{pmatrix} 4 & 5 & 6 \\ 0 & 1 & 2 \end{pmatrix} = \begin{pmatrix} 12+0 & 15+2 & 18+4 \\ 4+0 & 5+0 & 6+0 \end{pmatrix} = \begin{pmatrix} 12 & 17 & 22 \\ 4 & 5 & 6 \end{pmatrix}$$

Now, $$(AB)C = \begin{pmatrix} 12 & 17 & 22 \\ 4 & 5 & 6 \end{pmatrix}\begin{pmatrix} 1 & -4 & -1 \\ -2 & 5 & -3 \\ 3 & 6 & 5 \end{pmatrix}$$

$$= \begin{pmatrix} 12-34+66 & -48+85+132 & -12-51+110 \\ 4-10+18 & -16+25+36 & -4-15+30 \end{pmatrix} = \begin{pmatrix} 44 & 169 & 47 \\ 12 & 45 & 11 \end{pmatrix}_{2\times3}$$

Now, $$BC = \begin{pmatrix} 4 & 5 & 6 \\ 0 & 1 & 2 \end{pmatrix}\begin{pmatrix} 1 & -4 & -1 \\ -2 & 5 & -3 \\ 3 & 6 & 5 \end{pmatrix}$$

$$= \begin{pmatrix} 4-10+18 & -16+25+36 & -4-15+30 \\ 0-2+6 & 0+5+12 & 0-3+10 \end{pmatrix} = \begin{pmatrix} 12 & 45 & 11 \\ 4 & 17 & 7 \end{pmatrix}$$

Now, $$A(BC) = \begin{pmatrix} 3 & 2 \\ 1 & 0 \end{pmatrix}\begin{pmatrix} 12 & 45 & 11 \\ 4 & 17 & 7 \end{pmatrix} = \begin{pmatrix} 36+8 & 135+34 & 33+14 \\ 12+0 & 45+0 & 11+0 \end{pmatrix} = \begin{pmatrix} 44 & 169 & 47 \\ 12 & 45 & 11 \end{pmatrix}_{2\times3}$$

$\therefore$ **$(AB)\,C = A(BC)$** is verified.

Example 15. If $A = \begin{pmatrix} 5 \\ 4 \\ 3 \end{pmatrix}$ **and** $B = \begin{pmatrix} 3 \\ -1 \\ 2 \end{pmatrix}$, **find AB' and $A'B$.**

Solution : $A = \begin{pmatrix} 5 \\ 4 \\ 3 \end{pmatrix}_{3\times1}$ and $B' = (3 \;\; -1 \;\; 2)_{1\times3}$. The product AB' is defined and is of order 3×3.

Now, $$AB' = \begin{pmatrix} 5 \\ 4 \\ 3 \end{pmatrix} \cdot \downarrow (3 \;\; -1 \;\; 2) = \begin{pmatrix} \mathbf{15} & \mathbf{-5} & \mathbf{10} \\ \mathbf{12} & \mathbf{-4} & \mathbf{8} \\ \mathbf{9} & \mathbf{-3} & \mathbf{6} \end{pmatrix}_{3\times3}$$

Now, $A' = \overrightarrow{(5 \;\; 4 \;\; 3)}_{1\times3}$ and $B = \begin{pmatrix} 3 \\ -1 \\ 2 \end{pmatrix}_{3\times1}$; $A'B$ is defined and is of order 1×1

$$A'B = (5 \;\; 4 \;\; 3)\begin{pmatrix} 3 \\ -1 \\ 2 \end{pmatrix} = (15 - 4 + 6) = \mathbf{(17)_{1\times1}}$$

Example 16. If $A = \begin{pmatrix} 2 & -3 \\ 4 & 5 \end{pmatrix}$ and $B = \begin{pmatrix} 3 \\ 17 \end{pmatrix}$, find the matrix X such that $AX = B$.

Solution : A is of 2×2 order and B is of order 2×1. The product AX is of 2×1 implies X is of order 2×1.

Let the required matrix be $X = \begin{pmatrix} a \\ b \end{pmatrix}$

Now, by data, $AX = B \Rightarrow \begin{pmatrix} 2 & -3 \\ 4 & 5 \end{pmatrix} \begin{pmatrix} a \\ b \end{pmatrix} = \begin{pmatrix} 3 \\ 17 \end{pmatrix}$

$$\Rightarrow \begin{pmatrix} 2a - 3b \\ 4a + 5b \end{pmatrix} = \begin{pmatrix} 3 \\ 17 \end{pmatrix} \Rightarrow 2a - 3b = 3 \text{ and } 4a + 5b = 17$$

Solving these two equations we get, $a = 3$ and $b = 1 \Rightarrow \mathbf{X = \begin{pmatrix} 3 \\ 1 \end{pmatrix}}$

Example 17. Find x and y if $(x \quad y) \begin{pmatrix} 2 & 3 \\ 4 & -1 \end{pmatrix} = (2 \quad -4)$.

Solution : $(x \quad y) \begin{pmatrix} 2 & 3 \\ 4 & -1 \end{pmatrix} = (2 \quad -4) \Rightarrow (2x + 4y \quad 3x - y) = (2 \quad -4)$

$$\Rightarrow 2x + 4y = 2 \text{ and } 3x - y = -4$$

Solving these equations we get $\mathbf{x = -1}$ **and** $\mathbf{y = 1}$

Example 18. If $A = \begin{pmatrix} 2 & 4 \\ -1 & 2 \end{pmatrix}$ and $B = \begin{pmatrix} 1 & 0 \\ 2 & 1 \end{pmatrix}$ find $A^2 - B^2$ and show that $A^2 - B^2 \neq (A-B)(A+B)$.

Solution :

$$A^2 = \begin{pmatrix} 2 & 4 \\ -1 & 2 \end{pmatrix} \begin{pmatrix} 2 & 4 \\ -1 & 2 \end{pmatrix} = \begin{pmatrix} 4-4 & 8+8 \\ -2-2 & -4+4 \end{pmatrix} = \begin{pmatrix} 0 & 16 \\ -4 & 0 \end{pmatrix}$$

$$B^2 = \begin{pmatrix} 1 & 0 \\ 2 & 1 \end{pmatrix} \begin{pmatrix} 1 & 0 \\ 2 & 1 \end{pmatrix} = \begin{pmatrix} 1+0 & 0+0 \\ 2+2 & 0+1 \end{pmatrix} = \begin{pmatrix} 1 & 0 \\ 4 & 1 \end{pmatrix}$$

$$A^2 - B^2 = \begin{pmatrix} 0 & 16 \\ -4 & 0 \end{pmatrix} - \begin{pmatrix} 1 & 0 \\ 4 & 1 \end{pmatrix} = \begin{pmatrix} \mathbf{-1} & \mathbf{16} \\ \mathbf{-8} & \mathbf{-1} \end{pmatrix} \quad \text{.... (1)}$$

$$A - B = \begin{pmatrix} 2 & 4 \\ -1 & 2 \end{pmatrix} - \begin{pmatrix} 1 & 0 \\ 2 & 1 \end{pmatrix} = \begin{pmatrix} 1 & 4 \\ -3 & 1 \end{pmatrix}$$

$$A + B = \begin{pmatrix} 2 & 4 \\ -1 & 2 \end{pmatrix} + \begin{pmatrix} 1 & 0 \\ 2 & 1 \end{pmatrix} = \begin{pmatrix} 3 & 4 \\ 1 & 3 \end{pmatrix}$$

$$(A - B)(A + B) = \begin{pmatrix} 1 & 4 \\ -3 & 1 \end{pmatrix} \begin{pmatrix} 3 & 4 \\ 1 & 3 \end{pmatrix} = \begin{pmatrix} 3+4 & 4+12 \\ -9+1 & -12+3 \end{pmatrix} = \begin{pmatrix} \mathbf{7} & \mathbf{16} \\ \mathbf{-8} & \mathbf{-9} \end{pmatrix} \quad \text{.... (2)}$$

From (1) and (2) it follows that $\mathbf{A^2 - B^2 \neq (A - B)(A + B)}$.

Example 19. If $A = \begin{pmatrix} 1 & 2 & 2 \\ 2 & 1 & 2 \\ 2 & 2 & 1 \end{pmatrix}$ show that $A^2 - 4A - 5I = O$.

Solution :

$$A^2 = \begin{pmatrix} 1 & 2 & 2 \\ 2 & 1 & 2 \\ 2 & 2 & 1 \end{pmatrix}\begin{pmatrix} 1 & 2 & 2 \\ 2 & 1 & 2 \\ 2 & 2 & 1 \end{pmatrix} = \begin{pmatrix} 1+4+4 & 2+2+4 & 2+4+2 \\ 2+2+4 & 4+1+4 & 4+2+2 \\ 2+4+2 & 4+2+2 & 4+4+1 \end{pmatrix} = \begin{pmatrix} 9 & 8 & 8 \\ 8 & 9 & 8 \\ 8 & 8 & 9 \end{pmatrix}$$

$$A^2 - 4A = \begin{pmatrix} 9 & 8 & 8 \\ 8 & 9 & 8 \\ 8 & 8 & 9 \end{pmatrix} - \begin{pmatrix} 4 & 8 & 8 \\ 8 & 4 & 8 \\ 8 & 8 & 4 \end{pmatrix} = \begin{pmatrix} 5 & 0 & 0 \\ 0 & 5 & 0 \\ 0 & 0 & 5 \end{pmatrix}$$

$$A^2 - 4A - 5I = \begin{pmatrix} 5 & 0 & 0 \\ 0 & 5 & 0 \\ 0 & 0 & 5 \end{pmatrix} - \begin{pmatrix} 5 & 0 & 0 \\ 0 & 5 & 0 \\ 0 & 0 & 5 \end{pmatrix} = \begin{pmatrix} 0 & 0 & 0 \\ 0 & 0 & 0 \\ 0 & 0 & 0 \end{pmatrix} \Rightarrow A^2 - 4A - 5I = O$$

Example 20. If $A = \begin{pmatrix} a & b \\ c & d \end{pmatrix}$ and I is the unit matrix of order 2, show that

$$A^2 - (a+d)A = (bc - ad)I$$

Solution :

$$A^2 = \begin{pmatrix} a & b \\ c & d \end{pmatrix}\begin{pmatrix} a & b \\ c & d \end{pmatrix} = \begin{pmatrix} a^2 + bc & ab + bd \\ ac + dc & bc + d^2 \end{pmatrix}$$

Now, $$(a+d)A = \begin{pmatrix} a^2 + ad & ab + bd \\ ac + dc & ad + d^2 \end{pmatrix}$$

$$A^2 - (a+d)A = \begin{pmatrix} bc - ad & 0 \\ 0 & bc - ad \end{pmatrix} = (bc - ad)\begin{pmatrix} 1 & 0 \\ 0 & 1 \end{pmatrix} = (bc - ad)I$$

Example 21. Evaluate $(x \ \ y \ \ z)\begin{pmatrix} a & h & g \\ h & b & f \\ g & f & c \end{pmatrix}\begin{pmatrix} x \\ y \\ z \end{pmatrix}$.

Solution : Consider

$$\begin{pmatrix} a & h & g \\ h & b & f \\ g & f & c \end{pmatrix}_{3\times 3}\begin{pmatrix} x \\ y \\ z \end{pmatrix}_{3\times 1} = \begin{pmatrix} ax + hy + gz \\ hx + by + fz \\ gx + fy + cz \end{pmatrix}_{3\times 1}$$

Now, $$(x \ \ y \ \ z)_{1\times 3} \times \begin{pmatrix} ax + hy + gz \\ hx + by + fz \\ gx + fy + cz \end{pmatrix}_{3\times 1} = [(ax + hy + gz)x + (hx + by + fz)y + (gx + fy + cz)z]$$

$$= (ax^2 + by^2 + cz^2 + 2hxy + 2fyz + 2gzx)_{1\times 1}$$

Example 22. If $\begin{pmatrix} 1 & 0 & 0 \\ 0 & y & 0 \\ 0 & 0 & 1 \end{pmatrix}\begin{pmatrix} x \\ 1 \\ z \end{pmatrix} = \begin{pmatrix} 1 \\ 0 \\ 1 \end{pmatrix}$, find x, y and z.

Solution : By data, $\begin{pmatrix} 1 & 0 & 0 \\ 0 & y & 0 \\ 0 & 0 & 1 \end{pmatrix}_{3\times 3} \begin{pmatrix} x \\ 1 \\ z \end{pmatrix}_{3\times 1} = \begin{pmatrix} 1 \\ 0 \\ 1 \end{pmatrix}_{3\times 1} \Rightarrow \begin{pmatrix} x \\ y \\ z \end{pmatrix} = \begin{pmatrix} 1 \\ 0 \\ 1 \end{pmatrix} \Rightarrow$ $\mathbf{x = 1,\ y = 0,\ z = 1}$

Example 23. If $A = \begin{pmatrix} 1 & 2 \\ 2 & 4 \\ 5 & 6 \end{pmatrix}$ **and** $B = \begin{pmatrix} 1 & -2 & 5 \\ 2 & 4 & -6 \end{pmatrix}$, **show that** $(AB)' = B' \cdot A'$.

Solution : $A \cdot B = \begin{pmatrix} 1 & 2 \\ 2 & 4 \\ 5 & 6 \end{pmatrix} \begin{pmatrix} 1 & -2 & 5 \\ 2 & 4 & -6 \end{pmatrix} = \begin{pmatrix} 1+4 & -2+8 & 5-12 \\ 2+8 & -4+16 & 10-24 \\ 5+12 & -10+24 & 25-36 \end{pmatrix} = \begin{pmatrix} 5 & 6 & -7 \\ 10 & 12 & -14 \\ 17 & 14 & -11 \end{pmatrix}$

$$(AB)' = \begin{pmatrix} 5 & 10 & 17 \\ 6 & 12 & 14 \\ -7 & -14 & -11 \end{pmatrix}$$

$$B'A' = \begin{pmatrix} 1 & 2 \\ -2 & 4 \\ 5 & -6 \end{pmatrix} \begin{pmatrix} 1 & 2 & 5 \\ 2 & 4 & 6 \end{pmatrix} = \begin{pmatrix} 1+4 & 2+8 & 5+12 \\ -2+8 & -4+16 & -10+24 \\ 5-12 & 10-24 & 25-36 \end{pmatrix} = \begin{pmatrix} 5 & 10 & 17 \\ 6 & 12 & 14 \\ -7 & -14 & -11 \end{pmatrix}$$

Clearly, $\mathbf{(AB)' = B'A'}$

Exercise

I.

1. If $A = \begin{pmatrix} 0 & 2 & 4 \\ 2 & 1 & 3 \end{pmatrix}$ and $B = \begin{pmatrix} 7 & 6 & 3 \\ 1 & 4 & 5 \end{pmatrix}$ find $3A - B$.

2. Find $\begin{pmatrix} a^2+b^2 & b^2+c^2 \\ a^2+c^2 & a^2+b^2 \end{pmatrix} + \begin{pmatrix} 2ab & 2bc \\ -2ac & -2ab \end{pmatrix}$.

3. If $A = \begin{pmatrix} 2 & 4 \\ 3 & 2 \end{pmatrix}$, $B = \begin{pmatrix} 1 & 3 \\ -2 & 5 \end{pmatrix}$, $C = \begin{pmatrix} -2 & 5 \\ 3 & 4 \end{pmatrix}$, find $2A - B + 2C$.

4. Find the matrix A, if $2A + B = \begin{pmatrix} 1 & 0 \\ -3 & 2 \end{pmatrix}$ where $B = \begin{pmatrix} 3 & 2 \\ 1 & 4 \end{pmatrix}$.

5. If $A = \begin{pmatrix} 2 & 3 \\ 1 & 5 \end{pmatrix}$, find X, such that $A - 2X = \begin{pmatrix} 1 & 8 \\ 7 & -6 \end{pmatrix}$.

6. If $A + B + C = O$, where $A = \begin{pmatrix} 1 & -3 \\ 2 & -1 \\ 2 & 0 \end{pmatrix}$, $B = \begin{pmatrix} 1 & 2 \\ 0 & 1 \\ 1 & 3 \end{pmatrix}$ find C.

7. If $A = \begin{pmatrix} 3 & 1 & 2 \\ 4 & 3 & 5 \end{pmatrix}$, $B = \begin{pmatrix} 2 & -1 & 4 \\ 1 & 3 & 2 \end{pmatrix}$ verify $(A + B)' = A' + B'$.

8. If $A = \begin{pmatrix} 1 & -1 \\ -2 & 3 \end{pmatrix}$ and $B = \begin{pmatrix} 4 & 1 \\ 3 & -2 \end{pmatrix}$ find $A \cdot B$ and $B \cdot A$.

9. If $A = \begin{pmatrix} 1 & 2 & 3 \\ 2 & 3 & 4 \\ -1 & 1 & 2 \end{pmatrix}$ and $B = \begin{pmatrix} 0 & 2 & -1 \\ -1 & 3 & 4 \\ 0 & -2 & -3 \end{pmatrix}$ find $A \cdot B$ and $B \cdot A$.

10. Find $A \cdot B$, if $A = (1\ 3\ 8)$ and $B = \begin{pmatrix} 7 \\ 3 \\ 1 \end{pmatrix}$.

11. Find $A \cdot B$, if $A = \begin{pmatrix} 1 \\ 4 \\ 2 \end{pmatrix}$ and $B = (1\ \ 8\ \ 1)$.

12. Show that $(1\ \ 1\ \ 1) \begin{pmatrix} 1 & 0 & 0 \\ 0 & 1 & 0 \\ 0 & 0 & 1 \end{pmatrix} \begin{pmatrix} 4 \\ 4 \\ 4 \end{pmatrix} = (12)$.

13. If $A = \begin{pmatrix} 1 & i \\ -i & 1 \end{pmatrix}$ and $B = \begin{pmatrix} i & -1 \\ -1 & -i \end{pmatrix}$, find $A \cdot B$ and $B \cdot A$; $i^2 = -1$.

14. If $A = \begin{pmatrix} 2 & -3 & -5 \\ -1 & 4 & 5 \\ -1 & 3 & 4 \end{pmatrix}$ and $B = \begin{pmatrix} -1 & 3 & 5 \\ 1 & -3 & -5 \\ -1 & 3 & 5 \end{pmatrix}$, show that $A \cdot B = O$.

15. If $A = \begin{pmatrix} 1 & 2 & 3 \\ 1 & 2 & 3 \\ -1 & -2 & -3 \end{pmatrix}$, find A^2.

16. If $A = \begin{pmatrix} 1 & -1 \\ -1 & 1 \end{pmatrix}$, show that $A^2 = 2A$.

17. If $(2\ \ x\ \ 2) \begin{pmatrix} 1 \\ 3 \\ 5 \end{pmatrix} = (9)$, show that $x = -1$.

18. If $A = \begin{pmatrix} 3 & -1 & 2 \\ 1 & 3 & 2 \end{pmatrix}$, find $A \cdot A'$ and $A' \cdot A$.

19. If $A = \begin{pmatrix} 1 & -1 & 0 \\ 2 & 3 & 4 \\ 0 & 1 & 2 \end{pmatrix}$, $B = \begin{pmatrix} 2 & 2 & -4 \\ -4 & 2 & 4 \\ 2 & -1 & 5 \end{pmatrix}$, show that $BA = 6I$.

II.

1. If $A = \begin{pmatrix} 3 & 8 & 1 \\ 2 & -6 & 3 \\ 7 & 4 & -5 \end{pmatrix}$, $B = \begin{pmatrix} 4 & 0 & 2 \\ 6 & 2 & 3 \\ 1 & 3 & 2 \end{pmatrix}$, verify $3(A + B) = 3A + 3B$.

2. If $\begin{pmatrix} 3 & 2 \\ 4 & 1 \end{pmatrix} \begin{pmatrix} a & 1 \\ 5 & b \end{pmatrix} = \begin{pmatrix} 4 & 5 \\ -3 & 5 \end{pmatrix}$, show that $a = -2$, $b = 1$

3. If $\begin{pmatrix} 3 \\ 2 \\ 2 \end{pmatrix} (4\ \ 2) \begin{pmatrix} 3 \\ -1 \end{pmatrix} = \begin{pmatrix} x \\ y \\ z \end{pmatrix}$, find x, y, z.

4. If $\begin{pmatrix} x \\ y \end{pmatrix} = \begin{pmatrix} 2 & -1 \\ 3 & 4 \end{pmatrix} \begin{pmatrix} 2 \\ -1 \end{pmatrix}$, show that $x = 5$, $y = 2$.

5. If $\begin{pmatrix} x \\ y \\ z \end{pmatrix} = \begin{pmatrix} 1 & 2 & 3 \\ 3 & 4 & 5 \\ 6 & 7 & 8 \end{pmatrix} \begin{pmatrix} 2 \\ 3 \\ 4 \end{pmatrix}$, find x, y, z.

6. If $\begin{pmatrix} 1 & 3 \\ 1 & 2 \end{pmatrix} \begin{pmatrix} 2 & -1 & 0 \\ 2 & 1 & 4 \end{pmatrix} = \begin{pmatrix} 8 & x & 12 \\ y & 1 & z \end{pmatrix}$ find x, y and z.

7. Solve for x and y, $x\begin{pmatrix}2\\1\end{pmatrix}+y\begin{pmatrix}3\\5\end{pmatrix}+\begin{pmatrix}4\\6\end{pmatrix}=\begin{pmatrix}12\\17\end{pmatrix}$

8. Find A and B if

(a) $A+B=\begin{pmatrix}7&0\\2&5\end{pmatrix},\ A-B=\begin{pmatrix}3&0\\0&3\end{pmatrix}$ **(b)** $2A+B=\begin{pmatrix}2&3&1\\1&4&0\end{pmatrix},\ 3A+2B=\begin{pmatrix}4&6&1\\2&3&5\end{pmatrix}$

(c) $2A-3B=\begin{pmatrix}2&-4\\-12&1\end{pmatrix},\ A+5B=\begin{pmatrix}1&24\\33&7\end{pmatrix}$

9. Find $x, y, a, b,$ if $\begin{pmatrix}3x+4y&2&x-2y\\a+b&2a-b&-1\end{pmatrix}=\begin{pmatrix}2&2&4\\5&-5&-1\end{pmatrix}$

10. Find x and y, if $\begin{pmatrix}x+y&3\\-1&x-y\end{pmatrix}=\begin{pmatrix}4&3\\-1&8\end{pmatrix}$

11. If $A=\begin{pmatrix}3&-5\\-4&2\end{pmatrix}$, show that $A^2-5A=14I.$ **12.** If $A=\begin{pmatrix}1&2\\3&4\end{pmatrix}$, show that $A^2-5A=2I.$

13. If $A=\begin{pmatrix}3&1\\-1&2\end{pmatrix}$, show that $A^2-5A+7I=0.$

14. If $A=\begin{pmatrix}1&3&0\\1&1&0\\4&1&0\end{pmatrix}$, and $B=\begin{pmatrix}0&1&0\\1&0&0\\0&5&1\end{pmatrix}$ show that $AB\neq BA.$

15. If $A=\begin{pmatrix}1&2\\2&4\\5&6\end{pmatrix}$, and $B=\begin{pmatrix}1&2&5\\2&4&6\end{pmatrix}$ verify $(A\cdot B)'=B'\cdot A'$

16. If $A=\begin{pmatrix}3&-1&2\\1&3&2\\0&1&-1\end{pmatrix}$, and $B=\begin{pmatrix}1&2\\2&-1\\1&1\end{pmatrix}$, verify $(A\cdot B)'=B'\cdot A'.$

III. 1. If $A=\begin{pmatrix}3&1\\4&2\end{pmatrix}, B=\begin{pmatrix}2&-1\\3&1\end{pmatrix}$ and $C=\begin{pmatrix}2&3\\4&1\end{pmatrix}$ verify $A\cdot(B+C)=A\cdot B+A\cdot C.$

2. If $A=\begin{pmatrix}3&2\\1&0\end{pmatrix}, B=\begin{pmatrix}4&5&6\\0&1&2\end{pmatrix}$ and $C=\begin{pmatrix}1&-4&-1\\-2&5&-3\\3&6&5\end{pmatrix}$ verify $(A\cdot B)\cdot C=A\cdot(B\cdot C).$

3. If $A=\begin{pmatrix}1&1&1\\1&1&1\\1&1&1\end{pmatrix}$, show that $A^n=3^{n-1}\cdot A$, where n is a positive integer.

4. If $A=\begin{pmatrix}2&-1\\-1&2\end{pmatrix}$, and $B=\begin{pmatrix}1&4\\-1&1\end{pmatrix}$ show that $(A+B)^2\neq A^2+B^2+2AB.$

5. If $A=\begin{pmatrix}i&0\\0&-i\end{pmatrix}, B=\begin{pmatrix}0&-1\\1&0\end{pmatrix}$ and $C=\begin{pmatrix}0&i\\i&0\end{pmatrix}$ where $i^2=-1$ show that

(a) $A^2=B^2=C^2=-I$ **(b)** $AB=-BA=-C.$

6. If $A = \begin{pmatrix} \cos\theta & \sin\theta \\ -\sin\theta & \cos\theta \end{pmatrix}$, show that $A^n = \begin{pmatrix} \cos n\theta & \sin n\theta \\ -\sin n\theta & \cos n\theta \end{pmatrix}$ where n is a positive integer.

7. If $A = \begin{pmatrix} -1 & 1 & 0 \\ 3 & -3 & 3 \\ 5 & -5 & 5 \end{pmatrix}$, $B = \begin{pmatrix} 0 & 4 & 3 \\ 1 & -3 & -3 \\ -1 & 4 & 4 \end{pmatrix}$ show that $A^2 \cdot B^2 = A^2$.

8. If $A = \begin{pmatrix} 3 & -4 \\ 1 & -1 \end{pmatrix}$, show that $A^n = \begin{pmatrix} 1+2n & -4n \\ n & 1-2n \end{pmatrix}$ where n is a positive integer.

9. If $A = \begin{pmatrix} 0 & -\tan\frac{\theta}{2} \\ \tan\frac{\theta}{2} & 0 \end{pmatrix}$, show that $I + A = (I - A) \cdot \begin{pmatrix} \cos\theta & -\sin\theta \\ \sin\theta & \cos\theta \end{pmatrix}$

10. If $f(x) = x^2 - 5x + 7$, find $f(A)$, where $A = \begin{pmatrix} 3 & 1 \\ -1 & 2 \end{pmatrix}$.

Answers

I. **1.** $\begin{pmatrix} -7 & 0 & 9 \\ 5 & -1 & 4 \end{pmatrix}$ **2.** $\begin{pmatrix} (a+b)^2 & (b+c)^2 \\ (a-c)^2 & (a-b)^2 \end{pmatrix}$ **3.** $\begin{pmatrix} -1 & 15 \\ 14 & 7 \end{pmatrix}$

4. $\begin{pmatrix} -1 & -1 \\ -2 & -1 \end{pmatrix}$ **5.** $\begin{pmatrix} 1/2 & -5/2 \\ 3 & 11/2 \end{pmatrix}$ **6.** $\begin{pmatrix} -2 & 1 \\ -2 & 0 \\ -3 & -3 \end{pmatrix}$

8. $AB = \begin{pmatrix} 1 & 3 \\ 1 & -8 \end{pmatrix}$, $BA = \begin{pmatrix} 2 & -1 \\ 7 & -9 \end{pmatrix}$ **9.** $AB = \begin{pmatrix} -2 & 2 & -2 \\ -3 & 5 & -2 \\ -1 & -3 & -1 \end{pmatrix}$, $BA = \begin{pmatrix} 5 & 5 & 6 \\ 1 & 11 & 17 \\ -1 & -9 & -14 \end{pmatrix}$

10. (24) **11.** $\begin{pmatrix} 1 & 8 & 1 \\ 4 & 32 & 4 \\ 2 & 16 & 2 \end{pmatrix}$ **13.** $AB = O,\ B \cdot A = \begin{pmatrix} 2i & -2 \\ -2 & -2i \end{pmatrix}$

15. Zero matrix **18.** $AA' = \begin{pmatrix} 14 & 4 \\ 4 & 14 \end{pmatrix}$, $A'A = \begin{pmatrix} 10 & 0 & 8 \\ 0 & 10 & 4 \\ 8 & 4 & 8 \end{pmatrix}$

II. **3.** $x = 30,\ y = 20,\ z = 20$ **5.** $x = 20, y = 38, z = 65$

6. $x = 2,\ y = 6,\ z = 8$ **7.** $x = 1,\ y = 2$

8. (a) $A = \begin{pmatrix} 5 & 0 \\ 1 & 4 \end{pmatrix}$ $B = \begin{pmatrix} 2 & 0 \\ 1 & 1 \end{pmatrix}$; **(b)** $A = \begin{pmatrix} 0 & 0 & 1 \\ 0 & 5 & -5 \end{pmatrix}$, $B = \begin{pmatrix} 2 & 3 & -1 \\ 1 & -6 & 10 \end{pmatrix}$

(c) $A = \begin{pmatrix} 1 & 4 \\ 3 & 2 \end{pmatrix}$ $B = \begin{pmatrix} 0 & 4 \\ 6 & 1 \end{pmatrix}$ **9.** $x = 2,\ y = -1,\ a = 0,\ b = 5$ **10.** $x = 6,\ y = -2$

III. 10. $A^2 - 5A + 7I = 0$

Chapter 6

Determinants

6.1 Introduction

In this chapter we shall introduce the concept of "determinants" and its properties. Further we shall see its usefullness, in solving system of linear equations in two and three unknowns.

6.2 Determinants and its Expansion

(a) Determinants of second order

An arrangement denoted by $\begin{vmatrix} a & b \\ c & d \end{vmatrix}$, where a, b, c and d are any four numbers, is called a **second order determinant** and its value is defined to be $ad - bc$.

Thus, $\begin{vmatrix} a & b \\ c & d \end{vmatrix} = ad - bc$

The number $ad - bc$ is also called the expansion of the determinant.

We usually denote determinants by Δ or by Δ_1, Δ_2 etc.

Sometimes, the determinants are denoted by A, B, C etc.

Examples : (i) $A = \begin{vmatrix} 2 & 3 \\ 5 & 4 \end{vmatrix} = 2 \cdot 4 - 3 \cdot 5 = 8 - 15 = -7$

$$B = \begin{vmatrix} 2 & -3 \\ 1 & 4 \end{vmatrix} = 2 \cdot 4 - (-3) \cdot 1 = 8 + 3 = 1$$

$$C = \begin{vmatrix} 0 & -1 \\ 3 & 4 \end{vmatrix} = 0 - (-3) = 3$$

(b) Determinant of third order

The arrangement denoted by $\begin{vmatrix} a_1 & b_1 & c_1 \\ a_2 & b_2 & c_2 \\ a_3 & b_3 & c_3 \end{vmatrix}$, where a_i's, b_i's and c_i's are any numbers, is called a **third order determinant**. Its value is given by

$$\begin{vmatrix} a_1 & b_1 & c_1 \\ a_2 & b_2 & c_2 \\ a_3 & b_3 & c_3 \end{vmatrix} = a_1 \begin{vmatrix} b_2 & c_2 \\ b_3 & c_3 \end{vmatrix} - b_1 \begin{vmatrix} a_2 & c_2 \\ a_3 & c_3 \end{vmatrix} + c_1 \begin{vmatrix} a_2 & b_2 \\ a_3 & b_3 \end{vmatrix}$$

$$= a_1 (b_2c_3 - c_2b_3) - b_1 (a_2c_3 - c_2a_3) + c_1 (a_2b_3 - b_2a_3)$$

Observe that, we have written the value of the third order determinant (also called the expansion of the third order determinant), interms of second order determinant. Further we have considered the first row elements and each is multiplied by the second order determinant, obtained by deleting the row and the column in which the first row elements exist, with positive or negative signs alternatively.

For instance, the first element a_1 of the first row is multiplied by the 2nd order determinant obtained by deleting the row and the column in which a_1 appears (i.e., deleting the first row and the first columns. Similarly b_1 is multiplied by the 2nd order determinant obtained by deleting the first row and the second column; here we have taken the –ve sign.

A third order determinant can also be expanded by considering the elements of any row or any column and the value will be same, but we have to put proper signs before the elements of the row or column which is under consideration.

The rule to put the signs before the elements, while expanding the determinant is given by $(-1)^{i+j}$, here i stands for the row number and j stands for column number of the element in which it lies.

For example, consider the element c_2 in the above determinant. This element lies in the 2nd row and 3rd column. Thus the sign that is associated to this element is given by $(-1)^{2+3} = (-1)^5 = -1$, i.e., negative sign. The signs associated to different elements of the third order determinant is shown below.

$$\begin{vmatrix} + & - & + \\ - & + & - \\ + & - & + \end{vmatrix}$$

Expansion of the determinant of A with the help of 2nd column elements is given by

$$\begin{vmatrix} a_1 & b_1 & c_1 \\ a_2 & b_2 & c_2 \\ a_3 & b_3 & c_3 \end{vmatrix} = -b_1 \begin{vmatrix} a_2 & c_2 \\ a_3 & c_3 \end{vmatrix} + b_2 \begin{vmatrix} a_1 & c_1 \\ a_3 & c_3 \end{vmatrix} - b_3 \begin{vmatrix} a_1 & c_1 \\ a_2 & c_2 \end{vmatrix}$$

$$= -b_1 (a_2c_3 - c_2a_3) + b_2 (a_1c_3 - a_3c_1) - b_3 (a_1c_2 - c_1a_2)$$

For example,

$$\begin{vmatrix} 1 & 2 & 3 \\ 2 & 3 & 4 \\ 3 & 4 & 5 \end{vmatrix} = 1 \begin{vmatrix} 3 & 4 \\ 4 & 5 \end{vmatrix} - 2 \begin{vmatrix} 2 & 4 \\ 3 & 5 \end{vmatrix} + 3 \begin{vmatrix} 2 & 3 \\ 3 & 4 \end{vmatrix}$$

$$= (15 - 16) - 2 (10 - 12) + 3 (8 - 9) = -1 + 4 - 3 = \mathbf{0}$$

Again expanding by using the elements of second column we have,

$$\begin{vmatrix} 1 & 2 & 3 \\ 2 & 3 & 4 \\ 3 & 4 & 5 \end{vmatrix} = -2 \begin{vmatrix} 2 & 4 \\ 3 & 5 \end{vmatrix} + 3 \begin{vmatrix} 1 & 3 \\ 3 & 5 \end{vmatrix} - 4 \begin{vmatrix} 1 & 3 \\ 2 & 4 \end{vmatrix}$$

$$= -2 (10 - 12) + 3 (5 - 9) - 4 (4 - 6) = 4 - 12 + 8 = \mathbf{0}$$

Observe that the value of the determinant is same in both the cases.

Example 1. Find the value of (a) $\begin{vmatrix} 2 & 3 \\ -1 & 4 \end{vmatrix}$ (b) $\begin{vmatrix} \tan x & \sec x \\ \sec x & \tan x \end{vmatrix}$.

Solution : (a) $\begin{vmatrix} 2 & 3 \\ -1 & 4 \end{vmatrix} = 8 - (-3) = \mathbf{11}$ **(b)** $\begin{vmatrix} \tan x & \sec x \\ \sec x & \tan x \end{vmatrix} = \tan^2 x - \sec^2 x = \mathbf{1}$

Example 2. Find x if $\begin{vmatrix} x & 2 \\ 2 & x \end{vmatrix} = 0$.

Solution : By data, $\begin{vmatrix} x & 2 \\ 2 & x \end{vmatrix} = 0 \Rightarrow x^2 - 4 = 0 \Rightarrow x^2 = 4 \Rightarrow \mathbf{x = \pm 2}$

Example 3. Evaluate $\begin{vmatrix} a & b & c \\ b & c & a \\ c & a & b \end{vmatrix}$.

Solution : $\begin{vmatrix} a & b & c \\ b & c & a \\ c & a & b \end{vmatrix} = a(bc - a^2) - b(b^2 - ac) + c(ab - c^2)$

$= abc - a^3 - b^3 + abc + abc - c^3 = \mathbf{3abc - a^3 - b^3 - c^3}$

Example 4. Find x, if $\begin{vmatrix} 1 & 2 & 3 \\ 2 & x & 3 \\ 3 & 4 & 3 \end{vmatrix} = 0$.

Solution : $\begin{vmatrix} 1 & 2 & 3 \\ 2 & x & 3 \\ 3 & 4 & 3 \end{vmatrix} = 0 \Rightarrow 1(3x - 12) - 2(6 - 9) + 3(8 - 3x) = 0$

$\Rightarrow 3x - 12 + 6 + 24 - 9x = 0 \Rightarrow -6x + 18 = 0 \Rightarrow 6x = 18 \Rightarrow \mathbf{x = 3}$

6.3 Properties of Determinants

In this section we shall state few properties of determinants which are useful in working the problems.

I. The value of the determinant remains same if the rows and columns are interchanged.

II. If two rows (or columns) of a determinant is interchanged, the value of the determinant changes by a sign.

III. If in a determinant any two rows (or columns) are identical, then the value of the determinant is zero.

IV. If in a determinant the elements of any row (or column) are multiplied by the same scalar, say *k*, then the value of the new determinant is *k* times the given determinant.

Note : The property IV implies that, if there is a common factor among the elements of a row or a column, it can be taken outside the determinant.

V. If in a determinant the elements of any row (or column) is made up of a sum of two quantities, then the determinant can be expressed as the sum of two determinants of the same order.

VI. If each element of a row (or column) is multiplied by a scalar and added to the corresponding elements of any other row (or column), then the value of the determinant is unaltered.

VII. If in a determinant all the elements of any row (or column) are zeros then the value of the determinant is zero.

VIII. If all the elements on one side of the principal diagonal of a determinant are zeros, then the value of the determinant is the product of the elements of the principal diagonal.

For example, $\begin{vmatrix} 2 & -1 & 3 \\ 0 & 4 & 1 \\ 0 & 0 & 2 \end{vmatrix} = 2 \cdot 4 \cdot 2 = 16$; $\begin{vmatrix} 1 & 0 & 0 \\ 2 & -1 & 0 \\ 3 & 2 & 4 \end{vmatrix} = 1 \cdot (-1)\, 4 = -4$

The above properties of determinant are very useful in evaluating the determinant. We use the following notation for the operations applied on the rows and columns of a determinant.

If two rows say i^{th} and j^{th} rows are interchanged, then it is denoted by $R_i \leftrightarrow R_j$. If the ith and j^{th} column are interchanged it is denoted by $C_i \leftrightarrow C_j$.

If the element of ith row is multiplied by a scalar k, then it is denoted by $R_i \to k\,R_i$. Similar notation for columns.

If k times of the elements of ith row of a determinant are added to the corresponding elements of the jth row, then it is denoted by $R_i \to R_j + k\,R_i$. Similar notation for the columns also.

Example 1. Evaluate $\begin{vmatrix} \mathbf{8} & \mathbf{2} & \mathbf{1} \\ \mathbf{12} & \mathbf{3} & \mathbf{-5} \\ \mathbf{16} & \mathbf{4} & \mathbf{2} \end{vmatrix}$.

Solution : In the given determinant, the elements of the first column as a common factor 4. Thus we can write

$$\begin{vmatrix} 8 & 2 & 1 \\ 12 & 3 & -5 \\ 16 & 4 & 2 \end{vmatrix} = 4 \begin{vmatrix} 2 & 2 & 1 \\ 3 & 3 & -5 \\ 4 & 4 & 2 \end{vmatrix} = 0 \qquad (\because \text{two columns are identical})$$

Example 2. Show that $\begin{vmatrix} x & y & y \\ y & x & y \\ y & y & x \end{vmatrix} = (x + 2y)\,(x - y)^2$.

Solution : Let the given determinant be Δ. Consider $R_1 \to R_1 + R_2 + R_3$, we have,

$$\Delta = \begin{vmatrix} x+2y & x+2y & x+2y \\ y & x & y \\ y & y & x \end{vmatrix} = (x+2y) \begin{vmatrix} 1 & 1 & 1 \\ y & x & y \\ y & y & x \end{vmatrix}$$

$$= (x+2y)\begin{vmatrix} 1 & 0 & 0 \\ y & x-y & 0 \\ y & 0 & x-y \end{vmatrix} \quad \begin{matrix} C_2 \to C_2 - C_1 \\ C_3 \to C_3 - C_1 \end{matrix}$$

$$= (x+2y)\,[1(x-y)(x-y) - 0] = \mathbf{(x+2y)(x-y)^2}$$

Example 3. Show (without actual expansion) that each of the following determinants is zero.

$$\textbf{(a)} \begin{vmatrix} 1 & a & b+c \\ 1 & b & c+a \\ 1 & c & a+b \end{vmatrix} \qquad \textbf{(b)} \begin{vmatrix} 0 & a & -b \\ -a & 0 & c \\ b & -c & 0 \end{vmatrix}$$

Solution : (a) Let the given determinant be Δ. Consider $C_3 \to C_3 + C_2$, we get

$$\Delta = \begin{vmatrix} 1 & a & a+b+c \\ 1 & b & a+b+c \\ 1 & c & a+b+c \end{vmatrix};$$ Taking $a+b+c$ common from C_3, we get

$$\Delta = (a+b+c)\begin{vmatrix} 1 & a & 1 \\ 1 & b & 1 \\ 1 & c & 1 \end{vmatrix} = (a+b+c)\cdot 0 = \mathbf{0} \qquad (\because C_1 \text{ and } C_3 \text{ are identical})$$

(b) Let the given determinant be Δ.

$$\Delta = \begin{vmatrix} 0 & a & -b \\ -a & 0 & c \\ b & -c & 0 \end{vmatrix} = (-1)^3\begin{vmatrix} 0 & -a & b \\ a & 0 & -c \\ -b & c & 0 \end{vmatrix} \qquad \text{(multiplying each row by } -1)$$

$$= -\begin{vmatrix} 0 & a & -b \\ -a & 0 & c \\ b & -c & 0 \end{vmatrix} \qquad \text{(by interchanging rows and columns)}$$

$$\Rightarrow \qquad \Delta = -\Delta \quad \Rightarrow 2\Delta = 0 \quad \Rightarrow \mathbf{\Delta = 0}$$

Example 4. Show that $\begin{vmatrix} \mathbf{1} & \boldsymbol{a} & \boldsymbol{bc} \\ \mathbf{1} & \boldsymbol{b} & \boldsymbol{ca} \\ \mathbf{1} & \boldsymbol{c} & \boldsymbol{ab} \end{vmatrix} = \mathbf{(b-c)(c-a)(a-b)}$**.**

Solution : Consider $R_2 \to R_2 - R_1$, $R_3 \to R_3 - R_1$, we get

$$\text{LHS} = \begin{vmatrix} 1 & a & bc \\ 0 & b-a & c(a-b) \\ 0 & c-a & b(a-c) \end{vmatrix} = (a-b)(c-a)\begin{vmatrix} 1 & a & bc \\ 0 & -1 & c \\ 0 & 1 & -b \end{vmatrix}$$

Taking common $(a-b)$ and $(c-a)$ from R_2 and R_3

$$= (a-b)(c-a)\cdot 1\,[(b-c)] \qquad \text{(expanding along } C_1)$$

$$= \mathbf{(b-c)(c-a)(a-b)}$$

Example 5. Show that $\begin{vmatrix} \mathbf{1} & \mathbf{1} & \mathbf{1} \\ \boldsymbol{a} & \boldsymbol{b} & \boldsymbol{c} \\ \boldsymbol{a^2} & \boldsymbol{b^2} & \boldsymbol{c^2} \end{vmatrix} = \boldsymbol{(a-b)(b-c)(c-a)}$.

Solution : Consider $C_1 \to C_1 - C_3$, $C_2 \to C_2 - C_3$, we get

$$\text{LHS} = \begin{vmatrix} 0 & 0 & 1 \\ a-c & b-c & c \\ a^2-c^2 & b^2-c^2 & c^2 \end{vmatrix}.$$ Taking common $(a-c)$ and $(b-c)$ from C_1 and C_2

$$= (a-c)(b-c)\begin{vmatrix} 0 & 0 & 1 \\ 1 & 1 & c \\ a+c & b+c & c^2 \end{vmatrix}.$$ Now expanding along R_1, we get,

$$= (a-c)(b-c)[1(b+c-a-c)] = \boldsymbol{(a-b)(b-c)(c-a)} = \text{RHS}$$

Example 6. Show that $\begin{vmatrix} \mathbf{1} & \mathbf{1} & \mathbf{1} \\ \boldsymbol{a} & \boldsymbol{b} & \boldsymbol{c} \\ \boldsymbol{a^3} & \boldsymbol{b^3} & \boldsymbol{c^3} \end{vmatrix} = \boldsymbol{(a-b)(b-c)(c-a)(a+b+c)}$.

Solution : Consider, $C_1 \to C_1 - C_2$, $C_2 \to C_2 - C_3$

$$\text{LHS} = \begin{vmatrix} 0 & 0 & 1 \\ a-b & b-c & c \\ a^3-b^3 & b^3-c^3 & c^3 \end{vmatrix}$$ (Taking common $(a-b)$ and $(b-c)$ from C_1 and C_2)

$$= (a-b)(b-c)\begin{vmatrix} 0 & 0 & 1 \\ 1 & 1 & c \\ a^2+ab+b^2 & b^2+bc+c^2 & c^3 \end{vmatrix}$$

$$= (a-b)(b-c)[1(b^2+bc+c^2-a^2-ab-b^2)]$$

$$= (a-b)(b-c)[b(c-a)+(c^2-a^2)]$$

$$= (a-b)(b-c)[(c-a)(b+c+a)] = \boldsymbol{(a-b)(b-c)(c-a)(a+b+c)} = \text{RHS}$$

Example 7. Solve $\begin{vmatrix} \mathbf{3x-8} & \mathbf{3} & \mathbf{3} \\ \mathbf{3} & \mathbf{3x-8} & \mathbf{3} \\ \mathbf{3} & \mathbf{3} & \mathbf{3x-8} \end{vmatrix} = \mathbf{0}$.

Solution : Consider $R_1 \to R_1 + R_2 + R_3$ we get,

$$\begin{vmatrix} 3x-2 & 3x-2 & 3x-2 \\ 3 & 3x-8 & 3 \\ 3 & 3 & 3x-8 \end{vmatrix} = 0 \quad \Rightarrow \quad (3x-2)\begin{vmatrix} 1 & 1 & 1 \\ 3 & 3x-8 & 3 \\ 3 & 3 & 3x-8 \end{vmatrix} = 0$$

Consider $C_2 \to C_2 - C_1$ and $C_3 \to C_3 - C_1$ we get,

$$\Rightarrow \quad (3x-2)\begin{vmatrix} 1 & 0 & 0 \\ 3 & 3x-11 & 0 \\ 3 & 0 & 3x-11 \end{vmatrix} = 0$$

$$\Rightarrow \quad (3x-2)\,[1\,(3x-11)\,(3x-11)] = 0$$

$$\Rightarrow \quad (3x-2)\,(3x-11)\,(3x-11) = 0$$

$$\Rightarrow \quad x = \frac{2}{3}, \frac{11}{3} \text{ and } \frac{11}{3}$$

Example 8. Solve $\begin{vmatrix} 1 & -2 & x+3 \\ 1 & x-2 & 3 \\ x+1 & -2 & 3 \end{vmatrix} = 0.$

Solution : Consider $C_1 \to C_1 + C_2 + C_3$ we get,

$$\begin{vmatrix} x+2 & -2 & x+3 \\ x+2 & x-2 & 3 \\ x+2 & -2 & 3 \end{vmatrix} = 0 \quad \Rightarrow \quad (x+2)\begin{vmatrix} 1 & -2 & x+3 \\ 1 & x-2 & 3 \\ 1 & -2 & 3 \end{vmatrix} = 0$$

Consider $R_2 \to R_2 - R_1$ and $R_3 \to R_3 - R_1$ we get,

$$\Rightarrow \quad (x+2)\begin{vmatrix} 1 & -2 & x+3 \\ 0 & x & -x \\ 0 & 0 & -x \end{vmatrix} = 0$$

$$\Rightarrow \quad (x+2)\,[1\,(-x^2-0)] = 0$$

$$\Rightarrow \quad (x+2)\,(-x^2) = 0 \quad \Rightarrow \quad x = 0, -2$$

Exercise

I.

1. Evaluate the following

(a) $\begin{vmatrix} 4200 & 4201 \\ 4202 & 4203 \end{vmatrix}$ **(b)** $\begin{vmatrix} 5431 & 5433 \\ 5435 & 5437 \end{vmatrix}$ **(c)** $\begin{vmatrix} 500 & 503 \\ 506 & 509 \end{vmatrix}$ **(d)** $\begin{vmatrix} 4785 & 4787 \\ 4789 & 4791 \end{vmatrix}$ **(e)** $\begin{vmatrix} \cos\theta & -\sin\theta \\ -\sin\theta & \cos\theta \end{vmatrix}$

2. Evaluate the following

(a) $\begin{vmatrix} 101 & 102 & 103 \\ 104 & 105 & 106 \\ 107 & 108 & 109 \end{vmatrix}$ **(b)** $\begin{vmatrix} 1 & 6 & 7 \\ 2 & 3 & 0 \\ 0 & 1 & 4 \end{vmatrix}$ **(c)** $\begin{vmatrix} 12 & 0 & 0 \\ 14 & 3 & 0 \\ 2 & 2 & -3 \end{vmatrix}$

(d) $\begin{vmatrix} 1 & 3 & 5 \\ 2 & 6 & 10 \\ 31 & 11 & 39 \end{vmatrix}$ **(e)** $\begin{vmatrix} x & y & z \\ 1 & 2 & 3 \\ 2 & 4 & 6 \end{vmatrix}$ **(f)** $\begin{vmatrix} 81 & 82 & 83 \\ 84 & 85 & 86 \\ 87 & 88 & 89 \end{vmatrix}$

3. If $\begin{vmatrix} x & 12 \\ 12 & x \end{vmatrix} = 0$ find x. **4.** If $\begin{vmatrix} x & 3 \\ 8 & 4 \end{vmatrix} = 0$ find x. **5.** If $\begin{vmatrix} 2 & x & 3 \\ 4 & 1 & 6 \\ 1 & 2 & 7 \end{vmatrix} = 0$ find x.

6. If $\begin{vmatrix} 1 & 2 & 3 \\ 2 & x & 3 \\ 3 & 4 & 3 \end{vmatrix} = 0$ find x.

II.

1. Without actual expansion, show that

(a) $\begin{vmatrix} 1 & 5 & 7 \\ 5 & 25 & 35 \\ 12 & 20 & 24 \end{vmatrix} = 0$ **(b)** $\begin{vmatrix} 43 & 1 & 6 \\ 35 & 7 & 4 \\ 17 & 3 & 2 \end{vmatrix} = 0$ **(c)** $\begin{vmatrix} 1 & 1 & 1 \\ a & b & c \\ b+c & c+a & a+b \end{vmatrix} = 0$

(d) $\begin{vmatrix} x+y & y+z & z+x \\ x & y & z \\ 1 & 1 & 1 \end{vmatrix} = 0$ **(e)** $\begin{vmatrix} x & 4 & y+z \\ y & 4 & z+x \\ z & 4 & x+y \end{vmatrix} = 0$ **(f)** $\begin{vmatrix} a-b & b-c & c-a \\ x-y & y-z & z-x \\ p-q & q-r & r-p \end{vmatrix} = 0$

(g) $\begin{vmatrix} a-b & b-c & c-a \\ b-c & c-a & a-b \\ c-a & a-b & b-c \end{vmatrix} = 0$ **(h)** $\begin{vmatrix} \sin^2 x & \cos^2 x & 1 \\ \cos^2 x & \sin^2 x & 1 \\ -10 & 12 & 2 \end{vmatrix} = 0$ **(i)** $\begin{vmatrix} 0 & p-q & p-r \\ q-p & 0 & q-r \\ r-p & r-q & 0 \end{vmatrix} = 0$

(j) $\begin{vmatrix} 1 & \log_x y & \log_x z \\ \log_y x & 1 & \log_y z \\ \log_z x & \log_z y & 1 \end{vmatrix} = 0$ (x, y, z are positive) **(k)** $\begin{vmatrix} 2 & 2^2 & 2^3 \\ 2^4 & 2^5 & 2^6 \\ 2^7 & 2^8 & 2^9 \end{vmatrix} = 0$ **(l)** $\begin{vmatrix} 4 & 4^2 & 4^3 \\ 4^2 & 4^3 & 4^4 \\ 4^3 & 4^4 & 4^5 \end{vmatrix} = 0$

III.

1. Show that $\begin{vmatrix} 1 & 1 & 1 \\ bc & ca & ab \\ b+c & c+a & a+b \end{vmatrix} = (a-b)(b-c)(c-a)$.

2. Show that $\begin{vmatrix} x & y & p \\ p & x & q \\ p & q & x \end{vmatrix} = (x+p+q)(x-p)(x-q)$.

3. Show that $\begin{vmatrix} a & b+c & a^2 \\ b & c+a & b^2 \\ c & a+b & c^2 \end{vmatrix} = -(a+b+c)(a-b)(b-c)(c-a)$.

4. Show that $\begin{vmatrix} 1 & 1 & 1 \\ a^2 & b^2 & c^2 \\ bc & ca & ab \end{vmatrix} = (a+b+c)(a-b)(b-c)(c-a)$.

5. Show that $\begin{vmatrix} 1 & 1 & 1 \\ x^2 & y^2 & z^2 \\ x^3 & y^3 & z^3 \end{vmatrix} = (x-y)(y-z)(z-x)(xy+yz+zx)$.

6. Show that $\begin{vmatrix} 1+a^2 & ab & ac \\ ab & 1+b^2 & bc \\ ac & bc & 1+c^2 \end{vmatrix} = 1 + a^2 + b^2 + c^2$.

7. Show that $\begin{vmatrix} b+c & a & a \\ b & c+a & b \\ c & c & a+b \end{vmatrix} = 4abc$.

8. Solve

(a) $\begin{vmatrix} x+1 & x+2 & 3 \\ 3 & x+2 & x+1 \\ x+1 & z & x+3 \end{vmatrix} = 0$ **(b)** $\begin{vmatrix} x+1 & 2 & 3 \\ 1 & x+2 & 3 \\ 1 & 2 & x+3 \end{vmatrix} = 0$ **(c)** $\begin{vmatrix} x+2 & 3 & 4 \\ 2 & x+3 & 4 \\ 2 & 3 & x+4 \end{vmatrix} = 0$

(d) $\begin{vmatrix} 3-x & -6 & 3 \\ -6 & 3-x & 3 \\ 3 & 3 & -6-x \end{vmatrix} = 0$ **(e)** $\begin{vmatrix} x & -6 & -1 \\ 2 & -3x & x-3 \\ -3 & 2x & x+2 \end{vmatrix} = 0$ **(f)** $\begin{vmatrix} a+x & a-x & a-x \\ a-x & a+x & a-x \\ a-x & a-x & a+x \end{vmatrix} = 0$

(g) $\begin{vmatrix} x-2 & 2x-3 & 3x-4 \\ x-4 & 2x-9 & 3x-16 \\ x-8 & 2x-27 & 3x-64 \end{vmatrix} = 0$

Answers

I. 1. (a) -2 **(b)** -8 **(c)** -18 **(d)** -8 **(e)** $\cos 2\theta$

2. (a) 0 **(b)** -22 **(c)** -108 **(d)** 0 **(e)** 0 **(f)** 0

3. $x = \pm 12$ **4.** $x = 6$ **5.** $x = \frac{1}{2}$ **6.** $x = 3$

II. 8. (a) 0, 2, -3 **(b)** 0, -6 **(c)** 0, -9 **(d)** 0, 9, -9 **(e)** 2, -3, 1 **(f)** $3a$, 0 **(g)** 4

6.4 Solution of Linear Equations – Cramer's Rule

Consider the system of equations

$$\begin{aligned} a_1 x + b_1 y &= c_1 \\ a_2 x + b_2 y &= c_2 \end{aligned} \qquad \text{.... (1)}$$

in two variables x and y.

Let $\Delta = \begin{vmatrix} a_1 & b_1 \\ a_2 & b_2 \end{vmatrix}$, Δ is the determinant formed by taking the coefficients of x and y.

Consider, $x \cdot \Delta = x \begin{vmatrix} a_1 & b_1 \\ a_2 & b_2 \end{vmatrix} \Rightarrow x \cdot \Delta = \begin{vmatrix} a_1 x & b_1 \\ a_2 x & b_2 \end{vmatrix}$

$$\Rightarrow \quad x \cdot \Delta = \begin{vmatrix} a_1 x + b_2 y & b_1 \\ a_2 x + b_2 y & b_2 \end{vmatrix} \qquad C_1 \to C_1 + yC_2$$

$$\Rightarrow \quad x \cdot \Delta = \begin{vmatrix} c_1 & b_1 \\ c_2 & b_2 \end{vmatrix} = \Delta_1 \text{ (say)} \qquad \text{(from (1))}$$

If $\Delta \neq 0$, $\quad x\,\Delta = \Delta_1 \Rightarrow x = \dfrac{\Delta_1}{\Delta}$

Similarly, $\quad y = \dfrac{\Delta_2}{\Delta}$, where $\Delta_2 = \begin{vmatrix} a_1 & c_1 \\ a_2 & c_2 \end{vmatrix}$

This method of solving the system of equations is called **Cramer's Rule.**

The above method is also useful to solve the system of equation in three variables. Consider

$$\begin{aligned} a_1x + b_1y + c_1z &= d_1 \\ a_2x + b_2y + c_2z &= d_2 \\ a_3x + b_3y + c_3z &= d_3 \end{aligned} \qquad \text{.... (2)}$$

in three variables x, y and z.

Let Δ be the determinant formed by taking the coefficients of x, y and z in the equations

$$\Delta = \begin{vmatrix} a_1 & b_1 & c_1 \\ a_2 & b_2 & c_2 \\ a_3 & b_3 & c_3 \end{vmatrix}$$

Consider, $$x\,\Delta = \begin{vmatrix} a_1x & b_1 & c_1 \\ a_2x & b_2 & c_2 \\ a_3x & b_3 & c_3 \end{vmatrix}$$

$\Rightarrow$ $$x\,\Delta = \begin{vmatrix} a_1x + b_1y + c_1z & b_1 & c_1 \\ a_2x + b_2y + c_2z & b_2 & c_2 \\ a_3x + b_3y + c_3z & b_3 & c_3 \end{vmatrix} \quad C_1 \to yC_2 + zC_3$$

$\Rightarrow$ $$x\,\Delta = \begin{vmatrix} d_1 & b_1 & c_1 \\ d_2 & b_2 & c_2 \\ d_3 & b_3 & c_3 \end{vmatrix} \qquad \text{[from (2)]}$$

Let the right side determinant be Δ_1. Then we have

$$x\,\Delta = \Delta_1 \quad \Rightarrow \quad x = \frac{\Delta_1}{\Delta} \qquad (\text{if } \Delta \neq 0)$$

Similarly, $y = \dfrac{\Delta_2}{\Delta}$ and $z = \dfrac{\Delta_3}{\Delta}$ where $\Delta_2 = \begin{vmatrix} a_1 & d_1 & c_1 \\ a_2 & d_2 & c_2 \\ a_3 & d_3 & c_3 \end{vmatrix}$ and $\Delta_3 = \begin{vmatrix} a_1 & b_1 & d_1 \\ a_2 & b_2 & d_2 \\ a_3 & b_3 & d_3 \end{vmatrix}$

Working rule of solving the system of equations in three variables by Cramer's rule.

1. **Write the given system in order so that the constant terms will be on right side.**
2. **Write the determinant Δ formed by the coefficients of x, y and z in the system.**
3. **Write the determinant Δ_1, by replacing the first column in Δ by the constant terms in the equation.**
4. **Replace the second column in Δ by constant terms and write Δ_2.**
5. **Replace the third column in Δ by constant terms and write Δ_3.**

Write the solution as $x = \dfrac{\Delta_1}{\Delta}$, $y = \dfrac{\Delta_2}{\Delta}$, $z = \dfrac{\Delta_3}{\Delta}$

Note : If $\Delta = 0$, then the solution cannot be obtained by Cramer's rule.

Example 1. Solve $2x + 3y - 1 = 0$, $3x - y + 2 = 0$ by Cramer's Rule.

Solution : The given system of equations is $2x + 3y = 1$; $3x - y = -2$

Now,

$$\Delta = \begin{vmatrix} 2 & 3 \\ 3 & -1 \end{vmatrix} = 2(-1) - 9 = \mathbf{-11} \ ; \ \Delta_1 = \begin{vmatrix} 1 & 3 \\ -2 & -1 \end{vmatrix} = -1 + 6 = \mathbf{5} \ ; \ \Delta_2 = \begin{vmatrix} 2 & 1 \\ 3 & -2 \end{vmatrix} = -4 - 3 = \mathbf{-7}$$

Now, $x = \frac{\Delta_1}{\Delta} = -\frac{\mathbf{5}}{\mathbf{11}}$ and $y = \frac{\Delta_2}{\Delta} = \frac{\mathbf{7}}{\mathbf{11}}$

Verification : Put, $x = -\frac{5}{11}$ and $y = \frac{7}{11}$ in the equation, we get

$$\text{LHS} = -\frac{10}{11} + \frac{21}{11} = 1 = \text{RHS} \ ; \quad \text{LHS} = -\frac{15}{11} - \frac{7}{11} = -2 = \text{RHS}$$

Example 2. Solve using Cramer's rule $5y + 2x + z = -1$; $x + 7y - 6z = -18$; $3y + 6z = 9$

Solution : The given system of equations is $2x + 5y + z = -1$; $x + 7y - 6z = -18$; $0x + 3y + 6z = 9$

Now, $$\Delta = \begin{vmatrix} 2 & 5 & 1 \\ 1 & 7 & -6 \\ 0 & 3 & 6 \end{vmatrix}$$

$$= 2\,(42 + 18) - 5\,(6 - 0) + 1\,(3 - 0) = 120 - 30 + 3 = 120 - 27 = \mathbf{93}$$

$$\Delta_1 = \begin{vmatrix} -1 & 5 & 1 \\ -18 & 7 & -6 \\ 9 & 3 & 6 \end{vmatrix}$$

$$= -1\,(42 + 18) - 5\,(-108 + 54) + 1\,(-54 - 63) = -60 + 270 - 117 = \mathbf{93}$$

$$\Delta_2 = \begin{vmatrix} 2 & -1 & 1 \\ 1 & -18 & -6 \\ 0 & 9 & 6 \end{vmatrix}$$

$$= 2\,(-108 + 54) + 1\,(6 - 0) + (9 - 0) = -108 + 6 + 9 = \mathbf{-93}$$

$$\Delta_3 = \begin{vmatrix} 2 & 5 & -1 \\ 1 & 7 & -18 \\ 0 & 3 & 9 \end{vmatrix}$$

$$= 2\,(63 + 54) - 5\,(9 - 0) - 1\,(3 - 0) = 234 - 45 - 3 = 234 - 48 = \mathbf{186}$$

$$\therefore \quad x = \frac{\Delta_1}{\Delta} = \frac{93}{93} = 1, \quad y = \frac{\Delta_2}{\Delta} = -\frac{93}{93} = -1, \quad z = \frac{\Delta_3}{\Delta} = \frac{186}{93} = 2. \ \Rightarrow \quad \mathbf{x = 1, \quad y = -1, \quad z = 2}$$

Example 3. Solve using Cramer's rule $4x + y = 7$; $3y + 4z = 5$; $3z + 5x = 2$.

Solution : The given system of equations is $4x + y + 0z = 7$; $0x + 3y + 4z = 5$; $5x + 0y + 3z = 2$

Now, $$\Delta = \begin{vmatrix} 4 & 1 & 0 \\ 0 & 3 & 4 \\ 5 & 0 & 3 \end{vmatrix} = 4\,(9 - 0) - 1\,(0 - 20) = 36 + 20 = \mathbf{56}$$

$$\Delta_1 = \begin{vmatrix} 7 & 1 & 0 \\ 5 & 3 & 4 \\ 2 & 0 & 3 \end{vmatrix} = 7\,(9-0) - 1\,(15-8) = 63 - 7 = \mathbf{56}$$

$$\Delta_2 = \begin{vmatrix} 4 & 7 & 0 \\ 0 & 5 & 4 \\ 5 & 2 & 3 \end{vmatrix} = 4\,(15-8) - 7\,(0-20) = 28 + 140 = \mathbf{168}$$

$$\Delta_3 = \begin{vmatrix} 4 & 1 & 7 \\ 0 & 3 & 5 \\ 5 & 0 & 2 \end{vmatrix} = 4\,(6-0) - 1\,(0-25) + 7\,(0-15) \quad = 24 + 25 - 105 = \mathbf{-56}$$

$$\therefore \quad x = \frac{\Delta_1}{\Delta} = \frac{56}{56} = 1, \quad y = \frac{\Delta_2}{\Delta} = \frac{168}{56} = 3, \quad z = \frac{\Delta_3}{\Delta} = -\frac{56}{56} = -1. \Rightarrow \boldsymbol{x = 1}, \quad \boldsymbol{y = 3}, \quad \boldsymbol{z = -1}$$

is the solution.

Exercise

I.

1. Solve the following system of equations by Cramer's rule.

(a) $2x - 3y + 1 = 0$, $3x + y - 1 = 0$ **(b)** $4x + 2y = 3$, $3x - 4y = 5$

(c) $3x - 2y + 1 = 0$, $3x - 4y - 2 = 0$ **(d)** $5x - 7y = 2$, $7x - 5y = 3$

II.

1. Solve the following system of equations by Cramer's rule.

(a) $5x - y - 4z = 5$, $2x + 3y + 5z = 2$, $7x - 2y + 6z = 5$.

(b) $x + 2y - z = 1$, $3x + 5y - 2z = 5$, $2x + 6y + 3z = -2$.

(c) $x - y + 2z = 7$, $3x + 4y - 5z = -5$, $2x - y + 3z = 12$.

(d) $x - y + 2z = 3$, $2x + z = 1$, $3x + 2y + z = 4$.

(e) $x - y - 2z = 3$, $2x + y + z = 5$, $4x - y - 2z = 11$.

(f) $5x - y + z = 4$, $3x + 2y - 5z = 2$, $x + 3y - 2z = 5$.

(g) $2x + 3y + 3z = 5$, $x - 2y + z = -4$, $3x - y - 2z = 3$.

(h) $2x + 5y + z = -1$, $x + 7y - 6z = -18$, $3y + 6z = 9$.

(i) $x + y + z = 7$, $2x + 3y + 2z = 17$, $4x + 9y + z = 37$.

Answers

I. 1. (a) $\frac{2}{11}, \frac{5}{11}$ **(b)** $1, -\frac{1}{2}$ **(c)** $\frac{-4}{3}, \frac{-3}{2}$ **(d)** $\frac{11}{24}, \frac{1}{24}$

II.1. (a) 3, 2, –2 **(b)** 5, –2, 0 **(c)** 2, 1, 3 **(d)** –1, 2, 3 **(e)** $\frac{8}{3}, \frac{-1}{3}, 0$

(f) 1, 2, 1 **(g)** 1, 2, –1 **(h)** 1, –1, 2 **(i)** 2, 3, 2.

UNIT II

Trigonometry

Chapter 1

Trigonometric Ratios of Acute Angle

1.1 Introduction

The word trigonometry is derived from two Greek words, trigonon (meaning a triangle) and metron (meaning 'I measure'). Thus the word trigonometry literally means "measurement of triangles". But at present it has much wider scope and application in every branch of science. In every branch of higher mathematics, pure or applied, a knowledge of trigonometry is of the greatest value.

1.2 Angles and Measurement of Angles

Let a line through a point O, be initially in the position OX. If this revolves about the point O, to an another point position OP, from the position on OX, then the angle between OX and OP is measured by the amount of revolution, which the line has undergone in passing from its initial position OX to its position OP. We denote this angle by $X\hat{O}P$ and read as angle XOP. The point O is called the **vertex** or the **origin**, the line OX is called **initial line** or **initial side** and the line OP is called the **generating line** or **terminating side** or the **radius vector**.

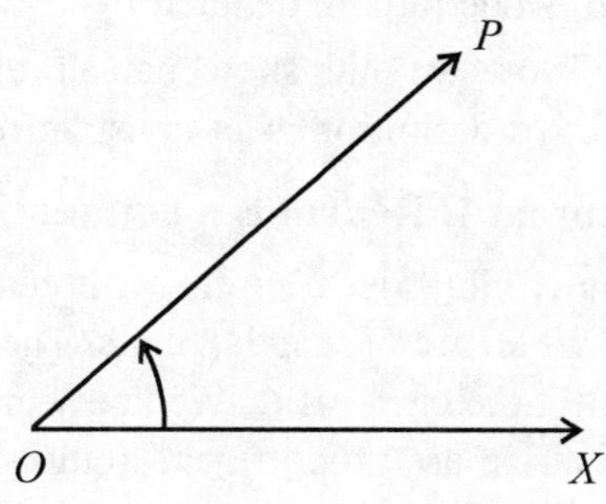

If the revolving line revolves in the anticlockwise direction from the initial line, to form an angle, then the angle described is said to be **positive**.

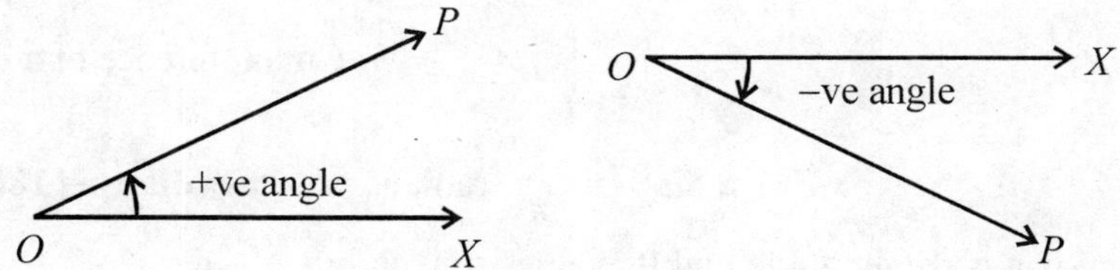

The angle described by the revolving line, by revolving in the clockwise direction from the initial line, is taken as **negative** or called negative angle.

1.3 Unit of Measurement of Angles

In general there are three units of measurement of an angle.

1. Sexagesimal system (English system) **2.** Centisimal system (French system)

3. Circular system or radian measure.

In sexagesimal system, the angle formed by the initial line by revolving one complete revolution, is divided into 360 equal parts. Each part is called **one degree** and it is denoted by 1°. The each part is again divided into 60 parts, and each part is called **one minute** and it is denoted by 1'. Further one minute is divided into 60 equal parts, and it is called **one second** and it is denoted by 1". Thus we have,

$$1 \text{ degree} = 60 \text{ minutes}$$

$$1 \text{ minute} = 60 \text{ seconds}$$

In this system, we call 90° = 1 right angle.

In centesimal system a right angle is divided into 100 equal parts and each part is called a **grade**, denoted by 1^g. Further **1^g = 100 minutes**, written 100'. Again 1 minute is equal to **100 seconds**, written as 100".

1.4 Radian Measure or Circular Measure

The unit of measurement used in circular system is called **Radian.** We shall give a formal definition.

Definition : A radian is the measure of an angle subtended at the centre of a circle by an arc of length equal to the radius of the circle. One radian is denoted by 1^c.

Let O be the centre of the circle of radius r units and AB be the arc of the circle, such that arc $AB = r$. Then the angle AOB = one radian, denoted by 1^c.

Now, we shall show that, a radian, constructed according to above definition is of constant magnitude.

Theorem 1. Radian is a constant angle.

Proof : Consider a circle with centre O and radius r units. Let AB an arc of length r units. Then $A\hat{O}B = 1^c$. Produce AO to meet the circle at C. We know that the angles at the centre of a circle are proportional to the arcs on which they stand.

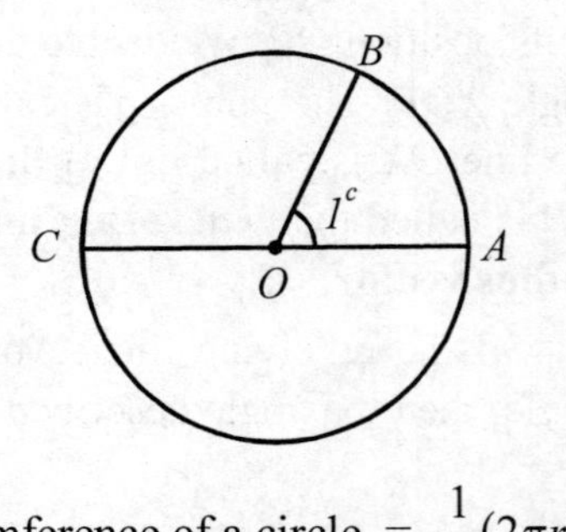

$$\therefore \quad \frac{A\hat{O}B}{A\hat{O}C} = \frac{\text{arc } AB}{\text{arc } ABC} \Rightarrow \frac{1 \text{ radian}}{180° \text{ degree}} = \frac{r}{\frac{1}{2}(2\pi r)}$$

$$\left(\because \overset{\frown}{A} = \frac{1}{2} \text{ Circumference of a circle } = \frac{1}{2}(2\pi r)\right)$$

$$\Rightarrow \quad 1 \text{ radian} = \frac{180}{\pi} \text{ radian} \Rightarrow \boldsymbol{\pi} \textbf{ radians = 180°}$$

Thus, 1 radian is a constant angle and its value is given by

1 radian = 57° 16' 22" (approximately)

Hence, the theorem.

Note : From the above theorem, we have a relation between degrees and radians. That is

$$\pi \text{ radians} = 180°. \quad \text{Also,} \quad 1° = \frac{\pi}{180} \text{ radians}$$

Radian measure of some common angles are given in the following table.

Degrees	15°	30°	45°	60°	75°	90°	180°	270°	360°
Radians	$\frac{\pi}{12}$	$\frac{\pi}{6}$	$\frac{\pi}{4}$	$\frac{\pi}{3}$	$\frac{5\pi}{12}$	$\frac{\pi}{2}$	π	$\frac{3\pi}{2}$	2π

Theorem 2. The length of an arc of a circle, which subtends an angle θ radian at the centre, is equal to $r\theta$, where r is the radius of the circle.

Proof : Let O be the centre of the circle of radius r. Let A and B are the points on the circle such that arc $AB = s$ and $A\hat{O}B = \theta^c$. Let $A\hat{O}P = 1$ radian.

Now we have, $\dfrac{A\hat{O}B}{A\hat{O}P} = \dfrac{\text{arc } AB}{\text{arc } AP} \Rightarrow \dfrac{\theta}{1} = \dfrac{s}{r} \Rightarrow \boldsymbol{s = r\theta}$

Thus, if s is length of the arc of a circle, θ radians is the angle subtended at the centre, and r is the radius of the circle, then we have, $s = r\theta$.

1.5 Introduction of Trigonometic Ratios

In this section we shall introduce six ratios called trigonometrical ratios for an acute angle and find relations between them.

1.6 Trigonometric Ratios

Let $P\hat{O}Q = \theta$, be any acute angle, B be any point on OP, one of the boundary lines and BC is the perpendicular to OQ. Thus we have a right angled triangle BOC. With respect to the angle θ, the side BC is called the *opposite side* and OC is called *adjecent side*. Of course, the side OB is the hypotenuse of the triangle. With reference to the angle θ, we define the following six ratios called trigonometrical ratios.

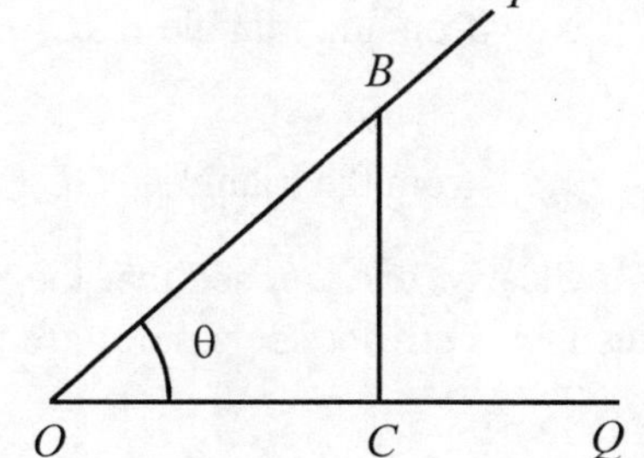

The ratio $\dfrac{BC}{OB} = \dfrac{\text{opposite side}}{\text{hypotenuse}}$ is called the sine θ

The ratio $\dfrac{OC}{OB} = \dfrac{\text{adjecent side}}{\text{hypotenuse}}$ is called the cosine θ

The ratio $\dfrac{BC}{OC} = \dfrac{\text{opposite side}}{\text{adjecent side}}$ is called the tangent θ

The ratio $\dfrac{OB}{BC} = \dfrac{\text{hypotenuse}}{\text{opposite side}}$ is called the cosecant θ

The ratio $\dfrac{OB}{OC} = \dfrac{\text{hypotenuse}}{\text{adjecent side}}$ is called the secant θ

The ratio $\dfrac{OC}{BC} = \dfrac{\text{adjecent side}}{\text{opposite side}}$ is called the cotangent θ

Instead of writing completely the words sine, cosine, tangent, cosecant, secant and cotangent, we shall abbreviate these and write more conveniently as below:

$$\sin\theta = \frac{BC}{OB} \quad \cos\theta = \frac{OC}{OB} \quad \tan\theta = \frac{BC}{OC} \quad \text{cosec}\,\theta = \frac{OB}{BC} \quad \sec\theta = \frac{OB}{OC} \quad \cot\theta = \frac{OC}{BC}$$

The above definition of trigonometrical ratios remains unaltered as long as the angle remains the same. That is trigonometrical ratios depends on the angle but not on the lengths of the sides of the right angled triangle.

This can be established as below:

Let $P\hat{O}Q$ be any acute angle, Let *B* and *D* be any two points on the boundary of *OP*. Draw *BC* and *DE* perpendiculars to *OQ*. Also, let *F* be any point on the boundary line *OQ* and *FG* be perpendicular from *F* onto *OP*.

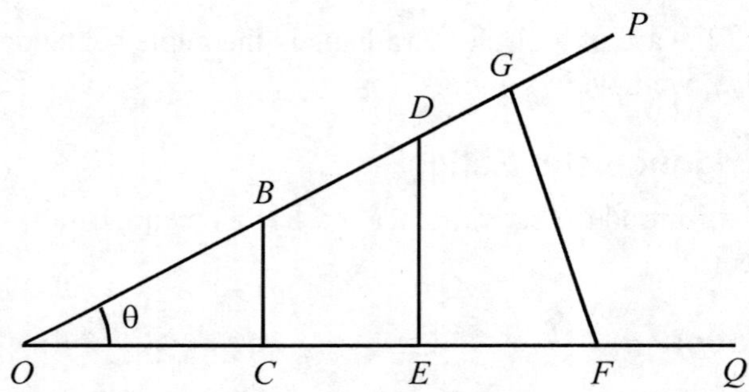

Now from the triangle *BOC*, $\sin\theta = \frac{BC}{OB}$,

from the triangle *DOE*, $\sin\theta = \frac{DE}{OD}$, from the triangle *FOG*, $\sin\theta = \frac{FG}{OF}$,

Clearly, one can see that the triangles *BOC*, *DOE* and *FOG* are similar triangles to each other. Thus their corresponding sides are proportional and therefore we have

$$\frac{BC}{OB} = \frac{DE}{OD} = \frac{FG}{OF}$$

Thus, the sine of the angle θ is the same, whether it is obtained from the triangle *BOC* or from the triangle *DOE* or from the triangle *FOG*.

A similar proof holds for each of other trigonometrical ratios.

Thus the trigonometrical ratios are independent of the lengths of the revolving line and depends only on the magnitude of the angle. Further a change is made in the value of the angle θ, then a consequent change can be seen in the values of the trigonometrical ratios of the angle θ. This we shall discuss it in a later chapter.

Note : We know that in every right angled triangle the hypotenuse is the greatest side. Thus it follows from the definition of trigonometric ratios, that those ratios which have the hypotenuse in the denominator can never be greater than unity, while those which have hypotenuse in the numerator can never be less than unity. Further, those ratios which do not involve the hypotenuse may have any numerical value. Thus we have the following results:

The sine and cosine of an angle can never be greater than 1.

The cosecant and secant of an angle can never be less than 1.

The tangent and cotangent of an angle may have any numerical value.

1.7 Relation Between the Trigonometric Ratios

Let ABC be a right angled triangle, right angled at B and $A\hat{C}B = \theta$.

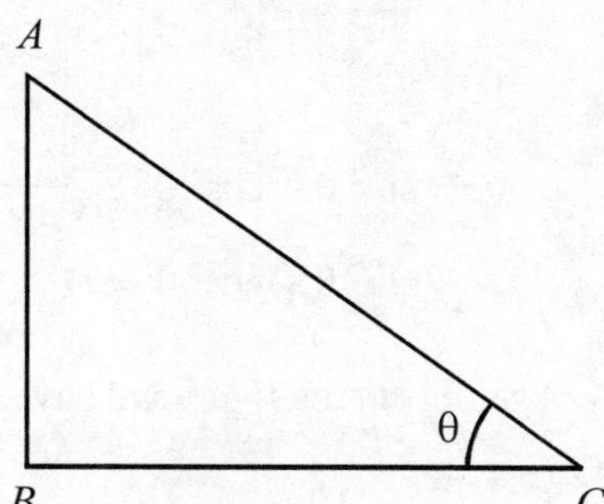

Then we have, $\sin\theta = \dfrac{AB}{AC}$ and $\operatorname{cosec}\theta = \dfrac{AC}{AB}$

$\Rightarrow \sin\theta \cdot \operatorname{cosec}\theta = \dfrac{AB}{AC}\cdot\dfrac{AC}{AB} = 1$

$$\Rightarrow \quad \mathbf{\sin\theta = \dfrac{1}{\operatorname{cosec}\theta} \quad and \quad \operatorname{cosec}\theta = \dfrac{1}{\sin\theta}}$$

Thus, $\sin\theta$ and $\operatorname{cosec}\theta$ are reciprocal to each other.

Similarly, we have

$$\mathbf{\cos\theta = \dfrac{1}{\sec\theta}, \quad \sec\theta = \dfrac{1}{\cos\theta} \quad and \quad \tan\theta = \dfrac{1}{\cot\theta}, \quad \cot\theta = \dfrac{1}{\tan\theta}}$$

That is $\cos\theta$ and $\sec\theta$ are reciprocal to each other, and $\tan\theta$ and $\cot\theta$ are reciprocal to each other.

Again from the figure we have,

$$\tan\theta = \dfrac{AB}{BC} \quad \Rightarrow \quad \tan\theta = \dfrac{(AB/AC)}{(BC/AC)} \quad \Rightarrow \tan\theta = \dfrac{\sin\theta}{\cos\theta}$$

Again, $$\cot\theta = \dfrac{BC}{AB} \quad \Rightarrow \quad \cot\theta = \dfrac{(BC/AC)}{(AB/AC)} \quad \Rightarrow \cot\theta = \dfrac{\cos\theta}{\sin\theta}$$

Thus we have, $\mathbf{\tan\theta = \dfrac{\sin\theta}{\cos\theta}}$ **and** $\mathbf{\cot\theta = \dfrac{\cos\theta}{\sin\theta}}$

Note: We frequently come across expressions containing the different powers of trigonometric ratios, like $(\sin\theta)^2$, $(\cos\theta)^3$, $(\sec\theta)^2$, etc., we write these different powers of trigonometric ratios as

$$\sin^2\theta, \quad \cos^3\theta, \quad \sec^2\theta \text{ etc.,}$$

and read as "sin squared θ", "cos cube θ", "secant squared θ" etc.

In the following section we shall derive few basics identities connecting trigonometric ratios.

1.8 Basic Identities

1. To prove that (a) $\mathbf{\sin^2\theta + \cos^2\theta = 1}$ **(b)** $\mathbf{1 + \tan^2\theta = \sec^2\theta}$ **(c)** $\mathbf{1 + \cot^2\theta = \operatorname{cosec}^2\theta}$

Proof : Let ABC be a right angled triangle and $A\hat{B}C = \theta$

(a) We have,

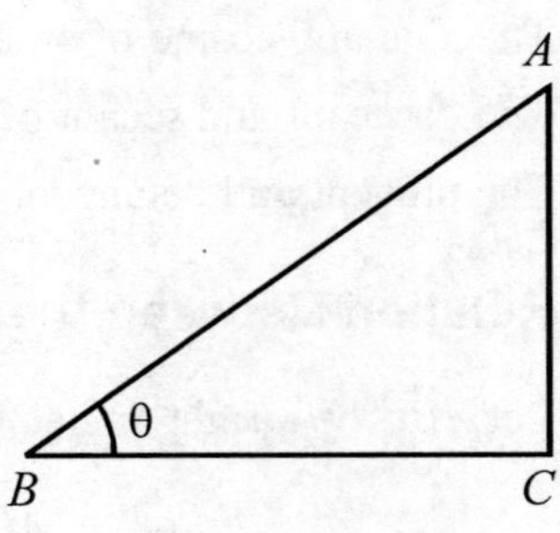

$$\sin\theta = \frac{AC}{AB} \text{ and } \cos\theta = \frac{BC}{AB}$$

Consider, $\sin^2\theta + \cos^2\theta = \frac{AC^2}{AB^2} + \frac{BC^2}{AB^2}$

$\Rightarrow \quad \sin^2\theta + \cos^2\theta = \frac{AC^2 + BC^2}{AB^2}$

$\Rightarrow \quad \sin^2\theta + \cos^2\theta = \frac{AB^2}{AB^2}$ $\quad (\because AB^2 = AC^2 + BC^2)$

$\Rightarrow \quad \mathbf{\sin^2\theta + \cos^2\theta = 1}$

(b) Again from the figure we have, $\tan\theta = \frac{AC}{BC}$ and $\sec\theta = \frac{AB}{BC}$

Now consider, $1 + \tan^2\theta = 1 + \frac{AC^2}{BC^2} \Rightarrow 1 + \tan^2\theta = \frac{BC^2 + AC^2}{BC^2}$

$\Rightarrow \quad 1 + \tan^2\theta = \frac{AB^2}{BC^2}$ $\quad (\because AB^2 = AC^2 + BC^2)$

$\Rightarrow \quad 1 + \tan^2\theta = \left(\frac{AB}{BC}\right)^2 = \sec^2\theta$

$\Rightarrow \quad \mathbf{1 + \tan^2\theta = \sec^2\theta}$

(c) Again from the figure, we have, $\cot\theta = \frac{BC}{AC}$ and $\text{cosec}\,\theta = \frac{AB}{AC}$

Now consider, $1 + \cot^2\theta = 1 + \frac{BC^2}{AC^2} \Rightarrow 1 + \cot^2\theta = \frac{AC^2 + BC^2}{AC^2}$

$\Rightarrow \quad 1 + \cot^2\theta = \frac{AB^2}{AC^2}$ $\quad (\because AC^2 + BC^2 = AB^2)$

$\Rightarrow \quad 1 + \cot^2\theta = \left(\frac{AB}{AC}\right)^2 = \text{cosec}^2\theta$

$\Rightarrow \quad \mathbf{1 + \cot^2\theta = cosec^2\theta}$

The following deductions from the above identities are useful.

1. $\mathbf{\sin^2\theta = 1 - \cos^2\theta}$, $\mathbf{\sin\theta = \sqrt{1 - \cos^2\theta}}$; $\mathbf{\cos^2\theta = 1 - \sin^2\theta}$, $\mathbf{\cos\theta = \sqrt{1 - \sin^2\theta}}$
2. $\mathbf{\sec^2\theta - \tan^2\theta = 1}$, $\mathbf{\sec\theta = \sqrt{1 + \tan^2\theta}}$; $\mathbf{\tan^2\theta = \sec^2\theta - 1}$, $\mathbf{\tan\theta = \sqrt{\sec^2\theta - 1}}$
3. $\mathbf{cosec^2\theta - \cot^2\theta = 1}$, $\mathbf{cosec\,\theta = \sqrt{1 + \cot^2\theta}}$; $\mathbf{\cot^2\theta = cosec^2\theta - 1}$, $\mathbf{\cot\theta = \sqrt{cosec^2\theta - 1}}$

Example 1. Prove the following identities

(a) $(1 - \cos^2\theta)\cdot \text{cosec}^2\theta = 1$ **(b)** $\tan\alpha \cdot \sqrt{1-\sin^2\alpha} = \sin\alpha$ **(c)** $(1-\cos^2\theta)(1+\tan^2\theta) = \tan^2\theta$

(d) $\sin^2\theta \cdot \cot^2\theta + \sin^2\theta = 1$ **(e)** $\text{cosec}^2\theta \cdot \tan^2\theta - 1 = \tan^2\theta$

Solution : (a) LHS $= (1-\cos^2\theta)\,\text{cosec}^2\theta = \sin^2\theta \cdot \text{cosec}^2\theta$ $(\because 1-\cos^2\theta = \sin^2\theta)$

$= (\sin\theta \cdot \text{cosec}\,\theta)^2 = \mathbf{1}$ $(\because \sin\theta\cdot\text{cosec}\,\theta = 1)$

$=$ RHS

(b) LHS $= \tan\alpha\sqrt{1-\sin^2\alpha} = \tan\alpha\cdot\sqrt{\cos^2\alpha}$ $(\because 1-\sin^2\alpha = \cos^2\alpha)$

$= \tan\alpha\cdot\cos\alpha = \dfrac{\sin\alpha}{\cos\alpha}\cdot\cos\alpha$ $\left(\because \tan\alpha = \dfrac{\sin\alpha}{\cos\alpha}\right)$

$= \mathbf{\sin\alpha} =$ RHS

(c) LHS $= (1-\cos^2\theta)\cdot(1+\tan^2\theta) = \sin^2\theta\,.\,\sec^2\theta$ $(\because 1+\tan^2\theta = \sec^2\theta)$

$= \dfrac{\sin^2\theta}{\cos^2\theta}$ $\left(\because \sec\theta = \dfrac{1}{\cos\theta}\right)$

$= \mathbf{\tan^2\theta}$ $\left(\because \dfrac{\sin\theta}{\cos\theta} = \tan\theta\right)$

$=$ RHS

(d) LHS $= \sin^2\theta\cdot\cot^2\theta + \sin^2\theta = \sin^2\theta\cdot\dfrac{\cos^2\theta}{\sin^2\theta} + \sin^2\theta$ $\left(\because \cot\theta = \dfrac{\cos\theta}{\sin\theta}\right)$

$= \cos^2\theta + \sin^2\theta = \mathbf{1}$ $(\because \cos^2\theta + \sin^2\theta = 1)$

$=$ RHS

(e) LHS $= \text{cosec}^2\theta\cdot\tan^2\theta - 1 = \text{cosec}^2\theta\cdot\dfrac{\sin^2\theta}{\cos^2\theta} - 1 = \dfrac{(\text{cosec}\,\theta\cdot\sin\theta)^2}{\cos^2\theta} - 1$

$= \dfrac{1}{\cos^2\theta} - 1$ $(\because \text{cosec}\,\theta\cdot\sin\theta = 1)$

$= \sec^2\theta - 1$ $\left(\because \sec\theta = \dfrac{1}{\cos\theta}\right)$

$= \mathbf{\tan^2\theta}$ $(\because 1+\tan^2\theta = \sec^2\theta)$

$=$ RHS

Example 2. Prove the following identities

(a) $\dfrac{1}{\cos^2\theta} - \dfrac{1}{\cot^2\theta} = 1$ **(b)** $\sec^4\alpha - 1 = 2\tan^2\alpha + \tan^4\alpha$

(c) $(\sec\theta\cdot\cot\theta)^2 - (\cos\theta\cdot\text{cosec}\,\theta)^2 = 1$ **(d)** $\tan^2\theta - \cot^2\theta = \sec^2\theta - \text{cosec}^2\theta.$

Solution : (a) LHS $= \dfrac{1}{\cos^2\theta} - \dfrac{1}{\cot^2\theta} = \sec^2\theta - \tan^2\theta = 1 = \text{RHS}$ $(\because \sec^2\theta - \tan^2\theta = 1)$

(b) LHS $= \sec^4\alpha - 1 = (\sec^2\alpha + 1)(\sec^2\alpha - 1) = (\sec^2\alpha + 1)\tan^2\alpha$ $(\because \sec^2\alpha - 1 = \tan^2\alpha)$

$= (1 + \tan^2\alpha + 1)\tan^2\alpha$ $(\because \sec^2\alpha = 1 + \tan^2\alpha)$

$= (2 + \tan^2\alpha)\tan^2\alpha = \mathbf{2\tan^2\alpha + \tan^4\alpha} = \text{RHS}$

(c) LHS $= (\sec\theta\cdot\cot\theta)^2 - (\cos\theta\cdot\operatorname{cosec}\theta)^2 = \left(\dfrac{1}{\cos\theta}\cdot\dfrac{\cos\theta}{\sin\theta}\right)^2 - \left(\cos\theta\cdot\dfrac{1}{\sin\theta}\right)^2$

$= \left(\dfrac{1}{\sin\theta}\right)^2 - \left(\dfrac{\cos\theta}{\sin\theta}\right)^2$

$= \operatorname{cosec}^2\theta - \cot^2\theta = \mathbf{1} = \text{RHS}$

(d) LHS $= \tan^2\theta - \cot^2\theta = (\sec^2\theta - 1) - (\operatorname{cosec}^2\theta - 1)$

$= \sec^2\theta - 1 - \operatorname{cosec}^2\theta + 1 = \mathbf{\sec^2\theta - \operatorname{cosec}^2\theta} = \text{RHS}$

Example 3. Prove the following identities

(a) $\tan^2\theta - \sin^2\theta = \tan^2\theta\cdot\sin^2\theta$

(b) $\dfrac{1+\cos\theta}{1-\cos\theta} = \dfrac{\tan^2\theta}{(\sec\theta - 1)^2}$

(c) $\sec^2\theta + \operatorname{cosec}^2\theta = \sec^2\theta\cdot\operatorname{cosec}^2\theta$

(d) $\dfrac{\operatorname{cosec}\theta + \cot\theta}{\operatorname{cosec}\theta - \cot\theta} = (\operatorname{cosec}\theta + \cot\theta)^2$

(e) $\dfrac{\sin\theta}{1+\cos\theta} + \dfrac{1+\cos\theta}{\sin\theta} = 2\operatorname{cosec}\theta$

(f) $\dfrac{1-\cos\theta}{\sin\theta} = \dfrac{\sin\theta}{1+\cos\theta}$

(g) $\dfrac{\tan\theta + \sec\theta - 1}{\tan\theta - \sec\theta + 1} = \dfrac{1+\sin\theta}{\cos\theta}$

Solution : (a) LHS $= \tan^2\theta - \sin^2\theta = \dfrac{\sin^2\theta}{\cos^2\theta} - \sin^2\theta = \dfrac{\sin^2\theta - \sin^2\theta\cdot\cos^2\theta}{\cos^2\theta}$

$= \dfrac{\sin^2\theta\,(1-\cos^2\theta)}{\cos^2\theta}$

$= \dfrac{\sin^2\theta}{\cos^2\theta}\cdot\sin^2\theta$ $(\because 1 - \cos^2\theta = \sin^2\theta)$

$= \mathbf{\tan^2\theta\cdot\sin^2\theta} = \text{RHS}$

(b) LHS $= \dfrac{1+\cos\theta}{1-\cos\theta} = \dfrac{(1+\cos\theta)(1-\cos\theta)}{(1-\cos\theta)(1-\cos\theta)} = \dfrac{1-\cos^2\theta}{(1-\cos\theta)^2}$

$= \dfrac{\sin^2\theta}{\left(1 - \dfrac{1}{\sec\theta}\right)^2}$ $(\because 1 - \cos^2\theta = \sin^2\theta)$

$$= \frac{\sin^2\theta \cdot \sec^2\theta}{(\sec\theta - 1)^2}$$

$$= \frac{\sin^2\theta \cdot \left(\frac{1}{\cos^2\theta}\right)}{(\sec\theta - 1)^2} \qquad \left(\because \sec\theta = \frac{1}{\cos\theta}\right)$$

$$= \frac{\mathbf{\tan^2\theta}}{\mathbf{(\sec\theta - 1)^2}} = \text{RHS}$$

(c) LHS $= \sec^2\theta + \text{cosec}^2\theta = \frac{1}{\cos^2\theta} + \frac{1}{\sin^2\theta} = \frac{\sin^2\theta + \cos^2\theta}{\cos^2\theta \cdot \sin^2\theta}$

$$= \frac{1}{\cos^2\theta \cdot \sin^2\theta} \qquad (\because \sin^2\theta + \cos^2\theta = 1)$$

$$= \mathbf{\sec^2\theta \cdot \text{cosec}^2\theta} = \text{RHS}$$

(d) LHS $= \frac{\text{cosec}\,\theta + \cot\theta}{\text{cosec}\,\theta - \cot\theta} = \frac{(\text{cosec}\,\theta + \cot\theta)(\text{cosec}\,\theta + \cot\theta)}{(\text{cosec}\,\theta - \cot\theta)(\text{cosec}\,\theta + \cot\theta)}$

$$= \frac{(\text{cosec}\,\theta + \cot\theta)^2}{\text{cosec}^2\theta - \cot^2\theta} = \mathbf{(\text{cosec}\,\theta + \cot\theta)^2} \quad (\because \text{cosec}^2\theta - \cot^2\theta = 1)$$

$$= \text{RHS}$$

(e) LHS $= \frac{\sin\theta}{1 + \cos\theta} + \frac{1 + \cos\theta}{\sin\theta} = \frac{\sin^2\theta + (1 + \cos\theta)^2}{(1 + \cos\theta) \cdot \sin\theta}$

$$= \frac{\sin^2\theta + \cos^2\theta + 2\cos\theta + 1}{(1 + \cos\theta) \cdot \sin\theta}$$

$$= \frac{2\cancel{(1 + \cos\theta)}}{\cancel{(1 + \cos\theta)} \cdot \sin\theta} \qquad (\because \sin^2\theta + \cos^2\theta = 1)$$

$$= \frac{2}{\sin\theta} = \mathbf{2\,\text{cosec}\,\theta} = \text{RHS}$$

(f) LHS $= \frac{1 - \cos\theta}{\sin\theta} = \frac{(1 - \cos\theta) \cdot (1 + \cos\theta)}{\sin\theta \cdot (1 + \cos\theta)} = \frac{1 - \cos^2\theta}{\sin\theta \cdot (1 + \cos\theta)}$

$$= \frac{\sin^2\theta}{\sin\theta \cdot (1 + \cos\theta)} \qquad (\because 1 - \cos^2\theta = \sin^2\theta)$$

$$= \frac{\mathbf{\sin\theta}}{\mathbf{1 + \cos\theta}} = \text{RHS}$$

(g) $\text{LHS} = \dfrac{\tan\theta + \sec\theta - 1}{\tan\theta - \sec\theta + 1}$

Writing, $1 = \sec^2\theta - \tan^2\theta$, in the numerator, we get

$$= \frac{(\tan\theta + \sec\theta) - (\sec^2\theta - \tan^2\theta)}{(1 - \sec\theta + \tan\theta)} = \frac{(\sec\theta + \tan\theta)\cdot[1 - (\sec\theta - \tan\theta)]}{(1 - \sec\theta + \tan\theta)}$$

$$= \frac{(\sec\theta + \tan\theta)\cdot\cancel{(1 - \sec\theta + \tan\theta)}}{\cancel{(1 - \sec\theta + \tan\theta)}}$$

$$= \sec\theta + \tan\theta = \frac{1}{\cos\theta} + \frac{\sin\theta}{\cos\theta} = \frac{1 + \sin\theta}{\cos\theta} = \text{RHS}$$

Example 4. If $x = a\cos\theta$, $y = b\sin\theta$, Prove that $\dfrac{x^2}{a^2} + \dfrac{y^2}{b^2} = 1$.

Solution : $x = a\cos\theta \Rightarrow \dfrac{x}{a} = \cos\theta$; $y = b\sin\theta \Rightarrow \dfrac{y}{b} = \sin\theta$

Now squaring and adding we get,

$$\frac{x^2}{a^2} + \frac{y^2}{b^2} = \cos^2\theta + \sin^2\theta \Rightarrow \frac{x^2}{a^2} + \frac{y^2}{b^2} = 1 \qquad (\because \sin^2\theta + \cos^2\theta = 1)$$

Example 5. If $x = r\cos\alpha\cos\beta$, $y = r\cos\alpha\sin\beta$, $z = r\sin\alpha$. Show that $x^2 + y^2 + z^2 = r^2$.

Solution : Consider,

$$\begin{aligned} x^2 + y^2 + z^2 &= r^2\cos^2\alpha\cos^2\beta + r^2\cos^2\alpha\sin^2\beta + r^2\sin^2\alpha \\ &= r^2\cos^2\alpha(\cos^2\beta + \sin^2\beta) + r^2\sin^2\alpha \\ &= r^2\cos^2\alpha + r^2\sin^2\alpha \qquad (\because \cos^2\beta + \sin^2\beta = 1) \\ &= r^2(\cos^2\alpha + \sin^2\alpha) \Rightarrow x^2 + y^2 + z^2 = r^2 \qquad (\because \cos^2\alpha + \sin^2\alpha = 1) \end{aligned}$$

Example 6. If $x = a\sec\theta$, $y = b\tan\theta$, Prove that $\dfrac{x^2}{a^2} - \dfrac{y^2}{b^2} = 1$.

Solution : $x = a\sec\theta \Rightarrow \dfrac{x}{a} = \sec\theta$; $y = b\tan\theta \Rightarrow \dfrac{y}{b} = \tan\theta$

Consider, $\dfrac{x^2}{a^2} - \dfrac{y^2}{b^2} = \sec^2\theta - \tan^2\theta \Rightarrow \dfrac{x^2}{a^2} - \dfrac{y^2}{b^2} = 1 \qquad (\because \sec^2\theta - \tan^2\theta = 1)$

Example 7. If $x = a\cos\theta + b\sin\theta$, $y = a\sin\theta - b\cos\theta$, Show that $x^2 + y^2 = a^2 + b^2$.

Solution : Consider, $x^2 = (a\cos\theta + b\sin\theta)^2$

$\Rightarrow$ $x^2 = a^2\cos^2\theta + 2ab\cos\theta\sin\theta + b^2\sin^2\theta$ .... (1)

Again consider, $y^2 = (a\sin\theta - b\cos\theta)^2$

$\Rightarrow$ $y^2 = a^2\sin^2\theta - 2ab\sin\theta\cos\theta + b^2\cos^2\theta$ .... (2)

Adding (1) and (2) we get

$$x^2 + y^2 = a^2(\cos^2\theta + \sin^2\theta) + b^2(\sin^2\theta + \cos^2\theta) \Rightarrow x^2 + y^2 = a^2 + b^2$$

Example 8. If $x = a\cos^2\theta + b\sin^2\theta$, show that $(x-a)(b-x) = (a-b)^2\sin^2\theta\cos^2\theta$

Solution : Consider,

$$x - a = a\cos^2\theta + b\sin^2\theta - a = b\sin^2\theta - a(1-\cos^2\theta) = b\sin^2\theta - a\sin^2\theta = (b-a)\sin^2\theta$$

Again consider,

$$b - x = b - a\cos^2\theta - b\sin^2\theta = b(1-\sin^2\theta) - a\cos^2\theta = b\cos^2\theta - a\cos^2\theta = (b-a)\cos^2\theta$$

Now $\quad (x-a)(b-x) = (b-a)\sin^2\theta \cdot (b-a)\cos^2\theta$

i.e., $\quad (x-a)(b-x) = \mathbf{(a-b)^2\sin^2\theta\cos^2\theta} \qquad (\because (a-b)^2 = (b-a)^2)$

Example 9. If $\sin\theta = \frac{3}{5}$ and θ is acute, find the value of $2\tan\theta + 3\sec\theta + 4\sec\theta\cdot\text{cosec}\,\theta$.

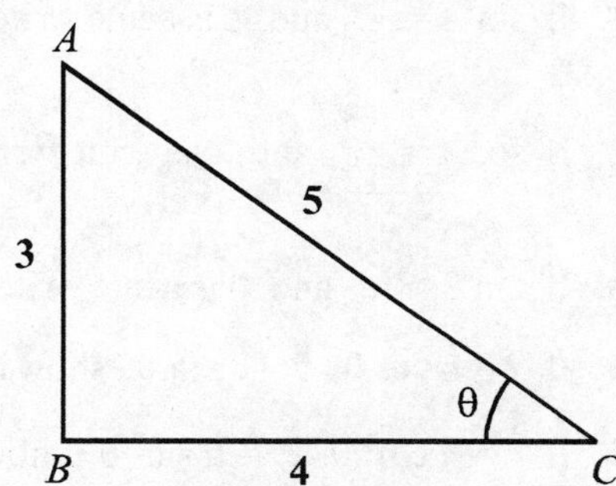

Solution : By data $\sin\theta = \frac{3}{5}$

Let in a right triangle ABC, $A\hat{C}B = \theta$

Now $\quad \sin\theta = \frac{3}{5}, \Rightarrow AB = 3$ and $AC = 5$

Now $\quad BC^2 = AC^2 - AB^2 \quad$ (by pythogoron theorem)

$\Rightarrow \quad BC^2 = 25 - 9 \Rightarrow BC^2 = 16 \Rightarrow BC = 4$

$\therefore \quad \tan\theta = \frac{3}{4}, \quad \sec\theta = \frac{5}{4}$ and $\text{cosec}\,\theta = \frac{5}{3}$

Now $\quad 2\tan\theta + 3\sec\theta + 4\sec\theta\cdot\text{cosec}\,\theta$

$$= 2\left(\frac{3}{4}\right) + 3\left(\frac{5}{4}\right) + 4\left(\frac{5}{4}\right)\left(\frac{5}{3}\right) = \frac{6}{4} + \frac{15}{4} + \frac{25}{3} = \frac{18+45+100}{12} = \mathbf{\frac{163}{12}}$$

Example 10. Express all the trigonometrical ratios of angle θ in terms of $\sin\theta$, where θ is acute.

Solution : We have, $\sin^2\theta + \cos^2\theta = 1 \Rightarrow \cos^2\theta = 1 - \sin^2\theta \Rightarrow \cos\theta = \sqrt{1-\sin^2\theta}$

$$\tan\theta = \frac{\sin\theta}{\cos\theta} \Rightarrow \tan\theta = \frac{\sin\theta}{\sqrt{1-\sin^2\theta}}; \quad \cot\theta = \frac{\cos\theta}{\sin\theta} \Rightarrow \cot\theta = \frac{\sqrt{1-\sin^2\theta}}{\sin\theta}$$

$$\sec\theta = \frac{1}{\cos\theta} \Rightarrow \sec\theta = \frac{1}{\sqrt{1-\sin^2\theta}}; \quad \sin\theta = \sin\theta \quad \text{and} \quad \text{cosec}\,\theta = \frac{1}{\sin\theta}$$

Example 11. Express $\tan\theta$ in terms of other trigonometrical ratios

Solution : We have,

$$\tan\theta = \frac{\sin\theta}{\cos\theta} \Rightarrow \tan\theta = \frac{\sin\theta}{\sqrt{1-\sin^2\theta}}; \quad \tan\theta = \frac{\sin\theta}{\cos\theta} \Rightarrow \tan\theta = \frac{\sqrt{1-\cos^2\theta}}{\cos\theta}$$

$$\tan^2\theta = \sec^2\theta - 1 \quad\Rightarrow\quad \tan\theta = \sqrt{\sec^2\theta - 1}$$

$$\cot^2\theta = \text{cosec}^2\theta - 1 \quad\Rightarrow\quad \frac{1}{\cot^2\theta} = \frac{1}{\text{cosec}^2\theta - 1} \Rightarrow \tan\theta = \frac{1}{\sqrt{\text{cosec}^2\theta - 1}}$$

Also $\tan\theta = \dfrac{1}{\cot\theta}$

Exercise

I.

1. If $\sin A = \dfrac{4}{5}$ and A is acute write other five trigonometrical ratios
2. If $\sec A = \dfrac{13}{12}$ and A is acute write other five trigonometrical ratios
3. If $\tan\theta = \dfrac{3}{4}$ and θ is acute write other five trigonometrical ratios
4. If $x = a\cos\theta,\ y = a\sin\theta$, show that $x^2 + y^2 = a^2$
5. If $x = a\cos^3\theta,\ y = a\sin^3\theta$, show that $x^{2/3} + y^{2/3} = a^{2/3}$
6. Prove the following identities
   - **(a)** $\sin A\cdot\cot A = \cos A$ **(b)** $\sin A\cdot\sec A = \tan A$ **(c)** $\cos A\cdot\text{cosec}\,A = \cot A$
   - **(d)** $\cot A\cdot\sec A\cdot\sin A = 1$ **(e)** $(1 - \sin^2 A)\sec^2 A = 1$ **(f)** $(1 - \cos^2 A)\,\text{cosec}^2 A = 1$
   - **(g)** $\cot^2\theta\,(1 - \cos^2\theta) = \cos^2\theta$ **(h)** $\sin^2 A\,(1 + \cot^2 A) = 1$ **(i)** $(\text{cosec}^2 A - 1)\tan^2 A = 1$
   - **(j)** $(1 - \cos^2 A)(1 + \cot^2 A) = 1$ **(k)** $\sin^2\theta\cdot\cot^2\theta + \sin^2\theta = 1$ **(l)** $\sin^2\theta\cdot\sec^2\theta = \sec^2\theta - 1$

II.

1. Express all the trigonometrical ratios of an acute angle A, in terms of
   (a) $\cos A$ **(b)** $\tan A$ **(c)** $\sec A$ **(d)** $\text{cosec}\,A$ **(e)** $\cot A$
2. Express $\cot A$ in terms of other trigonometrical ratios
3. Express $\sec A$ in terms of other trigonometrical ratios
4. If $x = ar\sin\theta\cos\phi,\ y = br\sin\theta\sin\phi,\ z = cr\cos\phi$ show that $\dfrac{x^2}{a^2} + \dfrac{y^2}{b^2} + \dfrac{z^2}{c^2} = r^2$
5. If $\cot\theta = \dfrac{4}{3}$ and θ is acute, show that $3\sin\theta + 4\cos\theta = 5$
6. If $\tan\theta = \dfrac{5}{12}$ and θ is acute, show that $3\sin\theta - 4\cos\theta = -\dfrac{33}{13}$
7. If $\cot\theta = \dfrac{5}{12}$ and θ is acute, show that $2\,\text{cosec}\,\theta - 4\sec\theta = \dfrac{247}{30}$
8. Prove the following identities.

   (a) $\dfrac{\tan A}{\sec A - 1} + \dfrac{\tan A}{\sec A + 1} = 2\,\text{cosec}\,A$ **(b)** $\dfrac{\sin\theta}{1 + \cos\theta} + \dfrac{1 + \cos\theta}{\sin\theta} = 2\,\text{cosec}\,\theta$

(c) $\frac{\sin^2\theta}{1-\cos\theta}-\frac{\cos^2\theta}{1-\sin\theta}=\cos\theta-\sin\theta$

(d) $\frac{\cos A}{1+\sin A}+\frac{\cos A}{1-\sin A}=2\sec A$

(e) $\frac{\sin\theta\cdot\tan\theta}{1-\cos\theta}=1+\sec\theta$

(f) $\sec\theta\,(1-\sin\theta)\,(\sec\theta+\tan\theta)=1$

(g) $\frac{\text{cosec}\,\theta+\cot\theta}{\text{cosec}\,\theta-\cot\theta}=(\text{cosec}\,\theta+\cot\theta)^2$

(h) $\frac{\tan\alpha+\cot\beta}{\cot\alpha+\tan\beta}=\frac{\tan\alpha}{\tan\beta}$

9. Prove the following identities.

(a) $\frac{\tan\theta-\cot\theta}{\sin\theta\cdot\cos\theta}=\sec^2\theta-\text{cosec}^2\theta$

(b) $\frac{\text{cosec}\,\theta}{\cot\theta+\tan\theta}=\cos\theta$

(c) $(1-\sin A+\cos A)^2=2\,(1-\sin A)\,(1+\cos A)$

(d) $\frac{\sin A+1-\cos A}{\sin A-1+\cos A}=\frac{1+\sin A}{\cos A}$

(e) $\frac{\cos\theta}{1-\tan\theta}+\frac{\sin\theta}{1-\cot\theta}=\sin\theta+\cos\theta$

(f) $\frac{\sec\theta+1}{\tan\theta}+\frac{\tan\theta}{\sec\theta+1}=2\,\text{cosec}\,\theta$

10. Prove the following identities.

(a) $\sqrt{\frac{1+\cos\theta}{1-\cos\theta}}=\text{cosec}\,\theta+\cot\theta$

(b) $\sqrt{\frac{\sec\theta+1}{\sec\theta-1}}=\cot\theta+\text{cosec}\,\theta$

(c) $\sqrt{\frac{1+\sin\theta}{1-\sin\theta}}=\sec\theta+\tan\theta$

(d) $\sqrt{\frac{1-\cos\theta}{1+\cos\theta}}=\text{cosec}\,\theta-\cot\theta$

11. If $\sin x+\sin^2 x=1$, show that $\cos^4 x+\cos^2 x=1$.

III.

1. If $\sin\theta=\frac{3}{5}$ and θ is acute show that $\frac{2\tan\theta}{1-\tan^2\theta}=\frac{24}{7}$
2. If $\cot\theta=\frac{5}{2}$ and θ is acute show that $\frac{5\cos\theta+2\sin\theta}{5\cos\theta-2\sin\theta}=\frac{29}{21}$
3. If $\tan\theta=\frac{a}{b}$ and θ is acute show that $\frac{a\sin\theta-b\cos\theta}{a\sin\theta+b\cos\theta}=\frac{a^2-b^2}{a^2+b^2}$
4. If $x=a\cos\theta+b\sin\theta$, $y=a\sin\theta-b\cos\theta$ show that $x^2+y^2=a^2+b^2$.
5. If $a=\sec\theta+\tan\theta$ show that $\sin\theta=\frac{a^2-1}{a^2+1}$.
6. If $a\sin^2\theta+b\cos^2\theta=c$ show that $\tan^2\theta=\frac{c-b}{a-c}$.
7. If $\tan\theta+\cot\theta=4$ show that $\tan^2\theta+\cot^2\theta=14$.
8. If $\sin\theta+\cos\theta=\sqrt{2}\cdot\cos\theta$ show that $\cos\theta-\sin\theta=\sqrt{2}\cdot\sin\theta$
9. If $\text{cosec}\,\theta-\sin\theta=a^3$, $\sec\theta-\cos\theta=b^3$ show that $a^2b^2(a^2+b^2)=1$

10. If $\sin\theta = \dfrac{a-b}{a+b}$, show that $\tan\theta + \sec\theta = \sqrt{\dfrac{a}{b}}$

11. If $\cos\theta = \dfrac{a-b}{a+b}$, show that $\cot\theta + \operatorname{cosec}\theta = \sqrt{\dfrac{a}{b}}$

12. If $\cos A = \sqrt{2}\cos B$, show that $1 + 2\tan^2 A = \tan^2 B$

1.9 Trigonometric ratios of certain standard angles

In the previous section we have seen the trigonometrical ratios of angles 0º, 90º, 180º, 270º and 360º, by considering the positions of the terminal side of the angle θ at these positions. In this section, we shall see the trigonometrical ratios of the angles 30º, 45º and 60º.

1. Trigonometrical ratios of 30º and 60º

Consider an equilateral triangle ABC of sides $2a$ units. Each angle of the triangle is 60º.

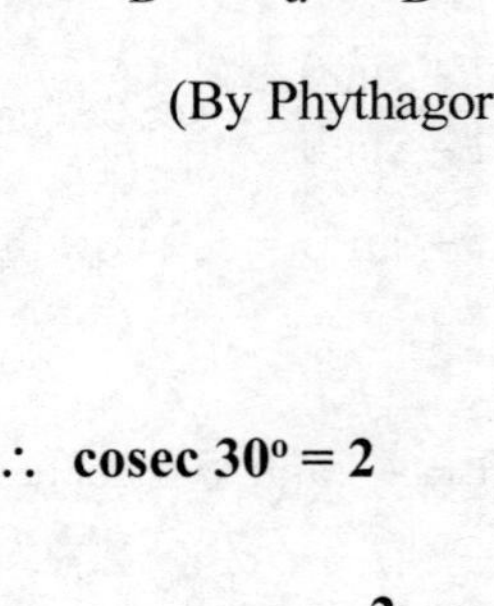

Draw AD perpendicular from A to BC.

Now, $\quad A\hat{B}D = 60^\circ$ and $B\hat{A}D = 30^\circ$.

Again $\quad BD = \dfrac{1}{2}BC \Rightarrow BD = a$

Now from the triangle ADB, we have

$$AD^2 = AB^2 - BD^2 \qquad \text{(By Phythagorus theorem)}$$

$$\Rightarrow \quad AD^2 = (2a)^2 - a^2 \Rightarrow AD^2 = 3a^2 \Rightarrow AD = \sqrt{3}\,a$$

Now from the right angled triangle ADB, we have

$$\sin 30^\circ = \frac{BD}{AB} = \frac{a}{2a} \Rightarrow \mathbf{\sin 30^\circ = \frac{1}{2}}; \quad \therefore \mathbf{\operatorname{cosec} 30^\circ = 2}$$

$$\cos 30^\circ = \frac{AD}{AB} = \frac{\sqrt{3}a}{2a} \Rightarrow \mathbf{\cos 30^\circ = \frac{\sqrt{3}}{2}}; \quad \therefore \mathbf{\sec 30^\circ = \frac{2}{\sqrt{3}}}$$

$$\tan 30^\circ = \frac{BD}{AD} = \frac{a}{\sqrt{3}} \Rightarrow \mathbf{\tan 30^\circ = \frac{1}{\sqrt{3}}}; \quad \therefore \mathbf{\cot 30^\circ = \sqrt{3}}$$

Again from the right angled triangle ADB, we have

$$\sin 60^\circ = \frac{AD}{AB} = \frac{\sqrt{3}a}{2a} \Rightarrow \mathbf{\sin 60^\circ = \frac{\sqrt{3}}{2}}; \quad \therefore \mathbf{\operatorname{cosec} 60^\circ = \frac{2}{\sqrt{3}}}$$

$$\cos 60^\circ = \frac{BD}{AB} = \frac{a}{2a} \Rightarrow \mathbf{\cos 60^\circ = \frac{1}{2}}; \quad \therefore \mathbf{\sec 60^\circ = 2}$$

$$\tan 60^\circ = \frac{AD}{BD} = \frac{\sqrt{3}a}{a} \Rightarrow \mathbf{\tan 60^\circ = \sqrt{3}}; \quad \therefore \mathbf{\cot 60^\circ = \frac{1}{\sqrt{3}}}$$

2. Trigonometrical ratios of 45°

Consider an isosceles right angled triangle ABC right angled at B.

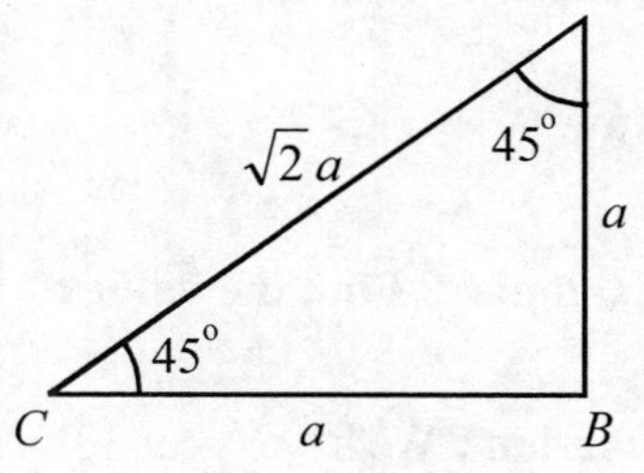

Clearly, $AB = BC = a$ (say)

Thus, $AC^2 = AB^2 + BC^2$

$\Rightarrow$ $AC^2 = 2a^2$

$\Rightarrow$ $AC = \sqrt{2}\,a$

Also $\hat{ACB} = 45°$ and $\hat{BAC} = 45°$

Now we have,

$$\sin 45° = \frac{AB}{AC} = \frac{a}{\sqrt{2}\,a} \Rightarrow \mathbf{\sin 45° = \frac{1}{\sqrt{2}}}; \qquad \therefore\ \mathbf{\operatorname{cosec} 45° = \sqrt{2}}$$

$$\cos 45° = \frac{BC}{AC} = \frac{a}{\sqrt{2}\,a} \Rightarrow \mathbf{\cos 45° = \frac{1}{\sqrt{2}}}; \qquad \therefore\ \mathbf{\sec 45° = \sqrt{2}}$$

$$\tan 45° = \frac{AB}{BC} = \frac{a}{a} \Rightarrow \mathbf{\tan 45° = 1}; \qquad \therefore\ \mathbf{\cot 45° = 1}$$

The following table is the summary of the trigonometrical ratios of certain standard angles up.

Measure of θ		**sin θ**	**cos θ**	**tan θ**	**cosec θ**	**sec θ**	**cot θ**
0°	0	0	1	0	undefined	1	undefined
30°	$\frac{\pi}{6}$	$\frac{1}{2}$	$\frac{\sqrt{3}}{2}$	$\frac{1}{\sqrt{3}}$	2	$\frac{2}{\sqrt{3}}$	$\sqrt{3}$
45°	$\frac{\pi}{4}$	$\frac{1}{\sqrt{2}}$	$\frac{1}{\sqrt{2}}$	1	$\sqrt{2}$	$\sqrt{2}$	1
60°	$\frac{\pi}{3}$	$\frac{\sqrt{3}}{2}$	$\frac{1}{2}$	$\sqrt{3}$	$\frac{2}{\sqrt{3}}$	2	$\frac{1}{\sqrt{3}}$
90°	$\frac{\pi}{2}$	1	0	undefined	1	undefined	0
180°	π	0	– 1	0	undefined	– 1	undefined
270°	$\frac{3\pi}{2}$	– 1	0	undefined	– 1	undefined	0
360°	2π	0	1	0	undefined	1	undefined
360° + θ	$2\pi + \theta$	sin θ	cos θ	tan θ	cosec θ	sec θ	cot θ

We have mentioned that tan 90°, cosec 90° etc., are undefined, for the reason that denominator of these ratios is zero. However we denote these as ∞ in the sence that a ratio $\frac{a}{b}$ $(a \neq 0)$ tends to ∞ as the denominator tends to 0.

Example 1. Find the values of (a) $\tan^2 60° + 2\tan^2 45°$ (b) $\cos\frac{\pi}{3} - \sin\frac{\pi}{6} - \cot^3\frac{\pi}{4}$

Solution : (a) We have, $\tan 60° = \sqrt{3}$, $\tan 45° = 1$

Now, $\tan^2 60° + 2\tan^2 45° = (\sqrt{3})^2 + 2(1)^2 = 3 + 2 = \mathbf{5}$

(b) We have, $\cos\frac{\pi}{3} = \cos 60° = \frac{1}{2}, \quad \sin\frac{\pi}{6} = \sin 30° = \frac{1}{2}, \quad \cot\frac{\pi}{4} = \cot 45° = 1$

Now, $\cos\frac{\pi}{3} - \sin\frac{\pi}{6} - \cot^3\frac{\pi}{4} = \frac{1}{2} - \frac{1}{2} - (1)^3 = \mathbf{-1}$

Example 2. Find the value of $\frac{4}{3}\cot^2 30° + 3\sin^2 60° - 2\,\text{cosec}^2 45° + \frac{1}{3}\tan^2 60° - \frac{1}{4}\cos 0°$

Solution : We have $\cot 30° = \sqrt{3}, \quad \sin 60° = \frac{\sqrt{3}}{2}, \quad \text{cosec}\, 45° = \sqrt{2}, \quad \tan 60° = \sqrt{3}, \quad \cos 0° = 1$

$$\therefore \quad \text{given expression} = \frac{4}{3}(\sqrt{3})^2 + 3\left(\frac{\sqrt{3}}{2}\right)^2 - 2\left(\sqrt{2}\right)^2 + \frac{1}{3}\left(\sqrt{3}\right)^2 - \frac{1}{4}\cdot 1$$

$$= 4 + \frac{9}{4} - 4 + 1 - \frac{1}{4} = \frac{9}{4} - \frac{1}{4} + 1 = 2 + 1 = \mathbf{3}$$

Example 3. Find the value of $3\tan^2 30° + \frac{4}{3}\cos^2 30° - \frac{1}{2}\sec^2 45° - \frac{1}{3}\sin^2 60°$.

Solution : The given expression $= 3\left(\frac{1}{\sqrt{3}}\right)^2 + \frac{4}{3}\left(\frac{\sqrt{3}}{2}\right)^2 - \frac{1}{2}\left(\sqrt{2}\right)^2 - \frac{1}{3}\left(\frac{\sqrt{3}}{2}\right)^2$

$$= 1 + \frac{4}{3}\cdot\frac{3}{4} - \frac{1}{2}\cdot 2 - \frac{1}{3}\cdot\frac{3}{4} = 1 + 1 - 1 - \frac{1}{4} = \mathbf{\frac{3}{4}}$$

Example 4. Find x, if $\dfrac{x\,\text{cosec}^2 30°\cdot\sec^2 45°}{8\cos 45°\sin 60°} = \tan^2 60° - \tan^2 30°$.

Solution : By data,

$$\frac{x\,\text{cosec}^2 30°\cdot\sec^2 45°}{8\cos 45°\sin 60°} = \tan^2 60° - \tan^2 30° \Rightarrow \frac{x\,(2)^2\cdot\left(\sqrt{2}\right)^2}{8\left(1/\sqrt{2}\right)\left(\sqrt{3}/2\right)} = \left(\sqrt{3}\right)^2 - \left(\frac{1}{\sqrt{3}}\right)^2$$

$$\Rightarrow \frac{8x\cdot 2\sqrt{2}}{8\sqrt{3}} = 3 - \frac{1}{3}$$

$$\Rightarrow \frac{2\sqrt{2}}{\sqrt{3}}\cdot x = \frac{8}{3} \Rightarrow x = \frac{8\sqrt{3}}{2\sqrt{2}\cdot 3} \Rightarrow \mathbf{x = \frac{4}{\sqrt{6}}}$$

Example 5. Find x from the equation, $x\sin 30°\cos^2 45° = \dfrac{\cot^2 30°\sec 60°\tan 45°}{\text{cosec}^2 45°\cdot\text{cosec}\,30°}$

Solution : The given expression becomes

$$x\cdot\frac{1}{2}\cdot\left(\frac{1}{\sqrt{2}}\right)^2 = \frac{\left(\sqrt{3}\right)^2\cdot 2\cdot 1}{\left(\sqrt{2}\right)^2\cdot 2} \Rightarrow x\cdot\frac{1}{4} = \frac{6}{4} \Rightarrow \mathbf{x = 6}$$

Example 6. Show that $\cos^2(\pi/4) - \cos^4(\pi/6) + \sin^4(\pi/6) + \sin^4(\pi/3) = \dfrac{9}{16}$

Solution : $\text{LHS} = \left(\dfrac{1}{\sqrt{2}}\right)^2 - \left(\dfrac{\sqrt{3}}{2}\right)^4 + \left(\dfrac{1}{2}\right)^4 + \left(\dfrac{\sqrt{3}}{2}\right)^4 = \dfrac{1}{2} - \dfrac{9}{16} + \dfrac{1}{16} + \dfrac{9}{16} = \mathbf{\dfrac{9}{16}} = \text{RHS}$

Exercise

I.

1. Find the values of

(a) $\sin^2\dfrac{\pi}{4} + \cos^2\dfrac{\pi}{4} - \tan^2\dfrac{\pi}{3}$

(b) $\text{cosec } 60^\circ - \sec 45^\circ + \cot 30^\circ$

(c) $(\tan^2 60^\circ - \sec^2 45^\circ)(\sin 60^\circ - \cos 30^\circ)$

(d) $\sin^2\dfrac{\pi}{4} \,.\, \cos^2\dfrac{\pi}{4} + \sin^2\dfrac{\pi}{3} \,.\, \cos^2\dfrac{\pi}{3}$

(e) $\cos\dfrac{\pi}{3} - \sin\dfrac{\pi}{6} - \cot^3\dfrac{\pi}{4}$

(f) $\tan^3 45^\circ + 4\cos^3 60^\circ$

(g) $2\,\text{cosec}^2\left(\dfrac{\pi}{4}\right) - 3\sec^2\left(\dfrac{\pi}{6}\right)$

(h) $2\sin 30^\circ \cos 30^\circ \cot 60^\circ$

(i) $\tan^2 60^\circ + 2\tan^2 45^\circ$

(j) $\cot\dfrac{\pi}{3} \tan\dfrac{\pi}{6} + \sec^2\dfrac{\pi}{4}$

(k) $\tan^2 45^\circ \sin 60^\circ \tan 30^\circ \tan^2 60^\circ$

(l) $\tan^2 60^\circ + 4\cos^2 45^\circ + 3\sec^2 30^\circ$

(m) $\dfrac{1}{2}\text{cosec}^2\dfrac{\pi}{3} + \sec^2\dfrac{\pi}{4} - 2\cot^2\dfrac{\pi}{3}$

(n) $\tan^2 30^\circ + 2\sin 60^\circ + \tan 45^\circ - \tan 60^\circ + \cos^2 30^\circ$

2. Find x, if $\sin 45^\circ \cos 60^\circ = x \tan 60^\circ \sec 30^\circ$

3. Find x, if $x\cos 45^\circ (\tan 45^\circ + \cot 45^\circ)^2 = 2$.

II.

1. Find the values of

(a) $\dfrac{4\sin^2 60^\circ \cos^3 60^\circ \sec^2 30^\circ}{2\,\text{cosec}^2 30^\circ - \dfrac{1}{2}\sin^2 60^\circ \tan^2 30^\circ}$

(b) $\sin^3 60^\circ \cot 30^\circ - 2\sec^2 45^\circ + 3\cos 60^\circ \tan^2 60^\circ$

(c) $\dfrac{\tan^2 60^\circ - 2\tan^2 45^\circ + \sec^2 30^\circ}{3\sin^2 45^\circ \sin 90^\circ + \cos^2 60^\circ \cos^3 0^\circ}$

(d) $\dfrac{\cos 0 \cdot \sin\left(\dfrac{\pi}{2}\right)\cos^2\left(\dfrac{\pi}{6}\right)\sec^2\left(\dfrac{\pi}{4}\right)}{\tan\left(\dfrac{\pi}{3}\right) + \cot\left(\dfrac{\pi}{3}\right)}$

(e) $\left(\dfrac{1-\cos(\pi/6)}{1+\cos(\pi/6)}\right) - \left(\dfrac{1-\cot 60^\circ}{1+\cot 60^\circ}\right)^2$

2. Find x, if

(a) $x\sin^2 45^\circ \cos^2 30^\circ = \dfrac{\tan 45^\circ \cot^2 30^\circ}{\sin^2 60^\circ + \text{cosec}^2 30^\circ}$

(b) $x \sin 30° \cdot \cos^2 45° = \dfrac{\cot^2 60° \sec 60° \tan 45°}{\text{cosec}^2 45° \text{ cosec } 60°}$

(c) $\tan^2 \dfrac{\pi}{4} - \cos^2 \dfrac{\pi}{3} = x \sin \dfrac{\pi}{4} \cos \dfrac{\pi}{4} \tan \dfrac{\pi}{3}$

(d) $\dfrac{x \sec^2 \dfrac{\pi}{3} \text{ cosec}^2 \dfrac{\pi}{4}}{8 \sin^2 \dfrac{\pi}{4} \cos^2 \dfrac{\pi}{4}} = \cot^2 \dfrac{\pi}{6} - \tan^2 \dfrac{\pi}{6}$

(e) $(3 \text{ cosec}^2 60° + 2 \cot 45°)\, x^2 - (4 \tan^2 60°)\, x = 5 \cos 90°$

(f) $(2x + 1) \cot^2 \left(\dfrac{\pi}{6}\right) - \sin^2 \left(\dfrac{\pi}{4}\right) = \cos^2 \left(\dfrac{\pi}{4}\right) \sec \left(\dfrac{\pi}{3}\right)$

3. Find the value of $\dfrac{\tan 60° + \tan 30° + \tan 45° - \tan 60° \tan 30° \tan 45°}{1 - \tan 60° \tan 30° - \tan 30° \tan 45° - \tan 45° \tan 60°}$

4. Show that $\sec^2 0, \sec^2 \dfrac{\pi}{6}, \sec^2 \dfrac{\pi}{4}, \sec^2 \dfrac{\pi}{3}$ are in H.P

5. If $\sin (A + B) = \dfrac{\sqrt{3}}{2}$ and $\tan (A - B) = 1$, find A and B which are both acute angles.

Answers

I. 1. (a) – 2 (b) $\dfrac{5 - \sqrt{6}}{\sqrt{3}}$ (c) 0 (d) $\dfrac{7}{16}$ (e) – 1 (f) $\dfrac{3}{2}$ (g) 0 (h) $\dfrac{1}{2}$ (i) 5 (j) $\dfrac{7}{3}$

(k) $\dfrac{3}{2}$ (l) 9 (m) 2 (n) $\dfrac{25}{12}$

2. $x = \dfrac{1}{4\sqrt{2}}$ 3. $\dfrac{1}{\sqrt{2}}$

II. 1. (a) $\dfrac{4}{63}$ (b) $\dfrac{13}{8}$ (c) $\dfrac{4}{3}$ (d) $\dfrac{3\sqrt{3}}{8}$ (e) 0

2. (a) $\dfrac{32}{19}$ (b) $\dfrac{2}{\sqrt{3}}$ (c) $\dfrac{\sqrt{3}}{2}$ (d) $\dfrac{2}{3}$ (e) $x = 0$, or 2 (f) $-\dfrac{1}{4}$

3. – 1 5. $A = 52.5°$, $B = 7.5°$

Chapter 2

Heights and Distances

2.1 Introduction

One of the important applications of trigonometry is in land surveying. It happens so many times that the direct measurement of certain distances, heights are angles are not possible. This has to be done by some technique and one of them is using the concept of trigonometrical ratios. We shall see some simple problems relating to problems of finding heights and distances. In the beginning we shall consider the angle of elevation and the angle of depression.

2.2 Angles of Elevation and Depression

Let P be the position of the object to be observed from the point O of observation. Let OA be a horizontal line in the same vertical plane as the object P and let OP be joined.

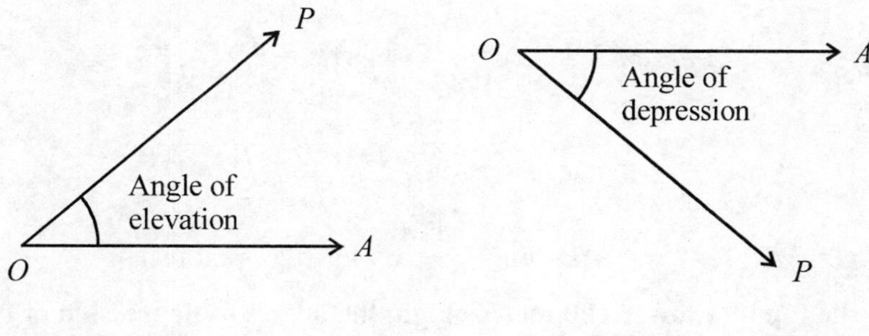

Fig 1. **Fig 2.**

In figure 1 the object P lies above the horizontal line OA, the angle AOP is called the **angle of elevation** of the object P as seen from the point O. The line OA is called the line of sight.

In the figure 2, the object P lies below the horizontal line OA, the angle AOP is called the **angle of depression** of the object P as seen from the point O. The line OA is called the line of sight.

Example 1. The angle of elevation of the top of a chimney at a distance of 100 metres is 30°, find its height.

Solution : Let AB be the chimney and C the point of observation. Fom the figure, we have

$$\tan 30^\circ = \frac{AB}{BC} \quad \Rightarrow \quad AB = BC \cdot \tan 30^\circ \Rightarrow AB = 100 \cdot \frac{1}{\sqrt{3}}$$

$\therefore$ the height of the chimney $= \dfrac{\mathbf{100}}{\sqrt{\mathbf{3}}}$ **metres.**

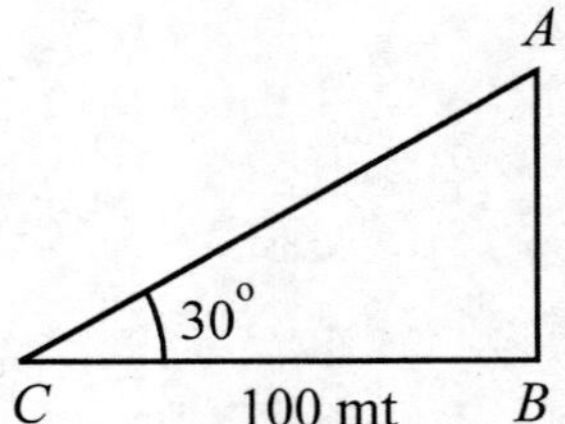

Example 2. A person is at the top of a tower 75 feet high. From there he observes a vertical pole and finds the angles of depressions of the top and the bottom of the pole which are 30° and 60° respectively. Find the height of the pole.

Solution : Let AB represent the tower and CD represent the vertical pole. Let AX be the horizontal drawn through A. By data, $X\hat{A}D = 60°$, $X\hat{A}C = 30°$. Also, $AB = 75$ feet. Let $CD = x$ feet.

Since AX is parallel to MC, we have $A\hat{C}M = X\hat{A}C = 30°$. Again, as AX is parallel to BD, we have $A\hat{D}B = X\hat{A}D = 60°$. Now from the right triangle ADB, we have

$$\tan 60° = \frac{AB}{BD} \Rightarrow BD = \frac{AB}{\tan 60°} \Rightarrow BD = \frac{75}{\sqrt{3}} \qquad (\because AB = 75)$$

Again from the right triangle AMC, we have

$$\tan 30° = \frac{AM}{MC} \Rightarrow AM = MC \cdot \tan 30° \Rightarrow AM = BD \cdot \frac{1}{\sqrt{3}} \qquad (\because MC = BD)$$

$$\Rightarrow AM = \frac{75}{\sqrt{3}} \cdot \frac{1}{\sqrt{3}} \qquad (\because BD = \frac{75}{\sqrt{3}})$$

$$\Rightarrow AM = 25$$

Now, $x = CD = MB \Rightarrow x = AB - AM \Rightarrow x = 75 - 25 =$ **50 feet**

Example 3. From the top of a tower 100 metres high, the angles of depression of two objects due east and west of the tower are 45° and 30° respectively. Find the distance between the objects.

Solution : Let AB be the tower. Let X and Y be the objects due east and west of the tower respectively. Let PAQ be the horizontal drawn through A.

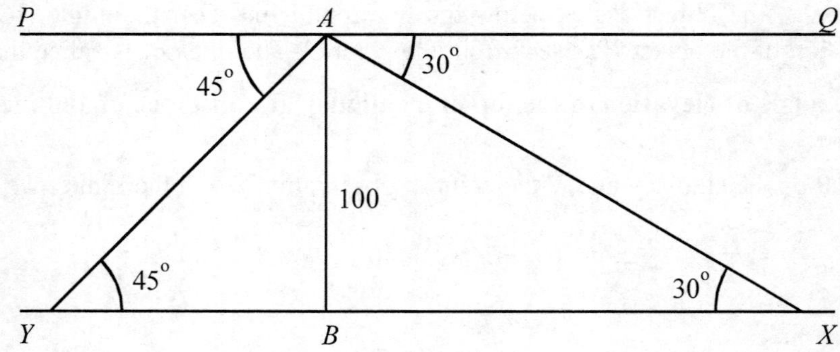

By data $Q\hat{A}X = 30^\circ$ and $P\hat{A}Y = 45^\circ$. Also $AB = 100$ metres. We have to find the lengths BX and BY and then XY. Since AQ is parallel to BX, we have $Q\hat{A}X = A\hat{X}B = 30^\circ$. Again AP is parallel to BY, we have, $P\hat{A}Y = A\hat{Y}B = 45^\circ$. From the right triangle ABX, we have

$$\tan 30^\circ = \frac{AB}{BX} \Rightarrow BX = \frac{AB}{\tan 30^\circ} \Rightarrow BX = \frac{100}{(1/\sqrt{3})} \quad (\because AB = 100)$$

$$\Rightarrow BX = \sqrt{3} \cdot 100 \text{ metres}$$

Again from the right triangle ABY, we have

$$\tan 45^\circ = \frac{AB}{BY} \Rightarrow BY = \frac{AB}{\tan 45^\circ} \Rightarrow BY = \frac{100}{1} \Rightarrow BY = 100 \text{ metres}$$

Now, $XY = BX + BY = 100\sqrt{3} + 100 = \mathbf{100\,(\sqrt{3} + 1)\ metres}$

Example 4. A person standing on the bank of a river observes that the angle of elevation of the top of a tree on the opposite bank is 60°, when he retires 40 feet from the bank he finds the angle to be 30°. Find the height of the tree and breadth of the river.

Solution : Let PQ represent the tree and B be the first point of observation and A be the second point of observation. Now BQ is the breadth of the river.

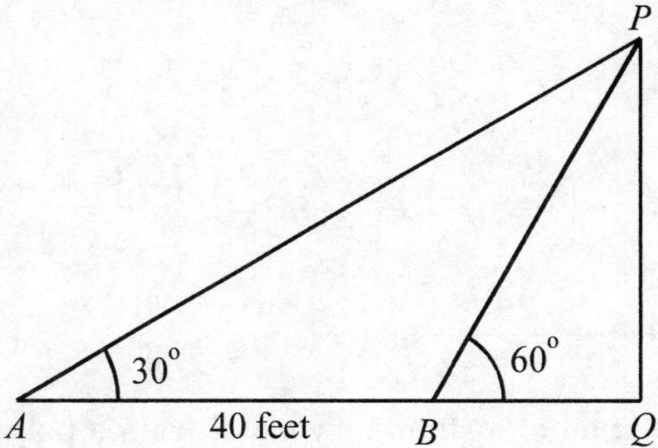

From the figure we have

$$\tan 60^\circ = \frac{PQ}{BQ} \Rightarrow PQ = BQ \cdot \sqrt{3} \quad \text{.... (1)}$$

Again from the figure we have

$$\tan 30^\circ = \frac{PQ}{AQ} \Rightarrow AQ = \frac{PQ}{\tan 30^\circ} \Rightarrow AQ = \sqrt{3} \cdot PQ \quad \left(\because \tan 30^\circ = \frac{1}{\sqrt{3}}\right)$$

$$\Rightarrow AQ = \sqrt{3}\,(\sqrt{3}\,BQ) \quad (\because PQ = \sqrt{3} \cdot BQ)$$

$$\Rightarrow AQ = 3\,BQ$$

Now, $BQ = AQ - AB \Rightarrow BQ = 3\,BQ - AB \Rightarrow 2\,BQ = 40$ $(\because AB = 40 \text{ feet})$

$\therefore$ Breadth of the river $= BQ =$ **20 feet**

Now, height of the tree $= PQ = \sqrt{3}\,BQ$ (from (1))

$\Rightarrow$ $PQ = \sqrt{3} \cdot 20 = \mathbf{20\sqrt{3}\ feet}$

Example 5. The angles of elevation of the top of a tower from the base and the top of a building are 60° and 45°. The building is 20 metres high. Find the height of the tower.

Solution : Let AB represent the building and CD represent the tower. Let BE be the horizontal drawn through the point B.

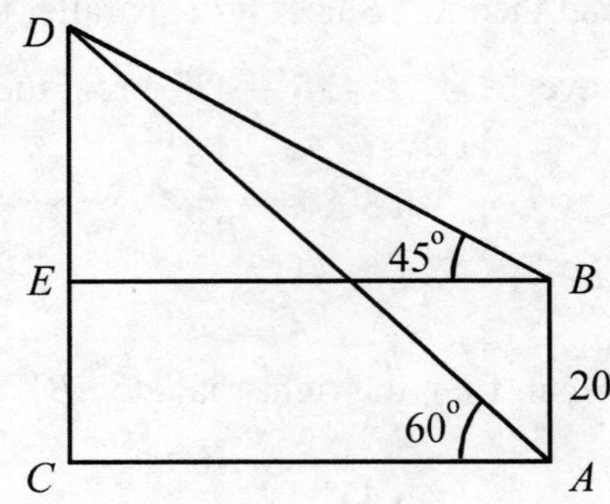

By data, $E\hat{B}D = 45°$, $C\hat{A}D = 60°$

Also, the height $AB = 20$ metres.

Now, from the right triangle ADC, we have

$$\tan 60° = \frac{DC}{AC} \Rightarrow AC = \frac{DC}{\sqrt{3}} \qquad (\because \tan 60° = \sqrt{3})$$

From the right triangle DEB, we have

$$\tan 45° = \frac{DE}{EB} \Rightarrow DE = EB \qquad (\because \tan 45° = 1)$$

$$\Rightarrow DE = AC \qquad (\because ED = \sqrt{3})$$

$$\Rightarrow DE = \frac{DC}{\sqrt{3}} \qquad \left(\because AC = \frac{DC}{\sqrt{3}}\right)$$

Now, $\quad DC = DE + EC \Rightarrow DC = \frac{DC}{\sqrt{3}} + AB \qquad (\because EC = AB)$

$$\Rightarrow DC = \frac{DC}{\sqrt{3}} + 20 \qquad (\because AB = 20)$$

$$\Rightarrow \left(1 - \frac{1}{\sqrt{3}}\right) DC = 20 \Rightarrow \left(\frac{\sqrt{3}-1}{\sqrt{3}}\right) DC = 20 \Rightarrow DC = \frac{20\sqrt{3}}{\sqrt{3}-1}$$

Thus, the height of the tower $= \mathbf{\frac{20\sqrt{3}}{\sqrt{3}-1}}$ **metres**

Example 6. The angle of elevation of a stationary cloud from a point 2500 cms above the lake is 30° and the angle of depression of its reflection in the lake is 60°. What is the height of the cloud above the lake.

Solution : Let D be the image of the cloud C, in the water.
Let AB be the surface of the lake. Then we have

$$CB = BD$$

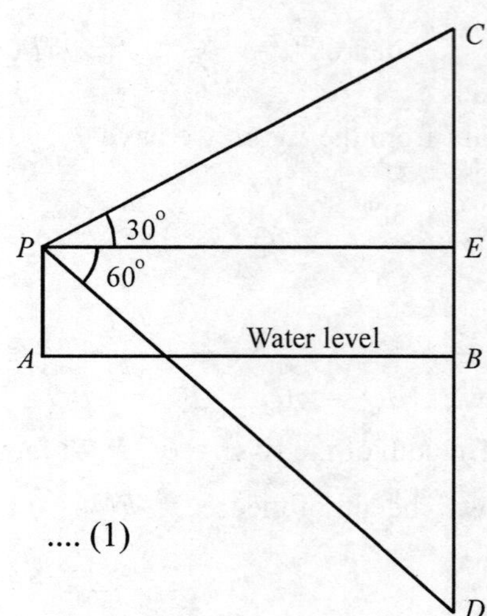

Let P be the position of observation.

Thus, by data $AP = 2500$ cms.

By data, $E\hat{P}C = 30°$, $E\hat{P}D = 60°$

From the right triangle PEC we have,

$$\tan 30° = \frac{CE}{PE} \Rightarrow PE = \frac{CE}{\tan 30°}$$

$$\Rightarrow PE = \sqrt{3} \cdot CE \qquad \text{.... (1)}$$

Again from the right triangle *PED*, we have

$$\tan 60^\circ = \frac{ED}{PE} \Rightarrow PE = \frac{ED}{\tan 60^\circ} \Rightarrow PE = \frac{ED}{\sqrt{3}} \quad \text{.... (2)}$$

From (1) and (2), we get,

$$\sqrt{3}\ CE = \frac{1}{\sqrt{3}}\ ED \Rightarrow 3\ CE = ED$$

$$\Rightarrow 3\ (BC - BE) = EB + BD$$

$$\Rightarrow 3\ (BC - 2500) = 2500 + BC \qquad (\because BD = BC \text{ and } BE = PA = 2500)$$

$$\Rightarrow 3\ BC - 7500 = 2500 + BC \Rightarrow 2\ BC = 10{,}000 \Rightarrow BC = 5000 \text{ cms}$$

Thus, the height of the cloud = **5000 cms.**

Exercise

I.

1. From a point on the ground, the angle of elevation of a top of a pole is 60°. If the pole is of height 30 metres find the distance between the point and the foot of the pole.
2. The angle of elevation of a top of a tower, from a point 100 metres from it, is 30°. Find the height of the tower.
3. From a ship's masthead 50 metres high, the angle of depression of a boat is observed to be 30°; find its distance from the ship.
4. Find the angle of elevation of the sun when the shadow of a tower 75 metres high is $25\sqrt{3}$ metres long.
5. What is the angle of elevation of the sun when the length of the shadow of a pole is $\sqrt{3}$ times the height of the pole?
6. The shadow of a tower standing on a level plane is found to be 60 feet longer when the sun's altitude is 30° than when it is 45°. Show that the height of the tower is $30\ (1 + \sqrt{3})$ feet.
7. The angles of elevation of the summit of a hill from the top and the bottom of a tower are 30° and 60° respectively. If the height of the tower is h, show that the height of the hill is $\frac{3h}{2}$.

II.

1. The angles of elevation of a mountain peak from two points due east of it and 500 metres aparts are found to be 45° and 60°. Find the height of the peak.
2. The angle of elevation of the top of the tower from a point on the same level as the foot of the tower is 30°. On advancing 150 metres towards the foot of the tower the angle of elevation increases to 60°. Find the height of the tower.
3. The angle of elevation of the top of a tower at a point *A* on the ground is 30°. On walking 20 metres towards the tower the angle of elevation is 60°. Find the height of the tower and its distance from *A*.
4. From the top of a light house 240 feet high, the angles of depression of the top of the mast of a ship and of its reflection in water are found to be 45° and 60° respectively. What is the height of the mast?
5. From the top of a tower the angles of depression of the top and the bottom of a building are 30° and 60°. If the height of the building is 50 metres, find the height of the tower.
6. From the top of a hill 300 feet high, the angles of depression of the top and the bottom of a vertical column are 45° and 60°. Find the height of the column.

7. From the top of a hill, the angles of depression of two consecutive kilometres stones on a level road leading to the foot of the hill are 30° and 60°. Find the height of the hill.

8. A person standing on the bank of a river observes the angle of elevation of the top of a telephone pole on the opposite bank to be 60°, when he retires 30 metres from the bank, he finds the angle of elevation to be 30°. Find the height of the telephone pole and breadth of the river.

9. From the top of a tower 50 feet high, the angle of elevation of the top of another tower is 45° and the angle of depression of its foot is 60°. Find the height of the second tower.

10. The angles of elevation of the top of a tower as seen from the top and bottom of a building are 30° and 45°. The height of the building is 10 metres. Find the height of the tower and the distance between the tower and the building.

11. The angle of elevation of an object from a point 200 metres above a lake is found to be 30° and the angle of depression of its image in the lake is 45°. Find the height of the object above the lake.

12. Two towers of height 14 metres and 25 metres stand on level ground. The angles of elevation of their tops from a point on the line joining their feet are respectively 45° and 60°. Find the distance between the towers.

13. A person standing in between the two posts observes the angles of elevation of their tops to be 60° each. Find the heights of the posts if the distance between the two posts is 25 metres and the height of one post is 1.5 times the other.

Answers

I. **1.** $10\sqrt{3}$ **2.** 57.73 mts **3.** 86.6 mts **4.** 60° **5.** 30°

II. **1.** $\dfrac{500\sqrt{3}}{\sqrt{3}-1}$ mts **2.** $75\sqrt{3}$ mts **3.** $10\sqrt{3}$ mts **4.** $240(2-\sqrt{3})$ feet **5.** 75 mts

6. $100\sqrt{3}(\sqrt{3}-1)$ ft **7.** $\dfrac{\sqrt{3}}{2}$ km **8.** 51.96 mts, 15 mts **9.** $\dfrac{50\sqrt{3}(\sqrt{3}+1)}{3}$

10. $5(\sqrt{3}+1)$, $5\sqrt{3}(\sqrt{3}+1)$ **11.** $200(2+\sqrt{3})$ mts **12.** $\dfrac{(25+14\sqrt{3})}{\sqrt{3}}$ mts **13.** $10\sqrt{3}$ and $15\sqrt{3}$

Chapter 3

Trigonometric Ratios of Compound Angles

3.1 Introduction

In this section we shall express the trigonometric ratios of the sum $A + B$ and difference $A - B$ of the angles A and B in terms of trigonometric ratios of A and B.

The angles of the form $A + B$, $A - B$, $A + B - C$ etc., are known as compound angles.

3.2 Ratios of Compound Angle

Theorem 1*. To prove geometricall, that

(a) $\sin(A+B) = \sin A \cos B + \cos A \sin B$ **(b)** $\cos(A+B) = \cos A \cos B - \sin A \sin B$

(c) $\tan(A+B) = \dfrac{\tan A + \tan B}{1 - \tan A \tan B}$

Proof* : Let the the radius vector start from OX and revolve through an angle A, and take the position OP and then revolve further through an angle B and take the final poisition OQ. Thus, $X\hat{O}P = \hat{A}$ and $P\hat{O}Q = \hat{B}$.

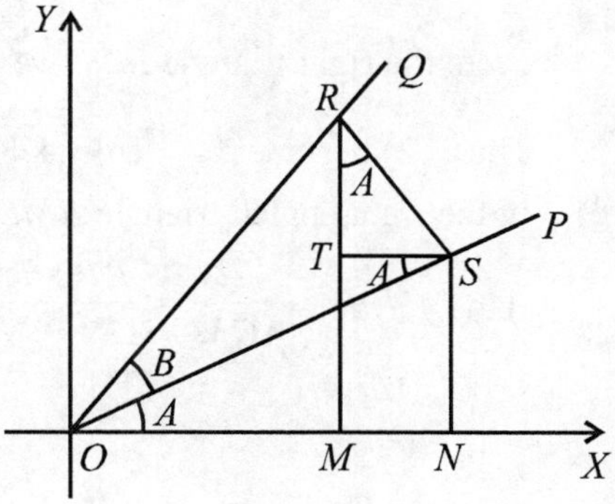

Clearly, $X\hat{O}Q = A + B$.

Let R be any point on OQ.

Draw RM perpendicular to OX
RS perpendicular to OP
SN perpendicular to OX
and ST perpendicular to RM.

Now, ST is parallel to OX. Thus

$$T\hat{S}O = S\hat{O}N \quad \text{(Alternative angle)}$$

$$\Rightarrow \quad T\hat{S}O = \hat{A}$$

Again, $$R\hat{S}T = 90^\circ - T\hat{S}O \quad (\because R\hat{S}O = 90^\circ)$$

$$\Rightarrow \quad R\hat{S}T = 90^\circ - \hat{A} \quad \Rightarrow \quad T\hat{R}S = \hat{A}$$

* Proof of the theorem may be avoided for the first reading.

(a) In the right angle triangle RMO, we have

$$\sin(A+B) = \frac{RM}{OR} = \frac{RT+TM}{OR} = \frac{RT+SN}{OR} \qquad (\because TM = SN)$$

$$\therefore \quad \sin(A+B) = \frac{RT}{OR} + \frac{SN}{OR} \Rightarrow \sin(A+B) = \frac{RT}{RS}\cdot\frac{RS}{OR} + \frac{SN}{OS}\cdot\frac{OS}{OR} \qquad \text{.... (1)}$$

From the right triangle RTS, we have $\frac{RT}{RS} = \cos A$

From the right triangle RSO, we have $\frac{RS}{OR} = \sin B$, $\frac{OS}{OR} = \cos B$

Again, from the right triangle SNO, we have $\frac{SN}{OS} = \sin A$.

Thus, (1) becomes

$\sin(A+B) = \cos A \sin B + \sin A \cos B$ i.e., $\mathbf{\sin(A+B) = \sin A \cos B + \cos A \sin B}$

(b) In the right angle triangle RMO, we have

$$\cos(A+B) = \frac{OM}{OR} = \frac{ON-MN}{OR} = \frac{ON-TS}{OR} \qquad (\because MN = TS)$$

$$\therefore \quad \cos(A+B) = \frac{ON}{OR} - \frac{TS}{OR} \Rightarrow \cos(A+B) = \frac{ON}{OS}\cdot\frac{OS}{OR} - \frac{TS}{RS}\cdot\frac{RS}{OR} \qquad \text{.... (2)}$$

From the right triangle ONS, we have $\frac{ON}{OS} = \cos A$

From the right triangle RSO, we have $\frac{OS}{OR} = \cos B$, $\frac{RS}{OR} = \sin B$

From the right triangle RTS, we have $\frac{TS}{RS} = \sin A$.

Thus, (2) becomes, $\mathbf{\cos(A+B) = \cos A \cos B - \sin A \sin B}$

(c) In the right angled triangle RMO, we have

$$\tan(A+B) = \frac{RM}{OM} = \frac{RT+TM}{ON-MN} \Rightarrow \tan(A+B) = \frac{RT+SN}{ON-TS} \qquad (\because TM = SN,\ MN = TS)$$

$$\Rightarrow \tan(A+B) = \frac{\frac{RT}{ON} + \frac{SN}{ON}}{1 - \frac{TS}{ON}}$$

$$\Rightarrow \tan(A+B) = \frac{\frac{RT}{ON} + \frac{SN}{ON}}{1 - \frac{TS}{RT}\cdot\frac{RT}{ON}} \qquad \text{.... (3)}$$

Clearly, the triangle RTS is similar to SNO. $\left(\because R\hat{T}S = S\hat{N}O = 90^\circ \ \ T\hat{R}S = \hat{A} = S\hat{O}N \ \therefore R\hat{S}T = O\hat{S}N\right)$.

Thus, the sides of the triangles are proportional.

i.e., $\dfrac{TS}{SN} = \dfrac{RT}{ON} = \dfrac{RS}{OS}$. Thus, (3) $\Rightarrow$ $\tan(A+B) = \dfrac{\dfrac{RS}{OS} + \dfrac{SN}{ON}}{1 - \dfrac{TS}{RT} \cdot \dfrac{RS}{OS}}$

But $\dfrac{RS}{OS} = \tan B$, in the triangle RSO.

In the right triangle SNO, $\dfrac{SN}{ON} = \tan A$ and in the right triangle RTS, $\dfrac{TS}{RT} = \tan B$

Thus, $$\mathbf{\tan(A+B) = \dfrac{\tan A + \tan B}{1 - \tan A \tan B}}$$

Note : 1. The results proved in the above theorem, hold good for angles A and B of any magnitude. Assuming the truth of the results for acute angle, we can establish them to be true universally.

Supposing A is an obtuse angle and B is acute. We can put $A = 90° + A_1$, where A_1 is an acute angle.

Clearly, $\sin A = \sin(90° + A_1) = \cos A_1$ and $\cos A = \cos(90° + A_1) = -\sin A_1$ .... (1)

Now consider, $\sin(A+B) = \sin[90° + (A_1 + B)] \Rightarrow \sin(A+B) = \cos(A_1 + B)$ .... (2)

Since A_1 and B are acute angle, we have,

$\cos(A_1 + B) = \cos A_1 \cos B - \sin A_1 \sin B$

$\Rightarrow$ $\cos(A_1 + B) = \sin A \cos B + \cos A \sin B$ (by using (1))

$\therefore$ (2) becomes, $\sin(A+B) = \sin A \cos B + \cos A \sin B$ which is the required result.

Similarly, $\cos(A+B) = \cos[90° + (A_1 + B)]$

$\Rightarrow$ $\cos(A+B) = -\sin(A_1 + B) = -[\sin A_1 \cos B + \cos A_1 \sin B]$

$\Rightarrow$ $\cos(A+B) = \cos A \cos B - \sin A \sin B$ (by using (1))

which is the required result.

In a similar way one can establish the results for other cases also. Hence, the theorem is true for all angles.

2. The formula for $\tan(A+B)$ can also be obtained by using the formulae of $\sin(A+B)$ and $\cos(A+B)$.

$$\tan(A+B) = \frac{\sin(A+B)}{\cos(A+B)} = \frac{\sin A \cos B + \cos A \sin B}{\cos A \cos B - \sin A \sin B}$$

Dividing both numerator and denominator by $\cos A \cos B$, we get, $\tan(A+B) = \dfrac{\tan A + \tan B}{1 - \tan A \tan B}$

Theorem 2. To prove geometrically,

(a) $\mathbf{\sin(A-B) = \sin A \cos B - \cos A \sin B}$ **(b)** $\mathbf{\cos(A-B) = \cos A \cos B + \sin A \sin B}$

(c) $\mathbf{\tan(A-B) = \dfrac{\tan A - \tan B}{1 + \tan A \tan B}}$.

Proof : Let the the radius vector start from OX and revolve through an angle A, and take the position OP. Again let it revolve in the opposite direction, and revolve through an angle B and finally take the poisition Q. Clearly, $X\hat{O}Q = A - B$.

Let R be any point on OQ.

Draw RM perpendicular to OX

RS perpendicular to OP

SN perpendicular to OX

and ST perpendicular to MR produced.

Now, ST is parallel to OX. Thus we have

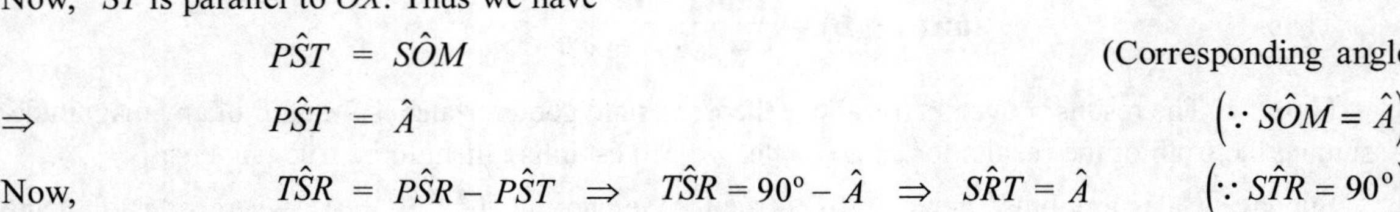

$$P\hat{S}T = S\hat{O}M \qquad \text{(Corresponding angle}$$

$$\Rightarrow \quad P\hat{S}T = \hat{A} \qquad \left(\because S\hat{O}M = \hat{A}\right)$$

Now, $$T\hat{S}R = P\hat{S}R - P\hat{S}T \Rightarrow T\hat{S}R = 90^\circ - \hat{A} \Rightarrow S\hat{R}T = \hat{A} \qquad \left(\because S\hat{T}R = 90^\circ\right)$$

(a) In the right angle triangle ROM, in which $R\hat{O}M = A - B$, we have

$$\sin(A - B) = \frac{RM}{OR} = \frac{TM - TR}{OR} = \frac{SN - TR}{OR} \qquad (\because TM = SN)$$

$$\Rightarrow \quad \sin(A - B) = \frac{SN}{OR} - \frac{TR}{OR} \Rightarrow \sin(A - B) = \frac{SN}{OS} \cdot \frac{OS}{OR} - \frac{TR}{SR} \cdot \frac{SR}{OR} \qquad \text{.... (1)}$$

Now, from the right triangle SNO, we have $\frac{SN}{OS} = \sin A$

from the right triangle OSR, we have $\frac{OS}{OR} = \cos B$, $\frac{SR}{OR} = \sin B$,

from the right triangle RTS, we have $\frac{TR}{SR} = \cos A$

Thus, (1) becomes, **$\sin(A - B) = \cos A \sin B - \sin A \cos B$** which is the required result.

(b) From the right angle triangle ROM, $R\hat{O}M = A - B$, we have

$$\cos(A - B) = \frac{OM}{OR} = \frac{ON + MN}{OR} = \frac{ON + TS}{OR} \qquad (\because NM = ST)$$

$$\therefore \quad \cos(A - B) = \frac{ON}{OR} + \frac{ST}{OR} \Rightarrow \cos(A - B) = \frac{ON}{OS} \cdot \frac{OS}{OR} + \frac{ST}{SR} \cdot \frac{SR}{OR} \qquad \text{.... (2)}$$

Now, from the right triangle SNO, we have $\frac{ON}{OS} = \cos A$

from the right triangle OSR, we have $\frac{OS}{OR} = \cos B$, $\frac{SR}{OR} = \sin B$,

from the right triangle RST, we have $\frac{ST}{SR} = \sin A$

Thus, (2) becomes, $\mathbf{\cos (A - B) = \cos A \cos B + \sin A \sin B}$ which is the required result.

(c) From the right angled triangle *ROM*, in which $R\hat{O}M = A - B$, we have

$$\tan (A - B) = \frac{RM}{OM} = \frac{TM - TR}{ON + NM} \Rightarrow \tan (A - B) = \frac{SN - TR}{ON + ST} \quad (\because TM = SN, NM = ST)$$

$$\Rightarrow \tan (A - B) = \frac{\frac{SN}{ON} - \frac{TR}{ON}}{1 + \frac{ST}{ON}}$$

$$\Rightarrow \tan (A - B) = \frac{\frac{SN}{ON} - \frac{TR}{ON}}{1 + \frac{ST}{TR} \cdot \frac{TR}{ON}} \quad .... (3)$$

Clearly, the triangle *STR* is similar to *SNO*.

Thus, the corresponding sides of the triangles are proportional. That is, $\frac{TR}{ON} = \frac{ST}{SN} = \frac{SR}{OS}$

$$\tan (A - B) = \frac{\frac{SN}{ON} - \frac{SR}{OS}}{1 - \frac{ST}{TR} \cdot \frac{SR}{OS}} \quad .... (3)$$

Now, from the right triangle *SNO*, $\frac{SN}{ON} = \tan A$

from the right triangle *RSO*, $\frac{SR}{OS} = \tan B$; from the right triangle *STR*, $\frac{ST}{TR} = \tan A$

Thus (3) becomes, $\mathbf{\tan (A - B) = \dfrac{\tan A - \tan B}{1 - \tan A \tan B}}$ which is the required result.

Note : 1. The results proved in the above theorem, hold good for angles *A* and *B* of any magnitude. This fact can be established, in the similar manner as we have done in the case of trigonometrical ratios of $A + B$. This we leave it to the readers, as an exercise.

2. Assuming the formulae for $\sin (A + B)$ and $\cos (A + B)$, we can deduce the formulae for $\sin (A - B)$ and $\cos (A - B)$. This can be done as below.

We have, $\sin (A + B) = \sin A \cos B + \cos A \sin B$

Replace *B* by $-B$, we get

$$\sin [A + (-B)] = \sin A \cos (-B) + \cos A \sin (-B)$$

$$\Rightarrow \sin [A - B)] = \sin A \cos B - \cos A \sin B \quad (\because \cos (-B) = \cos B, \sin (-B) = -\sin B)$$

Again, we have, $\cos (A + B) = \cos A \cos B - \sin A \sin B$

Replace *B* by $-B$, we get, $\cos [A + (-B)] = \cos A \cos (-B) - \sin A \sin (-B)$

$$\cos (A - B) = \cos A \cos B + \sin A \sin B$$

Also, we have, $\tan(A+B) = \dfrac{\tan A + \tan B}{1 - \tan A \tan B}$

Replace B by $-B$, we get, $\tan(A-B) = \dfrac{\tan A - \tan B}{1 + \tan A \tan B}$ $(\because \tan(-B) = -\tan B)$

Alternative method

We have, $\sin A = \sin[(A-B)+B] \Rightarrow \sin A = \sin(A-B)\cos B + \cos(A-B)\sin B$ .... (1)

Also, $\cos A = \cos[(A-B)+B] \Rightarrow \cos A = \cos(A-B)\cos B - \sin(A-B)\sin B$ .... (2)

Consider, (1) × $\cos B$ – (2) × $\sin B$, we get,

$$\sin A \cos B - \cos A \sin B = (\cos^2 B + \sin^2 B)\sin(A-B)$$

$$\Rightarrow \quad \sin(A-B) = \sin A \cos B - \cos A \sin B$$

Again consider, (1) × $\sin B$ + (2) × $\cos B$, we get,

$$\sin A \sin B + \cos A \cos B = (\sin^2 B + \cos^2 B)\cos(A-B)$$

$$\Rightarrow \quad \cos(A-B) = \cos A \cos B + \sin A \sin B$$

3. The formula for the $\tan(A-B)$ can be obtained by considering,

$$\tan(A-B) = \frac{\sin(A-B)}{\cos(A-B)} \Rightarrow \tan(A-B) = \frac{\sin A \cos B - \cos A \sin B}{\cos A \cos B + \sin A \sin B}$$

Dividing both numerator and denominator by $\cos A \cos B$ we get,

$$\Rightarrow \quad \tan A = \frac{\tan A - \tan B}{1 + \tan A \tan B}$$

Example 1. Show that $\dfrac{\tan 69^\circ + \tan 66^\circ}{1 - \tan 69^\circ \tan 66^\circ} = -1$.

Solution : LHS is of the form $\dfrac{\tan A + \tan B}{1 - \tan A \tan B}$, with $A = 69^\circ$ and $B = 66^\circ$

$$\therefore \quad \text{LHS} = \tan(69^\circ + 66^\circ) = \tan 135^\circ = \tan(180^\circ - 45^\circ) = -\tan 45^\circ = \mathbf{-1} = \text{RHS}$$

Example 2. Show that $\dfrac{\sqrt{3}\cos 23^\circ - \sin 23^\circ}{2} = \cos 53^\circ$.

Solution : $\text{LHS} = \dfrac{\sqrt{3}}{2}\cos 23^\circ - \dfrac{1}{2}\sin 23^\circ$

$= \cos 30^\circ \cos 23^\circ - \sin 30^\circ \sin 23^\circ$ $\left(\because \dfrac{\sqrt{3}}{2} = \cos 30^\circ, \dfrac{1}{2} = \sin 30^\circ\right)$

$= \cos(30^\circ + 23^\circ) = \mathbf{\cos 53^\circ} = \text{RHS}$

Example 3. Show that $\dfrac{\sin(A-B)}{\sin A\sin B}+\dfrac{\sin(B-C)}{\sin B\sin C}+\dfrac{\sin(C-A)}{\sin C\sin A}=0.$

Solution : Consider, $\dfrac{\sin(A-B)}{\sin A\sin B}=\dfrac{\sin A\cos B-\cos A\sin B}{\sin A\sin B}=\dfrac{\cos B}{\sin B}-\dfrac{\cos A}{\sin A}=\cot B-\cot A$

Similarly, $\dfrac{\sin(B-C)}{\sin B\sin C}=\cot C-\cot B$ and $\dfrac{\sin(C-A)}{\sin C\sin A}=\cot A-\cot C$

Thus, $\text{LHS}=\cot B-\cot A+\cot C-\cot B+\cot A-\cot C=\mathbf{0}=\text{RHS}$

Example 4. Show that $\sin^2(A+B)-\sin^2(A-B)=\sin 2A\sin 2B.$

Solution : We have from example 1; $\sin^2 A-\sin^2 B=\sin(A+B)\sin(A-B)$

Replacing A by $A+B$ and B by $A-B$, we get

$$\sin^2(A+B)-\sin^2(A-B)=\sin(A+B+A-B)\sin(A+B-A+B)$$

$$\Rightarrow\quad \sin^2(A+B)-\sin^2(A-B)=\sin 2A\cdot\sin 2B$$

which is the required expression.

Example 5. Show that $\cos^2\left(\dfrac{\pi}{4}+x\right)-\sin^2\left(\dfrac{\pi}{4}-x\right)$ **is independent of** x.

Solution : We know that (from the example 1), $\cos^2 A-\sin^2 B=\cos(A+B)\cdot\cos(A-B)$

Putting $A=\left(\dfrac{\pi}{4}+x\right)$ and $B=\left(\dfrac{\pi}{4}-x\right)$, we get,

$$\cos^2\left(\frac{\pi}{4}+x\right)-\sin^2\left(\frac{\pi}{4}-x\right)=\cos\left(\frac{\pi}{4}+x+\frac{\pi}{4}-x\right)\cdot\cos\left(\frac{\pi}{4}+x-\frac{\pi}{4}+x\right)$$

$$=\cos\frac{\pi}{2}\cdot\cos(2x)=\mathbf{0}\qquad\left(\because\ \cos\frac{\pi}{2}=0\right)$$

Example 6. If $\sin A=\dfrac{4}{5}$ **and** $\sin B=\dfrac{5}{13}$**, find the value of** $\sin(A+B)$, $\cos(A+B)$ **and** $\tan(A+B)$.

Solution : By data $\sin A=\dfrac{4}{5}$. From the adjacent figure (1) we have,

$$\cos A=\frac{3}{5}\quad\text{and}\quad\tan A=\frac{4}{3}$$

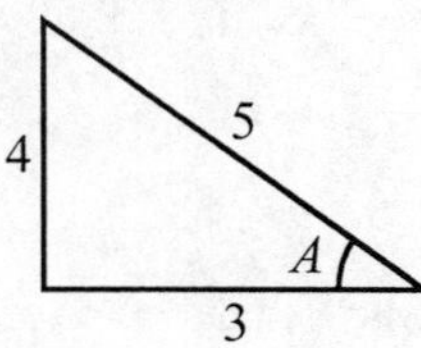

Fig (1)

Again by data $\sin B=\dfrac{5}{13}$. From the figure (2), we have,

$$\cos B=\frac{12}{13}\quad\text{and}\quad\tan B=\frac{5}{12}$$

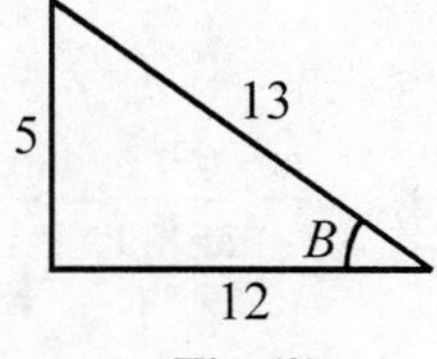

Fig (2)

Now, $\sin(A+B) = \sin A\cos B + \cos A\sin B = \frac{4}{5}\cdot\frac{12}{13} + \frac{3}{5}\cdot\frac{5}{13} = \frac{48+15}{65} = \frac{63}{65}$

$\cos(A+B) = \cos A\cos B - \sin A\sin B = \frac{3}{5}\cdot\frac{12}{13} - \frac{4}{5}\cdot\frac{5}{13} = \frac{36-20}{65} = \frac{16}{65}$

Again, $\tan(A+B) = \frac{\tan A + \tan B}{1+\tan A\tan B} = \frac{\frac{4}{3}+\frac{5}{12}}{1-\frac{4}{3}\cdot\frac{5}{12}} = \frac{48+15}{36-20} = \mathbf{\frac{63}{16}}$

Example 7. If $\tan(A-B) = \frac{7}{24}$ and $\tan A = \frac{4}{3}$, show that $A+B = \frac{\pi}{2}$.

Solution : By data, we have,

$$\Rightarrow \quad \tan(A-B) = \frac{7}{24} \quad \Rightarrow \quad \frac{\tan A - \tan B}{1+\tan A\tan B} = \frac{7}{24}$$

$$\Rightarrow \quad \frac{\frac{4}{3} - \tan B}{1+\left(\frac{4}{3}\right)\tan B} = \frac{7}{24} \qquad \left(\because \text{ by data } \tan A = \frac{4}{3}\right)$$

$$\Rightarrow \quad \frac{4-3\tan B}{3+4\tan B} = \frac{7}{24} \quad \Rightarrow \quad 96 - 72\tan B = 21 + 28\tan B$$

$$\Rightarrow \quad 100\tan B = 75 \quad \Rightarrow \quad \tan B = \frac{3}{4}$$

Now, $\tan(A+B) = \frac{\tan A + \tan B}{1-\tan A\tan B}$

$$\Rightarrow \quad \tan(A+B) = \frac{\frac{4}{3}+\frac{3}{4}}{1-\left(\frac{4}{3}\right)\left(\frac{3}{4}\right)} \Rightarrow \tan(A+B) = \frac{16+9}{12-12} = \infty \Rightarrow \mathbf{A+B = \frac{\pi}{2}}$$

Exercise

I.

1. Show that

(a) $\cos 47^\circ \sin 17^\circ - \sin 47^\circ \cos 17^\circ = \frac{1}{2}$

(b) $\sin(70^\circ + A)\cos(20^\circ - A) + \cos(20^\circ - A)\sin(70^\circ + A) = 1$

(c) $\frac{\tan 69^\circ + \tan 66^\circ}{1 - \tan 69^\circ \cdot \tan 66^\circ} = -1$

(d) $\cos 49^\circ 20' \cdot \cos 40^\circ 40' - \sin 49^\circ 20' \cdot \sin 40^\circ 40' = 0$

(e) $\tan(45^\circ + A)\tan(45^\circ - A) = 1$

(f) $\tan A + \tan B = \frac{\sin(A+B)}{\cos A\cos B}$

(g) $\sin(45^\circ + A) = \frac{1}{\sqrt{2}}(\sin A + \cos A)$ **(h)** $\tan(45^\circ - A) = \frac{1 - \tan A}{1 + \tan A}$

(i) $\tan(45^\circ + A) = \frac{1 + \tan A}{1 - \tan A}$ **(j)** $\tan 75^\circ + \cot 75^\circ = 4$

II.

1. Show that

(a) $\sin 75^\circ - \sin 15^\circ = \cos 105^\circ + \cos 15^\circ$ **(b)** $\sin A + \sin(120^\circ + A) - \sin(120^\circ - A) = 0$

(c) $\cos A + \cos(120^\circ + A) - \cos(120^\circ - A) = 0$ **(d)** $\frac{\sin 75^\circ - \sin 15^\circ}{\cos 75^\circ + \cos 15^\circ} = \frac{1}{\sqrt{3}}$

(e) $\frac{\cos 17^\circ + \sin 17^\circ}{\cos 17^\circ - \sin 17^\circ} = \tan 62^\circ$ **(e)** $\sin\left(\theta + \frac{\pi}{6}\right) + \cos\left(\theta + \frac{\pi}{3}\right) = \cos\theta$

2. If $\tan\alpha = \frac{1}{8}$, $\tan\beta = \frac{7}{9}$ find the value of $\alpha + \beta$ if $0^\circ < \alpha + \beta < 360^\circ$

3. If $\tan A = \frac{5}{6}$, $\tan B = \frac{1}{11}$ find $A + B$

4. If $\tan A = \frac{n}{n+1}$, $\tan B = \frac{1}{2n+1}$ show that $A + B = \frac{\pi}{4}$

5. If $\tan A = \frac{1}{2}$, $\tan(A + B) = \frac{7}{9}$ find $\tan B$

6. If $\cot\alpha = \frac{11}{3}$, $\tan\beta = \frac{-2}{11}$ find the value of $\tan(\alpha + \beta)$ given that, $\pi < \alpha < \frac{3\pi}{2}$, $\frac{\pi}{2} < \beta < \pi$

7. Show that $\sin\left(A + \frac{\pi}{4}\right) + \cos\left(A + \frac{\pi}{4}\right) = \sqrt{2}\cos A$

8. Show that, $\cos(A + B) + \sin(A + B) = (\cos A + \sin A)(\cos B - \sin B)$.

III.

1. Given $\sin A = \frac{4}{5}$, $\cos B = \frac{5}{12}$, A and B being acute, find $\sin(A - B)$.

2. If $\tan\alpha = \frac{3}{4}$, $\cot\beta = \frac{3}{4}$, find $\cos(\alpha - \beta)$.

3. If $\sin A = \frac{5}{13}$ and $\sin B = \frac{4}{5}$, find the value of $\cos(A - B)$.

4. If A and B are acute angle such that $\cos A = \frac{1}{7}$ and $\cos B = \frac{13}{14}$ show that $A - B = \frac{\pi}{3}$.

5. If $\tan(A + B) = 1$ and $\tan A = \frac{1}{2}$ find $\tan B$.

6. If $\sin A = \dfrac{8}{17}$ and $\sin B = \dfrac{5}{13}$ show that $\sin(A+B) = \dfrac{171}{221}$

7. Show that $\cos\left(\dfrac{\pi}{3}+A\right)\cos\left(\dfrac{\pi}{3}-A\right)-\sin\left(\dfrac{\pi}{3}+A\right)\sin\left(\dfrac{\pi}{3}-A\right)=-\dfrac{1}{2}$.

8. Show that $\cos^2 A + \cos^2(A+60^\circ) + \cos^2(A-60^\circ) = \dfrac{3}{2}$.

9. Show that $\tan 2A \cdot \tan 3A \cdot \tan 5A = \tan 5A - \tan 3A - \tan 2A$.

10. If $A+B+C = 180^\circ$ Show that $\tan 2A + \tan 2B + \tan 2C = \tan 2A \cdot \tan 2B \cdot \tan 2C$.

11. Show that $\sin(A-B)(\tan A + \tan B) = \sin(A+B)(\tan A - \tan B)$.

Answers

II. **2.** $45^\circ, 225^\circ$ **3.** $\dfrac{\pi}{4}$ **5.** $\dfrac{1}{5}$ **6.** $\dfrac{11}{127}$

III. **1.** $\dfrac{33}{65}$ **2.** $\dfrac{63}{65}$ **3.** $\dfrac{56}{65}$ **5.** $\dfrac{1}{3}$

3.3 Trigonometric Ratios of Multiple Angles and Submultiple Angles

The angles $2A$, $3A$, $4A$, etc., are called **multiple angles** whereas $\dfrac{A}{2}, \dfrac{A}{3}, \dfrac{3A}{2}$, etc., are called **submultiple angles.**

We shall find the trigonometrical ratios of multiple angles in terms of those of the angles

To show that

1. $\mathbf{\sin 2A = 2\sin A\cos A}$ **2.** $\mathbf{\cos 2A = \begin{cases}\cos^2 A - \sin^2 A\\ 1 - 2\sin^2 A\\ 2\cos^2 A - 1\end{cases}}$ **3.** $\mathbf{\tan 2A = \dfrac{2\tan A}{1-\tan^2 A}}$

Proof :

1. Consider, $\sin(A+B) = \sin A\cos B + \cos A\sin B$

Putting $B = A$, we get

$$\sin 2A = \sin A\cos A + \cos A\sin A \Rightarrow \mathbf{\sin 2A = 2\sin A\cos A}$$

2. Consider $\cos(A+B) = \cos A\cos B - \sin A\sin B$

Putting $B = A$, we get

$$\cos 2A = \cos A\cos A - \sin A\sin A \Rightarrow \mathbf{\cos 2A = \cos^2 A - \sin^2 A}$$

Again, $\cos 2A = 1 - \sin^2 A - \sin^2 A$ $(\because \cos^2 A = 1 - \sin^2 A)$

$\Rightarrow$ $\mathbf{\cos 2A = 1 - 2\sin^2 A}$

Again, $\cos 2A = \cos^2 A - \sin^2 A \Rightarrow \cos 2A = \cos^2 A - (1-\cos^2 A)$ $(\because \cos^2 A = 1 - \sin^2 A)$

$$\Rightarrow \mathbf{\cos 2A = 2\cos^2 A - 1}$$

3. Consider $\tan(A+B) = \dfrac{\tan A + \tan B}{1 - \tan A \cdot \tan A}$

Putting $B = A$, we get, $\tan 2A = \dfrac{\tan A + \tan A}{1 - \tan A \cdot \tan A} \Rightarrow \mathbf{\tan 2A = \dfrac{2\tan A}{1 - \tan^2 A}}$

Note : Since the trigonometrical ratios of compound angle $A + B$ are true for all values of A and B, any formulae derived from them are true for all values of A and B. In particular the above formulae for $\sin 2A$, $\cos 2A$ and $\tan 2A$ are true for all values of A.

To prove that

(a) $\sin 2A = \dfrac{2\tan A}{1+\tan^2 A}$ **(b)** $\cos 2A = \dfrac{1-\tan^2 A}{1+\tan^2 A}$ **(c)** $\cot A - \tan A = 2\cot 2A$

(d) $\cot A + \tan A = 2\operatorname{cosec} 2A$ **(e)** $\dfrac{1-\cos 2A}{1+\cos 2A} = \tan^2 A$ **(f)** $\dfrac{1+\cos 2A}{1-\cos 2A} = \cot^2 A$

Proof : (a) We have,

$$\sin 2A = 2\sin A\cos A \Rightarrow \sin 2A = \frac{2\sin A\cos A}{\cos^2 A + \sin^2 A} \qquad (\because \cos^2 A + \sin^2 A = 1)$$

$$\Rightarrow \sin 2A = \frac{2\left(\dfrac{\sin A}{\cos A}\right)}{1 + \left(\dfrac{\sin A}{\cos A}\right)^2} \qquad \left(\begin{array}{c}\text{dividing both numerator and}\\ \text{denominator by } \cos^2 A\end{array}\right)$$

$$\Rightarrow \mathbf{\sin 2A = \frac{2\tan A}{1+\tan^2 A}}$$

(b) We have,

$$\cos 2A = \cos^2 A - \sin^2 A \Rightarrow \cos 2A = \frac{\cos^2 A - \sin^2 A}{\cos^2 A + \sin^2 A} \qquad (\because \cos^2 A + \sin^2 A = 1)$$

$$\Rightarrow \cos 2A = \frac{1 - \left(\dfrac{\sin A}{\cos A}\right)^2}{1 + \left(\dfrac{\sin A}{\cos A}\right)^2} \qquad \left(\begin{array}{c}\text{dividing both numerator and}\\ \text{denominator by } \cos^2 A\end{array}\right)$$

$$\Rightarrow \mathbf{\cos 2A = \frac{1-\tan^2 A}{1+\tan^2 A}}$$

(c) We have,

$$\cot A - \tan A = \frac{\cos A}{\sin A} - \frac{\sin A}{\cos A} \Rightarrow \cot A - \tan A = \frac{\cos^2 A - \sin^2 A}{\sin A\cos A}$$

$$\Rightarrow \quad \cot A - \tan A = \frac{2(\cos 2A)}{2\sin A\cos A} \quad (\because \cos^2 A - \sin^2 A = \cos 2A)$$

$$\Rightarrow \quad \cot A - \tan A = \frac{2\cos 2A}{\sin 2A} = 2\cot 2A$$

$\therefore$ $\quad$ **$\cot A - \tan A = 2\cot 2A$**

(d) Consider,

$$\cot A + \tan A = \frac{\cos A}{\sin A} + \frac{\sin A}{\cos A} = \frac{\cos^2 A + \sin^2 A}{\sin A\cos A} = \frac{2\cdot(1)}{2\sin A\cos A} \quad (\because \cos^2 A + \sin^2 A = 1)$$

$$= \frac{2}{\sin 2A} = 2\operatorname{cosec} 2A$$

$\therefore$ $\quad$ **$\cot A + \tan A = 2\operatorname{cosec} 2A$**

(e) Consider,
$$\frac{1-\cos 2A}{1+\cos 2A} = \frac{1-(1-2\sin^2 A)}{1+(2\cos^2 A - 1)} = \frac{2\sin^2 A}{2\cos^2 A} = \tan^2 A$$

$$\therefore \quad \mathbf{\frac{1-\cos 2A}{1+\cos 2A} = \tan^2 A}$$

(f) Consider,
$$\frac{1+\cos 2A}{1-\cos 2A} = \frac{1+(2\cos^2 A - 1)}{1-(1-2\sin^2 A)} = \frac{\cos^2 A}{\sin^2 A} = \cot^2 A$$

$$\therefore \quad \mathbf{\frac{1+\cos 2A}{1-\cos 2A} = \cot^2 A}$$

3.4 Trigonometric Ratios of Angle 3*A*

To show that

1. $\sin 3A = 3\sin A - 4\sin^3 A$ $\quad$ **2.** $\cos 3A = 4\cos^3 A - 3\cos A$ $\quad$ **3.** $\tan 3A = \dfrac{3\tan A - \tan^3 A}{1 - 3\tan^2 A}$

Proof : 1. Consider,

$$\begin{aligned}\sin 3A &= \sin(2A + A) = \sin 2A\cos A + \cos 2A\sin A = 2\sin A\cos A\cos A + (1-2\sin^2 A)\cdot\sin A\\ &= 2\sin A(1-\sin^2 A) + \sin A - 2\sin^3 A\\ &= 2\sin A - 2\sin^3 A + \sin A - 2\sin^3 A\\ &= 3\sin A - 4\sin^3 A\end{aligned}$$

$\therefore$ $\quad$ **$\sin 3A = 3\sin A - 4\sin^3 A$**

2. Consider,

$$\begin{aligned}\cos 3A &= \cos(2A + A) = \cos 2A\cos A - \sin 2A\sin A = (2\cos^2 A - 1)\cos A - 2\sin A\cdot\cos A\cdot\sin A\\ &= 2\cos^3 A - \cos A - 2\cos A(1-\cos^2 A)\\ &= 2\cos^3 A - \cos A - 2\cos A + 2\cos^3 A\\ &= 4\cos^3 A - 3\cos A\end{aligned}$$

$\therefore$ $\quad$ **$\cos 3A = 4\cos^3 A - 3\cos A$**

3. Consider,

$$\tan 3A = \tan(2A + A) = \frac{\tan 2A + \tan A}{1 - \tan 2A \tan A} = \frac{\dfrac{2\tan A}{1-\tan^2 A} + \tan A}{1 - \dfrac{2\tan A}{1-\tan^2 A}\cdot \tan A} \qquad \left(\because \tan 2A = \frac{2\tan A}{1-\tan^2 A}\right)$$

$$= \frac{2\tan A + \tan A - \tan^3 A}{1 - \tan^2 A - 2\tan^2 A} = \frac{3\tan A - \tan^3 A}{1 - 3\tan^2 A}$$

$$\therefore \qquad \mathbf{\tan 3A = \frac{3\tan A - \tan^3 A}{1 - 3\tan^2 A}}$$

3.5 Trigonometric Ratios of Half Angles

We have the formulae for trigonometric ratios of the angles $2A$. In these by taking $2A = \theta$ $\left(\text{i.e., } A = \frac{\theta}{2}\right)$, we can get the formulae for the trigonometric ratios of the angle $\frac{\theta}{2}$.

1. $$\sin\theta = 2\sin\frac{\theta}{2}\cos\frac{\theta}{2} = \frac{2\tan\theta/2}{1+\tan^2\theta/2}$$

2. $$\cos\theta = \begin{cases} \cos^2\dfrac{\theta}{2} - \sin^2\dfrac{\theta}{2} \\ 1 - 2\sin^2\dfrac{\theta}{2} \\ 2\cos^2\dfrac{\theta}{2} - 1 \end{cases} ; \qquad \cos\theta = \frac{1-\tan^2\theta/2}{1+\tan^2\theta/2}$$

3. $$\tan\theta = \frac{2\tan\theta/2}{1-\tan^2\theta/2}$$

Example 1. Find the sine and cosine of the angles 18°, 36°, 54°, 72° and $22\frac{1}{2}^{\circ}$.

Solution : (a) To find sin 18° and cos 18°

Let $A = 18^\circ$. Then $5A = 90^\circ \Rightarrow 2A + 3A = 90^\circ$

Thus, we have, $2A = 90^\circ - 3A \Rightarrow \sin 2A = \sin(90^\circ - 3A)$

$$\Rightarrow \sin 2A = \cos 3A$$

$$\Rightarrow 2\sin A\cos A = 4\cos^3 A - 3\cos A$$

$$\Rightarrow 2\sin A = 4\cos^2 A - 3$$

[note $\cos A \neq 0$, for if $\cos A = 0 \Rightarrow A = 90^\circ$, which is not true]

$$\Rightarrow 2\sin A = 4(1 - \sin^2 A) - 3$$

$$\Rightarrow 4\sin^2 A + 2\sin A - 1 = 0$$

This is a quadratic equation in $\sin A$, Thus,

$$\sin A = \frac{-2 \pm \sqrt{4+16}}{8} = \frac{-2 \pm \sqrt{20}}{8} = \frac{-1 \pm \sqrt{5}}{4}$$

Thus $\quad \sin A = \dfrac{-1+\sqrt{5}}{4} \quad$ or $\quad \sin A = \dfrac{-1-\sqrt{5}}{4}$

Since $A = 18^\circ$ and $\sin 18^\circ$ is positive, we have, $\quad \mathbf{\sin 18^o = \dfrac{\sqrt{5}-1}{4}}$

Now, $\quad \cos 18^o = \sqrt{1-\sin^2 18^0} \Rightarrow \cos 18^o = \sqrt{1-\left(\dfrac{\sqrt{5}-1}{4}\right)^2}$

$$\Rightarrow \cos 18^o = \sqrt{\frac{16-\left(\sqrt{5}-1\right)^2}{16}} \Rightarrow \cos 18^o = \frac{\sqrt{10+2\sqrt{5}}}{4}$$

Thus, $\quad \mathbf{\cos 18^o = \dfrac{\sqrt{10+2\sqrt{5}}}{4}}$

(b) To find sin 36° and cos 36°

We have,

$$\cos 2A = 1-2\sin^2 A \Rightarrow \cos 36^o = 1-2\sin^2 18^o \Rightarrow \cos 36^o = 1-2\left(\frac{\sqrt{5}-1}{4}\right)^2$$

$$\Rightarrow \cos 36^o = 1-\frac{6-2\sqrt{5}}{8}$$

$$\Rightarrow \cos 36^o = \frac{2+2\sqrt{5}}{8}$$

$$\Rightarrow \cos 36^o = \frac{\sqrt{5}+1}{4}$$

$\therefore \quad \mathbf{\cos 36^o = \dfrac{\sqrt{5}+1}{4}}$

Now, $\quad \sin 36^o = \sqrt{1-\cos^2 36^0}$

$$\Rightarrow \sin 36^o = \sqrt{1-\left(\frac{\sqrt{5}+1}{4}\right)^2} \Rightarrow \sin 36^o = \sqrt{\frac{16-6-2\sqrt{5}}{16}} = \sqrt{\frac{10-2\sqrt{5}}{16}}$$

$\therefore \quad \mathbf{\sin 36^o = \dfrac{\sqrt{10-2\sqrt{5}}}{4}}$

(c) To find sin 54° and cos 54°

Consider, $$\mathbf{\sin 54^\circ} = \sin(90^\circ - 36^\circ) = \cos 36^\circ = \frac{\mathbf{\sqrt{5}+1}}{\mathbf{4}}$$

$$\mathbf{\cos 54^\circ} = \cos(90^\circ - 36^\circ) = \sin 36^\circ = \frac{\mathbf{\sqrt{10-2\sqrt{5}}}}{\mathbf{4}}$$

(d) To find sin 72° and cos 72°

Consider, $$\mathbf{\sin 72^\circ} = \sin(90^\circ - 18^\circ) = \cos 18^\circ = \frac{\mathbf{\sqrt{10+2\sqrt{5}}}}{\mathbf{4}}$$

$$\mathbf{\cos 72^\circ} = \cos(90^\circ - 18^\circ) = \sin 18^\circ = \frac{\mathbf{\sqrt{5}-1}}{\mathbf{4}}$$

(e) To find $\sin 22\frac{1}{2}^\circ$ and $\cos 22\frac{1}{2}^\circ$

We have, $\cos A = 1 - 2\sin^2\frac{A}{2}$. By taking $A = 45^\circ$, i.e., $\frac{A}{2} = 22\frac{1}{2}^\circ$, we get

$$\cos 45^\circ = 1 - 2\sin^2\left(22\frac{1}{2}^\circ\right) \Rightarrow 2\sin^2\left(22\frac{1}{2}^\circ\right) = 1 - \frac{1}{\sqrt{2}}$$

$$\Rightarrow \sqrt{2}\sin\left(22\frac{1}{2}^\circ\right) = \sqrt{\frac{\sqrt{2}-1}{\sqrt{2}}} \quad \text{(Taking square roots both sides)}$$

$$\Rightarrow \sin\left(22\frac{1}{2}^\circ\right) = \sqrt{\frac{\sqrt{2}-1}{2\sqrt{2}}}$$

$$\Rightarrow \sin\left(22\frac{1}{2}^\circ\right) = \sqrt{\frac{2-\sqrt{2}}{4}}$$

Thus, $$\mathbf{\sin\left(22\frac{1}{2}^\circ\right) = \sqrt{\frac{2-\sqrt{2}}{4}}}$$

Again, we have $\cos A = 2\cos^2\frac{A}{2} - 1$. By taking $A = 45^\circ$, i.e., $\frac{A}{2} = 22\frac{1}{2}^\circ$, we get

$$\cos 45^\circ = 2\cos^2\left(22\frac{1}{2}^\circ\right) - 1 \Rightarrow 2\cos^2\left(22\frac{1}{2}^\circ\right) = 1 + \cos 45^\circ$$

$$\Rightarrow 2\cos^2\left(22\frac{1}{2}^\circ\right) = 1 + \frac{1}{\sqrt{2}}$$

$$\Rightarrow \sqrt{2}\cos\left(22\frac{1}{2}^\circ\right) = \sqrt{\frac{\sqrt{2}+1}{\sqrt{2}}} \quad \text{(Taking square roots both sides)}$$

$$\Rightarrow \quad \cos\left(22\frac{1}{2}^{\circ}\right) = \sqrt{\frac{\sqrt{2}+1}{2\sqrt{2}}} = \sqrt{\frac{2+\sqrt{2}}{4}}$$

$$\therefore \quad \cos\left(22\frac{1}{2}^{\circ}\right) = \sqrt{\frac{2+\sqrt{2}}{4}}$$

Note : In establishing the results, in example 1, we have taken +ve square root, whenever a root sign involved. This is because of the fact that the trigonometric ratios involved are ratios of acute angle, which are positive.

The following results are useful and readers are advised to remember these

$$1 - \cos 2A = 2\sin^2 A, \quad 1 - \cos A = 2\sin^2\frac{A}{2}; \quad 1 + \cos 2A = 2\cos^2 A, \quad 1 + \cos A = 2\cos^2\frac{A}{2}$$

Exercise

I.

1. If $\sin A = \frac{1}{2}$, find $\sin 2A$, $\sin 3A$

2. If $\tan A = \frac{1}{2}$, find $\tan 3A$

3. If $\tan A = \frac{3}{4}$, find the values of $\sin 2A$, $\cos 2A$, $\tan 2A$.

4. Show that

   (a) $\cos^4 A - \sin^4 A = \cos 2A$

   (b) $\tan A + \cot A = 2\,\text{cosec}\, 2A$

   (c) $\sin(45^\circ + A)\sin(45^\circ - A) = \frac{1}{2}\cos 2A$

   (d) $\cot A - \tan A = 2\cot 2A$

   (e) $\frac{1 + \cos 2A}{\cot A} = \sin 2A$

II.

1. Show that

   (a) $\text{cosec}\, 2A + \cot 2A = \cot A$

   (b) $\frac{\cos 2A}{1 + \sin 2A} = \tan(45^\circ - A)$

   (c) $\frac{\sin 3A}{\sin A} - \frac{\cos 3A}{\cos A} = 2$

   (d) $\sin^6 A - \cos^6 A = -\cos 2A\left(1 - \frac{\sin^2 2A}{4}\right)$

   (e) $\sin^6 A + \cos^6 A = 1 - \frac{3}{4}\sin^2 2A$

   (f) $\cos 4A = 1 - 8\sin^2 A + 8\sin^4 A$

   (g) $\sqrt{2 + \sqrt{2 + 2\cos 4\theta}} = 2\cos\theta$

   (h) $\frac{\cos A}{1 + \sin A} = \tan\left(\frac{\pi}{4} - \frac{A}{2}\right)$

   (i) $\frac{\sin 3\theta}{1 + 2\cos 2\theta} = \sin\theta$

Chapter 4

Complex Numbers

4.1 Introduction

The readers are already familiar with the complex numbers expressed in the form of $a + ib$, where a and b are real numbers, and $i = \sqrt{-1}$. In this section we shall introduce complex numbers as an ordered pairs of real numbers and show that this representation, and the representation by $a + ib$ are equivalent. Further we shall see other forms of representation of complex numbers and few results associated to complex numbers.

4.2 Complex Numbers

Definition : A complex number is an ordered pair (a, b) of real numbers *a* and *b*.

For example, $(1, 2)$, $(-3, 0)$, $(0, -1)$, $\left(\sqrt{3}, \sqrt{2}\right)$ are complex numbers.

We shall develop the algebra of complex numbers as ordered pair of real numbers.

1. Equality of complex numbers

Two complex numbers (a, b) and (c, d) are said to be equal if and only if $a = c$ and $b = d$.

i.e.,
$$(a, b) = (c, d) \iff a = c \text{ and } b = d$$

2. Addition of two complex numbers

Let $z_1 = (a, b)$ and $z_2 = (c, d)$ be any two complex numbers. The sum $z_1 + z_2$ of the complex numbers z_1 and z_2 is defined by

$$z_1 + z_2 = (a, b) + (c, d) = (a + c, b + d)$$

3. Multiplication of two complex numbers

Let $z_1 = (a, b)$ and $z_2 = (c, d)$ be any two complex numbers. The product $z_1 \cdot z_2$ of the complex numbers is defined by

$$z_1 \cdot z_2 = (a, b) \cdot (c, d) = (ac - bd, ad + bc)$$

If $z_1 = (3, 2)$ and $z_2 = (-1, 2)$, then

$$z_1 + z_2 = (3 - 1, 2 + 2) = (2, 4) \text{ and } z_1 \cdot z_2 = (3, 2) \cdot (-1, 2) = (-3 - 4, 6 - 2) = (-7, 4)$$

4. Multiplication of a complex number by a scalar

Let $z = (a, b)$ be any complex number and k be any scalar. The scalar multiple of $k \cdot z$ of the complex number z by a scalar k, is defined by

$$k \cdot z = k \cdot (a, b) = (ka, kb)$$

If $z = (4, -5)$ then $3z = 3(4, -5) = (12, 15)$ and $\frac{1}{2}z = \left(2, -\frac{5}{2}\right)$

Supposing we denote the complex number (1, 0) by simply 1, then any complex number of the form $(a, 0)$ can be written simply as 'a'. For the reason,

$$(a, 0) = a\,(1, 0) = a \cdot 1 = a.$$

If we denote the complex number (0, 1) by i, then

$$i^2 = (0, 1) \cdot (0, 1) \;\Rightarrow\; i^2 = (-1, 0) \;\Rightarrow\; \mathbf{i^2 = -1}$$

Also we have, $\quad (0, b) = b\,(0, 1) = bi$

That is any complex number of the form $(0, b)$ can be written as bi. We call the complex number i as the **imaginary unit**.

Consider any complex number $z = (a, b)$

Now, $\quad z = (a, b) \;\Rightarrow\; z = (a, 0) + (0, b) \;\Rightarrow\; z = a + ib$

Thus, any complex number z can be expressed as $z = a + ib$, where a and b are real numbers and i is such that $i^2 = -1$. This form of representation of a complex number is called binomial form of representation.

Let $z = a + ib$ be a complex number. The real number a is called the **real part** and the real number b is called **imaginary part** of the complex number z. These are respectively denoted by Re(z) and Im(z)

i.e., $\quad z = a + ib \;\Rightarrow\; \text{Re}(z) = a, \;\text{ and }\; \text{Im}(z) = b$

The complex numbers whose imaginary part is zero are called **purely real numbers** and the complex numbers whose real part is zero are called **purely imaginary numbers**.

The algebra of complex numbers in the binomial notation is given below.

Equality : $\quad a + ib = c + id \Leftrightarrow a = c \text{ and } b = d$

Addition : $\quad (a + ib) + (c + id) = (a + c) + i\,(b + d)$

Multiplication : $\quad (a + ib) \cdot (c + id) = (ac - bd) + i\,(ad + bc)$

Scalar Multiple : $\quad k\,(a + ib) = ka + ikb$

Note : We have denoted $\sqrt{-1}$ by i; i.e., $i = \sqrt{-1}$

$\therefore \quad i^2 = 1, \quad$ Also, $\quad i^3 = i^2 \cdot i = -i, \quad i^4 = (i^2)^2 = (-1)^2 = 1$

Thus, we have, $\quad \mathbf{i^2 = -1}, \quad \mathbf{i^3 = -i}, \quad \mathbf{i^4 = 1}, \quad$ also $\quad \mathbf{\frac{1}{i} = -i}$

Now, any positive integral power of i will be either $-1, -i, i$ or 1.

In fact for any positive integer n

$$\mathbf{i^{4n} = 1, \qquad i^{4n+1} = i, \qquad i^{4n+2} = -1, \qquad i^{4n+3} = -i}$$

For example,

$$i^{435} = i^{4(108)+3} = -i\,; \qquad i^{397} = i^{4(99)+1} = i\,; \qquad i^{302} = i^{4(75)+2} = -1$$

The addition and multiplication of complex numbers satisfy the following properties:

Let z_1, z_2 and z_3 be any three complex numbers, then

1. (a) $\quad (z_1 + z_2) + z_3 = z_1 + (z_2 + z_3)$

(b) $\quad z_1 \cdot (z_2 \cdot z_3) = (z_1 \cdot z_2) \cdot z_3$

} Associative laws

2. (a) $z_1 + z_2 = z_2 + z_1$
(b) $z_1 \cdot z_2 = z_2 \cdot z_1$ } Commutative laws

3. $z_1 \cdot (z_2 + z_3) = z_1 \cdot z_2 + z_1 \cdot z_3$ } Distributive law

4.3 Conjugate of a Complex Number

To every complex number, we associate an another complex number, called the conjugate of the given complex number.

Definition : If $z = x + iy$ be any complex number, then the complex number $x - iy$ is called the conjugate of the complex number z and it is denoted by $\bar{z}$.

Thus, **if $z = x + iy$ then $\bar{z} = x - iy$**

Note : 1. The conjugate of a complex number, is obtained by changing the sign of the imaginary part.

2. If $\bar{z}$ is the conjugate of z, then z is the conjugate part of $\bar{z}$.

3. The conjugate of purely real number is itself and conversely.

Theorem : If z is any complex number, then

(a) $z + \bar{z}$ is purely a real number

(b) $z - \bar{z}$ is purely a imaginary number

(c) $z \cdot \bar{z}$ is a non zero positive real number

(d) If z_1 and z_2 are any two complex numbers, then $\overline{z_1 \pm z_2} = \bar{z}_1 \pm \bar{z}_2$

Proof : Let $z = x + iy$ then $\bar{z} = x - iy$

(a) $z + \bar{z} = (x + iy) + (x - iy) = x + \cancel{iy} + x - \cancel{iy} \Rightarrow z + \bar{z} = 2x \Rightarrow z + \bar{z} =$ purely real number

i.e., **$z + \bar{z} = 2$ (real part of z)**

(b) $z - \bar{z} = (x + iy) - (x - iy) = \cancel{x} + iy - \cancel{x} + iy \Rightarrow z - \bar{z} = 2iy =$ purely imaginary number

i.e., **$z - \bar{z} = 2i$ (imaginary part of z)**

(c) Consider, $z \cdot \bar{z} = (x + iy)(x - iy) \Rightarrow z \cdot \bar{z} = x^2 - (iy)^2$ $(\because (A + B)(A - B) = A^2 - B^2)$

$\Rightarrow z \cdot \bar{z} = x^2 + y^2$ $(\because i^2 = -1)$

$\Rightarrow z \cdot \bar{z} =$ a non zero real number

Hence, **$(x + iy)(x - iy) = x^2 + y^2$**

(d) Let $z_1 = x_1 + iy_1$ and $z_2 = x_2 + iy_2$. Consider,

$$z_1 + z_2 = (x_1 + x_2) + i(y_1 + y_2)$$

$$\Rightarrow \overline{z_1 + z_2} = (x_1 + x_2) - i(y_1 + y_2) \Rightarrow \overline{z_1 + z_2} = (x_1 - iy_1) + (x_2 - iy_2) \Rightarrow \overline{z_1 + z_2} = \bar{z}_1 + \bar{z}_2$$

Similarly, $\overline{z_1 - z_2} = \bar{z}_1 - \bar{z}_2$ Hence, $\overline{z_1 \pm z_2} = \bar{z}_1 \pm \bar{z}_2$

Now, we shall consider the quotient of two complex numbers. Let $z_1 = x_1 + iy_1$ and $z_2 = x_2 + iy_2$.

Now, $$\frac{z_1}{z_2} = \frac{x_1 + iy_1}{x_2 + iy_2} \Rightarrow \frac{z_1}{z_2} = \frac{(x_1 + iy_1)(x_2 - iy_2)}{(x_2 + iy_2)(x_2 - iy_2)}$$

(i.e., multiplying both numerator and denominator by the conjugate of $x_2 + iy_2$)

$$\Rightarrow \frac{z_1}{z_2} = \frac{(x_1x_2 + y_1y_2) + i(x_2y_1 - x_1y_2)}{(x_2^2 + y_2^2)} \Rightarrow \frac{z_1}{z_2} = \frac{x_1x_2 + y_1y_2}{x_2^2 + y_2^2} + i\frac{x_2y_1 - x_1y_2}{x_2^2 + y_2^2}$$

Clearly, the right hand side is of the form $a + ib$ and therefore $\frac{z_1}{z_2}$ is also a complex number, provided $z_2 \neq 0$.

Note : 1. To express a quotient of two complex number in the form of $a + ib$, multiply the numerator and denominator by the conjugate of the denominator complex number.

2. If $z \neq 0$, then the complex number $\frac{1}{z}$ is called multiplicative inverse of z.

Example 1. Express the following in $a + ib$ form

(a) $\frac{3 + 5i}{i - 1}$ (b) $\frac{3 - i}{2 + i} + \frac{3 + i}{2 - i}$.

Solution :

(a) Consider,

$$\frac{3 + 5i}{i - 1} = \frac{(3 + 5i)(-1 - i)}{(-1 - i)(-1 + i)} = \frac{(-3 + 5) + i(-5 - 3)}{(-1)^2 + (-1)^2} \quad (\because (x + iy)(x - iy) = x^2 + y^2)$$

$$= \frac{2 - 8i}{2} = \mathbf{1 - 4i}$$

(b) Consider,

$$\frac{3 - i}{2 + i} + \frac{3 + i}{2 - i} = \frac{(3 - i)(2 - i) + (3 + i)(2 + i)}{(2 + i)(2 - i)}$$

$$= \frac{(6 - 1) + i(-3 - 2) + (6 - 1) + i(2 + 3)}{4 + 1} = \frac{5 - 5i + 5 + 5i}{5} = \frac{10}{5} = 2 = \mathbf{2 + i0}$$

Example 2. Find the real and imaginary parts of

(a) $\frac{(1 + i)(1 + 2i)}{1 + 3i}$ (b) $\frac{(2 + 3i)}{1 + 2i} + \frac{1 + i}{2 + i}$.

Solution :

(a) Consider,

$$z = \frac{(1 + i)(1 + 2i)}{1 + 3i} = \frac{(1 - 2) + i(1 + 2)}{1 + 3i}$$

$$= \frac{-1 + 3i}{1 + 3i} = \frac{(-1 + 3i)(1 - 3i)}{(1 + 3i)(1 - 3i)} = \frac{(-1 + 9) + i(3 + 3)}{1 + 9}$$

$$= \frac{8 + 6i}{10} = \frac{8}{10} + \frac{6}{10}i \Rightarrow z = \mathbf{\frac{4}{5} + \frac{3}{5}i}$$

$\therefore$ **Re** $z = \mathbf{\frac{4}{5}}$ and **Im** $z = \mathbf{\frac{3}{5}}$

(b) Consider,

$$z_1 = \frac{2 + 3i}{1 + 2i} = \frac{(2 + 3i)(1 - 2i)}{(1 + 2i)(1 - 2i)} = \frac{(2 + 6) + i(3 - 4)}{1 + 4} = \frac{8 - i}{5} \Rightarrow z_1 = \mathbf{\frac{8}{5} - \frac{1}{5}i}$$

Again consider,

$$z_2 = \frac{1+i}{2+i} = \frac{(1+i)(2-i)}{(2+i)(2-i)} = \frac{(2+1)+i(2-1)}{4+1} = \frac{3+i}{5} \Rightarrow z_2 = \frac{3}{5} + \frac{1}{5}i$$

Now, $$z = \frac{(2+3i)}{1+2i} + \frac{1+i}{2+i} \Rightarrow z = z_1 + z_2 \Rightarrow z = \left(\frac{8}{5} + \frac{3}{5}\right) + i\left(-\frac{1}{5} + \frac{1}{5}\right) \Rightarrow z = \frac{11}{5} + 0i$$

$$\therefore \qquad \text{Re } z = \frac{11}{5}, \quad \text{Im } z = 0$$

Example 3. If $a + ib = \dfrac{1}{1 - \cos\theta + i\sin\theta}$, find a and b.

Solution : We know that , $1 - \cos\theta = 2\sin^2\dfrac{\theta}{2}$ and $\sin\theta = 2\sin\dfrac{\theta}{2}\cdot\cos\dfrac{\theta}{2}$

Now, $$1 - \cos\theta + i\sin\theta = 2\sin^2\frac{\theta}{2} + i\cdot 2\sin\frac{\theta}{2}\cdot\cos\frac{\theta}{2} = 2\sin\frac{\theta}{2}\left(\sin\frac{\theta}{2} + i\cos\frac{\theta}{2}\right)$$

$$\therefore \qquad a + ib = \frac{1}{2\sin\frac{\theta}{2}\left(\sin\frac{\theta}{2} + i\cos\frac{\theta}{2}\right)} = \frac{1}{2\sin\frac{\theta}{2}} \cdot \frac{\left(\sin\frac{\theta}{2} - i\cos\frac{\theta}{2}\right)}{\left(\sin\frac{\theta}{2} + i\cos\frac{\theta}{2}\right)\left(\sin\frac{\theta}{2} - i\cos\frac{\theta}{2}\right)}$$

$$= \frac{1}{2\sin\frac{\theta}{2}} \cdot \frac{\left(\sin\frac{\theta}{2} - i\cos\frac{\theta}{2}\right)}{\left(\sin^2\frac{\theta}{2} + \cos^2\frac{\theta}{2}\right)}$$

$$= \frac{1}{2\sin\frac{\theta}{2}} \cdot \left(\sin\frac{\theta}{2} - i\cos\frac{\theta}{2}\right) \qquad (\because \sin^2\alpha + \cos^2\alpha = 1)$$

$$= \frac{1}{2} - i\cot\frac{\theta}{2}$$

$$\therefore \qquad a = \frac{1}{2} \text{ and } b = -\cot\frac{\theta}{2}$$

Example 4. If $\dfrac{i^5 + i^6 + i^7 + i^8 + i^9}{1+i} = A + iB$, find A and B.

Solution : Consider,

$$\frac{i^5 + i^6 + i^7 + i^8 + i^9}{1+i} = \frac{0 + i^9}{1+i} \quad (\because i^n + i^{n+1} + i^{n+2} + i^{n+3} = 0, \text{ for any +ve integer } n)$$

$$= \frac{i}{1+i} = \frac{i(1-i)}{1+1} = \frac{1}{2} + \frac{1}{2}i = A + iB \quad \therefore A = \frac{1}{2}, \quad B = \frac{1}{2}$$

Example 5. Find the least positive integer n, for which $\left(\frac{1+i}{1-i}\right)^n = 1$.

Solution : Consider,

$$\frac{1+i}{1-i} = \frac{(1+i)(1+i)}{(1-i)(1+i)} = \frac{(1-1)+i(1+1)}{1+1} = \frac{2i}{2} = i$$

Now, $(i)^n = 1 \Rightarrow$ **$n = 4$** $(\because i^4 = 1)$. Thus the required least positive integer is 4.

Exercise

I.

1. If n is a positive integer find the value of $i^n + i^{n+1} + i^{n+2} + i^{n+3}$.
2. If $\dfrac{i^4 + i^9 + i^{16}}{2 - i^8 + i^{10} - i^{15}} = A + iB$ find A and B.
3. Find the value of $\dfrac{i^{592} + i^{590} + i^{588} + i^{586} + i^{584}}{i^{582} + i^{580} + i^{578} + i^{576} + i^{574}} + 1$
4. Express the following in $a + ib$ form and write their real and imaginary parts.

(a) $\dfrac{2}{6+5i}$ **(b)** $\dfrac{1-i}{1+i}$ **(c)** $\dfrac{3i}{3+4i}$ **(d)** $\dfrac{3}{3+4i}$

(e) $\dfrac{6}{4+3i}$ **(f)** $\dfrac{\sqrt{2}+i}{3i-2}$ **(g)** $\dfrac{2}{3+i} - \dfrac{3-i}{1+2i}$ **(h)** $\dfrac{(3+2i)^2}{4-3i}$

(i) $\dfrac{2}{4+3i} + \dfrac{3}{4-3i}$ **(j)** $\dfrac{(2+3i)^2}{(3+4i)^2}$ **(k)** $\dfrac{(1+i)(1+2i)}{1+3i}$ **(l)** $(1+i)^2 - (1-i)^2$

II.

1. Fin the least positive integer n for which $\dfrac{(1+i)^n}{(1-i)^{n-2}}$ is a real number.
2. Find the value of $\left(\dfrac{1+i}{1-i}\right)^{4n}$.
3. If $\left(\dfrac{1+i}{1-i}\right)^3 - \left(\dfrac{1-i}{1+i}\right)^3 = x + iy$ find x and y.
4. Express $\dfrac{1}{1+\cos\theta + i\sin\theta}$ in $A + iB$ form
5. Find the real values of x and y which satisfy the following equation

(a) $(2x - y) + i(3x + 2y + 1) = 0$ **(b)** $(x + 2y) + i(2x - 3y) = 5 - 4i$

(c) $(3 + i)x + (1 - 2i)y + 7i = 0$ **(d)** $(x + iy)(3 - 2i) = 12 + 5i$

(e) $(1 - i)x + (1 + i)y = 1 - 3i$ **(f)** $(1 + i)y^2 + 6 + i = (2 + i)x$

6. Find the multiplicative inverse of the following

(a) $a + ib$ **(b)** $\dfrac{1+i}{1-i}$ **(c)** $\dfrac{2}{2+3i}$ **(d)** $\dfrac{3+2i}{3-2i}$ **(e)** $\dfrac{(2+i)(3-i)}{1+i}$

Answers

I. 1. 0 **2.** $A = 1, B = -2$ **3.** 0

4. (a) $\frac{12}{61}, \frac{-10}{61}$ **(b)** 0, –1 **(c)** $\frac{12}{25}, \frac{9}{25}$ **(d)** $\frac{9}{25}, \frac{-12}{25}$ **(e)** $\frac{24}{25}, \frac{-18}{25}$

(f) $-\frac{2\sqrt{2}+3}{13}, -\left(\frac{2+3\sqrt{2}}{13}\right)$ **(g)** $\frac{2}{5}, \frac{4}{5}$ **(h)** $\frac{-16}{25}, \frac{63}{25}$ **(i)** $\frac{4}{5}, \frac{3}{25}$

(j) $\frac{323}{625}, \frac{36}{625}$ **(k)** $\frac{4}{5}, \frac{3}{5}$ **(l)** 0, 4

II.1. $n = 1$ **2.** 1 **3.** $x = 0, y = -2$ **4.** $\frac{1}{2} - \left(\frac{1}{2}\tan\frac{\theta}{2}\right)i$

5 (a) $x = -1/7,\ y = -2/7$ **(b)** $x = 1,\ y = 2$ **(c)** $x = -1, y = 3$

(d) $x = 2, y = 3$ **(e)** $x = 2, y = -1$ **(f)** $x = 5, y = \pm 2$

6 (a) $\frac{a - ib}{a^2 + b^2}$ **(b)** $-i$ **(c)** $1 + \frac{3}{2}i$ **(d)** $\frac{5}{13} - \frac{12}{13}i$ **(e)** $\frac{4}{25} + \frac{3}{25}i$

4.4 Polar Form of a Complex Numbers : Modulus and Amplitude of a Complex Number

The complex number expressed in the form

$$z = r(\cos\theta + i\sin\theta) \qquad \text{.... (1)}$$

is called polar form of a complex number.

Here r is a positive real number called the **modulus** of the complex number z and θ is called **amplitude** or **argument** of the complex number z. The complex number of the form (1), is also called **modulus–amplitude form**.

Any complex number of the form $a + ib$ can always be expressed in the polar form.

Let $z = a + ib$ be any complex number and $a + ib = r(\cos\theta + i\sin\theta)$

Equating the real and imaginary parts, we get

$$a = r\cos\theta, \qquad b = r\sin\theta$$

Squaring and adding, we get

$$r^2 = a^2 + b^2 \quad \Rightarrow \quad r = \sqrt{a^2 + b^2} \qquad \text{(We take +ve square root)}$$

Thus, **the modules of z, denoted by $|z|$, is given by $|z| = \sqrt{a^2 + b^2}$**

Now, $\cos\theta = \frac{a}{r} = \frac{a}{\sqrt{x^2 + y^2}}, \qquad \sin\theta = \frac{b}{r} = \frac{b}{\sqrt{x^2 + y^2}}$

These equations satisfy infinitely many values of θ, any two of which differ from each other by a multiple of 2π. However, we can find a unique value for θ such that $-\pi \le \theta \le \pi$. This value of θ is called the **principal value** of the amplitude (or argument) of z. The principal value of amplitude is generally denoted by **Amp** z (or **Arg** z).

The general value of the amplitude is given by $2n\pi + \theta$ where θ is the principal amplitude.

Note : 1. A complex number can be expressed in more general form

$$z = r\,[\cos(2n\pi + \theta) + i\sin(2n\pi + \theta)]$$

This is because, $\cos\theta + i\sin\theta$ remains unchanged if θ is replaced by $\theta + 2n\pi$, where n is an integer.

2. In what follows, unless otherwise stated, we only consider the principal amplitude of the complex number.

Given a complex number $z = a + ib$, we have $|z| = \sqrt{a^2 + b^2}$. The principal amplitude of z depends on the signs of the real and imaginary parts of z. The following table is useful in writing the principal amplitude of z:

a	b	Amp z
+	+	$\tan^{-1}\dfrac{b}{a}$
+	–	$-\tan^{-1}\dfrac{\|b\|}{a}$
–	+	$\pi - \tan^{-1}\dfrac{b}{\|a\|}$
–	–	$-\left(\pi - \tan^{-1}\left\|\dfrac{b}{a}\right\|\right)$

Example 1. Express the following complex numbers in the polar form

(a) $1 + i\sqrt{3}$ **(b)** $\sqrt{3} - i$ **(c)** $-4\sqrt{3} + 4i$ **(d)** $-3 - 4i$.

Solution : (a) Let $z = 1 + i\sqrt{3} = a + ib$, Here, $a = 1$ and $b = \sqrt{3}$

Now, $|z| = r = \sqrt{a^2 + b^2} \Rightarrow r = \sqrt{1+3} = 2$

Now, Amp $z = \tan^{-1}\dfrac{b}{a}$ $(\because a > 0,\ b > 0)$

$\Rightarrow$ Amp $z = \tan^{-1}\sqrt{3} \Rightarrow$ Amp $z = \dfrac{\pi}{3}$; $\therefore$ $1 + i\sqrt{3} = 2\left(\cos\dfrac{\pi}{3} + i\sin\dfrac{\pi}{3}\right)$

(b) Let $z = \sqrt{3} - i = a + ib$. Here, $a = \sqrt{3}$ and $b = -1$

Now, $|z| = r = \sqrt{a^2 + b^2} \Rightarrow r = \sqrt{3+1} = 2$

Now, Amp $z = -\tan^{-1}\dfrac{|b|}{a}$ $(\because a > 0,\ b < 0)$

$\Rightarrow$ Amp $z = -\tan^{-1}\dfrac{1}{\sqrt{3}} \Rightarrow$ Amp $z = -\dfrac{\pi}{6}$; $\therefore$ $\sqrt{3} - i = 2\left(\cos\left(-\dfrac{\pi}{6}\right) + i\sin\left(-\dfrac{\pi}{6}\right)\right)$

(c) Let $z = -4\sqrt{3} + 4i = a + ib$, Here, $a = -4\sqrt{3}$ and $b = 4$

Now, $|z| = r = \sqrt{a^2 + b^2} \Rightarrow r = \sqrt{48 + 16} = 8$

Now, $\quad \text{Amp } z = \pi - \tan^{-1}\dfrac{b}{|a|} \qquad (\because a < 0, b > 0)$

$\Rightarrow \quad \text{Amp } z = \pi - \tan^{-1}\dfrac{4}{4\sqrt{3}} \Rightarrow \text{Amp } z = \pi - \dfrac{\pi}{6} = \dfrac{5\pi}{6}$

$$\therefore \quad \mathbf{-4\sqrt{3} + 4i = 8\left(\cos\frac{5\pi}{6} + i\sin\frac{5\pi}{6}\right)}$$

(d) Let $z = -3 - 4i = a + ib$ Here, $a = -3$ and $b = -4$

Now, $\quad |z| = r = \sqrt{a^2 + b^2} \Rightarrow r = \sqrt{9 + 16} = 5$

Now, $\quad \text{Amp } z = -\left(\pi - \tan^{-1}\left|\dfrac{b}{a}\right|\right) \qquad (\because a < 0, b < 0)$

$\Rightarrow \quad \text{Amp } z = -\left(\pi - \tan^{-1}\dfrac{4}{3}\right) \Rightarrow \text{Amp } z = \tan^{-1}\dfrac{4}{3} - \pi = \theta$

$\therefore \quad \mathbf{-3 - 4i = 5(\cos\theta + i\sin\theta)}$ where $\theta = \tan^{-1}\dfrac{4}{3} - \pi$

Example 2. Express the following complex numbers in the polar form

(a) $-i$ (b) -7 (c) 3.

Solution : (a) Let $z = -i = 0 - i = a + ib \Rightarrow a = 0$ and $b = -1$

Clearly z can be written as

$$z = -i = 1\left(\cos\left(-\frac{\pi}{2}\right) + i\sin\left(-\frac{\pi}{2}\right)\right) \qquad \left(\because \cos\left(-\frac{\pi}{2}\right) = 0 \text{ and } \sin\left(-\frac{\pi}{2}\right) = -1\right)$$

(b) Let $z = -7 = -7 + 0i = a + ib \Rightarrow a = -7$ and $b = 0$

$\therefore \quad \mathbf{z = -7 = 7(\cos\pi + i\sin\pi)} \qquad (\because \cos\pi = -1, \sin\pi = 0)$

(c) Let $z = 3 = (3 + 0i) = a + ib \Rightarrow a = 3$ and $b = 0$

$\therefore \quad \mathbf{z = 3 = 3(\cos 0 + i\sin 0)} \qquad (\because \cos 0 = 1, \sin 0 = 0)$

The readers are advised to remember the following

(i) $\mathbf{1 = \cos 0 + i\sin 0,}$ **i.e.,** $\mathbf{|1| = 1, \quad \text{Amp } 1 = 0}$

(ii) $\mathbf{-1 = \cos\pi + i\sin\pi,}$ **i.e.,** $\mathbf{|-1| = 1, \quad \text{Amp }(-1) = \pi}$

(iii) $\mathbf{i = \cos\dfrac{\pi}{2} + i\sin\dfrac{\pi}{2},}$ **i.e.,** $\mathbf{|i| = 1, \quad \text{Amp }(i) = \dfrac{\pi}{2}}$

(iv) $\mathbf{-i = \cos\left(-\dfrac{\pi}{2}\right) + i\sin\left(-\dfrac{\pi}{2}\right),}$ **i.e.,** $\mathbf{|-i| = 1, \quad \text{Amp }(-i) = -\dfrac{\pi}{2}}$

4.5 Exponential Form of a Complex Numbers

We mention below, a result known as **Euler's formula** without proof.

$$\mathbf{e^{i\theta} = \cos\theta + i\sin\theta}$$

This is also known as **exponential** form of a complex number. If z is a complex number, such that $|z| = r$ and Amp $z = \theta$, then we have $z = r(\cos\theta + i\sin\theta) = re^{i\theta}$

Note: **1.** $e^{-i\theta} = \cos(-\theta) + i\sin(-\theta) \Rightarrow \boldsymbol{e^{-i\theta} = \cos\theta - i\sin\theta}$

2. $e^{2n\pi i} = \cos 2n\pi + i\sin 2n\pi \Rightarrow e^{2n\pi i} = 1 + i0 \Rightarrow \boldsymbol{e^{2n\pi i} = 1}$

3. $e^{i\theta} = \cos\theta + i\sin\theta,\ e^{-i\theta} = \cos\theta - i\sin\theta \Rightarrow \cos\theta = \dfrac{e^{i\theta} + e^{-i\theta}}{2}$, and $\sin\theta = \dfrac{e^{i\theta} - e^{-i\theta}}{2i}$

Expressing the complex numbers in the polar form or exponential form, it could help to establish certain results related to modulus and amplitude of the complex numbers, in a simple way. Thus, most of the times without loss of generosity we use these forms frequently.

Results Relating to Modulus and Amplitude

1. If $z = a + ib$, then $|z| = 0$ if and only if $z = 0$.

Proof : We have, $z = a + ib \Rightarrow |z| = \sqrt{a^2 + b^2}$

Now, $|z| = 0 \Leftrightarrow \sqrt{a^2 + b^2} = 0 \Leftrightarrow a^2 + b^2 = 0 \Leftrightarrow a = 0$ and $b = 0$ ($\because$ both a^2 and b^2 are +ve)

$$\Leftrightarrow a + ib = 0 \Leftrightarrow z = 0$$

$\therefore$ **$|z| = 0$ if and only if $z = 0$.**

2. If $z = a + ib$, then $z \cdot \bar{z} = |z|^2$.

Proof : Now, $z = a + ib,\ \bar{z} = a - ib$

Consider $z \cdot \bar{z} = (a + ib)(a - ib) \Rightarrow z \cdot \bar{z} = a^2 + b^2 \Rightarrow z \cdot \bar{z} = |z|^2$ ($\because |z| = \sqrt{a^2 + b^2}$)

3. If z_1 and z_2 are any two complex numbers then,

(a) $|z_1 \cdot z_2| = |z_1| \cdot |z_2|$, **amp** $(z_1 \cdot z_2) =$ **amp** $z_1 +$ **amp** z_2*

(b) $\left|\dfrac{z_1}{z_2}\right| = \dfrac{|z_1|}{|z_2|}$, **amp** $\left(\dfrac{z_1}{z_2}\right) =$ **amp** $z_1 -$ **amp** z_2* $(z_2 \neq 0)$

Proof : Let, $z_1 = r_1(\cos\theta_1 + i\sin\theta_1) = r_1 \cdot e^{i\theta_1}$ and $z_2 = r_2(\cos\theta_2 + i\sin\theta_2) = r_2 \cdot e^{i\theta_2}$

$\therefore$ $|z_1| = r_1,\ |z_2| = r_2$, amp $z_1 = \theta_1$ and amp $z_2 = \theta_2$

(a) Consider $z_1 \cdot z_2 = (r_1 \cdot e^{i\theta_1})(r_2 \cdot e^{i\theta_2}) \Rightarrow z_1 \cdot z_2 = (r_1 \cdot r_2) \cdot e^{i(\theta_1 + \theta_2)}$

$$\Rightarrow z_1 \cdot z_2 = (r_1 \cdot r_2) \cdot [\cos(\theta_1 + \theta_2) + i\sin(\theta_1 + \theta_2)]$$

$\Rightarrow$ $|z_1 \cdot z_2| = r_1 \cdot r_2 \Rightarrow |z_1 \cdot z_2| = |z_1| \cdot |z_2|$ ($\because |z_1| = r_1, |z_2| = r_2$)

Also, amp $(z_1 \cdot z_2) = \theta_1 + \theta_2 \Rightarrow$ **amp $(z_1 \cdot z_2) =$ amp $z_1 +$ amp z_2** ($\because \theta_1 =$ amp z_1, $\theta_2 =$ amp z_2)

(b) Consider

$$\frac{z_1}{z_2} = \frac{r_1 e^{i\theta_1}}{r_2 e^{i\theta_2}} \Rightarrow \frac{z_1}{z_2} = \left(\frac{r_1}{r_2}\right) e^{i(\theta_1 - \theta_2)} \Rightarrow \frac{z_1}{z_2} = \left(\frac{r_1}{r_2}\right)\left[\cos(\theta_1 - \theta_2) + i\sin(\theta_1 - \theta_2)\right]$$

---

* The results may not be valid if we just consider the principal values. That is, if θ_1 and θ_2 are principal values of amplitudes of z_1 and z_2, then $\theta_1 + \theta_2$ is not necessarily the principal value of amplitude of $(z_1 \cdot z_2)$ nor is $\theta_1 - \theta_2$ necessarily the principal value of the amplitude of $\left(\dfrac{z_1}{z_2}\right)$.

$$\therefore \quad \left|\frac{z_1}{z_2}\right| = \frac{r_1}{r_2} \Rightarrow \left|\frac{z_1}{z_2}\right| = \frac{|z_1|}{|z_2|} \qquad (\because |z_1| = r_1, |z_2| = r_2)$$

Also, $\text{amp}\left(\frac{z_1}{z_2}\right) = \theta_1 - \theta_2 \Rightarrow \textbf{amp}\left(\frac{z_1}{z_2}\right) = \textbf{amp } z_1 - \textbf{amp } z_2$ $\qquad (\because \theta_1 = \text{amp } z_1, \theta_2 = \text{amp } z_2)$

Note : The result (a) can be extended to finite number of complex numbers:
If $z_1, z_2, z_3, ..., z_n$ are any n complex numbers, then

(i) $|z_1, z_2, z_3, ..., z_n| = |z_1| \cdot |z_2| \cdot |z_3| \cdot ... \cdot |z_n|$

(ii) $\text{amp}(z_1, z_2, z_3, ..., z_n) = (\text{amp } z_1) + (\text{amp } z_2) + \cdot (\text{amp } z_3) + ... + (\text{amp } z_n)$

Example 1. Find the modulus and amplitude of the following complex number

(a) $\frac{1 + 3i}{2 + i}$ **(b)** $\frac{(3 + 2i)^2}{4 - 3i}$ **(c)** $\frac{1 + i\sqrt{3}}{\sqrt{3} + i}$.

Solution : (a) Let $z = \frac{1 + 3i}{2 + i} \Rightarrow z = \frac{(1 + 3i)(2 - i)}{(2 + i)(2 - i)}$

$$\Rightarrow z = \frac{(2 + 3) + i(6 - 1)}{4 + 1} \Rightarrow z = \frac{5}{5} + \frac{5}{5}i = 1 + i = a + ib$$

$$|z| = \sqrt{a^2 + b^2} \Rightarrow |z| = \sqrt{1 + 1} = \sqrt{2}$$

$$\Rightarrow \quad \text{amp } z = \tan^{-1}\frac{b}{a} \Rightarrow \text{amp } z = \tan^{-1} 1 = \frac{\pi}{4} \qquad (\because a > 0, b > 0)$$

(b) Let $z = \frac{(3 + 2i)^2}{4 - 3i} \Rightarrow z = \frac{(9 - 4) + i(12)}{4 - 3i}$

$$\Rightarrow z = \frac{5 + 12i}{4 - 3i} \Rightarrow z = \frac{(5 + 12i)(4 + 3i)}{(4 - 3i)(4 + 3i)}$$

$$\Rightarrow z = \frac{(20 - 36) + i(15 + 48)}{16 + 9} \Rightarrow z = \frac{-16 + 63i}{25} = \frac{-16}{25} + \frac{63}{25}i$$

Now, $z = \frac{-16}{25} + \frac{63}{25}i = a + ib; \Rightarrow a = \frac{-16}{25}, \quad b = \frac{63}{25}$

$$\therefore \quad |z| = \sqrt{a^2 + b^2} = \sqrt{\left(-\frac{16}{25}\right)^2 + \left(\frac{63}{25}\right)^2} = \sqrt{\frac{256 + 3969}{(25)^2}} = \frac{65}{25} = \mathbf{\frac{13}{5}}$$

Now, $\text{amp } z = \pi - \tan^{-1}\frac{b}{|a|} \Rightarrow \textbf{amp } z = \pi - \tan^{-1}\frac{63}{16}$ $\qquad (\because a < 0, b > 0)$

(c) Let $z = \frac{1 + i\sqrt{3}}{\sqrt{3} + i} \Rightarrow z = \frac{(1 + i\sqrt{3})(\sqrt{3} - i)}{(\sqrt{3} + i)(\sqrt{3} - i)}$

$$\Rightarrow \quad z = \frac{(\sqrt{3}+\sqrt{3}) + i\,(3-1)}{3+1} \quad \Rightarrow \quad z = \frac{2\sqrt{3}+2i}{4} = \frac{\sqrt{3}}{2} + \frac{1}{2}i$$

Now, $$z = \frac{\sqrt{3}}{2} + \frac{1}{2}i = a + ib; \quad a = \frac{\sqrt{3}}{2}, \quad b = \frac{1}{2}$$

$$\therefore \quad |z| = \sqrt{a^2+b^2} = \sqrt{\frac{3}{4}+\frac{1}{4}} = \mathbf{1}$$

Now, $$\text{amp } z = \tan^{-1}\left(\frac{b}{a}\right) \qquad (\because a > 0, b > 0)$$

$$\Rightarrow \quad \text{amp } z = \tan^{-1}\left(\frac{1/2}{\sqrt{3}/2}\right) = \tan^{-1}\frac{1}{\sqrt{3}} \Rightarrow \textbf{amp } z = \frac{\pi}{6} \qquad \left(\because \tan^{-1}\frac{1}{\sqrt{3}} = \frac{\pi}{6}\right)$$

Example 2. Express the following in the polar form

(a) $1 + \cos\theta + i\sin\theta$ **(b) $1 + \sin\theta + i\cos\theta$** **(c) $\dfrac{1}{1-\cos\theta + i\sin\theta}$.**

Solution : (a) Let $z = 1 + \cos\theta + i\sin\theta$

$$\Rightarrow \quad z = 2\cos^2\frac{\theta}{2} + i\cdot 2\sin\frac{\theta}{2}\cdot\sin\frac{\theta}{2} \Rightarrow z = 2\cos\frac{\theta}{2}\left(\cos\frac{\theta}{2} + i\sin\frac{\theta}{2}\right)$$

This is of the form $r(\cos\alpha + i\sin\alpha)$

$$\therefore \quad |z| = \mathbf{2\cos\frac{\theta}{2}} \quad \text{and} \quad \textbf{amp } z = \frac{\theta}{2}$$

(b) Let $z = 1 + \sin\theta + i\cos\theta \Rightarrow z = 1 + \cos\left(\frac{\pi}{2}-\theta\right) + i\sin\left(\frac{\pi}{2}-\theta\right)$

$$\Rightarrow \quad z = 1 + \cos\alpha + i\sin\alpha \qquad \left(\text{where } \alpha = \frac{\pi}{2} - \theta\right)$$

$$\Rightarrow \quad z = 2\cos^2\frac{\alpha}{2} + i2\sin\frac{\alpha}{2}\cdot\cos\frac{\alpha}{2}$$

$$\Rightarrow \quad z = 2\cos\frac{\alpha}{2}\left(\cos\frac{\alpha}{2} + i\sin\frac{\alpha}{2}\right)$$

$$\Rightarrow \quad z = 2\cos\frac{1}{2}\left(\frac{\pi}{2}-\theta\right)\left[\cos\frac{1}{2}\left(\frac{\pi}{2}-\theta\right) + i\sin\frac{1}{2}\left(\frac{\pi}{2}-\theta\right)\right]$$

$$\Rightarrow \quad z = 2\cos\left(\frac{\pi}{4}-\frac{\theta}{2}\right)\left[\cos\left(\frac{\pi}{4}-\frac{\theta}{2}\right) + i\sin\left(\frac{\pi}{4}-\frac{\theta}{2}\right)\right]$$

This is of the form $r(\cos\beta + i\sin\beta)$

$$\therefore \quad \boldsymbol{r} = \mathbf{2\cos\left(\frac{\pi}{4}-\frac{\theta}{2}\right)} \quad \text{and} \quad \boldsymbol{\beta} = \mathbf{\frac{\pi}{4}-\frac{\theta}{2}}$$

(c) Consider, $1 - \cos\theta + i\sin\theta = 2\sin^2\frac{\theta}{2} + i\,2\sin\frac{\theta}{2}\cdot\cos\frac{\theta}{2} = 2\sin\frac{\theta}{2}\left(\sin\frac{\theta}{2} + i\cos\frac{\theta}{2}\right)$

Now, $z = \dfrac{1}{1 - \cos\theta + i\sin\theta}$ $\Rightarrow$ $z = \dfrac{1}{2\sin\frac{\theta}{2}\left(\sin\frac{\theta}{2} + i\cos\frac{\theta}{2}\right)}$

$$\Rightarrow \quad z = \frac{1}{2\sin\frac{\theta}{2}}\left[\frac{\sin\frac{\theta}{2} - i\cos\frac{\theta}{2}}{\sin^2\frac{\theta}{2} + \cos^2\frac{\theta}{2}}\right] \Rightarrow z = \frac{1}{2}\operatorname{cosec}\frac{\theta}{2}\left[\cos\left(\frac{\pi}{2} - \frac{\theta}{2}\right) - i\sin\left(\frac{\pi}{2} - \frac{\theta}{2}\right)\right]$$

$$\Rightarrow \quad z = r(\cos\alpha + i\sin\alpha)$$

Here, $\boldsymbol{r = \frac{1}{2}\operatorname{cosec}\frac{\theta}{2}}$ and $\boldsymbol{\alpha = -\left(\frac{\pi}{2} - \frac{\theta}{2}\right)}$

Example 3. Show that modulus and amplitude of $e^{(x+iy)}$ are respectively e^x and y. Also find modulus and amplitude of $e^{\left(-1 + i\frac{\pi}{6}\right)}$.

Solution : We have, $e^{i\theta} = \cos\theta + i\sin\theta$.

Now, $e^{(x+iy)} = e^x \cdot e^{iy} \quad \Rightarrow \quad e^{x+iy} = e^x(\cos y + i\sin y)$

The modulus of $e^{x+iy} = e^x$ and amp $e^{x+iy} = y$.

Now, $z = e^{\left(-1 + i\frac{\pi}{6}\right)} = e^{-1} \cdot e^{i\frac{\pi}{6}} \Rightarrow z = e^{-1}\left(\cos\frac{\pi}{6} + i\sin\frac{\pi}{6}\right) \Rightarrow \boldsymbol{|z| = e^{-1}}$ and $\boldsymbol{\text{amp}\, z = \frac{\pi}{6}}$

Example 4. Show that $\dfrac{1 + \cos\theta + i\sin\theta}{1 + \cos\theta - i\sin\theta} = e^{i\theta}$.

Solution : Consider,

$$\frac{1 + \cos\theta + i\sin\theta}{1 + \cos\theta - i\sin\theta} = \frac{2\cos^2\frac{\theta}{2} + i\,2\sin\frac{\theta}{2}\cdot\cos\frac{\theta}{2}}{2\cos^2\frac{\theta}{2} - i\,2\sin\frac{\theta}{2}\cdot\cos\frac{\theta}{2}} = \frac{2\cos\frac{\theta}{2}\left(\cos\frac{\theta}{2} + i\sin\frac{\theta}{2}\right)}{2\cos\frac{\theta}{2}\left(\cos\frac{\theta}{2} - i\sin\frac{\theta}{2}\right)} = \frac{\left(\cos\frac{\theta}{2} + i\sin\frac{\theta}{2}\right)}{\left(\cos\frac{\theta}{2} - i\sin\frac{\theta}{2}\right)}$$

$$= \frac{e^{i\theta/2}}{e^{-i\theta/2}} = e^{i\left(\frac{\theta}{2} + \frac{\theta}{2}\right)} = e^{i\theta} = \text{R.H.S}$$

2nd Method:

$$\frac{1 + \cos\theta + i\sin\theta}{1 + \cos\theta - i\sin\theta} = \frac{1 + e^{i\theta}}{1 + e^{-i\theta}} \qquad \left(\begin{array}{l}\because \cos\theta + i\sin\theta = e^{i\theta} \\ \because \cos\theta - i\sin\theta = e^{-i\theta}\end{array}\right)$$

$$= \frac{1 + e^{i\theta}}{1 + \left(\frac{1}{e^{i\theta}}\right)} = \frac{e^{i\theta}\cancel{(1 + e^{i\theta})}}{\cancel{(e^{i\theta} + 1)}} = e^{i\theta} = \text{R.H.S.}$$

Example 5. Show that $2\cos\dfrac{\pi}{20} \cdot e^{-\frac{\pi}{20}i} = 1 + \cos\dfrac{\pi}{10} - i\sin\dfrac{\pi}{10}$.

Solution : Consider,

$$\text{L.H.S} = 2\cos\frac{\pi}{20}\left(\cos\frac{\pi}{20} - i\sin\frac{\pi}{20}\right) = 2\cos^2\frac{\pi}{20} - i\cdot\left(2\sin\frac{\pi}{20}\cdot\cos\frac{\pi}{20}\right)$$

$$= 1 + \cos\frac{\pi}{10} - i\sin\frac{\pi}{10} \qquad (\because 2\cos^2\theta = 1 + \cos 2\theta)$$

$$= \textbf{R.H.S}$$

Example 6. If $z_n = \cos\frac{\pi}{2^n} + i\sin\frac{\pi}{2^n}$, **show that** $z_1\cdot z_2\cdot z_3\cdot \ldots z_n \ldots$ **to** $\infty = -1$.

Solution : We have,

$$z_1 = \cos\frac{\pi}{2} + i\sin\frac{\pi}{2} = e^{i\frac{\pi}{2}}$$

$$z_2 = \cos\frac{\pi}{2^2} + i\sin\frac{\pi}{2^2} = e^{i\frac{\pi}{2^2}}$$

$$z_3 = \cos\frac{\pi}{2^3} + i\sin\frac{\pi}{2^3} = e^{i\frac{\pi}{2^3}}$$

....

....

Now, $z_1 \cdot z_2 \cdot z_3 \ldots$ to $\infty = e^{i\pi/2} + e^{i\pi/2^2} + e^{i\pi/2^3} + \ldots$ to ∞.

$\Rightarrow$ $z_1 \cdot z_2 \cdot z_3 \ldots$ to $\infty = e^{i\left(\frac{1}{2} + \frac{1}{2^2} + \frac{1}{2^3} + \ldots \text{ to } \infty\right)\pi}$

$\Rightarrow$ $z_1 \cdot z_2 \cdot z_3 \ldots$ to $\infty = e^{i\left(\frac{1/2}{1-1/2}\right)\pi}$ $\qquad \left(s_\infty \text{ of G.P.} = \frac{a}{1-r}\right)$

$\Rightarrow$ $z_1 \cdot z_2 \cdot z_3 \ldots$ to $\infty = e^{i\pi} = \cos\pi + i\sin\pi$

$\Rightarrow$ $\mathbf{z_1 \cdot z_2 \cdot z_3 \ldots}$ **to** $\mathbf{\infty = -1}$ $\qquad (\because \cos\pi = -1 \text{ and } \sin\pi = 0)$

Example 7. If $x + iy = \sqrt{\frac{a+ib}{c+id}}$, **show that** $x^2 + y^2 = \sqrt{\frac{a^2+b^2}{c^2+d^2}}$.

Solution : We have, $x + iy = \sqrt{\frac{a+ib}{c+id}}$. Squaring both the sides, we get

$$(x+iy)^2 = \frac{a+ib}{c+id}$$

Taking the modulus on both the sides, we get,

$$|(x+iy)\cdot(x+iy)| = \left|\frac{a+ib}{c+id}\right| \Rightarrow \sqrt{x^2+y^2}\cdot\sqrt{x^2+y^2} = \frac{\sqrt{a^2+b^2}}{\sqrt{c^2+d^2}} \quad \left(\because \left|\frac{z_1}{z_2}\right| = \frac{|z_1|}{|z_2|}, |z_1\cdot z_2| = |z_1|\cdot|z_2|\right)$$

$$\Rightarrow x^2 + y^2 = \sqrt{\frac{a^2+b^2}{c^2+d^2}}$$

Example 8. If $(1 + i)(1 + 2i)(1 + 3i) \ldots (1 + ni) = x + iy$, show that $2 \cdot 5 \cdot 10 \ldots (1 + n^2) = x^2 + y^2$.

Solution : We have,

$$(1 + i) \cdot (1 + 2i) \cdot (1 + 3i) \ldots (1 + ni) = x + iy$$

$$\Rightarrow \quad |(1 + i) \cdot (1 + 2i) \cdot (1 + 3i) \ldots (1 + ni)| = |x + iy|$$

$$\Rightarrow \quad |(1 + i)| \cdot |(1 + 2i)| \cdot |(1 + 3i)| \ldots |(1 + ni)| = \sqrt{x^2 + y^2}$$

$$\Rightarrow \quad \sqrt{1+1} \cdot \sqrt{1+4} \cdot \sqrt{1+9} \ldots\ldots \sqrt{1+n^2} = \sqrt{x^2 + y^2}$$

Squaring both the sides we get,

$$\mathbf{2 \cdot 5 \cdot 10 \ldots (1 + n^2) = x^2 + y^2}$$

Example 9. Show that $\dfrac{1 + \sin\theta + i\cos\theta}{1 - \sin\theta - i\cos\theta} = i(\tan\theta + \sec\theta)$.

Solution : Consider,

$$\frac{1 + \sin\theta + i\cos\theta}{1 - \sin\theta - i\cos\theta} = \frac{[(1 + \sin\theta) + i\cos\theta]\,[(1 - \sin\theta) + i\cos\theta]}{[(1 - \sin\theta) - i\cos\theta]\,[(1 - \sin\theta) + i\cos\theta]}$$

$$= \frac{(1 - \sin^2\theta - \cos^2\theta) + i\,[(1 + \sin\theta)\cos\theta + \cos(1 - \sin\theta)]}{(1 - \sin\theta)^2 + \cos^2\theta}$$

$$= \frac{\cancel{(\cos^2\theta - \cos^2\theta)} + i\,(\cancel{\sin\theta\cos\theta} + \cos\theta + \cos\theta - \cancel{\sin\theta\cos\theta})}{1 - 2\sin\theta + (\sin^2\theta + \cos^2\theta)}$$

$$= \frac{i\,\cancel{2}\cos\theta}{\cancel{2}\,(1 - \sin\theta)} = \frac{i\cos\theta\,(1 + \sin\theta)}{\cos^2\theta} = \frac{i\,(1 + \sin\theta)}{\cos\theta} = \mathbf{i\,(\tan\theta + \sec\theta) = R.H.S}$$

Exercise

I.

1. Express the following in polar form.

   (a) $1 + i$ **(b)** $1 - i$ **(c)** $-1 + i$ **(d)** $-1 - i$ **(e)** $(1 + i)^2$

   (f) $-\sqrt{3} - i$ **(g)** $\sqrt{3} + i$ **(h)** $\dfrac{2 - i}{1 + i}$ **(i)** $1 - \cos\theta + i\sin\theta$

   (j) $-\dfrac{1}{2} + i\dfrac{\sqrt{3}}{2}$ **(k)** $-i\sqrt{2} - \sqrt{2}$ **(l)** $1 + \cos\theta - i\sin\theta$ **(m)** $1 + \sin\dfrac{\pi}{15} + i\cos\dfrac{\pi}{15}$

   (n) $1 + \cos\dfrac{\pi}{10} + i\sin\dfrac{\pi}{10}$ **(o)** $1 + i\tan\alpha$

2. Find the modulus and amplitude of the following.

   (a) $\dfrac{1 - i}{1 + i\sqrt{3}}$ **(b)** $-i\sqrt{2} - \sqrt{2}$ **(c)** $3 + 4i$ **(d)** $-3 + 4i$ **(e)** $-4 - 4i$ **(f)** $2 - 3i$ **(g)** $-4 + 2i$

3. Find the modulus and amplitude of the following.

   (a) $e^{i\frac{\pi}{4}}$ **(b)** $e^{\left(2 + i\frac{\pi}{3}\right)}$ **(c)** $e^{\left(-1 + i\frac{\pi}{6}\right)}$ **(d)** $e^{\left(\sqrt{2} + \frac{\pi}{4}i\right)}$

4. Show that $e^{\left(\frac{\pi}{4}-i\frac{\pi}{4}\right)} \cdot e^{\left(\frac{\pi}{2}+i\frac{\pi}{2}\right)} = e^{3\frac{\pi}{4}} \cdot \left(\frac{1+i}{\sqrt{2}}\right)$.

II.

1. If $z_n = \cos\frac{\pi}{3^n} + i\sin\frac{\pi}{3^n}$ show that $z_1 \cdot z_2 \cdot z_3 \cdot \ldots z_n \ldots$ to $\infty = i$.

2. If $(a_1 + ib_1)(a_2 + ib_2) \ldots (a_n + ib_n) = A + iB$, show that

(a) $(a_1^2 + b_1^2)(a_2^2 + b_2^2) \ldots (a_n^2 + b_n^2) = A^2 + B^2$ **(b)** $\tan^{-1}\left(\frac{b_1}{a_1}\right)\tan^{-1}\left(\frac{b_2}{a_2}\right) \ldots \tan^{-1}\left(\frac{b_n}{a_n}\right) = \tan^{-1}\frac{B}{A}$

3. If $x + iy = \frac{1}{1+\cos\theta + i\sin\theta}$, show that $4x^2 = 1$.

4. If $(1 + 2i)(1 + 4i)(1 + 6i) \ldots (1 + 2ni) = a + ib$ Show that $5 \cdot 17 \cdot 37 \ldots (1 + 4n^2) = a^2 + b^2$.

Answers

I. 1. ($\cos\theta + i\sin\theta$ is denoted by **cis** θ)

(a) $\sqrt{2}\text{ cis}\frac{\pi}{4}$ **(b)** $\sqrt{2}\text{ cis}\left(-\frac{\pi}{4}\right)$ **(c)** $\sqrt{2}\text{ cis}\left(\frac{3\pi}{4}\right)$ **(d)** $\sqrt{2}\text{ cis}\left(-\frac{3\pi}{4}\right)$ **(e)** $2\text{ cis}\frac{\pi}{2}$

(f) $2\text{ cis}\left(-\frac{5\pi}{6}\right)$ **(g)** $2\text{ cis}\frac{\pi}{6}$ **(h)** $\sqrt{\frac{5}{2}}\text{ cis}(-\alpha), \left(\alpha = \tan^{-1}3\right)$

(i) $2\sin\frac{\theta}{2}\left[\cos\left(\frac{\pi}{2}-\frac{\theta}{2}\right) + i\sin\left(\frac{\pi}{2}-\frac{\theta}{2}\right)\right]$ **(j)** $\text{cis}\frac{2\pi}{3}$ **(k)** $2\text{cis}\left(-\frac{3\pi}{4}\right)$

(l) $2\cos\frac{\theta}{2}\left[\cos\left(-\frac{\theta}{2}\right) + i\sin\left(-\frac{\theta}{2}\right)\right]$ **(m)** $2\cos\frac{13\pi}{60}\text{cis}\frac{13\pi}{60}$ **(n)** $2\cos\frac{\pi}{20}\text{cis}\frac{\pi}{20}$ **(o)** $\sec\alpha \cdot \text{cis}\,\alpha$

2. (a) $\frac{1}{\sqrt{2}}, -\frac{7\pi}{12}$ **(b)** $2, -\frac{3\pi}{4}$ **(c)** $5, \tan^{-1}\frac{4}{3}$ **(d)** $5, \pi - \tan^{-1}\frac{4}{3}$

(e) $4\sqrt{2}, -\frac{3\pi}{4}$ **(f)** $\sqrt{13}, -\tan^{-1}\frac{3}{2}$ **(g)** $2\sqrt{5}, \pi - \tan^{-1}\frac{1}{2}$

3. (a) $1, \frac{\pi}{4}$ **(b)** $e^2, \frac{\pi}{3}$ **(c)** $e^{-1}, \frac{\pi}{6}$ **(d)** $e^{\sqrt{2}}, \frac{\pi}{4}$

4.6 Geometrical Representation of Complex Numbers

The complex numbers can be represented geometrically by means of points on a plane. The diagram in which this representation is carried out is called the **Argand diagram**, named after the French mathematician Argand, in 1806.

Choose a rectangular coordinate axes $X'OX$ and $Y'OY$, through any point O. Since to every complex number $z = x + iy$, we associate a pair (x, y) of real numbers, we can represent the complex number $z = x + iy$ by the point P whose co-ordinates are given by (x, y). Conversely to every point $Q(x', y')$ on the plane, we can associate a complex number $x' + i\,y'$.

The point $P(x, y)$ is called the "affix" of the complex number $z = x + iy$. The plane on which the complex numbers are represented is called **Argand plane** or **complex plane**.

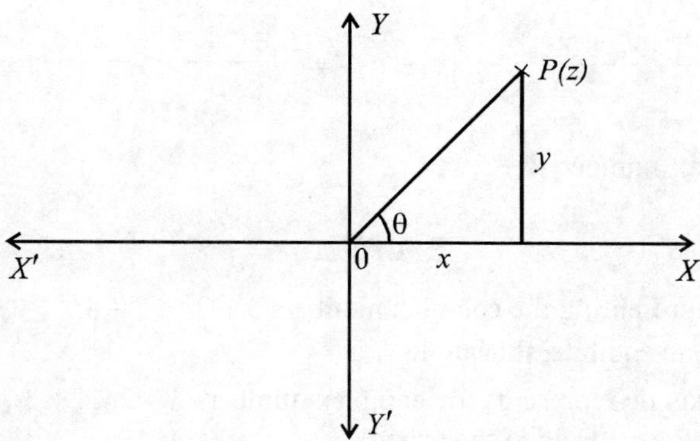

Clearly every point on the x–axis represents purely real numbers (i.e., complex numbers of the form $a + 0i$) and conversely. For this reason this axis is also called ***"real axis"***. Similarly, every point on the y–axis (except 0) represents purely imaginary numbers (i.e., of complex numbers of the form $0 + ib$) and conversely. For this reason this axis is also called ***"imaginary axis"***.

Supposing $z = x + iy = r(\cos\theta + i\sin\theta)$, is the polar form of z, then we have

$$x = r\cos\theta, \qquad y = r\sin\theta, \qquad |z| = \sqrt{x^2 + y^2}$$

Thus OP makes an angle θ with the x-axis and $OP = \sqrt{x^2 + y^2}$.

Thus to represent the complex number given in polar form $z = r(\cos\theta + i\sin\theta)$, we draw a line through O, making angle θ with the x-axis and take a point P on this line such that $OP = r$. This point P represents the complex number $z = r(\cos\theta + i\sin\theta)$.

Example 1. Show that the points *P, Q, R, S* representing the complex numbers $-1, 3i, 3 + 2i, 2 - i$ respectively on the Argand diagram are the vertices of a square.

Solution : Since the points P, Q, R and S represent the complex numbers $-1, 3i, 3 + 2i, 2 - i$, the co-ordinates of P, Q, R and S are respectively $(-1, 0), (0, 3), (3, 2), (2, -1)$. Consider

$$PQ = \sqrt{(-1-0)^2 + (0-3)^2} = \sqrt{10}\ ; \quad QR = \sqrt{(0-3)^2 + (3-2)^2} = \sqrt{10}$$

$$RS = \sqrt{(3-2)^2 + (2+1)^2} = \sqrt{10}\ ; \quad SP = \sqrt{(2+1)^2 + (-1-0)^2} = \sqrt{10}$$

Thus all the sides are equal.

Now,
$$PR = \sqrt{(-1-3)^2 + (0-2)^2} = \sqrt{16+4} = \sqrt{20}$$

$$QS = \sqrt{(0-2)^2 + (3+1)^2} = \sqrt{4+16} = \sqrt{20}$$

Thus, the diagonals are equal. Hence $PQRS$ is a square.

Example 2. If the vertices of the triangle *ABC* represents the complex number $3 - 2i$, $4 + 7i$ and $-1 - 3i$, find the complex number represented by the centroid of the triangle.

Solution : The vertices A, B and C represent the complex numbers $3 - 2i$, $4 + 7i$ and $-1 - 3i$ respectively.

$\therefore \qquad A = (3, -2),\ B = (4, 7), \quad C = (-1, -3)$

The centroid G of the triangle ABC is given by

$$G = \left(\frac{x_1 + x_2 + x_3}{3}, \frac{y_1 + y_2 + y_3}{3}\right) \Rightarrow G = \left(\frac{3+4-1}{3}, \frac{-2+7-3}{3}\right) = \left(2, \frac{2}{3}\right)$$

$\therefore$ G represents the complex number $\mathbf{2 + \frac{2}{3}}\boldsymbol{i}$.

Exercise

1. Show that the points representing the complex numbers $3 + 3i$, $-3 - 3i$, $-3\sqrt{3} + 3\sqrt{3}i$ on the Argand diagram are vertices of an equilateral triangle.
2. If the vertices of the triangle represents the complex numbers $3 - 2i$, $4 + 3i$, $5 + 5i$, find the complex number represented by the centroid of the triangle.
3. Show that the points representing the complex numbers, $2i$, $1 + i$, $4 + 4i$ and $3 + 5i$ on the Argand diagram are the vertices of a rectangle.
4. If the points A and B have affixes $4 - 5i$ and $-5 + 4i$ find the affix of the following :
   (a) the mid point of AB **(b)** the points of trisection of AB
   (c) the point which divides AB externally in the ratio 3 : 2.

4.7 DeMoivre's Theorem

In this section we shall see one of the most remarkable result, which enables us to extract any power of a complex number in a most simplified manner. This result is called **DeMoivre's theorem,** named after **Abraham DeMoivre** (1667 – 1754).

Theorem : (DeMoivre's Theorem)

If *n* is an integer then,

$$(\cos\theta + i\sin\theta)^n = \cos n\theta + i\sin n\theta.$$

If *n* is a fraction, then one of the values of $(\cos\theta + i\sin\theta)^n$ is $\cos n\theta + i\sin n\theta$

Proof : Case 1. Let n be a positive integer.

Consider

$$(\cos\theta_1 + i\sin\theta_1)(\cos\theta_2 + i\sin\theta_2) = e^{i\theta_1}\cdot e^{i\theta_2} = e^{i(\theta_1+\theta_2)} = \cos(\theta_1+\theta_2) + i\sin(\theta_1+\theta_2)$$

Similarly,

$$(\cos\theta_1 + i\sin\theta_1)(\cos\theta_2 + i\sin\theta_2)(\cos\theta_3 + i\sin\theta_3) = \cos(\theta_1+\theta_2+\theta_3) + i\sin(\theta_1+\theta_2+\theta_3)$$

In general,

$$(\cos\theta_1 + i\sin\theta_1)(\cos\theta_2 + i\sin\theta_2)\ ....\ (\cos\theta_n + i\sin\theta_n)$$
$$= \cos(\theta_1+\theta_2+...+\theta_n) + i\sin(\theta_1+\theta_2+...+\theta_n)$$

Now putting $\theta_1 = \theta_2 = \theta_3 = ... = \theta_n = \theta$, we get

$$(\cos\theta + i\sin\theta)(\cos\theta + i\sin\theta)\ ....\ (\cos\theta + i\sin\theta) = \cos(\theta+\theta+...+\theta) + i\sin(\theta+\theta+...+\theta)$$

$$\Rightarrow \qquad (\cos\theta + i\sin\theta)^n = \cos n\theta + i\sin n\theta$$

Thus the theorem is true when n is a positive integer.

Case 2. Let n be a negative integer. Let $n = -m$, where m is a positive integer.

Now,

$$(\cos\theta + i\sin\theta)^n = (\cos\theta + i\sin\theta)^{-m} \Rightarrow (\cos\theta + i\sin\theta)^n = \frac{1}{(\cos\theta + i\sin\theta)^m}$$

$$\Rightarrow \quad (\cos\theta + i\sin\theta)^n = \frac{1}{\cos m\theta + i\sin m\theta} \quad \text{(by case 1)}$$

$$\Rightarrow \quad (\cos\theta + i\sin\theta)^n = \frac{1}{e^{im\theta}}$$

$$\Rightarrow \quad (\cos\theta + i\sin\theta)^n = e^{-im\theta}$$

$$\Rightarrow \quad (\cos\theta + i\sin\theta)^n = \cos m\theta - i\sin m\theta$$

$$\Rightarrow \quad (\cos\theta + i\sin\theta)^n = \cos(-m\theta) + i\sin(-m\theta)$$

$$\Rightarrow \quad (\cos\theta + i\sin\theta)^n = \cos n\theta + i\sin n\theta \quad (\because -m = n)$$

Thus, the theorem is true where n is a negative integer.

Case 3. Let $n = 0$.

Consider, $(\cos\theta + i\sin\theta)^0 = 1 = \cos 0 + i\sin 0 \Rightarrow (\cos\theta + i\sin\theta)^0 = \cos 0\theta + i\sin 0\theta$

Thus, the theorem is true where $n = 0$. This proves the theorem for all integer n.

Case 4. Let n be a fraction, positive or negative.

Let $n = \frac{p}{q}$, where q is a positive integer and p is any integer positive or negative.

Since q is a positive integer we have,

$$\left(\cos\frac{\theta}{q} + i\sin\frac{\theta}{q}\right)^q = \cos\left(q\cdot\frac{\theta}{q}\right) + i\sin\left(q\cdot\frac{\theta}{q}\right) \Rightarrow \cos\theta + i\sin\theta = \left(\cos\frac{\theta}{q} + i\sin\frac{\theta}{q}\right)^q$$

Taking q^{th} root both sides, we have

$$\text{one of the values of, } (\cos\theta + i\sin\theta)^{1/q} = \cos\frac{\theta}{q} + i\sin\frac{\theta}{q}$$

$$\Rightarrow \quad \text{one of the values of, } (\cos\theta + i\sin\theta)^{p/q} = \left(\cos\frac{\theta}{q} + i\sin\frac{\theta}{q}\right)^p$$

$$\Rightarrow \quad \text{one of the values of, } (\cos\theta + i\sin\theta)^{p/q} = \cos\frac{p}{q}\theta + i\sin\frac{p}{q}\theta$$

$$\Rightarrow \quad \text{one of the values of, } (\cos\theta + i\sin\theta)^n = \cos n\theta + i\sin n\theta \quad \left(\because \frac{p}{q} = n\right)$$

Thus, if n is a fraction, then one of the values of $(\cos\theta + i\sin\theta)^n$ is $\cos n\theta + i\sin n\theta$.

This completes the proof of DeMoivre's theorem.

The following observations, which follows from the DeMoivre's theorem are useful

1. $(\cos\theta + i\sin\theta)^{-1} = \cos(-\theta) + i\sin(-\theta) = \cos\theta - i\sin\theta$

$$\therefore \quad \boldsymbol{\frac{1}{\cos\theta + i\cos\theta} = \cos\theta - i\sin\theta}$$

Similarly, $$\boldsymbol{\frac{1}{\cos\theta - i\cos\theta} = \cos\theta + i\sin\theta}$$

2. $(\cos\theta - i\sin\theta)^n = [\cos(-\theta) + i\sin(-\theta)]^n = \cos(-n\theta) + i\sin(-n\theta) = \cos n\theta - i\sin n\theta$

$$\boldsymbol{(\cos\theta - i\sin\theta)^n = \cos n\theta - i\sin n\theta}$$

3. Note that $(\sin\theta + i\cos\theta)^n \neq \sin n\theta + i\cos n\theta$

However we have, $(\sin\theta + i\cos\theta)^n = \left[\cos\left(\frac{\pi}{2}-\theta\right) + i\sin\left(\frac{\pi}{2}-\theta\right)\right]^n$

$$\Rightarrow \quad \mathbf{(\sin\theta + i\cos\theta)^n = \cos n\left(\frac{\pi}{2}-\theta\right) + i\sin n\left(\frac{\pi}{2}-\theta\right)}$$

We can also find $(\sin\theta + i\cos\theta)^n$, in the following way.

$$\sin\theta + i\cos\theta = i(\cos\theta - i\sin\theta) \qquad \text{(Note this step)}$$

$$\Rightarrow \quad (\sin\theta + i\cos\theta)^n = i^n(\cos\theta - i\sin\theta)^n$$

$$\therefore \quad \mathbf{(\sin\theta + i\cos\theta)^n = i^n(\cos n\theta - i\sin n\theta)}$$

Similarly, $\mathbf{(\sin\theta - i\cos\theta)^n = (-i)^n(\cos n\theta + i\sin n\theta)}$

We know that, $i^{4m+1} = i^{4m}\cdot i = 1\cdot i$ i.e., $i^{4m+1} = i$.

If n is of the form $4m + 1$

$(\sin\theta + i\cos\theta)^n = i^{4m+1}(\cos n\theta - i\sin n\theta) \Rightarrow (\sin\theta + i\cos\theta)^n = i(\cos n\theta - i\sin n\theta)$

$$\mathbf{(\sin\theta + i\cos\theta)^n = \sin n\theta + i\cos n\theta}$$

Thus,

> **If $n = 4m + 1$ (i.e., $4\,|\,n - 1$ or $n \equiv 1 \pmod 4$), then**
>
> $$\mathbf{(\sin\theta + i\cos\theta)^n = \sin n\theta + i\cos n\theta}$$
> $$\mathbf{(\sin\theta - i\cos\theta)^n = \sin n\theta - i\cos n\theta}$$

Example 1. Simplify

(a) $\mathbf{(\cos\theta - i\sin\theta)^5\cdot(\cos\theta + i\sin\theta)^2}$ **(b)** $\mathbf{\dfrac{(\cos\theta - i\sin\theta)^4}{(\cos\theta + i\sin\theta)^5}}$

(c) $\mathbf{(\sin 4\theta + i\cos 4\theta)^3\cdot(\sin 2\theta + i\cos 2\theta)^{-2}}$

Solution : (a) Consider

$$(\cos\theta - i\sin\theta)^5\cdot(\cos\theta + i\sin\theta)^2 = (\cos 5\theta - i\sin 5\theta)(\cos 2\theta + i\sin 2\theta)$$
$$= e^{-i(5\theta)}\cdot e^{i(2\theta)} = e^{i(-5\theta+2\theta)} = e^{-i(3\theta)} = \mathbf{(\cos 3\theta - i\sin 3\theta)}$$

(b) Consider,

$$\frac{(\cos\theta - i\sin\theta)^4}{(\cos\theta + i\sin\theta)^5} = \frac{\cos 4\theta - i\sin 4\theta}{\cos 5\theta + i\sin 5\theta} = \frac{e^{-i(4\theta)}}{e^{i(5\theta)}} = e^{i(-4\theta-5\theta)} = e^{-i(9\theta)} = \mathbf{\cos 9\theta - i\sin 9\theta}$$

(c) Consider, $(\sin 4\theta + i\cos 4\theta)^3\cdot(\sin 2\theta + i\cos 2\theta)^{-2}$

$$= i^3(\cos 4\theta - i\sin 4\theta)^3\cdot i^{-2}(\cos 2\theta - i\sin 2\theta)^{-2} \qquad \text{(Note this step)}$$
$$= -i(\cos 12\theta - i\sin 12\theta)\cdot(-1)(\cos(-4\theta) - i\sin(-4\theta)) \qquad (\because i^3 = -i, i^2 = 1)$$
$$= i(\cos 12\theta - i\sin 12\theta)(\cos 4\theta + i\sin 4\theta)$$
$$= i\,e^{-i12\theta}\cdot e^{i4\theta} = ie^{i(-12+4)\theta} = ie^{-i8\theta} = i(\cos 8\theta - i\sin 8\theta) = \mathbf{\sin 8\theta + i\cos 8\theta}$$

Example 2. Simplify

(a) $\mathbf{\dfrac{(\cos 7\theta - i\sin 7\theta)^{-2}(\cos 3\theta + i\sin 3\theta)^4}{(\cos\theta + i\sin\theta)^{-4}}}$ **(b)** $\mathbf{\dfrac{(\cos 3\alpha - i\sin 3\alpha)^3(\cos 5\beta + i\sin 5\beta)^{-2}}{(\cos 3\alpha + i\sin 3\alpha)^{-6}(\cos 4\beta - i\sin 4\beta)^5}}$

Solution : (a) Consider,

$$\frac{(\cos 7\theta - i\sin 7\theta)^{-2}\,(\cos 3\theta + i\sin 3\theta)^{4}}{(\cos\theta + i\sin\theta)^{-4}} = \frac{(\cos(-14\theta) - i\sin(-14\theta))\,(\cos 12\theta + i\sin 12\theta)}{(\cos(-4\theta) + i\sin(-4\theta))}$$

$$= \frac{(\cos 14\theta + i\sin 14\theta)\,(\cos 12\theta + i\sin 12\theta)}{(\cos 4\theta - i\sin 4\theta)}$$

$$= \frac{e^{i(14\theta)}\cdot e^{i(12\theta)}}{e^{-i(4\theta)}} = e^{i(14+12+4)\theta} = e^{i(30)\theta} = \mathbf{\cos 30\theta + i\sin 30\theta}$$

(b) Given expression $= \dfrac{(\cos 9\alpha - i\sin 9\alpha)\left[\cos(-10\beta) - i\sin(-10\beta)\right]}{\left[\cos(-18\alpha) - i\sin(-18\alpha)\right](\cos 20\beta - i\sin 20\beta)}$

$$= \frac{(\cos 9\alpha - i\sin 9\alpha)\,(\cos 10\beta - i\sin 10\beta)}{(\cos 18\alpha - i\sin 18\alpha)\,(\cos 20\beta - i\sin 20\beta)}$$

$$= \frac{e^{-i(9\alpha)}\cdot e^{-i(10\beta)}}{e^{-i(18\alpha)}\cdot e^{-i(20\beta)}} = e^{i(-9+18)\alpha}\cdot e^{i(-10+20)\beta} = e^{i(9\alpha)}\cdot e^{i(10\beta)}$$

$$= e^{i(9\alpha+10\beta)} = \mathbf{\cos(9\alpha + 10\beta) + i\sin(9\alpha + 10\beta)}$$

Example 3. Show that $(\sin 4\theta + i\cos 4\theta)^{-4} = \cos 16\theta + i\sin 16\theta$

Solution : Consider,

$$(\sin 4\theta + i\cos 4\theta)^{-4} = \left[\cos\left(\frac{\pi}{2} - 4\theta\right) + i\sin\left(\frac{\pi}{2} - 4\theta\right)\right]^{-4}$$

$$= \left[\cos(-4)\left(\frac{\pi}{2} - 4\theta\right) + i\sin(-4)\left(\frac{\pi}{2} - 4\theta\right)\right]$$

$$= \cos(2\pi - 16\theta) - i\sin(2\pi - 16\theta) \qquad (\because \cos(-\theta) = \cos\theta,\ \sin(-\theta) = \sin\theta)$$

$$= \mathbf{\cos 16\theta + i\sin 16\theta} = \text{R.H.S} \qquad \begin{pmatrix}\because \cos(2\pi-\alpha) = \cos\alpha \\ \sin(2\pi-\alpha) = -\sin\alpha\end{pmatrix}$$

2nd method:

Consider, $\sin 4\theta + i\cos 4\theta = i(\cos 4\theta - i\sin 4\theta)$

$\Rightarrow$ $(\sin 4\theta + i\cos 4\theta)^{-4} = (i)^{-4}(\cos 4\theta - i\sin 4\theta)^{-4}$

$\Rightarrow$ $\mathbf{(\sin 4\theta + i\cos 4\theta)^{-4} = \cos 16\theta + i\sin 16\theta}$ $\qquad (\because i^4 = 1)$

Example 4. If $x + \dfrac{1}{x} = 2\cos\theta$, show that one of the values of x is $\cos\theta + i\sin\theta$.

Solution : We have, $x + \dfrac{1}{x} = 2\cos\theta \Rightarrow x^2 + 1 = (2\cos\theta)x \Rightarrow x^2 - (2\cos\theta)x + 1 = 0$

$\Rightarrow$ $x^2 - (2\cos\theta)x + \cos^2\theta + \sin^2\theta = 0$ $\qquad (\because 1 = \cos^2\theta + \sin^2\theta)$

$\Rightarrow$ $(x - \cos\theta)^2 = -\sin^2\theta$

$\Rightarrow$ $x - \cos\theta = \pm\sin\theta\cdot\sqrt{-1} \Rightarrow x = \cos\theta \pm i\sin\theta$ $\qquad (\because i = \sqrt{-1})$

Thus, one of the values of x is $\mathbf{\cos\theta + i\sin\theta}$.

Example 5. If $x + \frac{1}{x} = 2\cos\theta$, show that $x^n + \frac{1}{x^n} = 2\cos n\theta$ and $x^n - \frac{1}{x^n} = 2i\sin n\theta$.

Solution : By data $x + \frac{1}{x} = 2\cos\theta$. Then we can take

$$x = \cos\theta + i\sin\theta \quad \text{and} \quad \frac{1}{x} = \cos\theta - i\sin\theta$$

$$\Rightarrow \quad x^n = (\cos\theta + i\sin\theta)^n \quad \text{and} \quad \frac{1}{x^n} = (\cos\theta - i\sin\theta)^n$$

$$\Rightarrow \quad x^n = \cos n\theta + i\sin n\theta \quad \text{and} \quad \frac{1}{x^n} = \cos n\theta - i\sin n\theta$$

Adding we get $\mathbf{x^n + \frac{1}{x^n} = 2\cos n\theta}$; On subtraction, $\mathbf{x^n - \frac{1}{x^n} = 2i\sin n\theta}$

Example 6. If $x = \text{cis}\,\alpha$, $y = \text{cis}\,\beta$ and $z = \text{cis}\,\gamma$,* show that

(a) $\frac{x^3y^4}{z^2} + \frac{z^2}{x^3y^4} = 2\cos(3\alpha + 4\beta - 2\gamma)$ **(b)** $\frac{x^3y^4}{z^2} - \frac{z^2}{x^3y^4} = 2i\sin(3\alpha + 4\beta - 2\gamma)$

Solution : We have,

$$x = \cos\alpha + i\sin\alpha \quad \Rightarrow \quad x^3 = \cos 3\alpha + i\sin 3\alpha = e^{i(3\alpha)}$$

$$y = \cos\beta + i\sin\beta \quad \Rightarrow \quad y^4 = \cos 4\beta + i\sin 4\beta = e^{i(4\beta)}$$

$$z = \cos\gamma + i\sin\gamma \quad \Rightarrow \quad z^2 = \cos 2\gamma + i\sin 2\gamma = e^{i(2\gamma)}$$

Consider,

$$\frac{x^3y^4}{z^2} = \frac{e^{i(3\alpha)}\cdot e^{i(4\beta)}}{e^{i(2\gamma)}} \Rightarrow \frac{x^3y^4}{z^2} = \frac{e^{i(3\alpha+4\beta)}}{e^{i(2\gamma)}} \Rightarrow \frac{x^3y^4}{z^2} = e^{i(3\alpha+4\beta-2\gamma)} \Rightarrow \frac{z^2}{x^3y^4} = e^{-i(3\alpha+4\beta-2\gamma)}$$

$$\therefore \quad \frac{x^3y^4}{z^2} = \cos(3\alpha + 4\beta - 2\gamma) + i\sin(3\alpha + 4\beta - 2\gamma) \quad \text{.... (1)}$$

and $$\frac{z^2}{x^3y^4} = \cos(3\alpha + 4\beta - 2\gamma) - i\sin(3\alpha + 4\beta - 2\gamma) \quad \text{.... (2)}$$

(a) Adding (1) and (2) we get, $\mathbf{\frac{x^3y^4}{z^2} + \frac{z^2}{x^3y^4} = 2\cos(3\alpha + 4\beta - 2\gamma)}$

(b) Consider (1) – (2) we get, $\mathbf{\frac{x^3y^4}{z^2} - \frac{z^2}{x^3y^4} = 2i\sin(3\alpha + 4\beta - 2\gamma)}$

Example 7. Evaluate $\left(\sqrt{3} - i\right)^9$.

Solution : Let $z = \sqrt{3} - i = a + ib$; $a = \sqrt{3}$, $b = -1$.

$$\therefore\ r = |z| = \sqrt{3+1} = 2; \quad \text{amp}\, z = \theta = -\tan^{-1}\frac{|b|}{a} \quad (\because a > 0, b < 0)$$

$$\Rightarrow \quad \text{amp}\, z = \theta = -\tan^{-1}\frac{1}{\sqrt{3}} = -\frac{\pi}{6}$$

---

* $\cos\theta + i\sin\theta$ is denoted by cis θ.

$$\therefore\ \sqrt{3}-i = 2\left[\cos\left(-\frac{\pi}{6}\right)+i\sin\left(-\frac{\pi}{6}\right)\right] \Rightarrow \sqrt{3}-i = 2\left[\cos\left(\frac{\pi}{6}\right)-i\sin\left(\frac{\pi}{6}\right)\right]$$

$$\Rightarrow \left(\sqrt{3}-i\right)^9 = 2^9\left[\cos 9\left(\frac{\pi}{6}\right)-i\sin 9\left(\frac{\pi}{6}\right)\right]$$

$$\Rightarrow \left(\sqrt{3}-i\right)^9 = 2^9\left[\cos\frac{3\pi}{2}-i\sin\frac{3\pi}{2}\right]$$

$$\Rightarrow \left(\sqrt{3}-i\right)^9 = 2^9\,(0-i(-1))\left(\because \cos\frac{3\pi}{2}=0,\ \sin\frac{3\pi}{2}=-1\right)$$

$$\Rightarrow \left(\sqrt{3}-i\right)^9 = 2^9\cdot i$$

Example 8. Show that $\left(-1+i\sqrt{3}\right)^{3n}+\left(-1-i\sqrt{3}\right)^{3n}=2^{3n+1}$.

Solution : Let $z=-1+i\sqrt{3}=a+ib$; $a=-1$, $b=\sqrt{3}$. $r=|z|=\sqrt{1+3}=2$

$$\theta = \text{amp}\, z = \pi-\tan^{-1}\left(\frac{b}{|a|}\right) \qquad (\because a<0,\ b>0)$$

$$\Rightarrow \quad \theta = \pi-\tan^{-1}\left(\sqrt{3}\right) \Rightarrow \theta=\pi-\frac{\pi}{3}=\frac{2\pi}{3}$$

$$\therefore \quad -1+i\sqrt{3} = 2\left(\cos\frac{2\pi}{3}+i\sin\frac{2\pi}{3}\right)$$

$$\Rightarrow \quad -1-i\sqrt{3} = 2\left(\cos\frac{2\pi}{3}-i\sin\frac{2\pi}{3}\right)$$

$$\Rightarrow \quad \left(-1+i\sqrt{3}\right)^{3n} = 2^{3n}\left(\cos\frac{6n\pi}{3}+i\sin\frac{6n\pi}{3}\right)$$

$$\Rightarrow \quad \left(-1-i\sqrt{3}\right)^{3n} = 2^{3n}\left(\cos\frac{6n\pi}{3}-i\sin\frac{6n\pi}{3}\right)$$

$$\Rightarrow \quad \left(-1+i\sqrt{3}\right)^{3n}+\left(-1-i\sqrt{3}\right)^{3n} = 2\cdot 2^{3n}\cdot\cos\left(\frac{6n\pi}{3}\right)$$

$$= 2^{3n+1}\,.\cos(2n\pi) = 2^{3n+1}\,.\,1 \qquad (\because \cos 2n\pi = 1)$$

$$\therefore \quad \left(-1+i\sqrt{3}\right)^{3n}+\left(-1-i\sqrt{3}\right)^{3n} = 2^{3n+1}$$

Note : $\left(-1+i\sqrt{3}\right)^{3n}-\left(-1-i\sqrt{3}\right)^{3n}=0 \qquad (\because \sin 2n\pi=0)$

Example 9. Evaluate $(1+i)^6+(1-i)^6$ and $(1+i)^6-(1-i)^6$.

Solution : Let $z=1+i=a+ib$; $a=1$, $b=1$. $r=|z|=\sqrt{1+1}=\sqrt{2}$;

$$\theta = \text{amp}\, z = \tan^{-1}\frac{b}{a} \qquad (a>0,\ b>0)$$

$$\Rightarrow \quad \theta = \tan^{-1}1 \Rightarrow \theta=\frac{\pi}{4}$$

$$\therefore \quad 1+i = \sqrt{2}\left(\cos\frac{\pi}{4}+i\sin\frac{\pi}{4}\right) \Rightarrow \quad 1-i=\sqrt{2}\left(\cos\frac{\pi}{4}-i\sin\frac{\pi}{4}\right)$$

$$\Rightarrow \quad (1+i)^6 = \left(\sqrt{2}^6\right)\left(\cos\frac{6\pi}{4} + i\sin\frac{6\pi}{4}\right)$$

$$\Rightarrow \quad (1-i)^6 = \left(\sqrt{2}^6\right)\left(\cos\frac{6\pi}{4} - i\sin\frac{6\pi}{4}\right)$$

Now, $(1+i)^6 + (1-i)^6 = 2\cdot\left(\sqrt{2}^6\right)\cdot\cos\frac{6\pi}{4} = 2\cdot 2^{6/2}\cdot\cos\frac{3\pi}{2} = 2^4 \cdot 0$ $\left(\because \cos\frac{3\pi}{2} = 0\right)$

$\therefore$ $$\mathbf{(1+i)^6 + (1-i)^6 = 0}$$

Again, $(1+i)^6 - (1-i)^6 = 2\cdot\left(\sqrt{2}^6\right)\cdot i\sin\frac{6\pi}{4} = 2\cdot 2^{6/2}\cdot i\sin\frac{3\pi}{2} = 2^4\, i\,(-1)$ $\left(\because \sin\frac{3\pi}{2} = -1\right)$

$\therefore$ $$\mathbf{(1+i)^6 + (1-i)^6 = -2^4\, i}$$

Example 10. Show that $\boldsymbol{(a+ib)^{\frac{m}{n}} + (a-ib)^{\frac{m}{n}} = 2\left(a^2+b^2\right)^{\frac{m}{2n}}\cos\left(\frac{m}{n}\tan^{-1}\frac{b}{a}\right)}$

Solution : Let $a + ib = r(\cos\theta + i\sin\theta)$. $\Rightarrow$ $r = \sqrt{a^2+b^2}$ and $\theta = \tan^{-1}\frac{b}{a}$

Now, $a + ib = r(\cos\theta + i\sin\theta)$; $a - ib = r(\cos\theta - i\sin\theta)$

$$\Rightarrow \quad (a+ib)^{\frac{m}{n}} = r^{\frac{m}{n}}\left(\cos\frac{m}{n}\theta + i\sin\frac{m}{n}\theta\right); \quad (a-ib)^{\frac{m}{n}} = r^{\frac{m}{n}}\left(\cos\frac{m}{n}\theta - i\sin\frac{m}{n}\theta\right)$$

$$\Rightarrow \quad (a+ib)^{\frac{m}{n}} + (a-ib)^{\frac{m}{n}} = 2\cdot r^{\frac{m}{n}}\cdot\cos\frac{m}{n}\theta$$

$$\Rightarrow \quad \boldsymbol{(a+ib)^{\frac{m}{n}} + (a-ib)^{\frac{m}{n}} = 2\left(a^2+b^2\right)^{\frac{m}{2n}}\cos\left(\frac{m}{n}\tan^{-1}\frac{b}{a}\right)} \quad \left(\because r = \sqrt{a^2+b^2},\ \theta = \tan^{-1}\frac{b}{a}\right)$$

Note : $\boldsymbol{(a+ib)^{\frac{m}{n}} - (a-ib)^{\frac{m}{n}} = 2i\left(a^2+b^2\right)^{\frac{m}{2n}}\sin\left(\frac{m}{n}\tan^{-1}\frac{b}{a}\right)}$

The above two results are useful to write directly, the values of the expressions like:

$(1+i)^6 + (1-i)^6 = 2\cdot(1+1)^{6/2}\cdot\cos(6\tan^{-1}1) = 2\cdot 2^3\cdot\left(\cos 6\cdot\frac{\pi}{4}\right) = 0$ $\left(\because \cos\frac{3\pi}{2} = 0\right)$

$$\begin{aligned}
\left(-1+i\sqrt{3}\right)^5 \left(-1-i\sqrt{3}\right)^5 &= 2i(1+3)^{5/2}\cdot\sin\left[5\left(\pi - \tan^{-1}\sqrt{3}\right)\right] \\
&= 2i(4)^{5/2}\cdot\sin\left[5\left(\pi - \frac{\pi}{3}\right)\right] \\
&= 2i\,2^5\cdot\sin\frac{10\pi}{3} = 2^6\, i\cdot\sin\left(2\pi + \frac{4\pi}{3}\right) \\
&= 2^6\, i\cdot\sin\left(\frac{4\pi}{3}\right) = 2^6\, i\cdot\sin\left(\pi + \frac{\pi}{3}\right) = 2^6\, i\left(-\sin\frac{\pi}{3}\right) = \boldsymbol{-2^5\sqrt{3}\, i}
\end{aligned}$$

Example 11. Show that $\boldsymbol{(1+\cos\theta + i\sin\theta)^n + (1+\cos\theta - i\sin\theta)^n = 2^{n+1}\cdot\cos^n\left(\frac{\theta}{2}\right)\cdot\cos\left(\frac{n\theta}{2}\right)}$.

Solution : Consider, $1 + \cos\theta + i\sin\theta = 2\cdot\cos^2\frac{\theta}{2} + i\cdot 2\sin\frac{\theta}{2}\cos\frac{\theta}{2}$

$$\Rightarrow \qquad 1 + \cos\theta + i\sin\theta = 2\cdot\cos\frac{\theta}{2}\left(\cos\frac{\theta}{2} + i\sin\frac{\theta}{2}\right)$$

$$\Rightarrow \qquad 1 + \cos\theta - i\sin\theta = 2\cdot\cos\frac{\theta}{2}\left(\cos\frac{\theta}{2} - i\sin\frac{\theta}{2}\right)$$

$$\left[\begin{array}{l} (1 + \cos\theta + i\sin\theta)^n = 2^n\cdot\cos^n\frac{\theta}{2}\left(\cos\frac{n\theta}{2} + i\sin\frac{n\theta}{2}\right) \\ (1 + \cos\theta - i\sin\theta)^n = 2^n\cdot\cos^n\frac{\theta}{2}\left(\cos\frac{n\theta}{2} - i\sin\frac{n\theta}{2}\right) \end{array}\right.$$

Adding, we get,

$$(1 + \cos\theta + i\sin\theta)^n + (1 + \cos\theta - i\sin\theta)^n = 2\cdot 2^n\cdot\cos^n\left(\frac{\theta}{2}\right)\cdot\cos\left(\frac{n\theta}{2}\right)$$

$$\therefore \qquad \mathbf{(1 + \cos\theta + i\sin\theta)^n + (1 + \cos\theta - i\sin\theta)^n = 2^{n+1}\cdot\cos^n\left(\frac{\theta}{2}\right)\cdot\cos\left(\frac{n\theta}{2}\right)}$$

Example 12. If $x = \cos\theta + i\sin\theta$, show that $\dfrac{x^{2n} - 1}{x^{2n} + 1} = i\tan n\theta$.

Solution : The expression $\dfrac{x^{2n} - 1}{x^{2n} + 1}$ can be written as

$$\frac{x^{2n} - 1}{x^{2n} + 1} = \frac{x^n - \dfrac{1}{x^n}}{x^n + \dfrac{1}{x^n}} \qquad \text{.... (1)}$$

Thus we shall find $x^n - \dfrac{1}{x^n}$ and $x^n + \dfrac{1}{x^n}$.

Now, $\qquad x = \cos\theta + i\sin\theta \quad \Rightarrow \quad \dfrac{1}{x} = \cos\theta - i\sin\theta$

$\Rightarrow \qquad x^n = \cos n\theta + i\sin n\theta \quad ; \quad \dfrac{1}{x^n} = \cos n\theta - i\sin n\theta$

$\therefore \qquad x^n + \dfrac{1}{x^n} = 2\cos n\theta \quad$ and $\quad x^n - \dfrac{1}{x^n} = 2i\sin n\theta$

Thus (1) $\Rightarrow \qquad \dfrac{x^{2n} - 1}{x^{2n} + 1} = \dfrac{2i\sin n\theta}{2\cos n\theta} \quad \Rightarrow \quad \mathbf{\dfrac{x^{2n} - 1}{x^{2n} + 1} = i\tan n\theta}$

Example 13. Show that $\left(\dfrac{1 + \cos\theta + i\sin\theta}{1 + \cos\theta - i\sin\theta}\right)^n = \cos n\theta + i\sin n\theta$.

Solution : Let, $z = \cos\theta + i\sin\theta \;\Rightarrow\; \dfrac{1}{z} = \cos\theta - i\sin\theta$

$$\frac{1+\cos\theta+i\sin\theta}{1+\cos\theta-i\sin\theta} = \left(\frac{1+z}{1+\frac{1}{z}}\right)$$

$$= \frac{z(1+z)}{(z+1)} = z = \cos\theta + i\sin\theta$$

$$\left(\frac{\mathbf{1+\cos\theta+\mathit{i}\sin\theta}}{\mathbf{1+\cos\theta-\mathit{i}\sin\theta}}\right)^{n} = \mathbf{\cos \mathit{n}\theta + \mathit{i}\sin \mathit{n}\theta}$$

Example 14. Simplify $\left(\dfrac{1+\sin\frac{\pi}{8}+i\cos\frac{\pi}{8}}{1+\sin\frac{\pi}{8}-i\cos\frac{\pi}{8}}\right)^{1/2}$.

Solution : Let $\sin\frac{\pi}{8} + i\cos\frac{\pi}{8} = z$ Clearly, $\sin\frac{\pi}{8} - i\cos\frac{\pi}{8} = \frac{1}{z}$

$$\therefore \quad \left(\frac{1+\sin\frac{\pi}{8}+i\cos\frac{\pi}{8}}{1+\sin\frac{\pi}{8}-i\cos\frac{\pi}{8}}\right)^{1/2} = \left(\frac{1+z}{1+\frac{1}{z}}\right)^{1/2} = \left(\frac{z(1+z)}{(z+1)}\right)^{1/2} = z^{1/2}$$

$$\therefore \quad \left(\frac{1+\sin\frac{\pi}{8}+i\cos\frac{\pi}{8}}{1+\sin\frac{\pi}{8}-i\cos\frac{\pi}{8}}\right)^{1/2} = \left(\sin\frac{\pi}{8}+i\cos\frac{\pi}{8}\right)^{1/2}$$

$$= \left(\cos\left(\frac{\pi}{2}-\frac{\pi}{8}\right) + i\sin\left(\frac{\pi}{2}-\frac{\pi}{8}\right)\right)^{1/2} = \left(\cos\frac{3\pi}{8} + i\sin\frac{3\pi}{8}\right)^{1/2}$$

$$\therefore \quad \left(\frac{\mathbf{1+\sin\frac{\pi}{8}+\mathit{i}\cos\frac{\pi}{8}}}{\mathbf{1+\sin\frac{\pi}{8}-\mathit{i}\cos\frac{\pi}{8}}}\right)^{1/2} = \mathbf{\cos\frac{3\pi}{16} + \mathit{i}\sin\frac{3\pi}{16}}$$

Example 15. Simplify $\left(\dfrac{1+\sin\theta+i\cos\theta}{1-\sin\theta-i\cos\theta}\right)^{7} = -i\cot^{7}\left(\dfrac{\pi}{4}-\dfrac{\theta}{2}\right)$.

Solution : Consider,

$$1+\sin\theta+i\cos\theta = 1+\cos\left(\frac{\pi}{2}-\theta\right) + i\sin\left(\frac{\pi}{2}-\theta\right)$$

$$= 1+\cos\alpha + i\sin\alpha \qquad \left(\text{where } \alpha = \frac{\pi}{2}-\theta\right)$$

$$= 2\cos^{2}\frac{\alpha}{2} + i\,2\sin\frac{\alpha}{2}\cos\frac{\alpha}{2} = 2\cos\frac{\alpha}{2}\left(\cos\frac{\alpha}{2} + i\sin\frac{\alpha}{2}\right)$$

Again consider

$$1 - \sin\theta - i\cos\theta = 1 - \cos\left(\frac{\pi}{2} - \theta\right) - i\sin\left(\frac{\pi}{2} - \theta\right)$$

$$= 1 - \cos\alpha - i\sin\alpha \qquad \left(\text{where } \alpha = \frac{\pi}{2} - \theta\right)$$

$$= 2\sin^2\frac{\alpha}{2} - i\,2\sin\frac{\alpha}{2}\cos\frac{\alpha}{2} = -i\,2\sin\frac{\alpha}{2}\left(\cos\frac{\alpha}{2} + i\sin\frac{\alpha}{2}\right)$$

(Note this step)

Now, $$\frac{1 + \sin\theta + i\cos\theta}{1 - \sin\theta - i\cos\theta} = \frac{2\cos\frac{\alpha}{2}\left(\cos\frac{\alpha}{2} + i\sin\frac{\alpha}{2}\right)}{-i\,2\sin\frac{\alpha}{2}\left(\cos\frac{\alpha}{2} + i\sin\frac{\alpha}{2}\right)} = -\frac{1}{i}\cot\frac{\alpha}{2} = \boldsymbol{i\cot\frac{\alpha}{2}} \qquad \left(\because \frac{1}{i} = -i\right)$$

$$\therefore \quad \left(\frac{1 + \sin\theta + i\cos\theta}{1 - \sin\theta - i\cos\theta}\right)^7 = i^7\cot^7\left(\frac{\alpha}{2}\right)$$

$$\Rightarrow \quad \boldsymbol{\left(\frac{1 + \sin\theta + i\cos\theta}{1 - \sin\theta - i\cos\theta}\right)^7 = -i\cot^7\left(\frac{\pi}{4} - \frac{\theta}{2}\right)} \qquad \left(\because i^7 = -i,\ \alpha = \frac{\pi}{2} - \theta\right)$$

Example 16. Show that $\dfrac{\left(\cos\frac{\pi}{6} - i\sin\frac{\pi}{6}\right)^{\frac{11}{2}}}{\left(\cos\frac{\pi}{6} + i\sin\frac{\pi}{6}\right)^{\frac{1}{2}}} = -1$**.**

Solution : Consider,

$$\text{L.H.S} = \frac{\cos\frac{11\pi}{12} - i\sin\frac{11\pi}{12}}{\cos\frac{\pi}{12} + i\sin\frac{\pi}{12}} = \frac{\cos\left(\pi - \frac{\pi}{12}\right) - i\sin\left(\pi - \frac{\pi}{12}\right)}{\cos\frac{\pi}{12} + i\sin\frac{\pi}{12}}$$

$$= \frac{-\cos\frac{\pi}{12} - i\sin\frac{\pi}{12}}{\cos\frac{\pi}{12} + i\sin\frac{\pi}{12}} \qquad (\because \cos(\pi - \theta) = -\cos\theta,\ \sin(\pi - \theta) = \sin\theta)$$

$$= \frac{-\left(\cos\frac{\pi}{12} + i\sin\frac{\pi}{12}\right)}{\left(\cos\frac{\pi}{12} + i\sin\frac{\pi}{12}\right)} = \mathbf{-1} = \text{R.H.S}$$

Example 17. If $a = \cos\alpha + i\sin\alpha$, $b = \cos\beta + i\sin\beta$, show that

(a) $\sin(\alpha - \beta) = \dfrac{1}{2i}\left(\dfrac{a}{b} - \dfrac{b}{a}\right)$ **(b)** $\cos(\alpha + \beta) = \dfrac{1}{2}\left(ab + \dfrac{1}{ab}\right)$

Solution : We have, $a = \cos\alpha + i\sin\alpha$; $b = \cos\beta + i\sin\beta$

(a) Consider $\frac{a}{b} = \frac{\cos\alpha + i\sin\alpha}{\cos\beta + i\sin\beta} = \frac{e^{i\alpha}}{e^{i\beta}} = e^{i(\alpha-\beta)} \Rightarrow \frac{a}{b} = \cos(\alpha-\beta) + i\sin(\alpha-\beta)$

$\Rightarrow \frac{b}{a} = \cos(\alpha-\beta) - i\sin(\alpha-\beta)$

$\therefore \quad \frac{a}{b} - \frac{b}{a} = 2i\sin(\alpha-\beta) \Rightarrow \mathbf{\sin(\alpha-\beta) = \frac{1}{2i}\left(\frac{a}{b} - \frac{b}{a}\right)}$

(b) Consider, $a\cdot b = (\cos\alpha + i\sin\alpha)(\cos\beta + i\sin\beta) \Rightarrow a\cdot b = e^{i\alpha}\cdot e^{i\beta} \Rightarrow a\cdot b = e^{i(\alpha+\beta)}$

$\therefore \quad a\cdot b = \cos(\alpha+\beta) + i\sin(\alpha+\beta) \Rightarrow \frac{1}{a\cdot b} = \cos(\alpha+\beta) - i\sin(\alpha+\beta)$

$\therefore \quad a\cdot b + \frac{1}{a\cdot b} = 2\cos(\alpha+\beta) \Rightarrow \mathbf{\cos(\alpha+\beta) = \frac{1}{2}\left(ab + \frac{1}{ab}\right)}$

Example 18. If $\cos\alpha + \cos\beta = 0$ and $\sin\alpha + \sin\beta = 0$. show that

(a) $\cos 2\alpha + \cos 2\beta = 2\cos(\pi + \alpha + \beta)$ (b) $\sin 2\alpha + \sin 2\beta = 2\sin(\pi + \alpha + \beta)$.

Solution : Let $a = \cos\alpha + i\sin\alpha = e^{i\alpha}$, $b = \cos\beta + i\sin\beta = e^{i\beta}$

Consider, $a + b = (\cos\alpha + i\sin\alpha) + (\cos\beta + i\sin\beta) \Rightarrow a + b = (\cos\alpha + \cos\beta) + i(\sin\alpha + \sin\beta)$

By data, $\cos\alpha + \cos\beta = 0$ and $\sin\alpha + \sin\beta = 0 \Rightarrow a + b = 0 + i0 \Rightarrow a + b = 0$

Now, $a^2 + b^2 = (a+b)^2 - 2ab \Rightarrow a^2 + b^2 = -2ab$ $(\because a + b = 0)$

$\Rightarrow (\cos\alpha + i\sin\alpha)^2 + (\cos\beta + i\sin\beta)^2 = -2e^{i\alpha}\cdot e^{i\beta}$

$\Rightarrow \cos 2\alpha + i\sin 2\alpha + \cos 2\beta + i\sin 2\beta = -2e^{i(\alpha+\beta)}$

$\Rightarrow (\cos 2\alpha + \cos 2\beta) + i(\sin 2\alpha + \sin 2\beta) = -2[\cos(\alpha+\beta) + i\sin(\alpha+\beta)]$

Equating the real and imaginary parts separately, we get,

$$\cos 2\alpha + \cos 2\beta = -2\cos(\alpha+\beta)$$

$$\sin 2\alpha + \sin 2\beta = -2\sin(\alpha+\beta)$$

$$\Rightarrow \left|\begin{array}{ll} \mathbf{\cos 2\alpha + \cos 2\beta = 2\cos(\pi + \alpha + \beta)} & (\because -\cos\theta = \cos(\pi + \theta)) \\ \mathbf{\sin 2\alpha + \sin 2\beta = 2\sin(\pi + \alpha + \beta)} & (\because -\sin\theta = \sin(\pi + \theta)) \end{array}\right.$$

Example 19. If $a = \cos\alpha + i\sin\alpha$, $b = \cos\beta + i\sin\beta$, $c = \cos\gamma + i\sin\gamma$ and $a + b + c = 0$, show that

(a) $\frac{1}{a} + \frac{1}{b} + \frac{1}{c} = 0$

(b) $\cos(\alpha+\beta) + \cos(\beta+\gamma) + \cos(\gamma+\alpha) = 0$, and $\sin(\alpha+\beta) + \sin(\beta+\gamma) + \sin(\gamma+\alpha) = 0$.

Solution : We have, $a = \cos\alpha + i\sin\alpha = e^{i\alpha}$, $b = \cos\beta + i\sin\beta = e^{i\beta}$, $c = \cos\gamma + i\sin\gamma = e^{i\gamma}$

By data, $a + b + c = 0 \Rightarrow (\cos\alpha + \cos\beta + \cos\gamma) + i(\sin\alpha + \sin\beta + \sin\gamma) = 0$

Equating the real and imaginary part separately to zero we get

$$\cos\alpha + \cos\beta + \cos\gamma = 0 \ ; \quad \sin\alpha + \sin\beta + \sin\gamma = 0 \qquad .... (1)$$

(a) Consider, $\frac{1}{a} + \frac{1}{b} + \frac{1}{c} = e^{-i\alpha} + e^{-i\beta} + e^{-i\gamma}$

$$= \cos\alpha - i\sin\alpha + \cos\beta - i\sin\beta + \cos\gamma - i\sin\gamma$$

$$= (\cos\alpha + \cos\beta + \cos\gamma) - i(\sin\alpha + \sin\beta + \sin\gamma) = 0 - i0 \quad \text{(from (1))}$$

$$\therefore \quad \boldsymbol{\frac{1}{a} + \frac{1}{b} + \frac{1}{c} = 0}$$

(b) Consider, $ab + bc + ca = abc\left(\frac{1}{a} + \frac{1}{b} + \frac{1}{c}\right) \Rightarrow ab + bc + ca = (abc)\,0 \quad \left(\because \frac{1}{a} + \frac{1}{b} + \frac{1}{c} = 0\right)$

$$\Rightarrow e^{i\alpha}\cdot e^{i\beta} + e^{i\beta}\cdot e^{i\gamma} + e^{i\gamma}\cdot e^{i\alpha} = 0$$

$$\Rightarrow e^{i(\alpha+\beta)} + e^{i(\beta+\gamma)} + e^{i(\gamma+\alpha)} = 0$$

$$\Rightarrow \cos(\alpha+\beta) + i\sin(\alpha+\beta) + \cos(\beta+\gamma) + i\sin(\beta+\gamma) + \cos(\gamma+\alpha) + i\sin(\gamma+\alpha) = 0$$

$$\Rightarrow [\cos(\alpha+\beta) + \cos(\beta+\gamma) + \cos(\gamma+\alpha)] + i[\sin(\alpha+\beta) + \sin(\beta+\gamma) + \sin(\gamma+\alpha)] = 0$$

Equating the real and imaginary parts to zero we get

$$\boldsymbol{\cos(\alpha+\beta) + \cos(\beta+\gamma) + \cos(\gamma+\alpha) = 0 \quad \text{and} \quad \sin(\alpha+\beta) + \sin(\beta+\gamma) + \sin(\gamma+\alpha) = 0.}$$

Example 20. If $\cos\alpha + \cos\beta + \cos\gamma = 0$, $\sin\alpha + \sin\beta + \sin\gamma = 0$, show that
(a) $\cos 3\alpha + \cos 3\beta + \cos 3\gamma = 3\cos(\alpha+\beta+\gamma)$ (b) $\sin 3\alpha + \sin 3\beta + \sin 3\gamma = 3\sin(\alpha+\beta+\gamma)$.

Solution : Let $a = \cos\alpha + i\sin\alpha = e^{i\alpha}$; $b = \cos\beta + i\sin\beta = e^{i\beta}$; $c = \cos\gamma + i\sin\gamma = e^{i\gamma}$

Now, $\quad a + b + c = (\cos\alpha + \cos\beta + \cos\gamma) + i(\sin\alpha + \sin\beta + \sin\gamma)$

But, $\quad \cos\alpha + \cos\beta + \cos\gamma = 0$ and $\sin\alpha + \sin\beta + \sin\gamma = 0$

$\therefore \quad a + b + c = 0 \Rightarrow a^3 + b^3 + c^3 = 3abc.$

$$\Rightarrow (\cos 3\alpha + i\sin 3\alpha) + (\cos 3\beta + i\sin 3\beta) + (\cos 3\gamma + i\sin 3\gamma) = 3e^{i\alpha}\cdot e^{i\beta}\cdot e^{i\gamma}.$$

$$\Rightarrow (\cos 3\alpha + \cos 3\beta + \cos 3\gamma) + i(\sin 3\alpha + \sin 3\beta + \sin 3\gamma) = 3e^{i(\alpha+\beta+\gamma)}.$$

$$\Rightarrow (\cos 3\alpha + \cos 3\beta + \cos 3\gamma) + i(\sin 3\alpha + \sin 3\beta + \sin 3\gamma) = 3[\cos(\alpha+\beta+\gamma) + i\sin(\alpha+\beta+\gamma)]$$

Equating the real and imaginary parts, we get

$$\boldsymbol{\cos 3\alpha + \cos 3\beta + \cos 3\gamma = 3\cos(\alpha+\beta+\gamma) \; ; \quad \sin 3\alpha + \sin 3\beta + \sin 3\gamma = 3\sin(\alpha+\beta+\gamma).}$$

Example 21. If α and β are the roots of $x^2 + 2x + 4 = 0$, show that $\alpha^6 + \beta^6 = 2^7$.

Solution : Consider the equation

$$x^2 + 2x + 4 = 0 \Rightarrow x^2 + 2x + 1 = -3 \Rightarrow (x+1)^2 = -3$$

$$\Rightarrow x + 1 = \pm i\sqrt{3} \qquad (\because \sqrt{-3} = i\sqrt{3})$$

$$\Rightarrow x = -1 \pm i\sqrt{3}$$

Let, $\alpha = -1 + i\sqrt{3}$, $\beta = -1 - i\sqrt{3}$

Now, repeat the procedure of example 8, with $n = 2$, we get

$$\Rightarrow \quad \boldsymbol{(-1 + i\sqrt{3})^6 + (-1 - i\sqrt{3})^6 = 2^{6+1} = 2^7.}$$

Time Saving Results (TSR).

The following are few results which are useful to answer the objective type questions quickly.

1. (i) $\sin\theta + i\cos\theta = i(\cos\theta - i\sin\theta)$; $\sin\theta - i\cos\theta = -i(\cos\theta + i\sin\theta)$

(ii) $(\sin\theta + i\cos\theta)^n = i^n(\cos n\theta - i\sin n\theta)$; $(\sin\theta - i\cos\theta)^n = (-i)^n(\cos n\theta + i\sin n\theta)$

Ex : $(\sin\theta + i\cos\theta)^7 = i^7(\cos 7\theta - i\sin 7\theta) = -i(\cos 7\theta - i\sin 7\theta) = \mathbf{-\sin 7\theta - i\cos 7\theta}$

$(\sin\theta - i\cos\theta)^6 = (-i)^6(\cos 6\theta + i\sin 6\theta) = \mathbf{-(\cos 6\theta + i\sin 6\theta)}$ $(\because i^6 = -1)$

2. $(\sin\theta \pm i\cos\theta)^n = \sin n\theta \pm i\cos n\theta$, **if** $4|(n-1)$, **i.e.,** $n \equiv 1 \pmod 4$

Ex : $(\sin\theta + \cos\theta)^{17} = \sin 17\theta + i\cos 17\theta$ $\quad\because 4|(17-1) - 1$ i.e., $4|16$

$(\sin\theta - i\cos\theta)9 = \sin 9\theta - i\cos 9\theta$ $\quad\because 4|(9-1) - 1$ i.e., $4|8$

3. (i) $(a+ib)^n + (a-ib)^n = 2\left(a^2+b^2\right)^{n/2}\cos\left(n\tan^{-1}\frac{b}{a}\right)$

(ii) $(a+ib)^n - (a-ib)^n = 2i\left(a^2+b^2\right)^{n/2}\sin\left(n\tan^{-1}\frac{b}{a}\right)$

Ex : $$\left(1+i\sqrt{3}\right)^{27} + \left(1-i\sqrt{3}\right)^{27} = 2(1+3)^{\frac{27}{2}}\cos\left(27\tan^{-1}\sqrt{3}\right)$$

$$= 2\cdot 2^{27}\cos\left(27\frac{\pi}{3}\right) = \mathbf{2^{28}\cdot(-1)} \qquad (\because \cos 9\pi = -1)$$

$$\left(\sqrt{3}+i\right)^{20} - \left(\sqrt{3}-i\right)^{20} = 2i(3+1)^{20/2}\sin\left(20\tan^{-1}\frac{1}{\sqrt{3}}\right)$$

$$= 2i\cdot 2^{20}\sin\left(\frac{10\pi}{3}\right) = 2^{21}i\cdot\sin\left(4\pi - \frac{2\pi}{3}\right)$$

$$= 2^{21}\cdot i\cdot\sin\left(\frac{2\pi}{3}\right) = \mathbf{2^{21}\cdot\sqrt{3}\,i} \qquad \left(\because \sin\frac{2\pi}{3} = \frac{\sqrt{3}}{2}\right)$$

Exercise

I.

1. Simplify the following

(a) $\left(\cos\frac{\pi}{5} + i\sin\frac{\pi}{5}\right)^5$ **(b)** $(\cos 3\pi + i\sin 3\pi)^9$ **(c)** $\left(\cos\frac{3\pi}{9} + i\sin\frac{3\pi}{9}\right)^{7/6}$

(d) $\left(\cos\frac{3\pi}{5} + i\sin\frac{3\pi}{5}\right)^5$ **(e)** $\left(\cos\frac{3\pi}{4} + i\sin\frac{3\pi}{4}\right)^{20}$ **(f)** $(\sin\theta + i\cos\theta)^{-4}$

(g) $(\sin 4\theta + i\cos 4\theta)^{-4}$ **(h)** $(\cos\theta - i\sin\theta)^5(\cos\theta + i\sin\theta)^2$

(i) $\dfrac{(\sin\theta + i\cos\theta)^4}{(\cos\theta - i\sin\theta)^4}$ **(j)** $(\sin 2\theta - i\cos 2\theta)^5$

II.

1. Simplify the following

(a) $\left(\dfrac{\cos 4\theta + i\sin 4\theta}{\cos 2\theta + i\sin 2\theta}\right)^3$ **(b)** $\left(\dfrac{\cos\frac{\pi}{4} + i\sin\frac{\pi}{4}}{\cos\frac{\pi}{3} + i\sin\frac{\pi}{3}}\right)^6$ **(c)** $\dfrac{\left(\cos\frac{\pi}{3} + i\sin\frac{\pi}{3}\right)^5}{\left(\cos\frac{\pi}{4} + i\sin\frac{\pi}{4}\right)^4}$

(d) $\dfrac{(\cos 4\theta + i\sin 4\theta)^3}{(\cos 2\theta - i\sin 2\theta)^4}$ **(e)** $\dfrac{(\cos\theta + i\sin\theta)^4}{(\sin\theta + i\cos\theta)^5}$ **(f)** $\dfrac{(\cos\theta + i\sin\theta)^7}{(\cos\theta - i\sin\theta)^4}$ **(g)** $\dfrac{\left(\sin\frac{\pi}{8} + i\cos\frac{\pi}{8}\right)^8}{\left(\sin\frac{\pi}{8} - i\cos\frac{\pi}{8}\right)^8}$

2. Simplify the following

(a) $\dfrac{(\cos 7\theta + i\sin 7\theta)^{-2}(\cos 3\theta + i\sin 3\theta)^4}{(\cos\theta + i\sin\theta)^{-4}}$ **(b)** $\dfrac{(\cos 5\theta - i\sin 5\theta)^2(\cos 7\theta + i\sin 7\theta)^{-3}}{(\cos 4\theta - i\sin 4\theta)^4(\cos\theta + i\sin\theta)^2}$

(c) $\dfrac{(\cos 2\theta + i\sin 2\theta)^2(\cos\theta - i\sin\theta)^3}{(\cos 2\theta - i\sin 2\theta)^{11}(\cos 3\theta - i\sin 3\theta)^5}$ **(d)** $\dfrac{(\cos\theta + i\sin\theta)^3(\cos\theta - i\sin\theta)^{-8}}{(\cos 4\theta + i\sin 4\theta)^5(\cos 3\theta + i\sin 3\theta)^2}$

(e) $\dfrac{(\cos 4\theta + i\sin 4\theta)^3(\cos 5\theta + i\sin 5\theta)^2}{(\cos 6\theta + i\sin 6\theta)^2(\cos\theta - i\sin\theta)^5}$ **(f)** $\dfrac{(\cos 2\alpha + i\sin 2\alpha)^3(\cos\beta + i\sin\beta)^{-4}}{(\cos\alpha - i\sin\alpha)^4(\cos 3\beta - i\sin 3\beta)^2}$

3. Find the modulus and amplitude of the following

(a) $\left(\sqrt{3} - i\right)^4$ **(b)** $(1 - i)^5$ **(c)** i^7 **(d)** i^4 **(e)** $\left(1 + i\sqrt{3}\right)^4$

(f) $(1 + \sin\alpha + i\cos\alpha)^5$ **(g)** $(1 - \cos\theta + i\sin\theta)^4$ **(h)** $(1 + \cos\theta - i\sin\theta)^5$

4. Show that

(a) $(1 - i)^9 = 16 - 16i$ **(b)** $\left(\sqrt{3} + i\right)^7 = -64\left(\sqrt{3} + i\right)$

(c) $(1 + i)^6 = -8i$ **(d)** $\left(-1 + i\sqrt{3}\right)^7 = 64\left(-1 + i\sqrt{3}\right)$

III.

1. Show that

(a) $\left(1 + i\sqrt{3}\right)^7 + \left(1 - i\sqrt{3}\right)^5 = 2^5$ **(b)** $\left(\sqrt{3} + i\right)^4 - \left(\sqrt{3} - i\right)^4 = 16\sqrt{3}i$

(c) $\left(\sqrt{3} + i\right)^6 - \left(\sqrt{3} - i\right)^6 = 0$ **(d)** $(1 + i)^n + (1 - i)^n = 2^{\frac{n+2}{2}}\cos\left(\dfrac{n\pi}{4}\right)$

(e) $(1 + i)^5 + (1 - i)^5 = -8$ **(f)** $\left(1 + i\sqrt{3}\right)^n + \left(1 - i\sqrt{3}\right)^n = 2^{n+1}\cos\left(\dfrac{n\pi}{3}\right)$

(g) $\left(1 + i\sqrt{3}\right)^7 + \left(1 - i\sqrt{3}\right)^7 = 128$ **(h)** $\left(1 + i\sqrt{3}\right)^5 + \left(1 - i\sqrt{3}\right)^5 = 32$

(i) $\left(\sqrt{3} + i\right)^6 + \left(\sqrt{3} - i\right)^6 = -2^7$ **(j)** $\left(\sqrt{3} + i\right)^5 + \left(\sqrt{3} - i\right)^5 = -32\sqrt{3}$

2. Show that

(a) $\left(\dfrac{1-\cos\dfrac{\pi}{10}+i\sin\dfrac{\pi}{10}}{1-\cos\dfrac{\pi}{10}-i\sin\dfrac{\pi}{10}}\right)^{10}=-1$ **(b)** $\left(\dfrac{1-\cos\theta+i\sin\theta}{1-\cos\theta-i\sin\theta}\right)^{n}=\cos(n\pi-n\theta)+i\sin(n\pi-n\theta)$

(c) $\left(\dfrac{1+\sin\dfrac{\pi}{8}+i\cos\dfrac{\pi}{8}}{1+\sin\dfrac{\pi}{8}-i\cos\dfrac{\pi}{8}}\right)^{1/2}=\cos\dfrac{3\pi}{16}+i\sin\dfrac{3\pi}{16}$ **(d)** $\left(\dfrac{1-i\tan\theta}{1+i\tan\theta}\right)^{n}=\dfrac{1-i\tan n\theta}{1+i\tan n\theta}$

3. If $x=\cos\alpha+i\sin\alpha,\ y=\cos\beta+i\sin\beta$, show that

(a) $xy+\dfrac{1}{xy}=2\cos(\alpha+\beta)$ **(b)** $xy-\dfrac{1}{xy}=2i\sin(\alpha+\beta)$

(c) $\dfrac{x^2}{y^3}-\dfrac{y^3}{x^2}=2\cos(2\alpha-3\beta)$ **(d)** $\dfrac{x^m}{y^n}-\dfrac{y^n}{x^m}=2i\sin(m\alpha-n\beta)$

(e) $\dfrac{x^m}{y^n}+\dfrac{y^n}{x^m}=2\cos(m\alpha-n\beta)$ **(f)** $\dfrac{x^6}{y^6}-\dfrac{y^6}{x^6}=2i\sin(6\alpha-6\beta)$

4. If $x=\cos\alpha+i\sin\alpha,\ y=\cos\beta+i\sin\beta,\ z=\cos\gamma+i\sin\gamma$, show that

(a) $\dfrac{x^5y^2}{z^3}+\dfrac{z^3}{x^5y^2}=2\cos(5\alpha+2\beta-2\gamma)$ **(b)** $\dfrac{x^3y^5}{z^4}-\dfrac{z^4}{x^3y^5}=2i\sin(3\alpha+5\beta-4\gamma)$

(c) $\dfrac{xy}{z}-\dfrac{z}{xy}=2i\sin(\alpha+\beta-\gamma)$.

5. If $a=\cos 2\alpha+i\sin 2\alpha$ and $b=\cos 2\beta+i\sin 2\beta$, show that

(a) $\sqrt{\dfrac{a}{b}}+\sqrt{\dfrac{b}{a}}=2\cos(\alpha-\beta)$ **(b)** $\sqrt{\dfrac{a}{b}}-\sqrt{\dfrac{b}{a}}=2i\sin(\alpha-\beta)$

6. If $x+\dfrac{1}{x}=2\cos\alpha$ and $y+\dfrac{1}{y}=2\cos\beta$ show that

(a) $x^my^n+\dfrac{1}{x^my^n}=2\cos(m\alpha+n\beta)$ **(b)** $x^my^n-\dfrac{1}{x^my^n}=2i\sin(m\alpha+n\beta)$

7. If $\cos\alpha+\cos\beta+\cos\gamma=0=\sin\alpha+\sin\beta+\sin\gamma$, show that

(a) $\cos 2\alpha+\cos 2\beta+\cos 2\gamma=0$
$\sin 2\alpha+\sin 2\beta+\sin 2\gamma=0$

(b) $\cos^2\alpha+\cos^2\beta+\cos^2\gamma=3/2$
$\sin^2\alpha+\sin^2\beta+\sin^2\gamma=3/2$

8. If $\cos\alpha+2\cos\beta+3\cos\gamma=0,\ \sin\alpha+2\sin\beta+3\sin\gamma=0$ show that
$\cos 3\alpha+8\cos 3\beta+27\cos 3\gamma=18\cos(\alpha+\beta+\gamma)$; $\sin 3\alpha+8\sin 3\beta+27\sin 3\gamma=18\sin(\alpha+\beta+\gamma)$

9. If $x^2-2x\cos\theta+1=0$, show that $x^{2n}-2x^n\cos n\theta+1=0$.

10. If α and β are the roots of the equation

(a) $x^2-2x+4=0$ show that $\alpha^n+\beta^n=2^{n+1}\cos\dfrac{n\pi}{3}$

(b) $x^2-2x+2=0$ show that $\alpha^n+\beta^n=2^{\frac{n+2}{2}}\cos\dfrac{n\pi}{4}$

Answers

I. 1. **(a)** -1 **(b)** -1 **(c)** $\text{cis}\ \frac{7\pi}{18}$ **(d)** -1 **(e)** -1 **(f)** $\text{cis}\ 4\theta$

(g) $\text{cis}\ 16\theta$ **(h)** $\text{cis}\ (-3\theta)$ **(i)** 1 **(j)** $\sin 10\ \theta - i \cos 10\ \theta$

II.1. **(a)** $\text{cis}\ 6\theta$ **(b)** i **(c)** $-\frac{1}{2} + i\frac{\sqrt{3}}{2}$ **(d)** $\text{cis}\ 20\theta$

(e) $\sin 9\theta - i \cos 9\theta$ **(f)** $\text{cis}\ 11\theta$ **(g)** 1

2. **(a)** $\text{cis}\ 2\theta$ **(b)** $\text{cis}\ (-17\theta)$ **(c)** $\text{cis}\ 44\theta$ **(d)** $\text{cis}\ (-15\theta)$ **(e)** $\text{cis}\ 15\theta$ **(f)** $\text{cis}\ (10\alpha + 2\beta)$

3. **(a)** $16, -\frac{2\pi}{3}$ **(b)** $4\sqrt{2}, -\frac{5\pi}{4}$ **(c)** $1, -\frac{\pi}{2}$ **(d)** $1, 0$ **(e)** $16, -\frac{2\pi}{3}$

(f) $32 \cos^{10}\left(\frac{\pi}{4} - \frac{\alpha}{2}\right), \left(\frac{\pi}{4} - \frac{5\alpha}{2}\right)$ **(g)** $16 \sin^4 \frac{\theta}{2}, -2\theta$ **(h)** $2^5 \cos\frac{\theta}{2}, -\frac{5\theta}{2}$

UNIT III

Analytical Geometry

- **Chapter 1** – Coordinate System in a Plane
- **Chapter 2** – Locus of a Point
- **Chapter 3** – Straight Line
- **Chapter 4** – Circles
- **Chapter 5** – Conic Section

Chapter 1

Coordinate System in a Plane

1.1 Introduction

In mathematics we have, in general, algebraic concept and geometrical concept. Each one can be understood independently. However the geometrical concepts can be transformed into algebraic one, and further the geometrical problems can be solved through algebraic methods. Study of geometry through algebra, is the subject called **Analytical Geometry** or **Coordinate Geometry.** The algebraic method of discussing the geometrical problems was first introduced by the French mathematician **Rane Descartes (1596 – 1650)**. To develop the subject matter first we represent each point in a plane by an ordered pair of real numbers called **coordinates** of the point. Then each geometrical curve can be expressed as an algebraic equation in the variables x and y and can be studied through algebraic methods.

1.2 Rectangular Cartesian Coordinate System

Let $X'OX$ and YOY' be two lines perpendicular to each other intersecting at O.

The point O is called the **origin**, the horizontal line $X'OX$ is called **x-axis** and the vertical line YOY' is called **y-axis**.

The origin O, together with the x-axis and y-axis, is called **rectangular Cartesian coordinate system**. The x and y axes are called **reference axes** or **coordinate axes**.

We shall follow the following conventions for the measurements of distances along the axes.

1. All the distances are measured from the origin along the x-axis or along the y-axis

2. The distances measured along OX and OY are taken as positive, where as those along OX' and OY' are taken as negative.

Let P be any point in the plane containing the coordinate axes. From P draw PM perpendicular to x-axis and PN perpendicular to y-axis. Let $OM = x$ and $ON = y$. For a given point P, these two distances are fixed. Thus we denote the point P by the ordered pair (x, y) and write $P = (x, y)$. The length $OM = x$ is called **x-coordinate** or **abscissa** of the point P and the length $ON = MP$ is called **y-coordinate** or **the ordinate** of the point P.

Note : For a given point, the x-coordinate (abscissa) of it is the distance of the point from y-axis where as the y-coordinate (ordinate) is the distance of the point from x-axis.

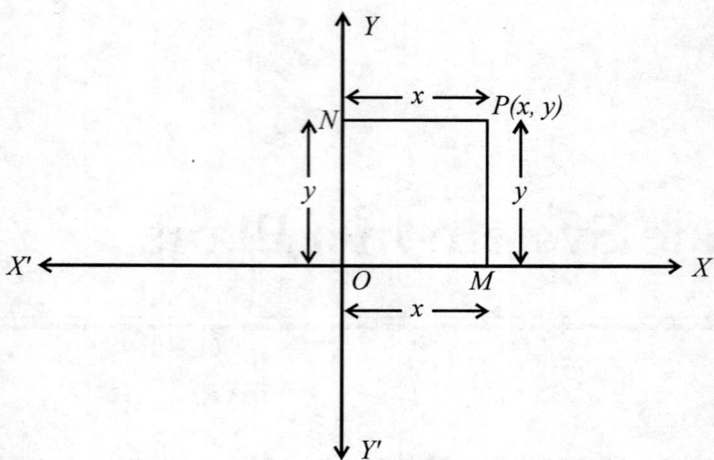

The coordinate axes $X'OX$ and YOY' divide the plane into four parts called **Quadrants**. The region XOY is called **first quadrant**, the region YOX' is called **second quadrant**, the region $X'OY'$ is called **third quadrant** and the region $Y'OX$ is called **fourth quadrant**.

According to our convention of signs of measurement of distances along the X and Y axes, the following table gives the signs of x and y coordinates of the points in different quadrants.

Quadrant	*x*-coordinate	*y*-coordinate
First	+	+
Second	–	+
Third	–	–
Fourth	+	–

Note : 1. The coordinates of the origin are (0,0).

2. The coordinates of any point on x-axis is of the form $(x, 0)$. That is for any point on the x-axis, the y coordinate is zero.

3. The coordinates of any point on y-axis is of the form $(0, y)$. That is for any point on the y-axis, the x-coordinate is zero.

Example 1. Mark the points $P = (2, 3)$, $Q = (5, -4)$, $R = (-3, -4)$ and $S = (-4, 5)$ on the plane.

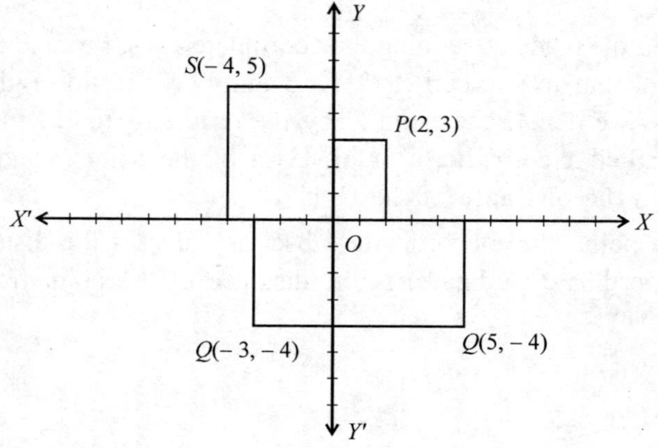

Example 2. Mark the points $A = (3, -4)$, $B = (-4, 2)$ on a plane. Find the reflection of A in the x-axis and that of B in y-axis

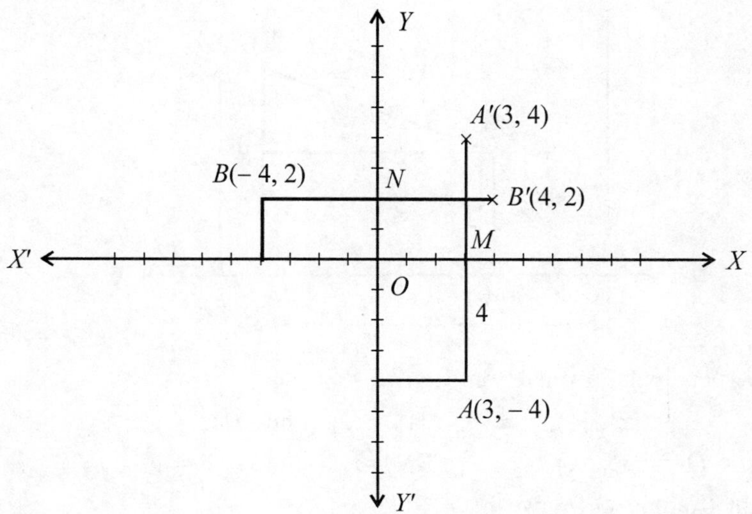

To find the image of $A(3, -4)$ in x-axis, produce AM to A' such that $MA' = 4$ units. We arrive at the point $A' = (3, 4)$ which is the image of A in x-axis.

Similarly to find the image of B in y-axis produce BN to B' such that $NB' = 4$ units, we arrive at the point $B' = (4, 2)$, which is the image of B in y-axis.

Note : Image of the point $A(x_1, y_1)$ in x-axis is $A' = (x_1, -y_1)$. Image of the point $A(x_1, y_1)$ in y-axis is $B' = (-x_1, y_1)$. For example, image of the point $P(2, 3)$ in x-axis is $P' = (2, -3)$ and the image of the point $Q = (-4, -5)$ in y-axis $Q' = (4, -5)$.

Exercise

1. Plot the points $A(3, 5)$, $B(-3, 4)$, $C(4, -1)$, $D(0, 2)$, $E(-4, 0)$.
2. Write the coordinates of images of the points $(3, 4)$, $(-7, 5)$, $(1, -3)$ in x-axis.
3. Write the coordinates of images of the points $(4, -3)$, $(-2, 5)$, $(-4, 0)$ in y-axis.
4. In which quadrant do the following points lie?
   (a) $(3, -5)$ **(b)** $(-4, 3)$ **(c)** $(-2, -1)$ **(d)** $(1, 4)$.
5. What are the distances of the following points from the coordinate axes
   (a) $(4, 3)$ **(b)** $(-3, 5)$ **(c)** $(-5, 2)$.

Answers

2. $(3, -4)$, $(-7, -5)$, $(1, 3)$ **3.** $(-4, -3)$, $(2, 5)$, $(4, 0)$

4. **(a)** 4th quadrant **(b)** 2nd quadrant **(c)** 3rd quadrant **(d)** 1st quadrant

5. **(a)** 4 units from y-axis, 3 units from x-axis **(b)** 3 units from y-axis, 5 units from x-axis
(c) 5 units from y-axis, 2 units from x-axis

1.3 Distance between Two Points (Distance Formula)

We know that the distance between the two points in the plane is the length of the line segment joining them.

To find the distance between the points $P(x_1, y_1)$ and $Q(x_2, y_2)$

Draw PM and QN perpendiculars to x-axis.

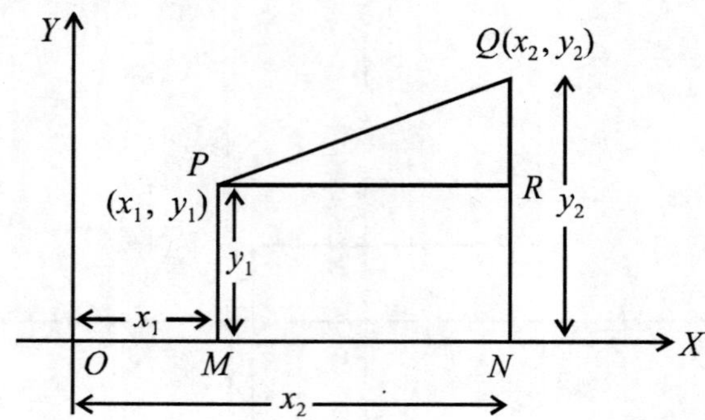

Let PR be perpendicular from P onto QN

Now $\quad P = (x_1, y_1) \Rightarrow OM = x_1$ and $PM = y_1$

$\quad Q = (x_2, y_2) \Rightarrow ON = x_2$ and $QN = y_2$

From the right triangle QRP, we have

$$PQ^2 = PR^2 + QR^2 \Rightarrow PQ^2 = MN^2 + QR^2 \quad (\because PR = MN)$$

$$\Rightarrow PQ^2 = (ON - OM)^2 + (QN - RN)^2$$

$$\Rightarrow PQ^2 = (ON - OM)^2 + (QN - PM)^2 \quad (\because RN = PM)$$

$$\Rightarrow PQ^2 = (x_2 - x_1)^2 + (y_2 - y_1)^2$$

Thus, the distance between the points $P(x_1, y_1)$ and $Q(x_2, y_2)$ is given by

$$|PQ| = \sqrt{(x_2 - x_1)^2 + (y_2 - y_1)^2} \quad .... (1)$$

Note : 1. The distance between the origin $O(0, 0)$ and any point $P(x_1, y_1)$ is given by

$$|OP| = \sqrt{(0 - x_1)^2 + (0 - y_1)^2} = \sqrt{x_1^2 + y_1^2}$$

2. The result proved above holds whatever may be the quadrant in which the points P and Q lie.

3. The formula (1) is known as **distance formula**.

Example 1. Find the distance between the points

(a) (2, 3) and (1, 3) **(b) (0, –2) and (–1, 0)**

(c) Origin and (–2, 3) **(d) (–3, 4) and (4, –5)**

Solution : (a) Let, $P = (2, 3) = (x_1, y_1)$ and $Q = (1, 3) = (x_2, y_2)$

Now, $|PQ| = \sqrt{(x_2 - x_1)^2 + (y_2 - y_1)^2} \Rightarrow |PQ| = \sqrt{(1-2)^2 + (3-3)^2} = \sqrt{1^2} = \mathbf{1}$

(b) Let, $A = (0, -2)$ and $B = (-1, 0)$

Now $\quad |AB| = \sqrt{(-1-0)^2 + (0+2)^2} = \sqrt{1+4} = \mathbf{\sqrt{5}}$

(c) Let $M = (-2, 3)$ and $O = (0, 0)$

Now, $\quad |OM| = \sqrt{(-2)^2 + 3^2} = \sqrt{4+9} = \mathbf{\sqrt{13}}$

(d) Let $R = (-3, 4)$ and $S = (4, -5)$

Now $|RS| = \sqrt{(4+3)^2 + (-5-4)^2} = \sqrt{49+81} = \mathbf{\sqrt{130}}$

Example 2. Find the value of '*a*' if the distance between the points (a, 2) and (3, 4) is $\sqrt{8}$ units.

Solution : Let $P = (a, 2)$ and $Q = (3, 4)$

By data, $|PQ| = \sqrt{8} \Rightarrow \sqrt{(3-a)^2 + (4-2)^2} = \sqrt{8} \Rightarrow (3-a)^2 + 2^2 = 8$

$\Rightarrow a^2 - 6a + 9 + 4 = 8$

$\Rightarrow a^2 - 6a + 5 = 0$

$\Rightarrow (a-5) \cdot (a-1) = 0$

$\Rightarrow \mathbf{a = 5}$ or $\mathbf{a = 1}$

Example 3. Find a point on the x-axis such that its distance from (3, –2) is 2 units.

Solution : Let $A = (3, -2)$. Let $B = (a, 0)$ be the point on the x-axis such that $|AB| = 2$ units (note that any point on the x-axis, y-coordinate is zero).

By data, $|AB| = 2 \Rightarrow AB^2 = 4 \Rightarrow (3-a)^2 + (-2-0)^2 = 4$

$\Rightarrow a^2 - 6a + 9 + 4 - 4 = 0$

$\Rightarrow a^2 - 6a + 9 = 0 \Rightarrow (a-3)^2 = 0 \Rightarrow \mathbf{a = 3}$

Thus the required point is (3, 0).

Example 4. Find a point on the x-axis which is equidistant from the points (7, 6) and (–3, 4).

Solution : Let $A = (7, 6)$, $B = (-3, 4)$.

Let $P(a, 0)$ be the point on the x-axis such that $PA = PB$.

Now, $PA = PB \Rightarrow PA^2 = PB^2 \Rightarrow (a-7)^2 + (0-6)^2 = (a+3)^2 + (0-4)^2$

$\Rightarrow a^2 - 14a + 49 + 36 = a^2 + 6a + 9 + 16$

$\Rightarrow -20a = -60 \Rightarrow \mathbf{a = 3}$

Thus the required point is **(3, 0)**.

Example 5. Show that the points (1, –1), (5, 2) and (9, 5) are collinear.

Solution : We know that the three points A, B and C taken in this order are collinear if and only if $AB + BC = AC$.

Let, $A = (1, -1)$, $B = (5, 2)$, $C = (9, 5)$.

Consider, $AB = \sqrt{(5-1)^2 + (2+1)^2} = \sqrt{16+9} = 5$

$BC = \sqrt{(9-5)^2 + (5-2)^2} = \sqrt{16+9} = 5$

$AC = \sqrt{(9-1)^2 + (5+1)^2} = \sqrt{64+36} = 10$

Clearly, $AB + BC = AC$

$\Rightarrow$ **The points A, B and C are collinear.**

Example 6. Show that the points (1, –1), (–1, 1) and $(-\sqrt{3}, -\sqrt{3})$ are the vertices of an equilateral triangle.

Solution : Let $A = (1, -1)$, $B = (-1, 1)$, $C = (-\sqrt{3}, -\sqrt{3})$

Consider, $AB = \sqrt{(-1-1)^2 + (1+1)^2} = \sqrt{4+4} = \sqrt{8}$

$BC = \sqrt{(-\sqrt{3}+1)^2 + (-\sqrt{3}-1)^2} = \sqrt{4-2\sqrt{3}+4+2\sqrt{3}} = \sqrt{8}$

$CA = \sqrt{(-\sqrt{3}-1)^2 + (-\sqrt{3}+1)^2} = \sqrt{4+2\sqrt{3}+4-2\sqrt{3}} = \sqrt{8}$

Thus, $AB = BC = CA$. Hence, the points A, B, C are the vertices of an **equilateral triangle**.

Example 7. Show that the points (2, 2), (6, 3) and (4, 11) form a right angled triangle.

Solution : Let $A = (2, 2)$, $B = (6, 3)$, $C = (4, 11)$

Consider, $AB = \sqrt{(6-2)^2 + (3-2)^2} = \sqrt{16+1} = \sqrt{17}$

$BC = \sqrt{(4-6)^2 + (11-3)^2} = \sqrt{4+64} = \sqrt{68}$

$AC = \sqrt{(4-2)^2 + (11-2)^2} = \sqrt{4+81} = \sqrt{85}$

Now, $AB^2 = 17$, $BC^2 = 68$ and $CA^2 = 85$

Clearly, $CA^2 = AB^2 + BC^2$

Hence, the points A, B, C form a **right angled triangle.**

Example 8. Show that the points (7, 9), (3, –7) and (–3, 3) are the vertices of a right angled isosceles triangle.

Solution : Let $A = (7, 9)$, $B = (3, -7)$, $C = (-3, 3)$

Now $AB^2 = (3-7)^2 + (-7-9)^2 = 16 + 256 = \mathbf{272}$

$BC^2 = (-3-3)^2 + (3+7)^2 = 36 + 100 = \mathbf{136}$

$CA^2 = (-3-7)^2 + (3-9)^2 = 100 + 36 = \mathbf{136}$

Clearly, $AB^2 = BC^2 + CA^2$

Thus, ABC is a **right angled triangle**.

Also, $BC^2 = CA^2 \Rightarrow |BC| = |CA| \Rightarrow ABC$ is an isosceles triangle. Hence, the given three points form a **right angled isosceles triangle**.

Example 9. Show that the points (–2, –1), (1, 0), (4, 3) and (1, 2) are the vertices of a parallelogram.

Solution : Let $A = (-2, -1)$, $B = (1, 0)$, $C = (4, 3)$ and $D = (1, 2)$

Consider, $AB = \sqrt{(1+2)^2 + (0+1)^2} = \sqrt{9+1} = \sqrt{10}$

$BC = \sqrt{(4-1)^2 + (3-0)^2} = \sqrt{9+9} = \sqrt{18}$

$CD = \sqrt{(1-4)^2 + (2-3)^2} = \sqrt{9+1} = \sqrt{10}$

and $DA = \sqrt{(1+2)^2 + (2+1)^2} = \sqrt{9+9} = \sqrt{18}$

We have, $AB = CD$ and $BC = DA$. That is opposites sides are equal. Thus $ABCD$ is a **parallelogram**.

Example 10. Show that the points (2, –2), (8, 4), (5, 7) and (–1, 1) are the vertices of a rectangle.

Solution : Let $A = (2, -2)$, $B = (8, 4)$, $C = (5, 7)$ and $D = (-1, 1)$

Consider,
$$AB = \sqrt{(8-2)^2 + (4+2)^2} = \sqrt{36+36} = \sqrt{\mathbf{72}}$$
$$BC = \sqrt{(5-8)^2 + (7-4)^2} = \sqrt{9+9} = \sqrt{\mathbf{18}}$$
$$CD = \sqrt{(-1-5)^2 + (1-7)^2} = \sqrt{36+36} = \sqrt{\mathbf{72}}$$
$$DA = \sqrt{(-1-2)^2 + (1+2)^2} = \sqrt{9+9} = \sqrt{\mathbf{18}}$$

Thus $AB = CD$ and $BC = DA$. That is opposites sides are equal. This shows that $ABCD$ is a parallelogram. In particular if we have to show that it is a rectangle, we have to show that the diagonals are equal.

$$AC = \sqrt{(5-2)^2 + (7+2)^2} = \sqrt{9+81} = \sqrt{\mathbf{90}}\ ;\ BD = \sqrt{(8+1)^2 + (4-1)^2} = \sqrt{81+9} = \sqrt{\mathbf{90}}$$

Clearly, the diagonals are equal. Therefore, $ABCD$ is a **rectangle**

Example 11. Show that the points (2, –1), (3, 4), (–2, 3) and (–3, –2) form a rhombus.

Solution : Let $A = (2, -1)$, $B = (3, 4)$, $C = (-2, 3)$ and $D = (-3, -2)$

Consider,
$$AB = \sqrt{(3-2)^2 + (4+1)^2} = \sqrt{1+25} = \sqrt{\mathbf{26}}$$
$$BC = \sqrt{(-2-3)^2 + (3-4)^2} = \sqrt{25+1} = \sqrt{\mathbf{26}}$$
$$CD = \sqrt{(-3+2)^2 + (-2-3)^2} = \sqrt{1+25} = \sqrt{\mathbf{26}}$$
$$DA = \sqrt{(-3-2)^2 + (-2+1)^2} = \sqrt{25+1} = \sqrt{\mathbf{26}}$$

Thus $AB = BC = CD = AD$. That is all the sides are equal. To show that $ABCD$ is a rhombus, we have to show that the diagonals are not equal.

Consider,
$$AC = \sqrt{(-2-2)^2 + (3+1)^2} = \sqrt{16+16} = \sqrt{\mathbf{32}}$$
$$BD = \sqrt{(-3-3)^2 + (-2-4)^2} = \sqrt{36+36} = \sqrt{\mathbf{72}}$$

Thus, $AC \neq BD$. That is diagonals are not equal. Therefore, $ABCD$ is a **rhombus**.

Example 12. The points (x, 2) is equidistant from (8, –2) and (2, –2). Find the value of x.

Solution : Let $A = (x, 2)$, $B = (8, -2)$, $C = (2, -2)$

By data, $AB = AC \Rightarrow AB^2 = AC^2$
$$\Rightarrow (8-x)^2 + (-2-2)^2 = (x-2)^2 + (2+2)^2$$
$$\Rightarrow x^2 - 16x + 64 + 16 = x^2 - 4x + 4 + 16$$
$$\Rightarrow -12x = -60 \Rightarrow \boldsymbol{x = 5}$$

Exercise

I.

1. **Find the distance between the following pair of points.**

   (i) $(1, 1), (-1, 1)$ **(ii)** $(-3, 7), (-8, 15)$ **(iii)** $(0, 0), (1, -3)$.

   (iv) $(a, 0), (0, b)$ **(v)** $(a + b, a + b), (a - b, a - b)$ **(vi)** $(at_1^2, 2at_1), (at_2^2, 2at_2)$

   (vii) $(\sqrt{3}, 1), (\sqrt{2}, \sqrt{2})$ **(viii)** (cos α, −□sin α), (−cos α, □sin α).

2. If the distance between the points $(a, 1)$ and $(8, a)$ is 5 units, show that $a = 5$ or 4.
3. Find the values of k, if the distance between $(k, -4)$ and $(-8, 2)$ be 10 units.
4. If the distance between $(a, 3)$ and $(4, 5)$ is $\sqrt{5}$ find a.
5. If the distance between $(a, 2)$ and $(3, 4)$ is $2\sqrt{2}$ show that a = 5 or 1.
6. Find a point on the x - axis, such that its distance from $(3, -2)$ is 2 units.
7. Find a point on the y - axis, such that its distance from $(2, -5)$ is $\sqrt{5}$ units.

II.

1. Show that the points $(7, 10)$ is equidistant from $(-10, -9)$, $(32, 5)$ and $(18, 33)$.
2. Show that the points $(11, 2)$ is equidistant from $(1, 2)$, $(5, -6)$, $(3, -4)$.
3. If the point (x, y) is equidistant from the points $(a + b, b - a)$, and $(b - a, a + b)$ show that $bx = ay$.
4. If $P(6, -1)$, $Q(1, 3)$ and $R(x, 8)$ are such that $PQ = QR$, find the value of x.
5. If the point (x, y) is equidistant from $(6, -1)$ and $(2, 3)$ find the relation between x and y.
6. If $A(a, 0)$ is equidistant from $(-4, 6)$ and $(14, -2)$ find x.
7. Show that the points A, B and C are collinear

   (a) $A = (-2, 3)$, $B = (1, 2)$, $C = (7, 0)$ **(b)** $A = (-2, 3)$, $B = (-6, 4)$, $C = (-10, 5)$

   (c) $A = (2, -4)$, $B = (4, -2)$, $C = (7, 1)$.

III. Essay Type Questions.

1. Show that the points $(1, 1)$, $(2, 3)$ and $(5, -1)$ form a right angled triangle.
2. Show that the points

   (a) $(2, 5)$, $(5, 2)$ and $(6, 6)$ are vertices of an isosceles triangle

   (b) $(7, 9)$, $(3, -7)$ and $(-3, 3)$ are vertices of a right angled triangle

   (c) $(3, 2)$, $(1, 0)$ and $(2 - \sqrt{3}, 1 + \sqrt{3})$ are vertices of an equilateral triangle

   (d) $(3, 2)$, $(6, 3)$ and $(4, 11)$ are vertices of a right angled triangle

   (e) $(7, 10)$, $(-2, 5)$ and $(3, -4)$ are vertices of an isosceles right angled triangle

   (f) $(4, 1)$, $(3, 4)$ and $(2, 1)$ are vertices of an isosceles triangle

   (g) $(2, 4)$, $(2, 6)$ and $(2 + \sqrt{3}, 5)$ are vertices of an equilateral triangle

   (h) $(2a, 4a)$, $(2a, 6a)$ and $(2a + \sqrt{3}a, 5a)$ are vertices of an equilateral triangle.

3. Show that the following points are the vertices of a parallelogram

   (a) $(-1, 0)$, $(3, 1)$, $(2, 2)$ and $(-2, 1)$ **(b)** $(1, -2)$, $(3, 6)$, $(5, 10)$ and $(3, 2)$

   (c) $(-1, 0)$, $(0, 3)$, $(1, 3)$ and $(0, 0)$.

4. Show that the following points are the vertices of a rectangle

(a) (2, –2), (8, 4), (5, 7), (–1, 1) **(b)** (1, 6), (5, 2), (12, 9), (8, 13)

(c) (3, 2), (11, 8), (8, 12), (0, 6) **(d)** (0, – 1), (–2, 3), (6, 7), (8, 3)

5. Show that the following points are the verteces of a square

(a) (2, 6), (5, 1), (0, –2) and (–3, 3) **(b)** (0, – 1), (2, 1), (0, 3) and (–2, 1)

(c) (3, 2), (0, 5), (–3, 2) and (0, – 1).

6. Show that the following points are the verteces of a rhombus

(a) (2, –3), (6, 5), (–2, 1) and (–6, –7) **(b)** (2, – 1), (3, 4), (–2, 3) and (–3, –2)

(c) (0, 5), (–2, –2), (5, 0) and (7, 7).

Answers

I. **1. (i)** $2\sqrt{2}$ **(ii)** $\sqrt{89}$ **(iii)** $\sqrt{10}$ **(iv)** $\sqrt{a^2+b^2}$ **(v)** $2\sqrt{2}\,b$

(vi) $a(t_2-t_1)\sqrt{(t_1+t_2)^2+4}$ **(vii)** $\sqrt{8-2\sqrt{2}(\sqrt{3}+1)}$ **(viii)** 2

3. 0 or –16 **4.** 3 or 5 **6.** (3, 0) **7.** (0, –6), (0, – 4)

II. **4.** $x = 5$ or -3 **5.** $x - y - 3 = 0$ **6.** $x = \dfrac{37}{9}$

1.4 Section Formula

In this section we shall derive a formula to find the coordinates of the point which divides the line joining the two points in the given ratio, internally or externally. First we shall consider the case of internal division.

1. To find the coordinates of the point which divides the line joining the points $A(x_1, y_1)$ and $B(x_2, y_2)$ internally in the ratio $l : m$.

Let $P(x, y)$ be the point on AB which divides AB in the ratio $l : m$, internally.

Thus, $$\frac{AP}{PB} = \frac{l}{m} \Rightarrow m \cdot (AP) = l \cdot (PB)$$

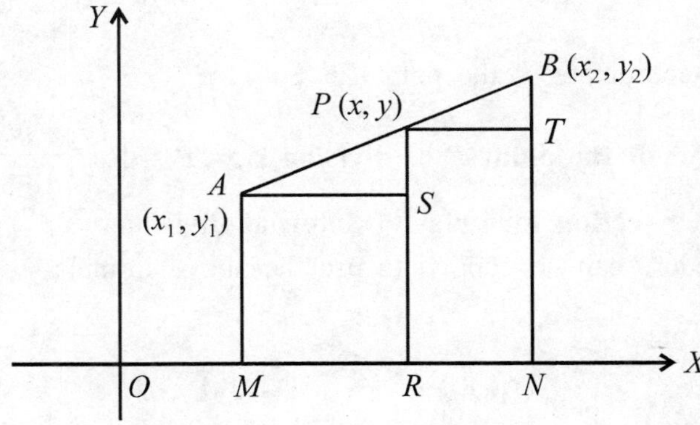

Draw perpendiculars AM, BN and PR from the points A, B and P on to the x-axis, respectively. Draw perpendiculars AS and PT from A on to PR and from P on to BN.

From the figure we have that the triangle ASP is similar to the triangle PTB (because $A\hat{S}P = P\hat{T}B = 90^0$, $P\hat{A}S = B\hat{P}T$, $A\hat{P}S = P\hat{B}T$)

$$\therefore \quad \frac{AS}{PT} = \frac{PS}{BT} = \frac{AP}{PB} \qquad \text{.... (1)}$$

Consider, $\frac{AS}{PT} = \frac{AP}{PB} \Rightarrow \frac{AS}{PT} = \frac{l}{m}$ $\qquad \left(\because \frac{AP}{PB} = \frac{l}{m}\right)$

$$\Rightarrow \frac{MR}{RN} = \frac{l}{m} \Rightarrow \frac{OR - OM}{ON - OR} = \frac{l}{m}$$

$$\Rightarrow \frac{x - x_1}{x_2 - x} = \frac{l}{m}$$

$$\Rightarrow m(x - x_1) = l(x_2 - x)$$

$$\Rightarrow mx + lx = lx_2 + mx_1$$

$$\Rightarrow (l + m)x = lx_2 + mx_1 \Rightarrow \boldsymbol{x = \frac{lx_2 + mx_1}{l + m}}$$

This gives us the x-coordinate of the point P.

Again from (1) consider,

$\frac{PS}{BT} = \frac{AP}{PB} \Rightarrow \frac{PS}{BT} = \frac{l}{m}$ $\qquad \left(\because \frac{AP}{PB} = \frac{l}{m}\right)$

$$\Rightarrow \frac{PR - SR}{BN - TN} = \frac{l}{m} \Rightarrow \frac{PR - AM}{BN - PR} = \frac{l}{m}$$

$$\Rightarrow \frac{y - y_1}{y_2 - y} = \frac{l}{m}$$

$$\Rightarrow m(y - y_1) = l(y_2 - y)$$

$$\Rightarrow my + ly = ly_2 + my_1$$

$$\Rightarrow (l + m)y = ly_2 + my_1 \Rightarrow \boldsymbol{y = \frac{ly_2 + my_1}{l + m}}$$

This gives us the y-coordinate of the point P.

Thus the coordinates of the point P of division is, $\boldsymbol{P = \left(\frac{lx_2 + mx_1}{l + m}, \frac{ly_2 + my_1}{l + m}\right)}$

This formula is called **section formula for internal division.**

The following procedure can be applied to use the above formula

$$\boldsymbol{l \quad : \quad m}$$

$$\boldsymbol{(x_1, y_1) \qquad (x_2, y_2)}$$

$$\boldsymbol{P = \left(\frac{lx_2 + mx_1}{l + m}, \frac{ly_2 + my_1}{l + m}\right)}$$

Corollary: The coordinates of the midpoint M of the line joining the points (x_1, y_1) and (x_2, y_2) is given by

$$M = \left(\frac{x_1 + x_2}{2}, \frac{y_1 + y_2}{2}\right)$$

Proof : The mid point M divides the line joining $A(x_1, y_1)$ and $B(x_2, y_2)$ in the ratio 1:1

Thus by putting $l = 1$ and $m = 1$, in the section formula for internal division we get

$$M = \left(\frac{x_1 + x_2}{2}, \frac{y_1 + y_2}{2}\right)$$

This formula is called **mid point formula.**

2. To find the coordinates of the point which divides the line joining the points $A(x_1, y_1)$ and $B(x_2, y_2)$ externally in the ratio $l : m$.

Let $P(x, y)$ be the point on AB which divides AB in the ratio $l : m$, externally.

$$\therefore \quad \frac{AP}{PB} = \frac{l}{m}$$

The adjoining figure is self explanatory.

From the figure we have that the triangle ASP and BTP are similar.

$$\therefore \quad \frac{AS}{BT} = \frac{PS}{PT} = \frac{AP}{PB} \quad \text{.... (1)}$$

Consider, $\dfrac{AS}{BT} = \dfrac{AP}{PB} \Rightarrow \dfrac{MR}{NR} = \dfrac{l}{m}$ $\quad \left(\because \dfrac{AP}{PB} = \dfrac{l}{m}\right)$

$$\Rightarrow \frac{OR - OM}{OR - ON} = \frac{l}{m} \Rightarrow \frac{x - x_1}{x - x_2} = \frac{l}{m}$$

$$\Rightarrow m(x - x_1) = l(x - x_2)$$

$$\Rightarrow x(l - m) = lx_2 - mx_1 \Rightarrow x = \frac{lx_2 - mx_1}{l - m}$$

This gives us the x-coordinate of the point P.

Again from (1) consider,

$\dfrac{PS}{PT} = \dfrac{AP}{PB} \Rightarrow \dfrac{PR - SR}{PR - TR} = \dfrac{l}{m}$ $\quad \left(\because \dfrac{AP}{PB} = \dfrac{l}{m}\right)$

$$\Rightarrow \frac{PR - AM}{PR - BN} = \frac{l}{m} \Rightarrow \frac{y - y_1}{y - y_2} = \frac{l}{m}$$

$$\Rightarrow m(y - y_1) = l(y - y_2)$$

$$\Rightarrow y(l - m) = ly_2 - my_1 \Rightarrow y = \frac{ly_2 - my_1}{l - m}$$

This gives us the y-coordinate of the point P of division.

Thus the coordinates of the point P of division is $P = \left(\dfrac{lx_2 - mx_1}{l - m}, \dfrac{ly_2 - my_1}{l - m}\right)$

This formula is called **section formula for external division.**

The following procedure will help to write the formula

$$\begin{array}{ccc} l & : & m \\ (x_1, y_1) & & (x_2, y_2) \end{array}$$

$$P = \left(\frac{lx_2 - mx_1}{l - m}, \frac{ly_2 - my_1}{l - m}\right)$$

As an application of the section formula, we can find the coordinates of the centroid of the triangle whose vertices are known.

3. To find the coordinates of the centroid of the triangle whose vertices are $A(x_1, y_1)$, $B(x_2, y_2)$ and $C(x_3, y_3)$.

Let G be the centroid of the triangle ABC. That is G is the point of intersection of the median of a triangle.

We know that the centroid G divides the median AD in the ratio 2 : 1 internally.

That is $$AG : GD = 2 : 1.$$

Now, D is the mid point of BC. Thus, $$D = \left(\frac{x_2 + x_3}{2}, \frac{y_2 + y_3}{2}\right)$$

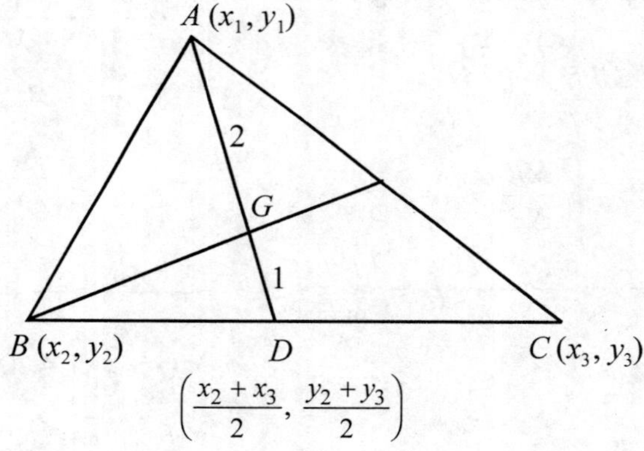

Now, $A = (x_1, y_1)$ and $D = \left(\frac{x_2 + x_3}{2}, \frac{y_2 + y_3}{2}\right)$ and $AG : GD = 2 : 1$.

Thus, we have, $$G = \left(\frac{2\left(\frac{x_2 + x_3}{2}\right) + 1.x_1}{2 + 1}, \frac{2\left(\frac{y_2 + y_3}{2}\right) + 1.y_1}{2 + 1}\right)$$

$$G = \left(\frac{x_1 + x_2 + x_3}{3}, \frac{y_1 + y_2 + y_3}{3}\right)$$

This is the coordinates of the centroid G of the triangle ABC.

Note : If $A(x_1, y_1)$ and $B(x_2, y_2)$ are the end points of the line segments and $P(x, y)$ is the point on AB (or AB produced) then the ratio $l : m$ in which P divides AB, can be obtained by using the formula

$$\frac{x - x_1}{x_2 - x} = \frac{l}{m} \quad \text{or} \quad \frac{y - y_1}{y_2 - y} = \frac{l}{m}$$

If the ratio thus obtained is +ve, then P divides AB internally in the ratio $l : m$. If the ratio thus obtained is –ve then P divides AB externally in the ratio $l : m$.

Example 1. Find the coordinates of the point P, which divides the line joining the points

(a) (1, –3) and (–3, 9) internally in the ratio 1 : 3

(b) (–2, 5) and (2, 5) externally in the ratio 2 : 5

(c) (–1, –4) and (8, 6) internally in the ratio 3 : 2

(d) (–1, 8) and (–2, 4) externally in the ratio 3 : 5.

Solution : (a) Let $A = (1, -3)$ and $B = (-3, 9)$

$$P = \left(\frac{1(-3)+3(1)}{1+3}, \frac{1(9)+3(-3)}{1+3}\right) = \left(\frac{-3+3}{4}, \frac{9-9}{4}\right) = \mathbf{(0,\ 0)}$$

1 : 3
(1, –3) (–3, 9)

(b) Let $A = (-2, 5)$ and $B = (2, 5)$

$$P = \left(\frac{2(2)-5(-2)}{2-5}, \frac{2(5)-5(5)}{2-5}\right) = \left(\frac{4+10}{-3}, \frac{10-25}{-3}\right) = \left(-\frac{\mathbf{14}}{\mathbf{3}}, \mathbf{5}\right)$$

2 : 5
(–2, 5) (2, 5)

(c) Let $A = (-1, -4)$ and $B = (8, 6)$

$$P = \left(\frac{3(8)+2(-1)}{3+2}, \frac{3(6)+2(-4)}{3+2}\right) = \left(\frac{24-2}{5}, \frac{18-8}{5}\right) = \left(\frac{\mathbf{22}}{\mathbf{5}}, \mathbf{2}\right)$$

3 : 2
(–1, –4) (8, 6)

(d) Let $A = (-1, 8)$ and $B = (-2, 4)$

$$P = \left(\frac{3(-2)-5(-1)}{3-5}, \frac{3(4)-5(8)}{3-5}\right) = \left(\frac{-6+5}{-2}, \frac{12-40}{-2}\right) = \left(\frac{\mathbf{1}}{\mathbf{2}}, \mathbf{14}\right)$$

3 : 5
(–1, 8) (–2, 4)

Example 2. The line joining the points (2, –2) and (–2, 4) is trisected. Find the points of trisection.

Solution : Let P and Q be the points of trisection

Clearly, $\frac{AP}{AB} = \frac{1}{2}$ and $\frac{AQ}{QB} = \frac{2}{1}$

Q B(–2, 4)
P
A(2, –2)

Now,
$$P = \left(\frac{1(-2)+2(2)}{1+2}, \frac{1(4)+2(-2)}{1+2}\right)$$

$$P = \left(\frac{-2+4}{3}, \frac{4-4}{3}\right) = \left(\frac{2}{3}, 0\right)$$

Again,
$$Q = \left(\frac{2(-2)+1(2)}{2+1}, \frac{2(4)+1(-2)}{2+1}\right)$$

2 : 1
(2, –2) (–2, 4)

$$Q = \left(\frac{-4+2}{3}, \frac{8-2}{3}\right) = \left(-\frac{2}{3}, 2\right)$$

Thus, the point of trisection are $\left(\frac{\mathbf{2}}{\mathbf{3}}, \mathbf{0}\right)$ and $\left(-\frac{\mathbf{2}}{\mathbf{3}}, \mathbf{2}\right)$.

Example 3. Find the ratio in which $P(2, 7)$ divides the line joining the points (8, 9) and (–7, 4).

Solution : Let $A = (x_1, y_1) = (8, 9)$, $B = (x_2, y_2) = (-7, 4)$ and $P = (x, y) = (2, 7)$.

The ratio $l : m$, in which P divides AB, is given by

$$\frac{x - x_1}{x_2 - x} = \frac{l}{m} \quad \text{or} \quad \frac{y - y_1}{y_2 - y} = \frac{l}{m}$$

$$\Rightarrow \quad \frac{2-8}{-7-2} = \frac{l}{m} \quad \text{or} \quad \frac{7-9}{4-7} = \frac{l}{m} \quad \Rightarrow \quad \frac{-6}{-9} = \frac{l}{m} \quad \text{or} \quad \frac{-2}{-3} = \frac{l}{m}$$

$\therefore$ The ratio is $l : m = \mathbf{2 : 3}$.

$\therefore$ P divides AB in the ratio 2 : 3, internally.

Note : Observe that both the equations gives us the same ratio. If, these equations give different ratios then it implies that the point P does not lie on the line joining AB; i.e., the points A, B, P are not collinear.

Alternate method:

Let P divide AB in the ratio $r : 1$.

Then, $$P = \left(\frac{-7r+8}{r+1}, \frac{4r+9}{r+1}\right)$$

$r : 1$

(8, 9) (−7, 4)

But, $P = (2, 7)$

$$\therefore \quad (2, 7) = \left(\frac{-7r+8}{r+1}, \frac{4r+9}{r+1}\right)$$

$$\Rightarrow \quad \frac{-7r+8}{r+1} = 2 \quad ; \quad \frac{4r+9}{r+1} = 7$$

$$\Rightarrow \quad -7r + 8 = 2r + 2 \quad ; \quad 4r + 9 = 7r + 7$$

$$\Rightarrow \quad 9r = 6 \quad ; \quad 3r = 2$$

$$\Rightarrow \quad r = \frac{2}{3} \quad ; \quad r = \frac{2}{3}$$

$\therefore$ The required ratio is **2 : 3**.

Example 4. Find the ratio in which $P(-1, -12)$ divides the line joining the points $A(3, 4)$ and $B(1, -4)$.

Solution : Let $A = (x_1, y_1) = (3, 4)$, $B = (x_2, y_2) = (1, -4)$ and $P = (x, y) = (-1, -12)$.

The ratio $l : m$ is given by

$$\frac{x - x_1}{x_2 - x} = \frac{l}{m} \Rightarrow \frac{-1-3}{1+1} = \frac{l}{m} \Rightarrow \frac{-4}{2} = \frac{l}{m} \Rightarrow \frac{l}{m} = \frac{-2}{1}$$

Thus the point P divides AB in the ratio **2 : 1 externally** ($\because$ of –ve sign).

Example 5. Find the ratio in which the line joining the points $(-4, 2)$ and $(8, 3)$ is divided by the y-axis. Also find the point of division.

Solution : Let $A = (x_1, y_1) = (-4, 2)$, $B = (x_2, y_2) = (8, 3)$

Let AB cut the y-axis at P and $P = (0, y)$. The ratio in which P divides AB is given by

$$\frac{0 - x_1}{x_2 - 0} = \frac{l}{m} \qquad (\because x = 0)$$

$$\Rightarrow \quad \frac{0+4}{8-0} = \frac{l}{m} \Rightarrow \frac{l}{m} = \frac{1}{2}$$

Thus the ratio 1 : 2. The coordinates of P are given by

$$P = \left(\frac{8-8}{1+2}, \frac{3+4}{1+2}\right) = \left(\mathbf{0}, \frac{\mathbf{7}}{\mathbf{3}}\right)$$

1 : 2
(−4, 2) (8, 3)

Example 6. Find the ratio in which the line joining the points (2, 5) and (1, 9) is divided by the *x*-axis. Also find the point of division.

Solution : Let $A = (x_1, y_1) = (2, 5)$, $B = (x_2, y_2) = (1, 9)$

Let AB cut the x-axis at P and $P = (x, 0)$. The ratio in which P divides AB is given by

$$\frac{0-y_1}{y_2-0} = \frac{l}{m} \qquad (\because y = 0)$$

$$\Rightarrow \quad \frac{-5}{9} = \frac{l}{m} \Rightarrow \frac{l}{m} = \frac{-5}{9}$$

Thus the point P divides AB is the ratio 5 : 9 externally. The coordinates of P are given by

$$P = \left(\frac{5(1)-9(2)}{5-9}, \frac{5(9)-9(5)}{5-9}\right)$$

5 : 9
(2, 5 (1, 9)

$$\therefore \quad P = \left(\frac{\mathbf{13}}{\mathbf{4}}, \mathbf{0}\right)$$

Example 7. Find the coordinates of the centroid of the triangle whose vertices are (−4, 6), (2, −2) and (2, 5).

Solution : Let $A = (-4, 6) = (x_1, y_1)$, $B = (2, -2) = (x_2, y_2)$, $C = (2, 5) = (x_3, y_3)$

The centroid G is given by,

$$G = \left(\frac{x_1+x_2+x_3}{3}, \frac{y_1+y_2+y_3}{3}\right) = \left(\frac{-4+2+2}{3}, \frac{6-2+5}{3}\right) = \mathbf{(0, 3)}$$

Example 8. If the centroid of the triangle *ABC* is (2, 3) and *A* = (4, 2) and *B* = (4, 5), find the coordinates of *C*.

Solution : By data $G = (2, 3)$, $A = (4, 2)$ and $B = (4, 5)$

Let $C = (h, k)$. We have,

$$G = \left(\frac{4+4+h}{3}, \frac{2+5+k}{3}\right) \Rightarrow (2,3) = \left(\frac{8+h}{3}, \frac{7+k}{3}\right) \Rightarrow \frac{8+h}{3} = 2, \frac{7+k}{3} = 3$$

$$\Rightarrow 8 + h = 6, \ 7 + k = 9$$

$$\Rightarrow h = -2, \quad k = 2$$

$\therefore$ The third vertex $\mathbf{C = (-2, 2)}$.

If *ABC* is a triangle with $A = (x_1, y_1)$, $B = (x_2, y_2)$ and $G = (h, k)$, then *C* is given by $C = (3h - (x_1 + x_2), 3k - (y_1 + y_2))$.

Example 9. The centroid of a triangle *ABC* is (3, 4) and *A* = (5, 4) if the vertices *B* and *C* lie respectively on the *x* and *y* axes, then find the coordinates of *B* and *C*.

Solution : By data $G = (3, 4)$, $A = (5, 4)$

The vertex *B* and *C* lies on *x* and *y* axes respectively. Let $B = (x, 0)$ and $C = (0, y)$.

Then we have,

$$G = \left(\frac{5+x+0}{3}, \frac{4+0+y}{3}\right) \Rightarrow (3,4) = \left(\frac{5+x}{3}, \frac{4+y}{3}\right)$$

$$\Rightarrow \frac{5+x}{3} = 3 \text{ and } \frac{4+y}{3} = 4 \Rightarrow x = 4 \text{ and } y = 8$$

Thus, the points *B* and *C* are **$B = (4, 0)$** and **$C = (0, 8)$**.

Example 10. If one end of the diameter of the circle is (3, 4) and the centre of the circle is (2, 3), find the coordinates of the other end.

Solution : Let $A = (3, 4)$, and $C = (2, 3)$. Let the other end of the diameter be $B = (h, k)$.

Clearly *C* is the midpoint of *AB*.

$$\Rightarrow (2, 3) = \left(\frac{3+h}{2}, \frac{4+k}{2}\right) \text{ (mid point formula)}$$

$$\Rightarrow \frac{3+h}{2} = 2 \text{ and } \frac{4+k}{2} = 3 \Rightarrow 3+h = 4 \text{ and } 4+k = 6 \Rightarrow \mathbf{h = 1, k = 2}$$

$\therefore$ **$B = (1, 2)$.**

Example 11. If $A = (3, 5), B = (-5, -4)$ and $C = (7, 10)$ are the vertices of a parallelogram, taken in the order. Find the coordinates of the fourth vertex.

Solution : We have $A = (3, 5)$, $B = (-5, -4)$ and $C = (7, 10)$.

Let the fourth vertex be $D = (h, k)$.

By data *ABCD* is a parallelogram. Thus the diagonals *AC* and *BD* bisects each other.

$\Rightarrow$ midpoint of AC = midpoint of BD

$$\Rightarrow \left(\frac{3+7}{2}, \frac{5+10}{2}\right) = \left(\frac{-5+h}{2}, \frac{-4+k}{2}\right)$$

$$\Rightarrow \left(5, \frac{15}{2}\right) = \left(\frac{h-5}{2}, \frac{k-4}{2}\right) \Rightarrow \frac{h-5}{2} = 5 \text{ and } \frac{15}{2} = \frac{k-4}{2} \Rightarrow h = 15 \text{ and } k = 19$$

$\therefore$ The fourth vertex D = **(15, 19)**

Note : The following result will be useful, again, to verify the answer in a problem of such type and also answer objective type questions.

> **If *A, B, C* are vertices of a parallelogram *ABCD,* taken in the order, then the fourth vertex *D* is given by**
>
> $$\mathbf{D = A - B + C}$$

That is if, $A = (x_1, y_1)$, $B = (x_2, y_2)$ and $C = (x_3, y_3)$ then $D = (x_1 - x_2 + x_3, y_1 - y_2 + y_3)$

For example, if $A = (3, 5)$, $B = (-5, -4)$ and $C = (7, 10)$ then $D = (3 + 5 + 7, 5 + 4 + 10) =$ **(15, 19)**

Exercise

I.

1. Find the coordinates of the point which divides the line joining the following pair of points, in the given ratio

   (a) (8, 9) and (–7, 4) internally in the ratio 2 : 3 (b) (1, –2) and (4, 7) internally in the ratio 1 : 2

   (c) (3, 4) and (–6, 2) externally in the ratio 3 : 2 (d) (–4, 4) and (1, 7) externally in the ratio 2 : 1

   (e) (p, q) and (q, p) internally in the ratio $p - q : p + q$.

2. Find the coordinates of the points of trisection of the line joining (5, –6) and (–7, 5).
3. Find the coordinates of the points of trisection of the line joining (3, –2) and (–3, –4)
4. Find the points which divide the line joining the points (12, 16) and (16, 12) into four parts
5. If one end of the diameter of a circle is (–1, 4) and the centre is (–2, –3), find the coordinates of the other end
6. If one end of the diameter of a circle is (3, –5) and the centre is (–3, 4), find the coordinates of the other end
7. Find the coordinates of the centroid of the triangle whose vertices are

   (a) (–4, 6), (2, –2) and (2, 5) (b) (4, 0), (6, –3) and (5, –5)

   (c) (4, 2), (4, 5) and (–2, 2) (d) $(a, 0)$, $(0, b)$ and (x, y).

8. The centroid of a triangle is (0, 3) and two vertices are (–4, 6) and (2, 5), find the third vertex.
9. In the triangle ABC, $A = (4, 2)$, $B = (4, 5)$ and the centroid $G = (2, 3)$, find C.
10. Find the coordinates of the point of trisection of the medians of the triangle whose vertices are (–2, 2), (–1, –3) and (5, 7). (Hint: find the centroid).

II.

1. In what ratio does the

   (a) point (–3, 7) divide the join of (–5, 11) and (4, –7)

   (b) point (–5, –20) divide the join of (4, 7) and (1, –2)

   (c) point (–5, 3) divide the join of (–3, –1) and (–8, 9)

   (d) point (1, 12) divide the join of (5, 6) and (7, 3)

2. In what ratio the line joining of the points

   (a) (2, –3) and (5, 6) is divided by x-axis (b) (2, –4) and (–3, 6) is divided by x-axis

   (c) (3, –6) and (–6, 8) is divided by y-axis (d) (–4, 2) and (8, 3) is divided by y-axis

   Find the points of division in all the cases.

Answers

I. 1. (a) (2, 7) (b) (2, 1) (c) (–24, –2) (d) (6, 10) (e) $\left(\dfrac{p^2 - q^2 + 2pq}{2p}, \dfrac{p^2 + q^2}{2p}\right)$

2. $\left(1, -\dfrac{7}{3}\right), \left(-3, \dfrac{4}{3}\right)$ 3. $\left(1, -\dfrac{8}{3}\right), \left(-1, -\dfrac{10}{3}\right)$ 4. (13, 15), (14, 14), (15, 13)

5. (–3, –10) 6. (–9, 13)

7. **(a)** (0, 3) **(b)** $\left(5, -\frac{8}{3}\right)$ **(c)** (2, 3) **(d)** $\left(\frac{a+x}{3}, \frac{b+y}{3}\right)$

8. (2, –2) 9. (–2, 2) 10. $\left(\frac{2}{3}, 2\right)$

II. 1. **(a)** 2 : 7 internally **(b)** 3 : 2 externally **(c)** 2 : 3 internally **(d)** 2 : 3 externally

2. **(a)** 1 : 2; (3, 0) **(b)** 2 : 3; (0, 0) **(c)** 1 : 2; $\left(0, -\frac{4}{3}\right)$ **(d)** 1 : 2; $\left(0, \frac{7}{3}\right)$

3. (–2, 1) 4. (–1, 1) 5. (–*b*, *b*) 6. (–3, –8) 8. $x = 4,\ y = 2$

1.5 Area of a Triangle

In this section we shall find a formula to find the area of a triangle whose vertices are known.

To show that the area of the triangle whose vertices are $A(x_1, y_1)$, $B(x_2, y_2)$ and $C(x_3, y_3)$ is

$$\Delta ABC = \frac{1}{2}\left[x_1(y_2 - y_3) + x_2(y_3 - y_1) + x_3(y_1 - y_2)\right]$$

We have $A = (x_1, y_1)$, $B = (x_2, y_2)$ and $C = (x_3, y_3)$.

Draw perpendiculars from *A, B* and *C, AL, BM* and *CN* onto *x*-axis

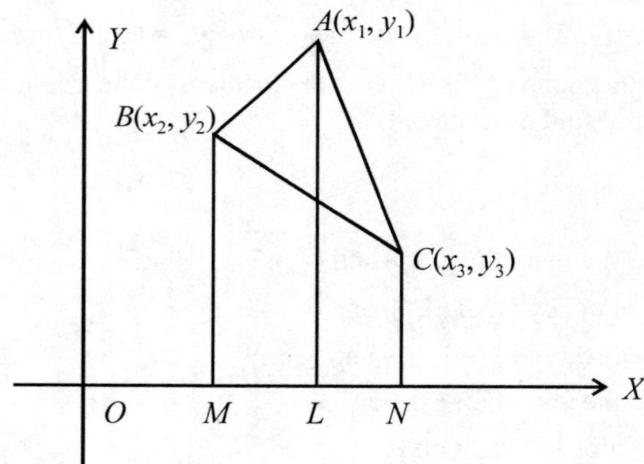

We have from the figure

Area of triangle *ABC* = (Area of Trapezium *BMLA*) + (Area of Trapezium *ALNC*) – (Area of Trapezium *BMNC*)

$$= \frac{1}{2}(BM + AL)(ML) + \frac{1}{2}(AL + CN)(LN) - \frac{1}{2}(BM + CN)(MN)$$

$$= \frac{1}{2}[(BM + AL)(OL - OM) + (AL + CN)(ON - OL) - (BM + CN)(ON - OM)]$$

$$= \frac{1}{2}[(y_2 + y_1)(x_1 - x_2) + (y_1 + y_3)(x_3 - x_1) - (y_2 + y_3)(x_3 - x_2)]$$

$$= \frac{1}{2}[x_1(y_2 + y_1 - y_1 - y_3) + x_2(-y_2 - y_1 + y_2 + y_3) + x_3(y_1 + y_3 - y_2 - y_3)]$$

$$\therefore \quad \Delta ABC = \frac{1}{2}[x_1(y_2 - y_3) + x_2(y_3 - y_1) + x_3(y_1 - y_2)]$$

Note : 1. The area of the triangle *ABC*, can also be written in the summation notation

$$\Delta ABC = \frac{1}{2}\sum x_1(y_2 - y_3)$$

2. It is a convention that the area of the triangle is taken to be positive, when the vertices *A, B, C* of the triangle taken in the anticlockwise direction, during the calculation of the area; otherwise it is taken to be negative.

3. If three points $A(x_1, y_1)$, $B(x_2, y_2)$ and $C(x_3, y_3)$ are collinear (i.e., lie on the same line), then the area *ABC* is zero and conversely. Thus the condition for three points *A, B, C* to be collinear is the area of the triangle *ABC* is zero.

Example 1. Find the area of the triangle whose vertices are (3, 4), (2, –1) and (4, –6).

Solution : Let $A = (x_1, y_1) = (3, 4)$, $B = (x_2, y_2) = (2, -1)$ and $C = (x_3, y_3) = (4, -6)$.

Now, $\Delta ABC = \frac{1}{2}[x_1(y_2 - y_3) + x_2(y_3 - y_1) + x_3(y_1 - y_2)]$

$$= \frac{1}{2}[3(-1 + 6) + 2(-6 - 4) + 4(4 + 1)] = \frac{1}{2}[15 - 20 + 20] = \mathbf{\frac{15}{2}} \textbf{ sq. units.}$$

Note : The area of the triangle *ABC*, where $A = (x_1, y_1)$, $B = (x_2, y_2)$ and $C = (x_3, y_3)$ can be calculated by writing the coordinates vertically as shown and multiplying the elements in the directions indicated, by taking positive signs while multiplying in the downward direction and negative signs while multiplying upwards directions.

$$\begin{matrix} x_1 & y_1 \\ x_2 & y_2 \\ x_3 & y_3 \\ x_1 & y_1 \end{matrix}$$

i.e., $\Delta ABC = \frac{1}{2}[(x_1y_2 + x_2y_3 + x_3y_1) - (x_1y_3 + x_3y_2 + x_2y_1)]$

For example, if $A = (3, 4)$, $B = (2, -1)$ and $C = (4, -6)$.

$$\begin{matrix} 3 & 4 \\ 2 & -1 \\ 4 & -6 \\ 3 & 4 \end{matrix}$$

$$\Delta ABC = \frac{1}{2}[(-3 - 12 + 16) - (-18 - 4 + 8)]$$

$$= \frac{1}{2}[1 + 14] = \mathbf{\frac{15}{2}} \textbf{ sq. units.}$$

Example 2. Show that the points (–5, 1), (5, 5) and (10, 7) are collinear.

Solution : Let $A = (-5, 1)$, $B = (5, 5)$ and $C = (10, 7)$.

We shall show that $\Delta ABC = 0$.

Now, $\Delta ABC = \frac{1}{2}[-5(5 - 7) + 5(7 - 1) + 10(1 - 5)] = \frac{1}{2}[10 + 30 - 40] = \mathbf{0}$

$\therefore$ **The points are collinear.**

Example 3. For what value of '*a*' the points (1, 4), (*a*, –2) and (–3, 16) are collinear.

Solution : Let $A = (1, 4)$, $B = (a, -2)$ and $C = (-3, 16)$. Let the points A, B, C be collinear.

$$\Rightarrow \quad \Delta ABC = 0 \quad \Rightarrow \quad \frac{1}{2}[1(-2-16) + a(16-4) - 3(4+2)] = 0$$

$$\Rightarrow \quad \frac{1}{2}[-18 + 12a - 18] = 0 \Rightarrow 12a - 36 = 0 \Rightarrow 12a = 36 \Rightarrow \boldsymbol{a = 3}$$

Example 4. If the point (x, y) lies on the line joining the points $(a, 0)$, and $(0, b)$, then show that $\frac{x}{a} + \frac{y}{b} = 1$.

Solution : Let $A = (x, y)$, $B = (a, 0)$ and $C = (0, b)$. By data A, B, C are collinear.

$$\Rightarrow \quad \Delta ABC = 0 \quad \Rightarrow \quad \frac{1}{2}[x(0-b) + a(b-y) + 0(y-0)] = 0$$

$$\Rightarrow \quad -bx + ab - ay = 0$$

$$\Rightarrow \quad ay + bx = ab \Rightarrow \frac{x}{a} + \frac{y}{b} = 1 \qquad \text{(dividing throughout by } ab)$$

Which is the required results.

Example 5. Find the area of the triangle, the coordinates of the mid points of whose sides are (5, 1), (0, –2) and (2, 6).

Solution : Let ABC be the triangle and $D(5, 1)$, $E(0, -2)$ and $F(2, 6)$ are the mid points of BC, CA and AB, respectively.

We know that the area $ABC = 4$ (area DEF)

Now, $$\Delta DEF = \frac{1}{2}[5(-2-6) + 0(6-1) + 2(1+2)] = \frac{1}{2}[-40+6] = -17$$

$\therefore \quad \Delta DEF = 17$ square units

$\therefore \quad \Delta ABC = 4(17) =$ **68 square units**

Example 6. Find the area of the quadrilateral whose vertices are (1, 1), (7, –3), (12, 2) and (7, 21).

Solution : Let $A = (x_1, y_1) = (1, 1)$, $B = (x_2, y_2) = (7, -3)$, $C = (x_3, y_3) = (12, 2)$ and $D = (x_4, y_4) = (7, 21)$.

The required area $ABCD$ = Area ABC + Area ACD

Now, Area $ABC = \frac{1}{2}[x_1(y_2 - y_3) + x_2(y_3 - y_1) + x_3(y_1 - y_2)]$

$$= \frac{1}{2}[1(-3-2) + 7(2-1) + 12(1+3)] = \frac{1}{2}[-5+7+48] = 25 \text{ sq. units}$$

Now, Area $ACD = \frac{1}{2}[x_1(y_3 - y_4) + x_3(y_4 - y_1) + x_4(y_1 - y_3)]$

$$= \frac{1}{2}[1(2-21) + 12(21-1) + 7(1-2)] = \frac{1}{2}[-19+240-7] = 107 \text{ sq. units}$$

$\therefore$ Area $ABCD = 25 + 107 =$ **132 sq. units.**

Example 7. Find the area of the quadrilateral whose vertices are $A(1, 1)$, $B(3, 4)$, $C(5, -2)$ and $D(4, -7)$.

Solution : Consider $A = (1, 1)$, $B = (3, 4)$, $C = (5, -2)$.

$$\text{Area } ABC = \frac{1}{2}[1(4+2) + 3(-2-1) + 5(1-4)] = \frac{1}{2}[6 - 9 - 15] = -\frac{18}{2} = -9$$

$\therefore$ $\Delta ABC = 9$ sq. units. (taking +ve sign)

Again Consider, $A = (1, 1)$, $C = (5, -2)$, $D = (4, -7)$.

$$\text{Area } ACD = \frac{1}{2}[1(-2+7) + 5(-7-1) + 4(1+2)] = \frac{1}{2}[5 - 40 + 12] = -\frac{23}{2}$$

$\therefore$ $\Delta ACD = \frac{23}{2}$ sq. units. (taking +ve sign)

$\therefore$ Area $ABCD = \Delta ABC + \Delta ACD = 9 + \frac{23}{2} = \mathbf{\frac{41}{2}}$ **sq. units.**

Exercise

I.

1. Find the area of the triangle whose vertices are
   - **(a)** $(6, 3)$, $(-3, 5)$, $(4, -2)$
   - **(b)** $(0, 0)$, $(-2, 3)$, $(10, 7)$
   - **(c)** $(a, 0)$, $(0, b)$, (x, y)
   - **(d)** $(5, 2)$, $(-9, -3)$, $(-3, -5)$
   - **(e)** $(a, c + a)$, (a, c), $(-a, c - a)$
   - **(f)** $(at_1^2, 2at_1)$, $(at_2^2, 2at_2)$, $(at_3^2, 2at_3)$
2. Show that the following points are collinear
   - **(a)** $(1, -1)$, $(2, 1)$, $(4, 5)$
   - **(b)** $(-5, -4)$, $(1, 2)$, $(3, 4)$
   - **(c)** $(4, -5)$, $(1, 1)$, $(-2, 7)$
   - **(d)** $(a. b + c)$, $(b, c + a)$, $(c, a + b)$
3. If the area of the triangle whose vertices are (x, y), $(1, 2)$ and $(2, 1)$ is 6, show that $x + y = 15$.

Answers

I. **1. (a)** $\frac{49}{2}$ **(b)** 22 **(c)** $\frac{1}{2}(bx + ay - ab)$ **(d)** 29 **(e)** a^2 **(f)** $a^2(t_1 - t_2)(t_2 - t_3)(t_3 - t_1)$

Chapter 2

Locus of a Point

2.1 Introduction

Earlier we have mentioned that the study of analytical geometry is the study of geometry through algebraic methods. To this end, in the previous chapter, we have expressed certain basic geometrical concepts in terms of algebraic relations involving the coordinates of the points. The most important aspect in analytical geometry is to express the plane curve in terms of algebraic equation, called the **equation of the curve**. For this purpose we must give a formal definition of a curve. It can be defined as the path traced by a moving point under a given geometrical condition(s). This is known as *locus* of a point. In this chapter, we shall see the procedure of finding the algebraic equation of the *locus* of a point.

2.2 Locus of a Point

Definition. The locus of a point, is the path traced by a point, which is moving under given geometrical condition(s).

For example, the path traced by a point which moves such that its distance from the fixed point is always same, is a curve called a circle.

An equation is said to represent the locus of a point, if the coordinates of every point on the path satisfy the equation and conversely those coordinates which satisfy the equation are the coordinates of some point on the locus.

Working rule to find the equation of a locus

Step 1: Take the coordinates of any point on the locus as $P(x, y)$.

Step 2: Apply the geometrical condition to it with which the path is traced.

Step 3: Convert the geometrical condition to algebraic equation, using the formulae derived earlier.

Step 4: Simplify the equation thus obtained, if needed. The resultant is the equation of the locus.

Example 1. Find the equation of the locus of the point which moves such that its distance from the point (2, 3) is twice its distance from (–2, 2).

Solution : Let $A = (2, 3)$ and $B = (-2, 2)$. Let $P(x, y)$ be any point on the locus.

Thus by data, $PA = 2\ PB \quad \Rightarrow \quad PA^2 = 4\ PB^2$

$$\Rightarrow \quad (x - 2)^2 + (y - 3)^2 = 4\ [(x + 2)^2 + (y - 2)^2]$$

$$\Rightarrow \quad x^2 - 4x + 4 + y^2 - 6y + 9 = 4\,(x^2 + 4x + 4 + y^2 - 4y + 4)$$

$$\Rightarrow \quad \mathbf{3x^2 + 3y^2 + 20x - 10y + 19 = 0}$$

This is the equation of the locus.

Example 2. Find the equation of the locus of the point which moves such that it is equidistant from the points (1, 2) and (–2, 3).

Solution : Let $A = (1, 2)$ and $B = (-2, 3)$. Let $P(x, y)$ be any point on the locus.

Thus we have, $PA = PB \Rightarrow PA^2 = PB^2$

$$\Rightarrow \quad (x - 1)^2 + (y - 2)^2 = (x + 2)^2 + (y - 3)^2$$

$$\Rightarrow \quad x^2 - 2x + 1 + y^2 - 4y + 4 = x^2 + 4x + 4 + y^2 - 6y + 9$$

$$\Rightarrow \quad 6x - 2y + 8 = 0 \quad \Rightarrow \mathbf{3x - y + 4 = 0}$$

This is the equation of the locus.

In fact the locus is the perpendicular bisector of the line joining the points (1, 2) and (–2, 3).

Example 3. Find the equation of the perpendicular bisector of the line joining $A(3, -2)$ and $B(4, 1)$.

Solution : The perpendicular bisector of the line joining A and B is the locus of the point which moves such that it is equidistant from A and B.

Now, we have $A = (3, -2)$ and $B = (4, 1)$. Let $P(x, y)$ be any point on the perpendicular bisector.

Thus we have, $PA = PB \Rightarrow PA^2 = PB^2$

$$\Rightarrow \quad (x - 3)^2 + (y + 2)^2 = (x - 4)^2 + (y - 1)^2$$

$$\Rightarrow \quad x^2 - 6x + 9 + y^2 + 4y + 4 = x^2 - 8x + 16 + y^2 - 2y + 1$$

$$\Rightarrow \quad 2x + 6y - 4 = 0 \quad \Rightarrow \mathbf{x + 3y - 2 = 0}$$

This is the equation of the locus.

Example 4. Find the equation of the locus of the point which moves such that it is equidistant from the point (4, 2) and the x-axis.

Solution : Let $A = (4, 2)$ and $P(x, y)$ be any point on the locus.

Thus by data, PA = Distance of P from x-axis

We know that distance of any point from x-axis is its y-coordinate.

$$\therefore \quad \sqrt{(x-4)^2 + (y-2)^2} = y \quad \Rightarrow \quad (x - 4)^2 + (y - 2)^2 = y^2$$

$$\Rightarrow \quad x^2 - 8x + 16 + y^2 - 4y + 4 = y^2$$

$$\Rightarrow \quad \mathbf{x^2 - 8x - 4y + 20 = 0}$$

This is the equation of the locus.

Example 5. Find the equation of the locus of the point which moves such that the sum of the squares of its distances from the coordinate axes is 5.

Solution : Let $P(x, y)$ be any point on the locus.

Now, distance of P from X-axis = y-coordinate of $P = y$

distance of P from Y-axis = x-coordinate of $P = x$

By data, $\qquad x^2 + y^2 = 5$

This is the equation of the locus.

Example 6. Find the equation of the locus of the point which moves such that the area of the triangle formed with the points (1, 2) and (2, –1) is equal to +3 units.

Solution : Let $P(x, y)$ be any point on the locus. Let $A = (1, 2)$ and $B = (2, -1)$.

By data, $\quad \Delta PAB = 3 \quad \Rightarrow \quad \frac{1}{2}[x(2 + 1) + 1(-1 - y) + 2(y - 2)] = 3$

$\Rightarrow \quad 3x - 1 - y + 2y - 4 = 6 \quad \Rightarrow \quad \mathbf{3x + y - 11 = 0}$

This is the equation of the locus.

Example 7. Find the equation of the locus of the point which moves such that the sum of its distance from (0, 4) and (0, –4) is 10 units.

Solution : Let $A = (0, 4)$ and $B = (0, -4)$. Let $P(x, y)$ be any point on the locus.

By data, $\qquad PA + PB = 10$

$$\Rightarrow \quad \sqrt{x^2 + (y-4)^2} + \sqrt{x^2 + (y+4)^2} = 10 \quad \Rightarrow \quad \sqrt{x^2 + (y-4)^2} = 10 - \sqrt{x^2 + (y+4)^2}$$

Squaring both sides, we get

$$x^2 + (y - 4)^2 = 100 + x^2 + (y + 4)^2 - 20\sqrt{x^2 + (y+4)^2}$$

$$\Rightarrow \quad (y - 4)^2 = 100 + (y + 4)^2 - 20\sqrt{x^2 + (y+4)^2}$$

$$\Rightarrow \quad y^2 - 8y + 16 = 100 + y^2 + 8y + 16 - 20\sqrt{x^2 + (y+4)^2}$$

$$\Rightarrow \quad -16y - 100 = -20\sqrt{x^2 + (y+4)^2} \quad \Rightarrow \quad 4y + 25 = 5\sqrt{x^2 + (y+4)^2}$$

Again squaring both sides, we get

$$16y^2 + 200y + 625 = 25(x^2 + y^2 + 8y + 16) \quad \Rightarrow \quad 25x^2 + 9y^2 = 225 \quad \Rightarrow \quad \mathbf{\frac{x^2}{9} + \frac{y^2}{25} = 1}$$

This is the equation of the locus.

Example 8. A point P moves such that $PA^2 = 3 \cdot PB^2$. If $A = (5, 0)$ and $B = (-5, 0)$ find the equation of the locus of P.

Solution : Let $P = (x, y)$.

By data, $\quad PA^2 = 3\,PB^2 \quad \Rightarrow \quad (x - 5)^2 + y^2 = 3[(x + 5)^2 + y^2]$

$\Rightarrow \quad x^2 - 10x + 25 + y^2 = 3(x^2 + 10x + 25 + y^2)$

$\Rightarrow \quad 2x^2 + 2y^2 + 40x + 50 = 0 \quad \Rightarrow \quad \mathbf{x^2 + y^2 + 20x + 25 = 0}$

This is the equation of the locus.

Exercise

I.

1. **Find the equation of the locus of the point which moves such that**

   (a) its distance from the x-axis is always 5 units.

(b) its distance from the y-axis is always 3 units.

(c) its distance from the coordinate axes which is in the ratio 5 : 3.

(d) sum of its distances from the coordinate axes is 4.

(e) product of its distances from the coordinate axes is 16.

(f) its distance from (1, –2) is 3.

(g) its distance from the x-axis is twice its distance from y-axis.

(h) square of its distance from (2, 3) is 3.

(i) sum of the square of its distances from the coordinate axes is 2.

II.

1. Find the equation of the locus of the point which moves such that

(a) its distance from (2, 3) is equal to its distance from (1, –2).

(b) its distance from (3, –4) is twice its distance from (1, 2).

(c) it is equidistant from the points (3, –4) and (2, 3).

(d) it is equidistant from the points (5, –2) and (–3, 1).

(e) it is equidistant from (4, 2) and the x-axis.

(f) it is equidistant from (2, 4) and the y-axis.

(g) its distance from the origin is twice its distance from (2, 3).

(h) ratio of its distances from (2, –3) and (4, –2) is 2 : 3.

(i) ratio of its distances from (1, –2) and (2, –3) is 2 : 1.

(j) the sum of the squares of its distances from (1, –3) and (2, – 1) is 4.

(k) the sum of the squares of its distances from the coordinate axes is equal to square of its distance from (1, –2).

III.

1. Find the equation of the locus of the point P which moves such that the area of the triangle PAB is +3 units; where $A = (-1, 2)$ and $B = (2, -1)$.

2. If $A = (1, 1)$ and $B = (-2, 3)$ are two fixed points, find the locus of a point P so that the area of the triangle $PAB = +9$ square units.

3. Find the equation of the locus of the point P, which moves such that the area of the triangle PAB is +4 units, where $A = (0, -2)$ and $B = (2, 1)$.

4. Find the equation of the locus of the point P, which moves such that $A\hat{P}B = 90^0$, where $A = (-2, 0)$ and $B = (2, 0)$.

5. If $A = (2, 3)$ and $B = (-1, 5)$ $A\hat{P}B = 90^0$, find the locus of P.

6. Find the equation of the locus of the point which moves such that sum of its distances from (3, 0) and (–3, 0) is 9.

7. Find the equation of the locus of the point which moves such that the sum of its distances from (0, 2) and (0, –2) is 6.

8. If $A = (ae, 0)$ and $B = (-ae, 0)$ and P moves such that $PA + PB = 2a$, show that the equation of the locus of P is $\frac{x^2}{a^2} + \frac{y^2}{b^2} = 1$, where $b^2 = a^2 (1 - e^2)$

Answers

I. **1.** **(a)** $y = 5$ **(b)** $x = 3$ **(c)** $5x = 3y$

(d) $x + y = 4$ **(e)** $xy = 16$ **(f)** $x^2 + y^2 - 2x + 4y - 4 = 0$

(g) $y = 2x$ **(h)** $x^2 + y^2 - 4x - 6y + 10 = 0$ **(i)** $x^2 + y^2 = 2$

II. **1.** **(a)** $x + 5y - 4 = 0$ **(b)** $3x^2 + 3y^2 - 2x - 24y - 5 = 0$ **(c)** $x - 7y - 6 = 0$

(d) $16x - 6y - 19 = 0$ **(e)** $x^2 - 8x - 4y + 20 = 0$ **(f)** $y^2 - 4x - 8y + 20 = 0$

(g) $3x^2 + 3y^2 - 16x - 24y + 52 = 0$ **(h)** $5x^2 + 5y^2 - 4x + 38y + 37 = 0$

(i) $3x^2 + 3y^2 - 14x + 20y + 47 = 0$ **(j)** $2x^2 + 2y^2 - 6x + 8y + 11 = 0$ **(k)** $2x - 4y - 5 = 0$

III. **1.** $x + y - 3 = 0$ **2.** $2x + 3y + 13 = 0$ **3.** $3x - 2y + 4 = 0$ **4.** $x^2 + y^2 = 4$

5. $x^2 + y^2 - x - 8y + 13 = 0$ **6.** $20x^2 + 36y^2 = 405$ **7.** $\dfrac{x^2}{5} + \dfrac{y^2}{9} = 1$

Chapter 3

Straight Line

3.1 Introduction

In this chapter, we shall derive the equation of a line in various forms. To this end we have to introduce the concept of the *slope* of a line, which associates a unique real number to a line.

3.2 Slope or Gradient of a Line

Given a line, it makes a fixed angle with the x-axis. This angle measured in the anticlockwise direction from the x-axis, is taken as positive, where as measured in the clockwise direction is taken as negative. This angle θ measured in anticlockwise direction is also called **inclination** of the line.

Note : Clearly, inclination of a line lies between 0° and 180°.

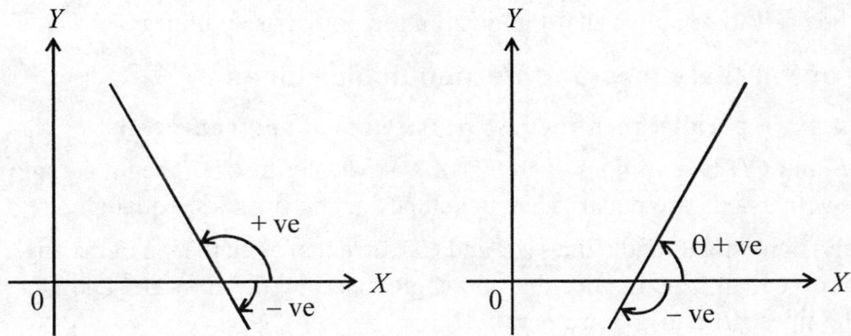

The *Steepness* of a line to the x-axis is described by a real number and is called ***slope*** of a line. The concept of the ***slope*** of the line is very important and fundamental.

Definition. (Slope of a line). The slope of a line is the tangent of the angle made by the line with the x-axis in the positive direction.

That is if θ is the angle made by the line with the x-axis in the positive direction, then $\tan\theta$ is the slope of the line.

For example, if a line makes an angle 45° with the x-axis in the positive direction, then $\tan 45° = 1$ is the slope of the line.

If a line makes an angle 150° with the x-axis, then the slope of the line is given by

$$\tan 150° = \tan(180° - 30°) = -\tan 30° = -\left(\frac{1}{\sqrt{3}}\right)$$

Following observations follows directly from the definition of the slope of a line.

1. Let θ is the angle made by the line with the x-axis in the positive direction. Now,

Slope of the line is positive $\Leftrightarrow$ $\tan\theta$ is positive

$\Leftrightarrow$ θ lies between 0° and 90° $\Leftrightarrow$ θ is acute

Now, Slope of the line is negative $\Leftrightarrow$ $\tan\theta$ is negative

$\Leftrightarrow$ θ lies between 90° and 180° $\Leftrightarrow$ θ is obtuse.

2. From the above observation, we have

(a) if the slope of a line is positive, then as we move from left to right along the line, then the line will "*rise*"

(b) if the slope of a line is negative, then as we move from left to right along the line, then the line will "*fall*".

3. If a line is parallel to x-axis, then the angle made by it with x-axis is zero, hence the slope is zero. In particular the slope of x-axis is zero.

Thus, the slope of any line parallel to x-axis is zero and in particular the slope of x-axis is zero.

4. A line parallel to y-axis, makes the angle 90° with the x-axis and $\tan 90^\circ$ is not defined. Thus, the slope of the line which is perpendicular to x-axis, we write the "Slope is ∞" (as we denote $\tan 90^\circ$ by ∞).

5. If three points A, B and C are collinear then the slope of AB = slope of BC and conversely. This fact can be used to establish that the given three points are collinear.

3.3 Slopes of Parallel Lines and Perpendicular Lines

1. If two lines are parallel then their slopes are equal and conversely.

Proof : Let AB and CD be two lines such that AB is parallel to CD. Then the angle of inclinations of AB and CD with x-axis are equal. Thus the slopes of the lines are equal (figure 1).

Conversely, if the slopes of the lines AB and CD are equal, then the tan of the angle of inclinations of AB and CD are equal. This implies that the angle made by the lines AB and CD with x-axis are equal. Thus, the lines AB and CD are parallel.

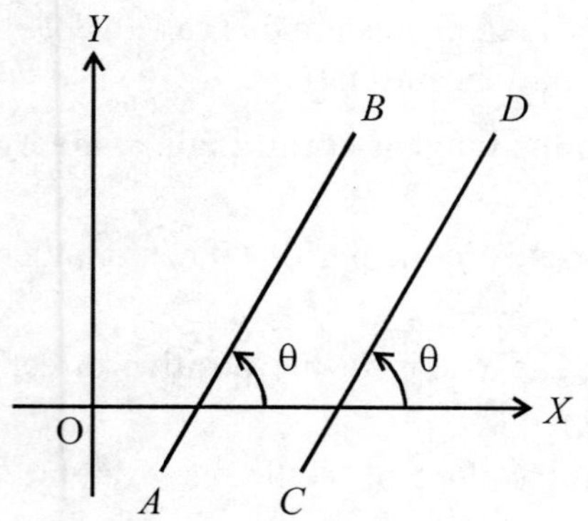

Fig. 1

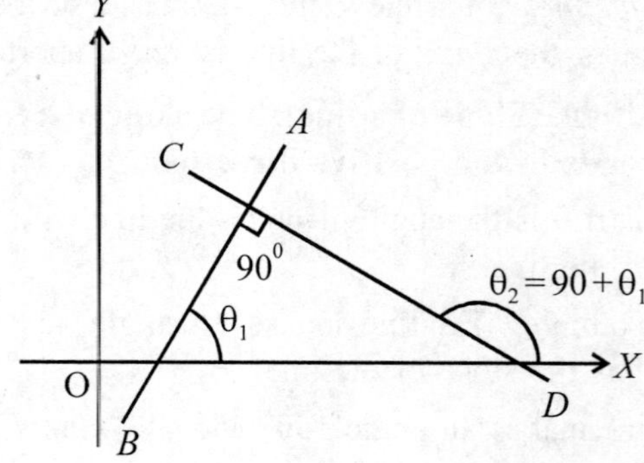

Fig. 2

2. If two lines are perpendicular to each other then the product of their slopes is –1 and conversely.

Proof : Let AB and CD be two lines such that AB is perpendicular to CD (figure 2). Let the angle made by lines AB and CD with x-axis, in the positive direction be θ_1 and θ_2 respectively.

Then, slope of $AB = \tan\theta_1 = m_1$ and slope of $CD = \tan\theta_2 = m_2$

From the figure 2, we have,

$$\theta_2 = 90 + \theta_1 \qquad \left(\because XDC = D\hat{B}A + B\hat{A}D\right)$$

$$\Rightarrow \tan\theta_2 = \tan(90 + \theta_1) \Rightarrow \tan\theta_2 = -\cot\theta_1 \qquad (\because \tan(90 + A) = -\cot A)$$

$$\Rightarrow \tan\theta_2 = -\frac{1}{\tan\theta_1} \Rightarrow \tan\theta_1 \cdot \tan\theta_2 = -1 \Rightarrow \boldsymbol{m_1 \cdot m_2 = 1}$$

Thus the product of the slopes = –1.

Conversely, let the product of the slopes of the lines AB and CD be –1.

i.e., $m_1 \cdot m_2 = -1 \Rightarrow \tan\theta_1 \cdot \tan\theta_2 = -1 \Rightarrow \tan\theta_2 = -\cot\theta_1$

$\Rightarrow \tan\theta_2 = \tan(90^\circ + \theta_1)$ or $\tan(\theta_1 - 90^\circ)$

$\Rightarrow \theta_2 = 90^\circ + \theta_1$ or $\theta_2 = \theta_1 - 90^\circ$

$\Rightarrow \theta_1$ and θ_2 differ by 90°

$\Rightarrow$ Angle between AB and CD are perpendicular to each other

Note : If the slope of the line l is m then the slope of any line perpendicular to it is given by

$$-\frac{1}{m} \text{ (i.e., negative reciprocal of } m\text{)}$$

3.4 Slope of the Line Joining Two Points

In this section, we shall find a formula to find the slope of the line joining two points.

Theorem. The slope of the line joining the points (x_1, y_1) and (x_2, y_2) is given by

$$\boldsymbol{m = \frac{y_2 - y_1}{x_2 - x_1}}.$$

Proof : Let $P = (x_1, y_1)$ and $Q = (x_2, y_2)$ be such that $x_1 \neq x_2$ (i.e., the line PQ is not parallel to y-axis).

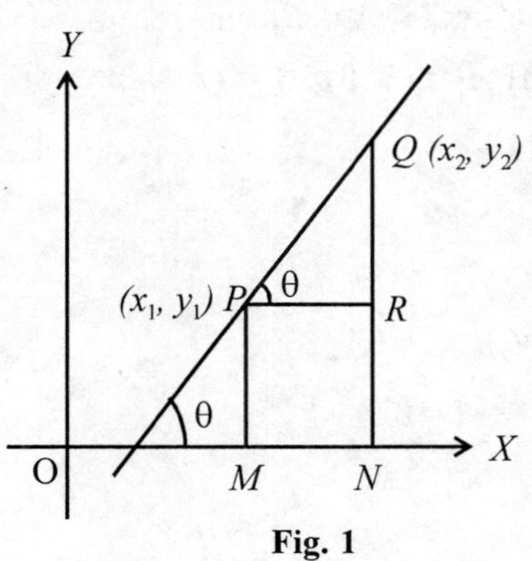

Fig. 1

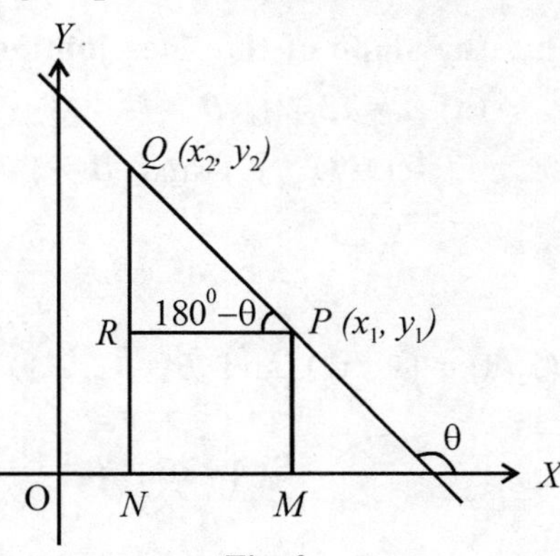

Fig. 2

Let θ be the inclination of the line joining *PQ* with the *x*-axis.

Let *PM* and *QN* be perpendiculars drawn from *P* and *Q* onto *x*-axis. Let *PR* be perpendicular from *P* onto *QN*.

For Fig. 1: $Q\hat{P}R = \theta$ For Fig. 2 : $Q\hat{P}R = 180^\circ - \theta$

$\Rightarrow \quad \tan\theta = \frac{QR}{PR} \qquad \tan(180^\circ - \theta) = \frac{QR}{PR}$

$\Rightarrow \quad \tan\theta = \frac{QN - RN}{MN} \qquad \tan(180^\circ - \theta) = \frac{QN - RN}{NM}$

$\Rightarrow \quad \tan\theta = \frac{QN - RN}{ON - OM} \qquad \tan(180^\circ - \theta) = \frac{QN - RN}{OM - ON}$

$\Rightarrow \quad \tan\theta = \frac{y_2 - y_1}{x_2 - x_1} \qquad -\tan\theta = \frac{y_2 - y_1}{x_1 - x_2} \quad (\because \tan(180^\circ - \theta) = -\tan\theta)$

$\Rightarrow \quad \boldsymbol{\tan\theta = \frac{y_2 - y_1}{x_2 - x_1}} \qquad \boldsymbol{\tan\theta = \frac{y_2 - y_1}{x_2 - x_1}} \quad (\because x_1 - x_2 = -(x_2 - x_1))$

Thus, the slope $m = \tan\theta$ of the line joining $P(x_1, y_1)$ and $Q(x_2, y_2)$ is given by

$$\boldsymbol{\tan\theta = m = \frac{y_2 - y_1}{x_2 - x_1}}$$

Note : The slope of a line not parallel to *y*-axis can be also be understood as below :

Supposing a point moves along a line from one position to another. Then there will be a change in its distance from the *x*-axis and as well in its distance from *y*-axis. The change in its distance from *x*-axis is called *"Vertical separation"* or *"rise"* and the change in its distance from *y*-axis is called *"Horizontal separation"* or *"run"*. Clearly, the slope of a line is the ratio of *vertical separation* to *horizontal separation* i.e., **the slope of a line is *rise* per unit *run*.**

If $P = (x_1, y_1)$ and $Q = (x_2, y_2)$, then clearly

Vertical separation $= y_2 - y_1$ (i.e., *rise*)

Horizontal separation $= x_2 - x_1$ (i.e., *run*)

$$\therefore \quad \textbf{Slope} = \frac{y_2 - y_1}{x_2 - x_1} = \frac{\textbf{rise}}{\textbf{run}}$$

Example 1. Find the slope of the lines joining the points *A* and *B*, where

(a) $A = (3, 2)$, $B = (-4, 3)$, (b) $A = (5, 0)$, $B = (2, -1)$.

Solution : (a) $A = (3, 2) = (x_1, y_1)$ and $B = (-4, 3) = (x_2, y_2)$.

$$\text{Slope of } AB = \frac{y_2 - y_1}{x_2 - x_1} = \frac{3 - 2}{-4 - 3} = -\frac{1}{7}$$

(b) $A = (5, 0) = (x_1, y_1)$ and $B = (2, -1) = (x_2, y_2)$.

$$\text{Slope of } AB = \frac{y_2 - y_1}{x_2 - x_1} = \frac{-1 - 0}{2 - 5} = \frac{1}{3}$$

Example 2. Find the value of x so that the slope of the line joining the points (2, 5) and (x, 3) is 2.

Solution : Let $A = (2, 5)$ and $B = (x, 3)$

By data slope of $AB = 2 \Rightarrow \dfrac{3-5}{x-2} = 2 \Rightarrow -2 = 2(x-2) \Rightarrow 2x = 2 \Rightarrow \mathbf{x = 1}$

Example 3. Show that the points (1, 1), (3, –2) and (5, –5) are collinear using the concept the slope of the line.

Solution : Let $A = (1, 1)$ and $B = (3, -2)$ and $C = (5, -5)$

If the points A, B, and C are collinear, then slope of AB = slope of BC.

Now, Slope of $AB = \dfrac{-2-1}{3-1} = -\dfrac{\mathbf{3}}{\mathbf{2}}$, Slope of $BC = \dfrac{-5+2}{5-3} = -\dfrac{\mathbf{3}}{\mathbf{2}}$

Clearly, slope of AB = slope of BC. Thus, **the points are collinear**.

Example 4. Find the value of a, if the points (–1, 4) and (2, 5) and (3, a) are collinear.

Solution : Let $A = (-1, 4)$, $B = (2, 5)$ and $C = (3, a)$. By data the points A, B and C are collinear.

$\Rightarrow$ slope of AB = slope of BC

$$\Rightarrow \frac{5-4}{2+1} = \frac{a-5}{3-2} \Rightarrow \frac{1}{3} = \frac{a-5}{1} \Rightarrow 3a - 15 = 1 \Rightarrow 3a = 16 \Rightarrow \mathbf{a = \frac{16}{3}}$$

Example 5. Show that the line joining the points (2, 3) and (4, 2) is perpendicular to the line joining the points (5, 3) and (6, 5).

Solution : Let, $A = (2, 3)$, $B = (4, 2)$, $C = (5, 3)$, $D = (6, 5)$.

Now, Slope of $AB = \dfrac{2-3}{4-2} = -\dfrac{\mathbf{1}}{\mathbf{2}} = m_1$, Slope of $CD = \dfrac{5-3}{6-5} = \mathbf{2} = m_2$

Clearly, slope of $m_1 \cdot m_2 = \left(-\dfrac{1}{2}\right) \cdot 2 = -1 \Rightarrow$ **Line AB is perpendicular to line CD.**

Example 6. Show that the line joining the points (2, –3) and (–5, 1) is parallel to the line joining the points (7, –1) and (0, 3).

Solution : Let, $A = (2, -3)$, $B = (-5, 1)$, $C = (7, -1)$ and $D = (0, 3)$.

Now, Slope of $AB = \dfrac{1+3}{-5-2} = -\dfrac{\mathbf{4}}{\mathbf{7}} = m_1$ and Slope of $CD = \dfrac{3+1}{0-7} = -\dfrac{\mathbf{4}}{\mathbf{7}} = m_2$

Clearly, $m_1 = m_2 \Rightarrow$ **The line AB is parallel to CD.**

Example 7. If the line joining the points (2, 3) and (5, 4) is parallel to the line joining the points (3, k) and (4, 2) find the value of k.

Solution : Let $A = (2, 3)$, $B = (5, 4)$, $C = (3, k)$ and $D = (4, 2)$. By data AB is parallel to CD.

$$\Rightarrow \text{slope of } AB = \text{slope of } BC \Rightarrow \frac{4-3}{5-2} = \frac{2-k}{4-3}$$

$$\Rightarrow \frac{1}{3} = \frac{2-k}{1} \Rightarrow 6 - 3k = 1 \Rightarrow 3k = 5 \Rightarrow \mathbf{k = \frac{5}{3}}$$

Exercise

I.

1. Find the slope of the line joining the points
   (a) (2, 3) and (–1, 2) **(b)** (–2, 4) and (3, –2) **(c)** (2, 11) and (3, 2)
   (d) $\left(\frac{1}{2}, \frac{3}{2}\right)$ and $\left(-2, \frac{1}{2}\right)$ **(e)** (2, 0) and (0, –3) **(f)** (1, –1) and (2, 1)
2. Find the slope of the line perpendicular to the line joining the points
   (a) (3, 2) and (–3, 1) **(b)** (–1, 2) and (2, –3)
3. If the line *AB* is perpendicular to *CD* and the slope of *AB* is $-\frac{4}{3}$, find the slope of *CD*.
4. If the slope of the line *AB* is 2/3 and the line *CD* is perpendicular to *AB*, then write the slope of *CD*.
5. Find the value of *k*, so that the slope of the line joining
   (a) *A* (2, –1) and *B* (1, *k*) is 2/3 **(b)** *A* (3, 2) and *B* (*k*, –1) is 3/5
   (c) *A* (2, –1) and *B* (0, *k*) is –1.

II.

1. Show that the following points are collinear
   (a) (2, –4), (4, –2) and (7, 1) **(b)** (3, 5), (7, 1) and (10, –2)
   (c) (–1, 6), (3, 10) and (8, 15) **(d)** $(a, b + c)$, $(b, c + a)$ and $(c, a + b)$.
2. Find the value of '*k*', if the points (*k*, 3), (–6, 4) and (–10, 5) are collinear.
3. State whether the two lines *AB* and *CD* are parallel, perpendicular or neither, where
   (a) $A = (2, -4)$, $B = (4, -2)$ and $C = (3, 4)$, $D = (-3, 5)$
   (b) $A = (-5, 7)$, $B = (0, -2)$ and $C = (1, -3)$, $D = (4, -\frac{4}{3})$
   (c) $A = (-5, 2)$, $B = (5, -2)$ and $C = (3, 6)$, $D = (1, 1)$
   (d) $A = (-1, 0)$, $B = (5, 2)$ and $C = (3, 1)$, $D = (9, 3)$
   (e) $A = (6, 5)$, $B = (3, 2)$ and $C = (-2, 9)$, $D = (-5, 6)$.
4. If the line joining the points (3, 2) and (2, –3) is parallel to the line joining the points (4, 3) and (2, *k*), find *k*.
5. If $A = (-3, 4)$, $B = (2, -3)$, $C = (3, k)$, $D = (2, -3)$ and the line *AB* is perpendicular to the line *CD*, then find *k*.
6. The line joining *A* (–3, 4) and *B* (2, –1) is parallel to the line joining *C* (1, –2) and *D* (0, *k*), find *k*.
7. Find *k* if *AB* is perpendicular to *CD*, where $A = (k, 1)$, $B = (3, 4)$, $C = (5, 1)$ and $D = (1, -1)$.
8. If *A* (4, 4), *B* (3, –2) and *C* (–3, 16) are the vertices of the triangle *ABC*, find
   (a) slope of the median *AD*, **(b)** slope of the altitude of *BE* from *B* to *AC*,
   (c) slope any line parallel to *BC*.
9. Using the concept of the slope, show that the points
   (a) *A* (4, –2), *B* (–4, 4), *C* (10, 6) are the vertices of a right angled triangle.
   (b) *A* (–3, 2), *B* (3, 4), *C* (5, –2) and *D* (–1, –4) are the vertices of parallelogram.

Answers

I. **1.** **(a)** $\frac{1}{3}$ **(b)** $-\frac{6}{5}$ **(c)** -9 **(d)** $\frac{2}{5}$ **(e)** $\frac{3}{2}$ **(f)** 2

2. **(a)** -6 **(b)** $\frac{3}{5}$ **3.** $\frac{3}{4}$ **4.** $-\frac{3}{2}$ **5.** **(a)** $-\frac{5}{3}$ **(b)** -2 **(c)** 1.

II. **2.** $a = -2$

3. **(a)** neither **(b)** perpendicular **(c)** perpendicular **(d)** parallel **(e)** parallel

4. $k = -7$ **5.** $k = -\frac{16}{7}$ **6.** $k = -1$ **7.** $k = \frac{9}{2}$ **8.** **(a)** $-\frac{3}{4}$ **(b)** $\frac{7}{12}$ **(c)** -3

3.5 Standard Forms of Equation of a Straight Line

In this section we shall see the equation of a line in different form.

Equation of a line means an algebraic equation in the variables x and y which is satisfied by the x and y coordinates of every point on the line and not by those of any other point.

To derive the equation of a line we must establish a common geometrical property (from the given data about the line) which should be true for every point on that line. The algebraic form of this geometric property is the equation of the line.

There are different forms of equation by which a line can be represented, depending upon the facts known about the line. To start with we shall consider the equations of coordinate axes and the lines parallel to coordinate axes.

I. Equation to coordinate axes and lines parallel to coordinate axes

(a) Equation of x-axis

We know that the y-coordinate (i.e., ordinate) of each point on the x-axis is zero. This is the common property that every point on the x-axis will satisfy. Thus, **the equation of x-axis is $y = 0$.**

(b) Equation of y-axis

Again we know that the x-coordinate (i.e., abscissa) of each point on the y-axis is zero. Thus, **the equation of y-axis is $x = 0$.**

(c) Equation of a line parallel to x-axis

Consider a line parallel to x-axis, which is at a distance 'a' units from the x-axis (Figure (i)).

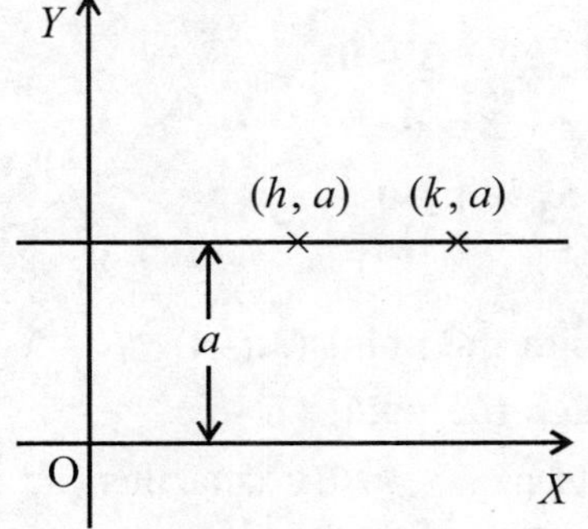

Fig. (i)

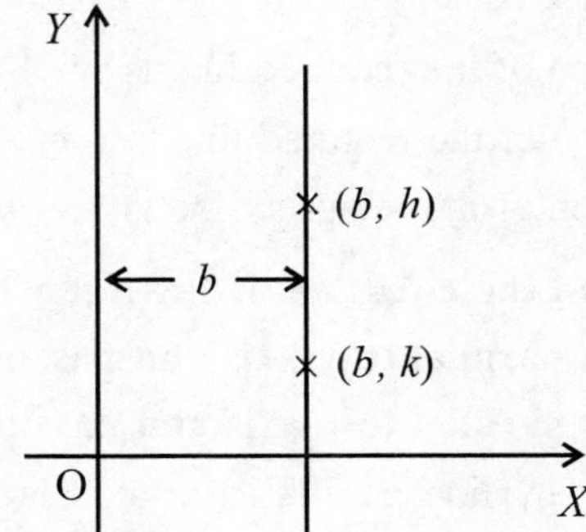

Fig. (ii)

Then clearly, the y-coordinate (ordinate) of every point on this line is 'a'. This is the common property that every point will satisfy.

Thus, **the equation of the line parallel to x-axis at a distance 'a' from it, is given by $y = a$.**

Conversely, the equation of the form $y = a$, represent a line parallel to x-axis, which is at a distance 'a' units from x-axis. If 'a' is positive, then the line is above the x-axis and it lies below x-axis if 'a' is negative.

For example, equation of the line parallel to x-axis which is 5 units above the x-axis is $y = 4$ or $y - 4 = 0$.

The equation $y + 7 = 0$, represent the line parallel to x-axis, which is 7 units below the x-axis.

(d) Equation of a line parallel to y-axis

Consider a line parallel to y-axis, which is at a distance 'b' units from the y-axis (Figure (ii)).

Then clearly, the x-coordinate (abscissa) of every point on this line is 'b'. This is the common property that every point will satisfy.

Thus, **the equation of the line parallel to y-axis at a distance 'b' from it, is given by $x = b$.**

Conversely, the equation of the form $x = b$, represent a line parallel to y-axis, which is at a distance 'b' units from y-axis. If 'b' is positive, then the line is to the right side of y-axis and it will be to the left side of y-axis if 'b' is negative.

For example, equation of the line parallel to y-axis which is 3 units to the right side of y-axis is given by $x = 3$.

The equation $2x + 5 = 0$, which can be written as $x = -\frac{5}{2}$, represent a line parallel to y-axis, which is $\frac{5}{2}$ units to the left side of the y-axis.

Example 1. Write the equation of

(a) the line parallel to x-axis which is 4 units above it

(b) the line parallel to x-axis which is 3/2 units below it

(c) the line parallel to y-axis which is 2 units to left of it

(d) the line parallel to y-axis which is 4/5 units to right of it.

Solution : **(a)** Equation of the line parallel to x-axis is of the form $y = k$. Thus, the required equation is $y = 4$ or $\mathbf{y - 4 = 0}$.

(b) Equation of the required line is $y = -3/2$ i.e., $\mathbf{2y + 3 = 0}$.

(c) Equation of the required line is $x = -2$ i.e., $\mathbf{x + 2 = 0}$.

(d) Equation of the required line is $x = 4/5$ i.e., $\mathbf{5x - 4 = 0}$.

Example 2. Find the equation of the line which is

(a) parallel to x-axis and passing through the point (3, –4)

(b) parallel to y-axis and passing through the point (2, –3).

Solution : (a) Any line parallel to x-axis is of the form $y = k$. By data this must pass through (3, –4). Thus, $-4 = k$. Hence, the equation of the line is $y = -4$ i.e., $\mathbf{y + 4 = 0}$.

(b) Any line parallel to y-axis is of the form $x = k$. By data this must pass through $(2, -3)$. Thus, $2 = k$. Thus, the equation of the line is $\boldsymbol{x = 2}$.

Example 3. Find the equation of the line which is equidistant from the line $x = -5$ and $x = 7$.

Solution : The equations represent lines parallel to y-axis. The required line is equidistant from these lines. Thus, it is also parallel to y-axis.

Now, distance of the required line from y-axis $= \frac{1}{2}(-5+7) = 1$.

$\therefore$ The equation of the line is $\boldsymbol{x = 1}$.

II. Equation of a line in slope-point form

To find the equation of a line whose slope is m and passing through the point (x_1, y_1)

Let $A = (x_1, y_1)$. Let $P(x, y)$ be any point on the line.

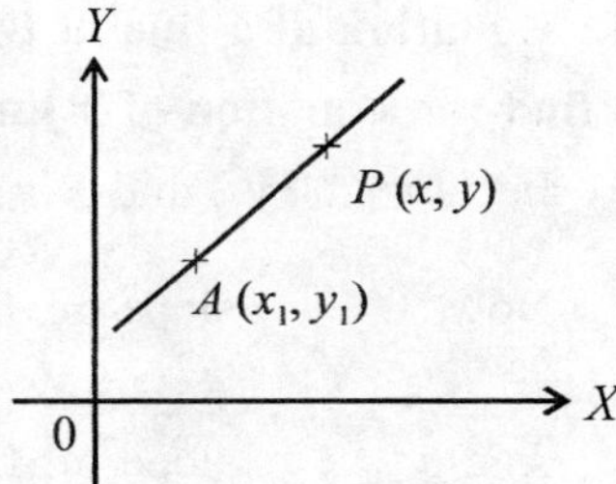

Now, slope of $AP = \dfrac{y - y_1}{x - x_1}$

But by data the slope of the line is m.

Thus, $\dfrac{y - y_1}{x - x_1} = m$

This relation is true for any point on the line.

Thus, the equation of the line whose slope is m and passing through (x_1, y_1) is given by

$$\boldsymbol{y - y_1 = m(x - x_1)}$$

This form of the equation of a line is called ***slope-point form.***

Note : Equation of the line passing through the origin and having slope m is given by $\boldsymbol{y = mx}$.

Example 4. Find the equation of the line through $(-2, 3)$ with slope $\frac{1}{2}$.

Solution : Let $(-2, 3) = (x_1, y_1)$ and $m = 1/2$. The equation of the line is given by

$$y - y_1 = m(x - x_1) \Rightarrow y - 3 = \frac{1}{2}(x + 2) \Rightarrow 2y - 6 = x + 2 \Rightarrow \boldsymbol{x - 2y + 8 = 0}$$

Example 5. Find the equation of the line passing through $(5, -2)$ and making an angle 150° with the x-axis in the positive direction.

Solution : By data the inclination of the line is 150°. Thus,

$$\text{slope of the line is } m = \tan 150^\circ \Rightarrow m = \tan(180^\circ - 30^\circ)$$

$$\Rightarrow m = -\tan 30^\circ \Rightarrow m = -\frac{1}{\sqrt{3}}$$

Let $(5, -2) = (x_1, y_1)$. Then the equation of the line is

$$y - y_1 = m(x - x_1) \Rightarrow y + 2 = -\frac{1}{\sqrt{3}}(x - 5)$$

$$\Rightarrow \sqrt{3}y + 2\sqrt{3} = -x + 5 \Rightarrow \boldsymbol{x + \sqrt{3}y + (2\sqrt{3} - 5) = 0}$$

Example 6. Find the equation of the line passing through (–2, 1) perpendicular to the line whose slope is $\frac{1}{2}$.

Solution : Let $m_1 = \frac{1}{2}$. Let the slope of the required line be m_2. By data $m_1 \cdot m_2 = -1$

($\because$ if the lines are perpendicular then the product of their slopes = –1).

Now, $m_1 \cdot m_2 = -1 \Rightarrow m_2 \cdot \frac{1}{2} = -1 \Rightarrow \boldsymbol{m_2 = -2}$

Let $(-2, 1) = (x_1, y_1)$. Thus the equation of the required line is

$$y - y_1 = m_1 (x - x_1) \Rightarrow y - 1 = -2 (x + 2)$$

$$\Rightarrow y - 1 = -2x - 4 \Rightarrow \mathbf{2x + y + 3 = 0}$$

III. Equation of a line in two points form

To find the equation of a line passing through the points (x_1, y_1) and (x_2, y_2)

Let $A = (x_1, y_1)$ and $B = (x_2, y_2)$. Let $P(x, y)$ be any point on the line joining A and B.

Now, slope of $AP = \frac{y - y_1}{x - x_1}$

Again, slope of $AB = \frac{y_2 - y_1}{x_2 - x_1}$

$P(x, y)$

$B(x_2, y_2)$

$A(x_1, y_1)$

Since the points A, B and P are collinear.

We have, slope of AP = slope of $AB \Rightarrow \frac{y - y_1}{x - x_1} = \frac{y_2 - y_1}{x_2 - x_1}$

This relation is true for all the points on the line.

Thus, the equation of the line joining the points (x_1, y_1) and (x_2, y_2) is given by

$$\frac{y - y_1}{x - x_1} = \frac{y_2 - y_1}{x_2 - x_1}$$

This form of equation of a line is called ***two-point form.***

Note : The equation of the line passing through origin and any point (x_1, y_1) is given by

$$\frac{y - y_1}{x - x_1} = \frac{0 - y_1}{0 - x_1} \Rightarrow yx_1 - x_1y_1 = xy_1 - x_1y_1 \Rightarrow y_1x - x_1y = 0$$

Example 7. Find the equation of the line passing through the points (2, 5) and (–3, 2).

Solution : Let $(x_1, y_1) = (2, 5)$ and $(x_2, y_2) = (-3, 2)$. Then equation of the line is given by

$$\frac{y - y_1}{x - x_1} = \frac{y_2 - y_1}{x_2 - x_1} \Rightarrow \frac{y - 5}{x - 2} = \frac{2 - 5}{-3 - 2}$$

$$\Rightarrow \frac{y - 5}{x - 2} = \frac{3}{5} \Rightarrow 5y - 25 = 3x - 6 \Rightarrow \mathbf{3x - 5y + 19 = 0}$$

IV. Equation of the line in the intercept form

Let a line cut the x-axis at A and y-axis at B. Let $OA = a$ and $OB = b$. Then the length $OA = a$ is called the **x-intercept** of the line and the length $OB = b$ is called the **y-intercept** of the line. If the line passes through the origin then both x and y intercepts are zero.

To find the equation of a line which makes intercepts 'a' and 'b' on the x- and y-axes, respectively.

Let the line cut the x-axis at A and y-axis at B. By data $OA = a$ and $OB = b$. Thus, $A = (a, 0)$ and $B = (0, b)$. Hence the required line is the lining joining A and B. Its equation can be obtained by using two point form of the equation of a line.

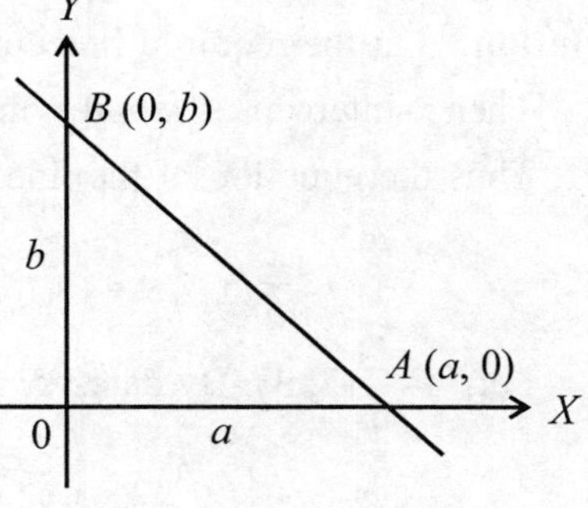

Thus the equation of AB is

$$\frac{y-0}{x-a} = \frac{0-b}{a-0} \Rightarrow \frac{y}{x-a} = -\frac{b}{a}$$

$$\Rightarrow ay = -bx + ab \Rightarrow bx + ay = ab$$

Dividing throughout by ab, we have, $\boldsymbol{\frac{x}{a} + \frac{y}{b} = 1}$

This form of equation of the line is called ***intercept form.***

Example 8. Find the equation of the line whose intercepts on the x-axis and y-axis are 3 and -4.

Solution : Let $a = 3$ and $b = -4$. The equation of the line is given by

$$\frac{x}{a} + \frac{y}{b} = 1 \Rightarrow \frac{x}{3} + \frac{y}{-4} = 1 \Rightarrow -4x + 3y = -12 \Rightarrow \mathbf{4x - 3y - 12 = 0}$$

Example 9. Find the equation of the line passing through $(-4, 5)$ and have intercepts on the axes, equal in magnitude and opposite in sign.

Solution : Let the x-intercept $= a$. Then y-intercept $= -a$. The equation of the line will be of the form

$$\frac{x}{a} + \frac{y}{-a} = 1 \Rightarrow x - y = a \quad \text{.... (1)}$$

By data this line must pass through $(-4, 5)$ $\Rightarrow -4 - 5 = a \Rightarrow a = -9$

Putting this value of a in (1) we get the equation of the line $x - y = -9 \Rightarrow \mathbf{x - y + 9 = 0}$

Example 10. Find the equation of the line passing through $(-2, 6)$ and the sum of intercepts on the coordinate axes is 5.

Solution : Let $a = x$-intercept and $b = y$-intercept of the line. By data $a + b = 5 \Rightarrow b = 5 - a$

Now, the equation of the line is given by,

$$\frac{x}{a} + \frac{y}{b} = 1 \Rightarrow \frac{x}{a} + \frac{y}{5-a} = 1 \quad \text{.... (1)}$$

$$\Rightarrow (5-a)x + ay = a(5-a) \quad \text{.... (2)}$$

By data this must pass through $(-2, 6)$. Thus equation (2) must satisfy $x = -2$ and $y = 6$.

$$\Rightarrow (5-a)(-2) + 6a = 5a - a^2 \Rightarrow -10 + 2a + 6a = 5a - a^2$$

$$\Rightarrow a^2 + 3a - 10 = 0$$

$$\Rightarrow (a+5)(a-2) = 0 \Rightarrow \boldsymbol{a = -5} \text{ or } \boldsymbol{a = 2}$$

Putting these values in (1), we get

$$\frac{x}{-5} + \frac{y}{10} = 1 \text{ and } \frac{x}{2} + \frac{y}{3} = 1 \Rightarrow -2x + y = 10 \text{ and } 3x + 2y = 6$$

i.e., $\mathbf{2x - y + 10 = 0}$ and $\mathbf{3x + 2y - 6 = 0}$

Example 11. A line passes through the point (2, 3) and this point bisects the portion of the line intercepted between the coordinate axes. Find the equation of the line.

Solution : Let the required line cut the x-axis at A $(a, 0)$ and the y-axis at B $(0, b)$.

Then x-intercept $= a$ and y-intercept $= b$.

Thus the equation of the line is given by

$$\frac{x}{a} + \frac{y}{b} = 1 \qquad .... (1)$$

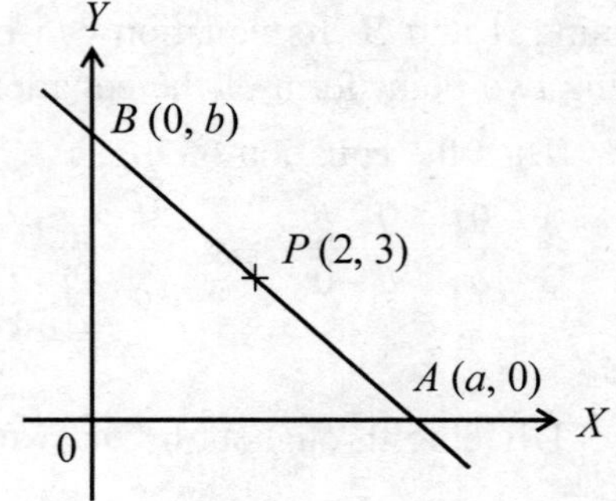

Let, $P = (2, 3)$. By data, $P =$ midpoint of AB.

$$\Rightarrow \qquad (2, 3) = \left(\frac{a}{2}, \frac{b}{2}\right)$$

$$\Rightarrow \qquad \frac{a}{2} = 2 \text{ and } \frac{b}{2} = 3 \Rightarrow a = 4 \text{ and } b = 6.$$

Thus, the equation of the line is, $\frac{x}{4} + \frac{y}{6} = 1 \Rightarrow \mathbf{3x + 2y = 12}$

Example 12. Find the equation of the line which passes through the point (–5, 4) and is such that the portion of it between the axes is divided by the point in the ratio 2 : 1.

Solution : Let $P = (-5, 4)$. Let the line cut the x-axis at A $(a, 0)$ and the y-axis at B $(0, b)$.

Then the equation of the line is given by, $\frac{x}{a} + \frac{y}{b} = 1$ .... (1)

By data P divides AB in the ratio 2 : 1.

$$\therefore \qquad P = \left(\frac{2(0)+1\cdot a}{2+1}, \frac{2\cdot b+1(0)}{2+1}\right) \qquad \begin{matrix} 2 : 1 \\ (a, 0) \quad (0, b) \end{matrix}$$

$$\Rightarrow \qquad (-5, 4) = \left(\frac{a}{3}, \frac{2b}{3}\right) \Rightarrow \frac{a}{3} = -5 \text{ and } \frac{2b}{3} = 4 \Rightarrow \mathbf{a = -15 \text{ and } b = 6}$$

Thus equation (1) becomes, $\frac{x}{-15} + \frac{y}{6} = 1 \Rightarrow \mathbf{2x - 5y + 30 = 0}$

Example 13. A line passes through the point P (–4, 1) and cuts and x- and y-axes at A and B respectively. If P divides AB in the ratio 1 : 2, find the equation of the line.

Solution : Let $A = (a, 0)$ and $B = (0, b)$. Then x-intercept $= a$ and y-intercept $= b$. Thus the equation of the line will be of the form,

$$\frac{x}{a} + \frac{y}{b} = 1 \qquad .... (1)$$

Now, by data P $(-4, 1)$ divides AB in the ratio 1 : 2.

$$P = \left(\frac{1(0)+2\cdot a}{1+2}, \frac{2\cdot b+2(0)}{1+2}\right) \qquad \begin{matrix} 1 : 2 \\ A(a, 0) \quad B(0, b) \end{matrix}$$

$$\Rightarrow \qquad (-4, 1) = \left(\frac{2a}{3}, \frac{b}{3}\right) \Rightarrow \frac{2a}{3} = -4 \text{ and } \frac{b}{3} = 1 \Rightarrow a = -6 \text{ and } b = 3$$

Putting these values in equation (1) we get

$$\frac{x}{-6} + \frac{y}{3} = 1 \Rightarrow -x + 2y = 6 \Rightarrow \mathbf{x - 2y + 6 = 0}$$

This is the required equation.

V. Equation of a line in slope-intercept form

To derive the equation of a line whose slope is *m* and making an intercept '*c*' on *y*-axis

Let the line cut the y-axis at A. By data y-intercept of the line is 'c'. Thus, $OA = c$ and $A = (0, c)$.

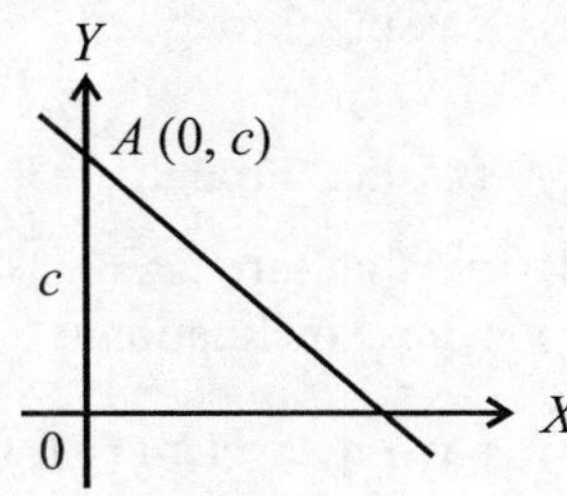

Thus the required line passes through the point $(0, c)$ and has slope m. Therefore, the equation of the line is

$$y - c = m(x - 0) \quad \text{(slope point formula)}$$

i.e., $y = mx + c$

Hence, the equation of the line whose slope is *m* and making an intercept '*c*' on *y*-axis is

$$\mathbf{y = mx + c}$$

This form of equation is called ***slope intercept* form.**

Note : Equation of the line passing through the origin and having slope m is $\mathbf{y = mx}$.

Example 14. Find the equation of the line whose *y*-intercept is –2 and slope is $\frac{3}{2}$.

Solution : Let $m = \frac{3}{2}$ and $c = -2$. The equation of the line is of the form $y = mx + c$.

i.e., $$y = \frac{3}{2}x - 2 \Rightarrow 2y = 3x - 4 \Rightarrow \mathbf{3x - 2y - 4 = 0}$$

Example 15. Find the equation of the line whose *y*-intercept is 3 units and makes an angle of 120° with the positive direction of *x*-axis.

Solution : Let $c = 3$. By data the angle of inclination is 120°.

Thus, slope $= m = \tan 120^\circ \Rightarrow m = \tan(180^\circ - 60^\circ) = -\tan(60^\circ) = -\sqrt{3}$.

Thus the equation of the line is,

i.e., $$y = -\sqrt{3}\,x + 3 \quad \text{or} \quad \mathbf{\sqrt{3}x + y - 3 = 0}$$

Exercise

I.

1. Find the equation of the line

(a) parallel to x-axis, which is 3 units above it

(b) parallel to x-axis, which is $2\frac{1}{2}$ units below it

(c) parallel to y-axis, which is 4 unit left of it

(d) parallel to y-axis, which is $\frac{1}{2}$ unit right of it

(e) parallel to x-axis and passing through $(3, -2)$

(f) parallel to x-axis and passing through $(2, -1)$
(g) perpendicular to x-axis and passing through $(-1, -2)$
(h) parallel to y-axis and passing through $(2, 0)$
(i) equidistant from the lines $y = 8$ and $y = -2$
(j) equidistant from the lines $x = -3$ and $x = 5$.

2. Find the equation of the line
(a) passing through $(2, -1)$, with slope -2
(b) passing through the origin, with slope $-\frac{3}{2}$
(c) passing through $(-3, -2)$, with slope $-\frac{2}{3}$
(d) passing through $(-2, 1)$ and making an angle of $135°$ with the positive direction of x-axis
(e) whose inclination is $30°$ and passing through $(0, -4)$
(f) passing through $(-1, -1)$ and perpendicular to the line whose slope is $-\frac{2}{5}$
(g) passing through $(2, -1)$ and parallel to the line whose slope is -1.

3. Find the equation of the line passing through the points
(a) $(2, 1)$ and $(4, -3)$
(b) $(-2, 5)$ and $(1, 0)$
(c) $(0, -4)$ and $(-2, -1)$
(d) $(1, 2)$ and $(-2, 5)$

4. Find the equation of the line whose x-intercept and y-intercept are respectively
(a) 2 and -3
(b) $\frac{3}{2}$ and $\frac{2}{5}$
(c) -4 and -2

5. Find the equation of the line
(a) with slope 3 and y-intercept is 2
(b) with slope $-\frac{1}{2}$ and y-intercept is -4
(c) with y-intercept 2 and parallel to the line whose slope is 2
(d) with y-intercept -3 and perpendicular to the line whose slope is $-\frac{1}{2}$.

II.

1. Find the equation of the line which passes through the point $(-2, 5)$ and cutting off equal intercepts on the axes.
2. Find the equation of the line passing through the point $(5, 6)$ and cutting off intercepts on the axes which are equal in magnitude but opposite in signs.
3. Find the equation of the line passing through $(4, -5)$ and the intercept on y-axis exceeds intercept on the x-axis by 3.
4. If a line $\frac{x}{a} + \frac{y}{b} = 1$ passes through the point $(7, -3)$ and $(-8, 12)$ find the values of a and b.
5. Find the equation of the line passing through $(-2, 2)$ and the sum of the intercepts on the coordinate axes is 3.
6. Find the equation of the line which passes through the point $(4, 6)$ such that the portion of it between the axes is bisected at this point.
7. A line passes through $(4, 3)$ and the portion of the line intercepted between the axes is bisected at this point; find its equation.

8. Find the equation of the line such that the segment between the axes is bisected at the point $(-3, 2)$.
9. Find the equation of the median through A, of the triangle ABC, where $A = (1, 4)$, $B = (2, -3)$ and $C = (-1, -2)$.

III.

1. Find the equation of a line which passes through the point $(-4, 1)$ and the portion of it between the axes is divided by this point in the ratio $1 : 2$.
2. Find the equation of a line which passes through the point $(3, -2)$ and cut off the positive intercept on x and y axes which are in the ratio $4 : 3$.
3. In what ratio is the line joining the points $(2, 3)$ and $(4, -5)$ divided by the line joining $(6, 8)$ and $(-3, 2)$.
4. A line passes through $(7, 9)$ and the portion of it between the axes is divided by this point in the ratio $3 : 1$. Find the equation of the line.
5. The area of the triangle formed by the coordinate axes and a line is 6 sq. units and the length of the hypotenuse is 5 units. Find the equation of the line.
6. Find the equation of the line at a distance of 3 units from the origin such that the perpendicular from the origin to the line makes an angle α such that $\tan\alpha = \frac{5}{12}$ with the positive direction of x-axis.

Answers

I. **1.** **(a)** $y = 3$ **(b)** $2y + 5 = 0$ **(c)** $x + 4 = 0$ **(d)** $2x - 1 = 0$ **(e)** $y + 2 = 0$
(f) $y + 1 = 0$ **(g)** $x + 1 = 0$ **(h)** $x - 2 = 0$ **(i)** $y - 3 = 0$ **(j)** $x - 1 = 0$.

2. **(a)** $2x + y - 3 = 0$ **(b)** $3x + 2y = 0$ **(c)** $2x + 3y + 12 = 0$ **(d)** $x + y + 1 = 0$
(e) $x - \sqrt{3}y - 4\sqrt{3} = 0$ **(f)** $5x - 2y + 3 = 0$ **(g)** $x + y - 1 = 0$.

3. **(a)** $2x + y - 5 = 0$ **(b)** $5x + 3y - 5 = 0$ **(c)** $3x + 2y + 8 = 0$ **(d)** $x + y - 3 = 0$.

4. **(a)** $3x - 2y - 6 = 0$ **(b)** $4x + 15y - 6 = 0$ **(c)** $x + 2y + 4 = 0$

5. **(a)** $y = 3x + 2$ **(b)** $x + 2y + 8 = 0$ **(c)** $y = 2x + 2$ **(d)** $y = 2x - 3$

6. **(a)** $\sqrt{3}x + y - 10 = 0$ **(b)** $x - y + 4\sqrt{2} = 0$ **(c)** $x - \sqrt{3}y + 4 = 0$ **(d)** $x + \sqrt{3}y - 10 = 0$

II. **1.** $x + y - 3 = 0$ **2.** $x - y + 1 = 0$ **3.** $5x + 2y - 10 = 0$, $x + 2y + 6 = 0$ **4.** $a = 4$, $b = 4$
5. $x + 2y - 2 = 0$, $2x - y + 6 = 0$ **6.** $3x + 2y - 24 = 0$ **7.** $3x + 4y - 24 = 0$
8. $2x - 3y + 12 = 0$ **9.** $13x - y - 9 = 0$

III. **1.** $x - 2y + 6 = 0$ **2.** $3x + 4y - 1 = 0$ **3.** $1 : 5$ externally **4.** $3x + 7y - 84 = 0$
5. $3x + 4y = 12$, $3x + 4y + 12 = 0$; $4x + 3y = 12$, or $4x + 3y + 12 = 0$
6. $12x + 5y = 39$.

3.6 Equation of a Line in General Form

In the previous section we have seen the different forms of equation of a line. Further all these forms were of the form $ax + by + c = 0$, where a, b and c are real constants. Now we shall see that the converse is also true. That is we shall show that the equation of the form $ax + by + c = 0$ always represents a line.

Definition. An equation of the form $ax + by + c = 0$, where a, b and c are real constants and a and b are not simultaneously zero is called general linear equation in the variables x and y or the general first degree equation in the variables x and y.

For example, $2x - 3y - 4 = 0$, $2x + 7 = 0$, $4y - 5 = 0$ are the first degree equations in x and y.

Theorem. A linear equation $ax + by + c = 0$ always represent a straight line and conversely.

Case 2: The line is not parallel to y–axis.

In this case the equation of the line will be of the form, $y = mx + c$ or $mx - y + c = 0$

Again this is also of the form $ax + by + c = 0$. Hence, the proof of the theorem

Note : The equation $ax + by + c = 0$ is called the ***general form*** of the equation of a line.

Consider the general equation of a line

$$Ax + By + C = 0 \quad \text{.... (1)}$$

This equation can be written as

$$By = -Ax - C \Rightarrow y = \left(-\frac{A}{B}\right)x + \left(-\frac{C}{B}\right) \quad (B \neq 0)$$

This equation is of the form $y = mx + c$, the slope-intercept form. Thus

$$\text{Slope of line (1)} = -\left(\frac{A}{B}\right) \quad \text{and} \quad y\text{–intercept of (1)} = -\left(\frac{C}{B}\right)$$

Again if $A \neq 0$ and $B \neq 0$, the equation (1) can be written as, $\dfrac{x}{-(C/A)} + \dfrac{y}{-(C/B)} = 1$

This equation is of the form $\dfrac{x}{a} + \dfrac{y}{b} = 1$, the intercept form of the equation of a line. Thus

$$x\text{–intercept of (1)} = -\left(\frac{C}{A}\right) \quad \text{and} \quad y\text{–intercept of (1)} = -\left(\frac{C}{B}\right)$$

Thus we have the following important results.

For the equation $ax + by + c = 0$ of a line, we have

$$\textbf{Slope} = -\frac{a}{b} = -\frac{\textbf{coefficient of } x}{\textbf{coefficient of } y}$$

$$x\textbf{–intercept} = -\frac{c}{a} = -\frac{\textbf{constant term}}{\textbf{coefficient of } x}$$

$$y\textbf{–intercept} = -\frac{c}{b} = -\frac{\textbf{constant term}}{\textbf{coefficient of } y}$$

Note : If the line $ax + by + c = 0$ cut the coordinate axes at A and B then we have $OA = -\dfrac{c}{a}$ and $OB = -\dfrac{c}{b}$. Thus

$$\boxed{\textbf{Area of } OAB = \left|\frac{1}{2}\, OA \cdot OB\right| = \frac{1}{2}\left|\frac{c^2}{ab}\right|}$$

The results of the following two theorems will help us to write the equation of any line parallel to given line and the equation of any line perpendicular to given line.

Theorem. The equation of any line parallel to the line $ax + by + c = 0$ is of the form $ax + by + \lambda = 0$, where λ is a constant.

3.7 Intersection of Two Lines

Consider two lines whose equations are

$$a_1 x + b_1 y + c_1 = 0 \quad \text{.... (1)}$$

and

$$a_2 x + b_2 y + c_2 = 0 \quad \text{.... (2)}$$

Supposing these two lines are parallel. Then their slopes are equal. That is

$$-\frac{a_1}{b_1} = -\frac{a_2}{b_2} \Rightarrow a_1 b_2 = a_2 b_1 \Rightarrow a_1 b_2 - a_2 b_1 = 0$$

Thus if $a_1 b_2 - a_2 b_1 = 0$ $\left(i.e., \frac{a_1}{b_1} = \frac{a_2}{b_2}\right)$ then the lines are parallel and they do not intersect.

If the lines are not parallel, then two lines (1) and (2) intersect at one point. Let the coordinates of the point of intersection of the lines (1) and (2) be (x_1, y_1). Thus both (1) and (2) must satisfy $x = x_1$ and $y = y_1$. Therefore, we have

$$a_1 x_1 + b_1 y_1 + c_1 = 0 ; \quad a_2 x_1 + b_2 y_1 + c_2 = 0$$

Solving these two equations we get x_1 and y_1.

We shall use the method of cross multiplication to solve. We get

$$\frac{x_1}{b_1 c_2 - c_1 b_2} = \frac{-y_1}{a_1 c_2 - c_1 a_2} = \frac{1}{a_1 b_2 - b_1 a_2} \Rightarrow \frac{x_1}{b_1 c_2 - c_1 b_2} = \frac{y_1}{c_1 a_2 - a_1 c_2} = \frac{1}{a_1 b_2 - b_1 a_2}$$

$$\therefore \quad x_1 = \frac{b_1 c_2 - c_1 b_2}{a_1 b_2 - b_1 a_2}, \quad y_1 = \frac{c_1 a_2 - a_1 c_2}{a_1 b_2 - b_1 a_2}$$

Thus, the coordinates of the point of intersection are $\left(\frac{b_1 c_2 - c_1 b_2}{a_1 b_2 - b_1 a_2}, \frac{c_1 a_2 - a_1 c_2}{a_1 b_2 - b_1 a_2}\right)$

Remember that $a_1 b_2 - b_1 a_2 \neq 0$

Thus, to find the coordinates of point of intersection of two non-parallel lines we solve the equations of the lines, simultaneously and the values of x and y so obtained are the coordinates of the point of intersection.

Example 1. Find the point of intersection of the lines $3x + 2y - 5 = 0$ and $4x - y - 3 = 0$.

Solution : We have,

$$3x + 2y - 5 = 0 \quad \text{.... (1)}$$

and

$$4x - y - 3 = 0 \quad \text{.... (2)}$$

To find the point of intersection, we solve (1) and (2) simultaneously.

Now,

$$\begin{array}{cccc} x & y & 1 & \\ 2 & -5 & 3 & 2 \\ -1 & -3 & 4 & -1 \end{array}$$

$$\frac{x}{-6-5} = \frac{y}{-20+9} = \frac{1}{-3-8} \Rightarrow \frac{x}{-11} = \frac{y}{-11} = \frac{1}{-11}$$

$$\Rightarrow x = \frac{-11}{-11}, \; y = \frac{-11}{-11} \Rightarrow x = 1, \quad y = 1$$

Thus the point of intersection is **(1, 1)**.

Example 2. Show that the lines $5x - 11y + 1 = 0$, $6x + 13y - 25 = 0$ and $x - 2y = 0$ are concurrent. Also find the point of concurrency.

Solution : We have,

$$5x - 11y + 1 = 0 \quad .... (1)$$

$$6x + 13y - 25 = 0 \quad .... (2)$$

$$x - 2y = 0 \quad .... (3)$$

To show that the lines are concurrent, we shall find the point of intersection of (1) and (2) and show that it lies on (3).

To solve (1) and (2) we have

$$\frac{x}{275-13} = \frac{-y}{-125-6} = \frac{1}{65+66} \Rightarrow \frac{x}{262} = \frac{y}{131} = \frac{1}{131} \Rightarrow x = 2, \; y = 1$$

Thus (1) and (2) intersect at **(2, 1)**

To show that it lies on the line (3), we shall put $x = 2$, $y = 1$ in (3), we get

$$2 - 2 = 0 \Rightarrow 0 = 0$$

Thus the lines (1), (2) and (3) meet at the (2, 1). Thus the lines are concurrent and point of concurrency = **(2, 1)**.

Example 3. For what values of k are the lines $x - 2y + 1 = 0$, $2x - 5y + 3 = 0$ and $5x - 9y + k = 0$ are concurrent.

Solution : We have,

$$x - 2y + 1 = 0 \quad .... (1)$$

$$2x - 5y + 3 = 0 \quad .... (2)$$

$$5x - 9y + k = 0 \quad .... (3)$$

First we shall find the point of intersection of (1) and (2). To this end consider,

$$\frac{x}{-6+5} = \frac{-y}{3-2} = \frac{1}{-5+4} \Rightarrow \frac{x}{-1} = \frac{y}{-1} = \frac{1}{-1} \Rightarrow x = 1, \; y = 1$$

$\therefore$ (1) and (2) intersect at (1, 1). By data (1), (2) and (3) are concurrent

$\Rightarrow$ (1, 1) lies on the line (3) $\Rightarrow$ $5 - 9 + k = 0 \Rightarrow$ **$k = 4$**

Example 4. Find the point of intersection of the lines $\frac{x}{a} + \frac{y}{b} = 1$ and $\frac{x}{b} + \frac{y}{a} = 1$.

Solution : The equations of the given lines can be written as

$$bx + ay - ab = 0 \quad \text{and} \quad ax + by - ab = 0$$

Solving, by the method of cross multiplication we have

$$\frac{x}{-a^2b+ab^2}=\frac{-y}{-ab^2+a^2b}=\frac{1}{b^2-a^2} \Rightarrow \frac{x}{ab(b-a)}=\frac{y}{ab(b-a)}=\frac{1}{b^2-a^2}$$

$$\Rightarrow x=\frac{ab(b-a)}{b^2-a^2}, \quad y=\frac{ab(b-a)}{b^2-a^2}$$

$$\Rightarrow x=\frac{ab}{b+a}, \quad y=\frac{ab}{b+a}$$

$\therefore$ point of intersection is $\left(\frac{ab}{b+a}, \frac{ab}{b+a}\right)$.

Example 5. The lines $ax+2y+1=0$, $bx+3y+1=0$ and $cx+4y+1=0$ are concurrent. Show that a, b, c are in A.P.

Solution : We have,

$$ax+2y+1=0 \quad \text{.... (1)}$$

$$bx+3y+1=0 \quad \text{.... (2)}$$

$$cx+4y+1=0 \quad \text{.... (3)}$$

Solving (1) and (2), we have,

$$\frac{x}{2-3}=\frac{-y}{a-b}=\frac{1}{3a-2b} \Rightarrow \frac{x}{-1}=\frac{y}{b-a}=\frac{1}{3a-2b} \Rightarrow x=\frac{-1}{3a-2b}, \quad y=\frac{b-a}{3a-2b}$$

Thus the point of intersection of (1) and (2) is $\left(-\frac{1}{3a-2b}, \frac{b-a}{3a-2b}\right)$.

By data this must lie on the line (3)

$$\therefore \quad c\left(\frac{-1}{3a-2b}\right)+4\left(\frac{b-a}{3a-2b}\right)+1=0 \Rightarrow -c+4b-4a+3a-2b=0$$

$$\Rightarrow -a+2b-c=0$$

$$\Rightarrow 2b=a+c \Rightarrow b=\frac{a+c}{2}$$

This shows that ***a, b, c* are in A.P.**

Exercise

I.

1. Find the point of intersection of lines

(a) $3x+4y+1=0$, $2x-y-3=0$ **(b)** $2x-3y+8=0$, $4x+5y=6$

(c) $3x+2y-5=0$, $4x-y=13$ **(d)** $2x+y=3$, $3x-y=7$

II.

1. Show that the following lines are concurrent

(a) $3x-4y+5=0$, $7x-8y+5=0$, $4x+5y=45$

(b) $2x-3y-7=0$, $3x-4y-13=0$, $8x-11y-33=0$

(c) $2x + 3y - 5 = 0$, $x - 3y + 2 = 0$, $7x - 3y - 4 = 0$

(d) $2x + 3y - 13 = 0$, $x + 2y - 8 = 0$, $3x - y - 3 = 0$

2. For what value of 'a' will the lines $3x + 4y + 1 = 0$, $ax + 2y - 3 = 0$, $2x - y - 3 = 0$ be concurrent.

3. For what value of 'a' the three lines $2y + 1 = 0$, $5x - 9y + a = 0$, $2x - 5y + 3 = 0$ are concurrent.

4. For what value of 'k' will the lines $3x - 4y + 5 = 0$, $7x - 8y + 5 = 0$, $4x + 5y + k = 0$ are concurrent.

5. Show that the lines $(a - b)x + (b - c)y = c - a$,
$(b - c)x + (c - a)y = a - b$ and $(c - a)x + (a - b)y = b - c$ are concurrent.

Answers

I. **1.** **(a)** $(1, -1)$ **(b)** $(-1, 2)$ **(c)** $\left(\frac{31}{11}, -\frac{19}{11}\right)$ **(d)** $(2, -1)$

II. **2.** $a = 5$ **3.** $\frac{37}{4}$ **4.** -45

3.8 Angle Between the Lines

In this section we shall derive a formula to find the angle between the lines whose slopes are known.

To find the angle between the lines whose slopes are m_1 and m_2

Let the lines whose slopes are m_1 and m_2 be AB and CD.

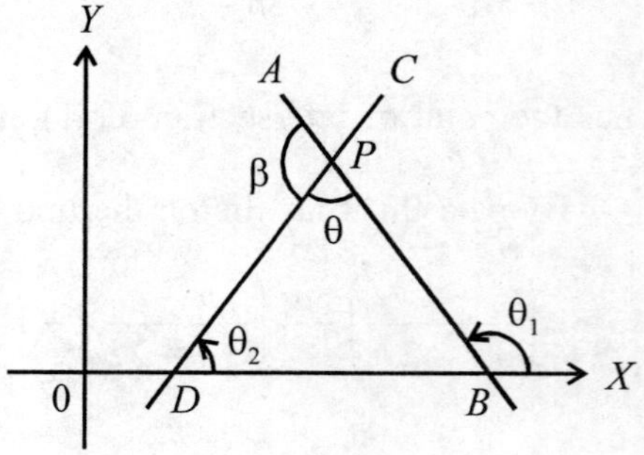

Let AB and CD intersect at P.

Let θ_1 and θ_2 be the angle made by the lines AB and CD respectively with the x-axis, in the positive direction.

$\therefore$ $m_1 = \tan\theta_1$ and $m_2 = \tan\theta_2$

Let θ be the acute angle between the lines.

From the figure, we have

$$\theta = \theta_1 - \theta_2 \Rightarrow \tan\theta = \tan(\theta_1 - \theta_2)$$

$$\Rightarrow \tan\theta = \frac{\tan\theta_1 - \tan\theta_2}{1 + \tan\theta_1 \tan\theta_2} \Rightarrow \tan\theta = \frac{m_1 - m_2}{1 + m_1 m_2}$$

Supposing, we consider the obtuse angle β between the lines.

Then, we have, $\beta = \pi - \theta$

$$\Rightarrow \tan\beta = \tan(\pi - \theta) \Rightarrow \tan\beta = -\tan\theta \Rightarrow \tan\beta = -\left(\frac{m_1 - m_2}{1 + m_1 m_2}\right) = \left(\frac{m_2 - m_1}{1 + m_1 m_2}\right)$$

From the above two facts, we conclude that :

The angle θ between the lines whose slopes are m_1 and m_2 is given by, $\tan\theta = \frac{m_1 - m_2}{1 + m_1 m_2}$

and θ will be acute or obtuse according as the r.h.s. is positive or negative.

Thus, the acute angle θ between the lines whose slopes are m_1 and m_2 is given by

$$\tan\theta = \left|\frac{m_1 - m_2}{1 + m_1 m_2}\right|$$

We shall consider two particular cases.

Case 1. Let the two lines be parallel to each other.

That is θ = 0. Thus, $\tan 0 = \frac{m_1 - m_2}{1 + m_1 m_2} \Rightarrow 0 = \frac{m_1 - m_2}{1 + m_1 m_2} \Rightarrow m_1 - m_2 = 0 \Rightarrow m_1 = m_2$

Thus, if two lines are parallel then their slopes are equal.

Case 2. Let the two lines are perpendicular to each other.

That is $\theta = 90^\circ \Rightarrow \cot\theta = \cot 90^\circ \Rightarrow \cot\theta = 0$

Now, $\tan\theta = \frac{m_1 - m_2}{1 + m_1 m_2} \Rightarrow \cot\theta = \frac{1 + m_1 m_2}{m_1 - m_2} \Rightarrow 0 = \frac{1 + m_1 m_2}{m_1 - m_2}$ $(\because \cot\theta = 0)$

$\Rightarrow 1 + m_1 m_2 = 0 \Rightarrow m_1 m_2 = -1$

Thus, if two lines are perpendicular to each other then the product of their slopes is –1.

Note : If the equations of the lines are

$$a_1 x + b_1 y + c_1 = 0 ; \qquad a_2 x + b_2 y + c_2 = 0$$

then their slopes are given by, $m_1 = -\frac{a_1}{b_1}$ and $m_2 = -\frac{a_2}{b_2}$

The angle θ between them is given by

$$\tan\theta = \frac{-\frac{a_1}{b_1} - \left(-\frac{a_2}{b_2}\right)}{1 + \left(-\frac{a_1}{b_1}\right)\left(-\frac{a_2}{b_2}\right)} \Rightarrow \tan\theta = \frac{a_2 b_1 - a_1 b_2}{a_1 a_2 + b_1 b_2}$$

Then acute angle is given by, $\tan\theta = \left|\frac{a_2 b_1 - a_1 b_2}{a_1 a_2 + b_1 b_2}\right|$

Also, the lines will be parallel if, $a_2 b_1 - a_1 b_2 = 0 \Rightarrow \frac{a_1}{a_2} = \frac{b_1}{b_2}$

and perpendicular, if, $a_1 a_2 + b_1 b_2 = 0$

Example 1. Find the acute angle between the lines $9x + 3y - 5 = 0$ and $2x + 4y + 3 = 0$.

Solution : We have, $9x + 3y - 5 = 0$ .... (1)

and $2x + 4y + 3 = 0$ .... (2)

Slope of (1) $= -\frac{9}{3} = -3 = m_1$, slope of (2) $= -\frac{2}{4} = -\frac{1}{2} = m_2$

The acute angle between the lines is given by

$$\tan\theta = \left|\frac{m_1 - m_2}{1 + m_1 m_2}\right| \Rightarrow \tan\theta = \left|\frac{-3 + \frac{1}{2}}{1 + \frac{3}{2}}\right| \Rightarrow \tan\theta = \left|-\frac{5}{5}\right| \Rightarrow \tan\theta = 1 \Rightarrow \theta = 45^\circ$$

Example 2. If the acute angle between the lines $4x - y + 7 = 0$ and $kx - 5y - 9 = 0$ is 45°, find k.

Solution : We have $4x - y + 7 = 0$ .... (1)

and $kx - 5y - 9 = 0$ .... (2)

$$\text{Slope of (1)} = -\left(\frac{4}{-1}\right) = 4 = m_1, \qquad \text{slope of (2)} = -\left(\frac{k}{-5}\right) = \left(\frac{k}{5}\right) = m_2$$

By data the angle between the (1) and (2) is 45°.

$$\therefore \quad \tan 45^\circ = \left|\frac{4 - \frac{k}{5}}{1 + \frac{4k}{5}}\right| \Rightarrow 1 = \left|\frac{20 - k}{5 + 4k}\right| \qquad (\because \tan 45^\circ = 1)$$

$$\Rightarrow \frac{20 - k}{5 + 4k} = \pm 1 \qquad (\because |a| = b \Rightarrow a = \pm b)$$

$$\Rightarrow 20 - k = \pm (5 + 4k)$$

$$\Rightarrow 20 - k = 5 + 4k \quad \text{or} \quad 20 - k = -(5 + 4k)$$

$$\Rightarrow \mathbf{k = 3} \quad \text{or} \quad \mathbf{k = -\frac{25}{3}}$$

Example 3. The angle between the lines is 45° and the slope of one of the lines is $\frac{1}{2}$. Find the slope of the another line.

Solution : Let $m_1 = \frac{1}{2}$ and m be the slope of another line.

$$\therefore \text{ We have, } \tan 45^\circ = \left|\frac{\frac{1}{2} - m}{1 + (m/2)}\right|$$

$$1 = \left|\frac{1 - 2m}{2 + m}\right| \Rightarrow \frac{1 - 2m}{2 + m} = \pm 1 \Rightarrow 1 - 2m = \pm (2 + m)$$

$$\Rightarrow 1 - 2m = 2 + m \quad \text{or} \quad 1 - 2m = -(2 + m)$$

$$\Rightarrow \mathbf{m = -\frac{1}{3}} \quad \text{or} \quad \mathbf{m = 3}$$

Exercise

I.

1. Find the acute angle between the lines

(a) $\sqrt{3}x + y + 6 = 0,\quad \sqrt{3}x - y + 11 = 0$ **(b)** $3x + y + 7 = 0,\quad x + 2y + 9 = 0$

(c) $x + \sqrt{3}y = 1,\quad \sqrt{3}x = y + 2$ **(d)** $x - 2y + 3 = 0,\quad 3x + y - 1 = 0$

(e) $x \cos \alpha_1 + y \sin \alpha_1 = p_1,\quad x \cos \alpha_2 + y \sin \alpha_2 = p_2$

(f) $\frac{x}{a} + \frac{y}{b} = -1$ and $\frac{x}{a} - \frac{y}{b} = 1$

II.

1. Find the angle between the lines OB and CD, where $O = (0, 0)$, $B = (2, 3)$, $C = (2, -2)$ and $D = (3, 5)$.
2. If the acute angle between the lines $3x + y + 7 = 0$ and $x + ky + 9 = 0$ is 45° find the value of k.
3. Find the value of a, such that the angle θ between the lines $x + y = 3$ and $2x + ay - 1 = 0$ is given by $\tan \theta = \frac{1}{5}$.
4. Find the angle between the diagonals of a parallelogram whose vertices taken in order are $A\,(1, -2)$, $B\,(2, 0)$, $C\,(1, 6)$ and $D\,(0, 4)$.

Answers

I. 1. **(a)** 60° **(b)** 45° **(c)** 90° **(d)** $\tan^{-1}\frac{7}{2}$ **(e)** $\alpha_1 - \alpha_2$ **(f)** 90°

II. 1. $\tan^{-1}\left(\frac{11}{23}\right)$ 2. $k = 2$ or $-\frac{1}{2}$ 3. $a = 3$ or $\frac{4}{3}$ 4. $\tan^{-1}\left(\frac{1}{2}\right)$

Chapter 4

Circles

4.1 Introduction

In this chapter we shall derive the equation of a circle, and few results connected to circles, including the concept of radical axis of two circles and orthogonal circles.

4.2 Definition and Equation of a Circle

Definition : A circle is the locus of a point, which moves on a plane such that its distance from the fixed point is always a given constant.

The fixed point is called the **centre** and the given constant distance is called the **radius** of the circle.

We shall derive the equation of the circle with the centre at (x_1, y_1) and radius r units.

Theorem : The equation of the circle with centre at (x_1, y_1) and radius equal to r units is given by

$$(x - x_1)^2 + (y - y_1)^2 = r^2$$

Proof : Let $P(x_1, y_1)$ be the centre and $Q(x, y)$ be any point on the circle.
By the definition of a circle,

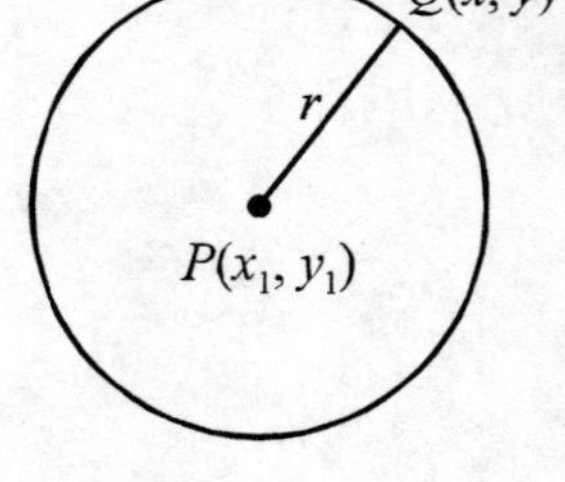

$$|PQ| = r \text{ units} \quad \Rightarrow \quad PQ^2 = r^2$$

$$\Rightarrow \quad (x - x_1)^2 + (y - y_1)^2 = r^2$$

which is the equation of the circle with centre at (x_1, y_1) and radius equal to r units.

Note : The equation of the circle with centre at the origin and radius equal to r units is given by $\mathbf{x^2 + y^2 = r^2}$ $\quad (\because x_1 = 0, y_1 = 0)$

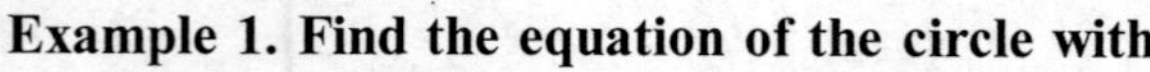

Example 1. Find the equation of the circle with

(a) Centre = (4, 2), radius = 3 (b) Centre = (–1, 2), radius = $\frac{1}{2}$ (c) Centre = $\left(\frac{3}{2}, -\frac{1}{2}\right)$, radius = 4.

Solution : (a) The equation of the circle is

$$(x - 4)^2 + (y - 2)^2 = 3^2 \Rightarrow x^2 - 8x + 16 + y^2 - 4y + 4 - 9 = 0 \Rightarrow \mathbf{x^2 + y^2 - 8x - 4y + 11 = 0}$$

(b) The equation of the circle is

$$(x + 1)^2 + (y - 2)^2 = \left(\frac{1}{2}\right)^2 \Rightarrow x^2 + 2x + 1 + y^2 - 4y + 4 - \frac{1}{4} = 0 \Rightarrow \mathbf{4x^2 + 4y^2 + 8x - 16y + 19 = 0}$$

(c) The equation of the circle is

$$\left(x - \frac{3}{2}\right)^2 + \left(y + \frac{1}{2}\right)^2 = 16 \Rightarrow x^2 - 3x + \frac{9}{4} + y^2 + y + \frac{1}{4} - 16 = 0 \Rightarrow \mathbf{4x^2 + 4y^2 - 12x + 4y - 54 = 0}$$

If we know the co–ordinates of the centre and the length of the radius, we can write the equation of the circle. Thus by finding these from the given data, we can write the equation of the circle.

Example 2. Find the equation of the circle with centre at (2, 1) and passing through the origin.

Solution : The centre $C = (2, 1)$. By data the circle passes through the origin $O\ (0, 0)$.

Thus, radius $= OC = \sqrt{(2-0)^2 + (1-0)^2} = \sqrt{4+1} = \sqrt{5}$

The equation of the circle with centre $C\ (2, 1)$ and radius $\sqrt{5}$ is given by

$$(x-2)^2 + (y-1)^2 = 5 \Rightarrow x^2 - 4x + 4 + y^2 - 2y + 1 = 5 \Rightarrow \mathbf{x^2 + y^2 - 4x - 2y = 0}$$

Example 3. Find the equation of the circle two of whose diameters are $2x + y - 5 = 0$ and $x + 2y - 7 = 0$ and passing through the point (1, –3).

Solution : The equation of the diameters are

$$2x + y - 5 = 0 \quad \text{.... (1)}$$

$$x + 2y - 7 = 0 \quad \text{.... (2)}$$

The centre of the required circle is the point of intersection of the diameters

Solving (1) and (2) we have,

$$\frac{x}{-7+10} = \frac{-y}{-14+5} = \frac{1}{4-1}$$

$$\Rightarrow \frac{x}{3} = \frac{y}{9} = \frac{1}{3} \Rightarrow \mathbf{x = 1,\ y = 3}$$

$$a_1x + b_1y + c_1 = 0$$
$$a_2x + b_2y + c_2 = 0$$
$$\Rightarrow \frac{x}{b_1c_2 - c_1b_2} = \frac{-y}{a_1c_2 - c_1a_2} = \frac{1}{a_1b_2 - b_1a_2}$$

Thus, the centre $C = (1, 3)$. Again by data the circle passes through $A\ (1, -3)$

$$\therefore \quad \text{radius} = CA = \sqrt{(1-1)^2 + (3+3)^2} \Rightarrow \text{radius} = 6$$

Hence, the equation of the circle with centre (1, 3) and radius = 6 is

$$(x-1)^2 + (y-3)^2 = 36 \Rightarrow x^2 - 2x + 1 + y^2 - 6y + 9 - 36 = 0 \Rightarrow \mathbf{x^2 + y^2 - 2x - 6y - 26 = 0}$$

Theorem : The equation of the circle described on the line joining the points (x_1, y_1) and (x_2, y_2) as a diameter is given by $x^2 + y^2 - (x_1 + x_2)x - (y_1 + y_2)y + (x_1x_2 + y_1y_2) = 0$

Proof : Let $A = (x_1, y_1)$ and $B = (x_2, y_2)$ be the be the ends of the diameter.

Let $P\,(x, y)$ be any point on the circle described on AB as diameter. Then $A\hat{P}B = 90°$.

($\because$ angle in a semi circle)

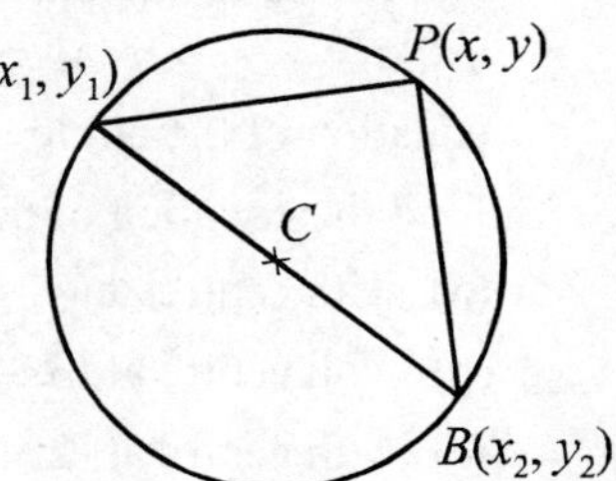

$\Rightarrow$ AP is perpendicular to PB.

$$\Rightarrow \quad (\text{Slope of } AP)\ (\text{Slope of } PB) = -1$$

$$\Rightarrow \quad \left(\frac{y - y_1}{x - x_1}\right)\left(\frac{y - y_2}{x - x_2}\right) = -1$$

$$\Rightarrow \quad (y - y_1)(y - y_2) = -(x - x_1)(x - x_2)$$

$$\Rightarrow \quad (x - x_1)(x - x_2) + (y - y_1)(y - y_2) = 0$$

$\Rightarrow \quad \mathbf{x^2 + y^2 - (x_1 + x_2)x - (y + y_1)y + (x_1x_2 + y_1y_2) = 0}$

This is the equation of the required circle.

Example 4. Find the equation of the circle described on the line joining the points (3, 4) and (1, –2) as a diameter.

Solution : Let $A = (x_1, y_1) = (3, 4)$, $B = (x_2, y_2) = (1, -2)$

Equation of the required circle is given by,

$$x^2 + y^2 - (x_1 + x_2)x - (y_1 + y_2)y + (x_1x_2 + y_1y_2) = 0$$

i.e., $$x^2 + y^2 - (3 + 1)x - (4 - 2)y + (3 - 8) = 0 \Rightarrow \mathbf{x^2 + y^2 - 4x - 2y - 5 = 0}$$

Example 5. Find the equation of the point circle with centre at (4, –5).

Solution : A circle with radius 0 is called a "point circle"

The equation of the circle with centre at (4, –5) and radius = 0 is

$$(x - 4)^2 + (y + 5)^2 = 0 \quad \Rightarrow \mathbf{x^2 + y^2 - 8x + 10y + 41 = 0}$$

Exercise

I.

1. Find the equation of the circle with centre at C and radius r, where

(a) $C = (2, 3)$, $r = 4$ **(b)** $C = (-1, -3)$, $r = \frac{1}{2}$ **(c)** $C = (2, 0)$, $r = \frac{3}{2}$

(d) $C = \left(\frac{2}{3}, \frac{1}{3}\right)$, $r = 2$ **(e)** $C = (0, -1)$, $r = 4$ **(f)** $C = (0, 0)$, $r = 4$

(g) $C = (3, 4)$, $r = 6$

2. Find the centre of the circle, two of whose diameters are

(a) $x + y = 2$, and $x - y = 0$ **(b)** $2x - 3y = 1$, and $3x - 2y = 2$

(c) $x + 2y - 1 = 0$, and $3x + y + 2 = 0$ **(d)** $x = 0$, and $y = 2x + 3$.

(e) $y = 0$, and $y = x - 5$. **(f)** x – axis and y – axis.

3. Write the equation of the point circle, with centre at

(a) (4, –5) **(b)** (–3, 2) **(c)** (1, 0)

II.

1. Find the equation of the circle

(a) two of the diameters are $x + y = 6$ and $x + 2y = 4$ and its radius is 10 units.

(b) two of whose diameters are $x + y = 3$ and $2x - y - 3 = 0$ and radius 3 units.

(c) two of whose diameters are $x - 2y - 4 = 0$ and $2x - y - 1 = 0$ and radius 4 units.

(d) two of whose diameters are $x + y = 4$ and $x - y = 2$ and passing through the point (2, – 1).

(e) two of whose diameters are $x = 4$ and $y = 3$ and passing through (1, 1).

2. The lines $2x - 3y = 5$ and $3x - 4y = 7$ are the diameters of a circle of area 154 sq. units. Find the equation of the circle.

3. Find the equation of the circle

(a) with centre at (–2, 1) and passing through the origin

(b) with centre at (3, –2) and passing through (3, 2)

(c) with centre at (2, 1) and passing through (0, – 1)

4. Find the equation of the circle described on the line joining the points A and B as a diameter, where
   (a) $A = (-5, 1), B = (1, 3)$ (b) $A = (2, 0), B = (0, 2)$ (c) $A = (5, 0), B = (0, 5)$
   (d) $A = (3, -6), B = (-2, 5)$ (e) $A = (3, 4), B = (1, -2)$ (f) $A = (a, 0), B = (0, b)$
   (g) $A = (a \cos\theta, b \sin\theta), B = (b \cos\theta, a \sin\theta)$.

Answers

I. (a) $x^2 + y^2 - 4x - 6y - 3 = 0$ (b) $4x^2 + 4y^2 + 8x + 24y + 39 = 0$
(c) $4x^2 + 4y^2 - 16x + 7 = 0$ (d) $9x^2 + 9y^2 - 12x - 6y - 31 = 0$
(e) $x^2 + y^2 + 2y - 15 = 0$ (f) $x^2 + y^2 = 16$ (g) $x^2 + y^2 - 6x - 8y - 11 = 0$

2. (a) $(1, 1)$ (b) $\left(\frac{4}{5}, \frac{1}{5}\right)$ (c) $(-1, 1)$ (d) $(0, 3)$ (e) $(5, 0)$ (f) $(0, 0)$

3. (a) $x^2 + y^2 - 8x + 10y + 41 = 0$ (b) $x^2 + y^2 + 6x - 4y + 13 = 0$ (c) $x^2 + y^2 - 2x + 1 = 0$

II.

1. (a) $x^2 + y^2 - 16x + 4y - 32 = 0$ (b) $x^2 + y^2 - 4x - 2y - 4 = 0$
(c) $9x^2 + 9y^2 + 12x + 42y - 91 = 0$ (d) $x^2 + y^2 - 6x - 2y + 5 = 0$ (e) $x^2 + y^2 - 8x - 6y + 12 = 0$

2. $x^2 + y^2 - 2x + 2y - 47 = 0$

3. (a) $x^2 + y^2 + 4x - 2y = 0$ (b) $x^2 + y^2 - 6x + 4y - 3 = 0$ (c) $x^2 + y^2 - 4x - 2y - 3 = 0$

4. (a) $x^2 + y^2 + 4x - 4y - 2 = 0$ (b) $x^2 + y^2 - 2x - 2y = 0$ (c) $x^2 + y^2 - 5x - 5y = 0$
(d) $x^2 + y^2 - x + y - 36 = 0$ (e) $x^2 + y^2 - 4x - 2y - 5 = 0$ (f) $x^2 + y^2 - ax - by = 0$
(g) $x^2 + y^2 - (a + b)\cos\theta \cdot x - (a + b)\sin\theta \cdot y + ab = 0$

4.3 General Equation of a Circle

The equation of the circle with the centre (x_1, y_1) and radius equal to r is given by

$$(x - x_1)^2 + (y - y_1)^2 = r^2$$

This equation can be written as $x^2 + y^2 - 2x_1x - 2y_1y + (x_1^2 + y_1^2 - r^2) = 0$

This equation is

(a) a second degree equation in x and y.

(b) one in which there is no second degree term containing xy

(c) one in which the coefficients of x^2 and y^2 are equal and equal to 1.

Now, we shall see that any equation of this type represents a circle.

Theorem : The second degree equation $x^2 + y^2 + 2gx + 2fy + c = 0$ always represents a circle, with centre at $(-g, -f)$ and the radius equal to $\sqrt{g^2 + f^2 - c}$.

Proof : Consider the equation

$$x^2 + y^2 + 2gx + 2fy + c = 0 \quad \text{.... (1)}$$

$$\Rightarrow \quad x^2 + y^2 + 2gx + 2fy = -c$$

Adding both sides $g^2 + f^2$, we get,

$$x^2 + 2gx + g^2 + y^2 + 2fy + f^2 = g^2 + f^2 - c$$

$$\Rightarrow \quad (x + g)^2 + (y + f)^2 = g^2 + f^2 - c$$

$$\Rightarrow \qquad [x - (-g)]^2 + [y - (-f)]^2 = \left(\sqrt{g^2 + f^2 - c}\right)^2 \qquad \text{.... (2)}$$

This equation is of the form $(x - x_1)^2 + (y - y_1)^2 = r^2$ .... (3)

which represents a circle with centre at (x_1, y_1) and radius r.

Thus, the equation (2), hence (1) represents a circle.

Comparing the equations (3) and (2) we have, the centre of the circle (1) at $(-g, -f)$ and radius $\sqrt{g^2 + f^2 - c}$.

Hence, the equation $x^2 + y^2 + 2gx + 2fy + c = 0$ represents a circle with, centre = $(-g, -f)$ and radius $\sqrt{g^2 + f^2 - c}$.

Definition : The equation of the circle in the form

$$x^2 + y^2 + 2gx + 2fy + c = 0$$

is called the standard form of the equation of the circle.

Note : 1. If $g^2 + f^2 - c > 0$, then $\sqrt{g^2 + f^2 - c}$ is a real number. In this case the circle is called a ***real*** circle.

2. If $g^2 + f^2 - c = 0$, then the radius of the circle will be zero. In this case the circle is called a ***point*** circle.

3. If $g^2 + f^2 - c < 0$, then $\sqrt{g^2 + f^2 - c}$ is an imaginary number. In this case the circle is called an ***imaginary*** circle.

Consider the general second degree equation $ax^2 + 2hxy + by^2 + 2gx + 2fy + c = 0$

This equation represents a circles, if $a = b$ and $h = 0$.

That is the equation,

$$ax^2 + ay^2 + 2gx + 2fy + c = 0$$

always represents a circle. This equation is called **general equation of the circle.**

In particular if $a = b = 1$, we get the standard form of the equation of the circle.

Consider the standard form of the equation of the circle

$$x^2 + y^2 + 2gx + 2fy + c = 0$$

Now, centre = $(-g, -f)$ and radius = $\sqrt{g^2 + f^2 - c}$

Clearly,

$$\textbf{x - coordinate of the centre} = -\left(\frac{\textbf{coff. of } x}{2}\right)$$

$$\textbf{y - coordinate of the centre} = -\left(\frac{\textbf{coff. of } y}{2}\right)$$

$$\textbf{radius} = \sqrt{\left(\frac{\textbf{coff. of } x}{2}\right)^2 + \left(\frac{\textbf{coff. of } y}{2}\right)^2 - \textbf{constant term}}$$

Example 1. Find the centre and radius of the following circles

(a) $x^2 + y^2 - 2x + 6y - 20 = 0$ (b) $3x^2 + 3y^2 - 6x + 4y + 1 = 0$ (c) $2x^2 + 2y^2 + 6x - 10y + 9 = 0$

Solution : (a) The equation of the circle is given by

$$x^2 + y^2 - 2x + 6y - 20 = 0$$

Here, $2g = -2,\ 2f = 6,\ c = -20 \Rightarrow g = -1,\ f = 3,\ c = -20$

$\therefore$ $C = (-g, -f) = \mathbf{(1, -3)}$, radius $= \sqrt{g^2 + f^2 - c} = \sqrt{1 + 9 + 20} = \mathbf{\sqrt{30}}$

(b) The equation of the circle is given by $3x^2 + 3y^2 - 6x + 4y + 1 = 0$

In this equation, the coefficient of x^2 and y^2 are not 1. Thus, we have to divide, throughout by 3. We get,

$$x^2 + y^2 - 2x + \frac{4}{3}y + \frac{1}{3} = 0$$

Here, $2g = -2,\ 2f = \frac{4}{3},\ c = \frac{1}{3} \Rightarrow g = -1,\ f = \frac{2}{3},\ c = \frac{1}{3}$

$\therefore$ $C = (-g, -f) = \left(\mathbf{1, -\frac{2}{3}}\right)$, radius $= \sqrt{g^2 + f^2 - c} = \sqrt{1 + \frac{4}{9} - \frac{1}{3}} = \mathbf{\frac{\sqrt{10}}{3}}$

(c) The equation of the given circle is $2x^2 + 2y^2 + 6x - 10y + 9 = 0$

Here also we have to make the coefficients of x^2 and y^2 as 1. To this end we shall divide throughout by 2. We get

$$x^2 + y^2 + 3x - 5y + \frac{9}{2} = 0$$

Here, $2g = 3,\ 2f = -5,\ c = \frac{9}{2} \Rightarrow g = \frac{3}{2},\ f = -\frac{5}{2},\ c = \frac{9}{2}$

$\therefore$ $C = (-g, -f) = \left(\mathbf{-\frac{3}{2}, \frac{5}{2}}\right)$, and radius $= \sqrt{g^2 + f^2 - c} = \sqrt{\frac{9}{4} + \frac{25}{4} - \frac{9}{2}} = \frac{4}{2} = \mathbf{2}$

Note : The students are adviced to practice to write the coordinates of the centre and the length of the radius directly by observing the coefficients of x and y, and the constant term, in the standard form of the equation of the circle.

For example, consider the circle $x^2 + y^2 - 2x + 4y + 2 = 0$. We have,

$$C = (-g, -f) = \left(-\left(-\frac{2}{2}\right), -2\right) = \mathbf{(1, -2)}\ ;$$

and radius $= \sqrt{g^2 + f^2 - c} = \sqrt{(-1)^2 + 2^2 - 2} = \sqrt{1 + 4 - 2} = \mathbf{\sqrt{3}}$

Example 2. If the radius of the circle $x^2 + y^2 - 2x + 3y + k = 0$ is $\frac{5}{2}$, find k.

Solution : The given equation is $x^2 + y^2 - 2x + 3y - k = 0$.

By data, radius $= \frac{5}{2} \Rightarrow \sqrt{\left(-\frac{2}{2}\right)^2 + \left(\frac{3}{2}\right)^2 - k} = \frac{5}{2}$

$$\Rightarrow \sqrt{1+\frac{9}{4}-k} = \frac{5}{2}$$

$$\Rightarrow \frac{13}{4} - k = \frac{25}{4} \Rightarrow k = \frac{13}{4} - \frac{25}{4} \Rightarrow k = -\frac{12}{4} = \mathbf{-3}$$

Example 3. If one end of the diameter of the circle $x^2 + y^2 - 4x - 2y - 5 = 0$ is (3, 4), find the coordinate of the other end.

Solution : The equation of the circle is $x^2 + y^2 - 4x - 2y - 5 = 0$

Now, centre $C = (-g, -f) = (2, 1)$. One end of the diameter is $A = (3, 4)$. Let the other end be $B = (h, k)$

$$\therefore \quad C = \text{midpoint of } AB \quad \Rightarrow \quad (2, 1) = \left(\frac{3+h}{2}, \frac{4+k}{2}\right)$$

$$\Rightarrow \quad 2 = \frac{3+h}{2}, \quad 1 = \frac{4+k}{2}$$

$$\Rightarrow \quad 4 = 3 + h, \quad 2 = 4 + k \quad \Rightarrow h = 1, \quad k = -2$$

$\therefore$ the other end of the diameter is **(1, –2).**

Example 4. If $x^2 + y^2 - 14x + by + 53 = 0$ represents a point circle, find 'b'.

Solution : By data $x^2 + y^2 - 14x + by + 53 = 0$ represents a point circle. Thus, the radius = 0.

$$\Rightarrow \quad \sqrt{g^2 + f^2 - c} = 0 \quad \Rightarrow \quad \sqrt{(-7)^2 + \left(\frac{b}{2}\right)^2 - 53} = 0$$

$$\Rightarrow \quad 49 + \frac{b^2}{4} - 53 = 0 \Rightarrow \frac{b^2}{4} = 4 \Rightarrow b^2 = 16 \Rightarrow \mathbf{b = \pm 4}$$

Example 5. Write one of the diameters of the circle $x^2 + y^2 + 4x - 2y = 0$.

Solution : The equation of the circle is $x^2 + y^2 + 4x - 2y = 0$

Diameter is a line through the centre of the circle. Now, the centre $C = (-2, 1)$

Any line through (–2, 1) is of the form $y - 1 = m(x + 2)$ .... (1)

[Any line through (x_1, y_1) is of the form $y - y_1 = m(x - x_1)$]

Now, for different values of m, (1) represents a diameter of the circle. Let us put $m = 0$, we get $\mathbf{y - 1 = 0}$, which is one of the diameters.

[Similarly, by putting other values form, we get other diameters - for $m = 1$, we get $x - y + 3 = 0$, for $m = \frac{1}{2}$, we get $x - 2y + 4 = 0$, etc.]

Consider the equation of the circle $x^2 + y^2 + 2gx + 2fy + c = 0$ .... (1)

Any circle concentric with this circle, will have the same centre. Thus, in its equation the coefficients of x^2 and y^2 will be same as that of (1). Hence, we have the following:

The equation of the circle concentric with the circle $x^2 + y^2 + 2gx + 2fy + c = 0$ is of the form $x^2 + y^2 + 2gx + 2fy + k = 0$.

For different values of k, we get different circles concentric with the given circle.

Example 6. Find the equation of the circle concentric with the circle $x^2 + y^2 - 6x - 2y + 4 = 0$ and passing through the point (3, 2).

Solution : The equation of the circle concentric with

$$x^2 + y^2 - 6x - 2y + 4 = 0$$

is of the form $\qquad x^2 + y^2 - 6x - 2y + k = 0 \qquad$ .... (1)

By data this must pass through (3, 2). Thus, the equation (1) must satisfy $x = 3,\ y = 2$.

$$\Rightarrow \quad 9 + 4 - 18 - 4 + k = 0 \Rightarrow \mathbf{k = 9}$$

Hence, the equation of the required circle is

$$\mathbf{x^2 + y^2 - 6x - 2y + 9 = 0}$$

Example 7. Find the equation of the circle whose centre is (–3, 1) and passing through the centre of the circle $x^2 + y^2 + 2x - 4y + 4 = 0$.

Solution : Let $C = (-3, 1)$. The centre of the given circle $x^2 + y^2 + 2x - 4y + 4 = 0$ is $A = (-1, 2)$ $(\because 2g = 2, 2f = -4)$

The required circle passes through A. Thus,

$$\text{radius} = CA \Rightarrow \text{radius} = \sqrt{(-3+1)^2 + (1-2)^2} = \sqrt{5}$$

Now, the equation of the circle with centre (–3, 1) and radius $\sqrt{5}$ units is given by

$$(x + 3)^2 + (y - 1)^2 = (\sqrt{5})^2 \Rightarrow \mathbf{x^2 + y^2 + 6x - 2y + 5 = 0}$$

2nd method.

By data the centre of the required circle is (–3, 1), $\therefore\ -g = -3$ and $-f = 1 \Rightarrow g = 3,\ f = -1$

Thus, the required circle is of the form $x^2 + y^2 + 6x - 2y + k = 0$.

This must pass through the centre of $x^2 + y^2 + 2x - 4y + 4 = 0$. i.e., it must pass through (–1, 2).

$$\Rightarrow \quad 1 + 4 - 6 - 4 + k = 0 \Rightarrow k = 5$$

Putting $k = 5$, we get the required equation, $\mathbf{x^2 + y^2 + 6x - 2y + 5 = 0}$.

Note : The standard form of the equation of the circle has three parameters g, f and c. Thus to find the equation of a circle, we have to find corresponding values of g, f and c. Hence, we require at least three conditions to determine the equation of a circle.

Example 8. Find the equation of the circle passing through the points (0, 0) and (1, 1), and has its centre on x–axis.

Solution : Let the equation of the required circle be

$$x^2 + y^2 + 2gx + 2fy + c = 0 \qquad \text{.... (1)}$$

Using the conditions in the problem, we shall determine the values of g, f and c.

By data the circle passes through the point (0, 0). Thus the equation (1) must satisfy $x = 0$ and $y = 0$.

$$\Rightarrow \quad 0 + 0 + 2g \cdot 0 + 2f \cdot 0 + c = 0 \Rightarrow \mathbf{c = 0.}$$

Again by data the circle passes through the point (1, 1). Thus the equation (1) must satisfy $x = 1$ and $y = 1$.

$$\Rightarrow \quad 1 + 1 + 2g + 2f + c = 0 \Rightarrow 2g + 2f + 2 = 0. \qquad (\because c = 0)$$

$$\Rightarrow g + f + 1 = 0. \qquad \text{.... (2)}$$

Again by data, the centre $(-g, -f)$ of (1) lies on x – axis

$\Rightarrow$ y-coordinate of the centre $= 0 \quad \Rightarrow \quad f = 0.$

Putting $f = 0$ in (2), we get $g + 1 = 0 \quad \Rightarrow \quad g = -1.$

Putting $g = -1, f = 0, c = 0$ in (1) we get the equation of the circle, $\mathbf{x^2 + y^2 - 2x = 0}$

Example 9. Find the equation of the circle passing through the points $(-1, 2)$ and $(3, -2)$, and has its centre on $x = 2y$.

Solution : Let the equation of the required circle be

$$x^2 + y^2 + 2gx + 2fy + c = 0$$

By data this passes through the point $(-1, 2)$ and $(3, -2)$

$$\Rightarrow \quad -2g + 4f + c + 5 = 0 \qquad .... (1)$$

$$\text{and} \quad 6g - 4f + c + 13 = 0 \qquad .... (2)$$

Again by data the centre $(-g, -f)$ lies on $x = 2y$.

$$\Rightarrow \quad -g = -2f \Rightarrow g = 2f. \qquad .... (3)$$

Solving the equations (1), (2) and (3) we get, g, f, and c.

Consider (1) – (2), we get $\quad -8g + 8f - 8 = 0 \quad \Rightarrow \quad g - f + 1 = 0.$

Putting $g = 2f$, we get, $\quad 2f - f + 1 = 0 \quad \Rightarrow \mathbf{f = -1}$; Now, $g = 2f \quad \Rightarrow \quad \mathbf{g = -2}$

From (1) we have, $\quad c = 2g - 4f - 5 \Rightarrow c = -4 + 4 - 5 \Rightarrow \mathbf{c = -5.}$

Thus the equation of the required circle is $\mathbf{x^2 + y^2 - 4x - 2y - 5 = 0.}$

Example 10. Find the equation of the circle passing through the points $(5, 3)$, $(1, 5)$ and $(3, -1)$.

Solution : Let the equation of the circle be

$$x^2 + y^2 + 2gx + 2fy + c = 0 \qquad .... (1)$$

By data this passes through the point (5, 3), (1, 5) and (3, –1). Thus, we have,

$$10g + 6f + c + 34 = 0 \qquad .... (2)$$

$$2g + 10f + c + 26 = 0 \qquad .... (3)$$

$$6g - 2f + c + 10 = 0 \qquad .... (4)$$

Solving these three equations we get f, g and c. To this end consider,

$$(2) - (3) \Rightarrow \quad 8g - 4f + 8 = 0 \quad \Rightarrow \quad 2g - f + 2 = 0 \qquad .... (5)$$

$$(3) - (4) \Rightarrow \quad -4g + 12f + 16 = 0 \quad \Rightarrow \quad g - 3f - 4 = 0 \qquad .... (6)$$

Solving (5) and (6), we get,

$$\frac{g}{4+6} = \frac{f}{-8-2} = \frac{1}{-6+1} \quad \Rightarrow \quad \frac{g}{10} = \frac{f}{10} = \frac{1}{-5} \Rightarrow \mathbf{g = -2, \quad f = -2}$$

Now, (1) $\Rightarrow \quad c = -10g - 6f - 34 \Rightarrow c = 20 + 12 - 34 \Rightarrow \mathbf{c = -2}$

Thus, the equation of the required circle is $\mathbf{x^2 + y^2 - 4x - 4y - 2 = 0}$

Example 11. Show that the following points $(2, 0)$, $(-1, 3)$, $(-2, 0)$ and $(1, -1)$ are concyclic.

Solution : We shall find the equation of the circle passing through the points (2, 0), (–1, 3) and (–2, 0) and then we shall show that the fourth point (1, –1) lies on it.

Let the equation of the circle passing through (2, 0), (–1, 3) and (–2, 0), be

$$x^2 + y^2 + 2gx + 2fy + c = 0$$

Since it passes through (2, 0), (–1, 3) and (–2, 0), we have,

$$4g + c + 4 = 0 \quad .... (1)$$

$$-2g + 6f + c + 10 = 0 \quad .... (2)$$

$$-4g + c + 4 = 0 \quad .... (3)$$

Solving these equations we get f, g and c.

Consider, (1) – (3) we get, $8g = 0 \Rightarrow$ $\mathbf{g = 0}$. Putting $g = 0$ in (1) we get, $\mathbf{c = -4}$.

Now (2) $\Rightarrow 6f = 2g - c - 10 \Rightarrow 6f = +4 - 10 \Rightarrow$ $\mathbf{f = -1}$

Thus, the equation of the circle is $x^2 + y^2 - 2y - 4 = 0$

Now, putting $x = 1$ and $y = -1$, in l.h.s, we get, $1 + 1 + 2 - 4 = 4 - 4 = 0 =$ r.h.s

Thus, the point (1, –1) lies on the circle. **Hence, the given four points are concyclic.**

Example 12. Find the equation of the diameter of the circle $x^2 + y^2 + 2x + 4y - 4 = 0$ which when produced passes through the point (–6, 2).

Solution : The equation of the circle is $x^2 + y^2 + 2x + 4y - 4 = 0$

The centre, $C = (-1, -2)$ $(\because g = 1, f = 2)$

The required diameter is the line joining the points C (–1, –2) and the point (–6, 2).

Its equation is, $\dfrac{y+2}{x+1} = \dfrac{2+2}{-6+1}$ (two points form of the eqn. to line)

$$\frac{y+2}{x+1} = \frac{4}{-5} \Rightarrow -5y - 10 = 4x + 4 \Rightarrow \mathbf{4x + 5y + 14 = 0}$$

Example 13. Show that the line $2x + y = 4$ passes through the centre of the circle $x^2 + y^2 - 3x - 2y + 2 = 0$.

Solution : The centre of the circle $x^2 + y^2 - 3x - 2y + 2 = 0$ is $C = \left(\frac{3}{2}, 1\right)$ $\left(\because g = \frac{-3}{2}, f = -1\right)$

Put, $x = \frac{3}{2}$, $y = 1$ in l.h.s of $2x + y = 4$, we get, $2\left(\frac{3}{2}\right) + 1 = 3 + 1 = 4 =$ r.h.s

Thus, $\left(\frac{3}{2}, 1\right)$ **lies on $2x + y = 4$.**

Example 14. Show that the length of the chord of the circle $x^2 + y^2 + 2gx + 2fy + c = 0$ intercepted by (i) x–axis is $2\sqrt{g^2 - c}$ (ii) y – axis is $2\sqrt{f^2 - c}$.

Solution : The equation of the circle is $x^2 + y^2 + 2gx + 2fy + c = 0$

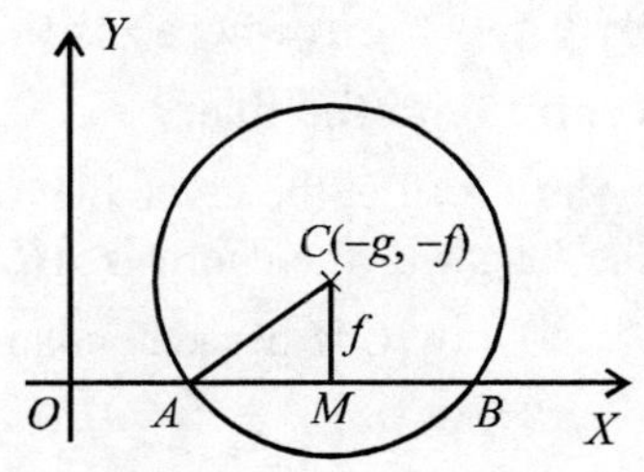

(i) Let the circle cut the x–axis at A and B.

Draw perpendicular CM from the centre onto AB.

Clearly M in the Midpoint of AB.

Now, x-intercept $= 2AM$

$\Rightarrow$ x-intercept $= 2\sqrt{AC^2 - CM^2}$ $(\because AMC$ is a right triangle)

$\Rightarrow$ x-intercept $= 2\sqrt{g^2 + \cancel{f^2} - c - \cancel{f^2}}$ $(\because AC =$ radius, $CM = -f)$

$\Rightarrow$ **x-intercept** $= 2\sqrt{g^2 - c}$

(ii) Let the circle cut the y–axis at D and E. Draw perpendicular CN from the centre on DE.

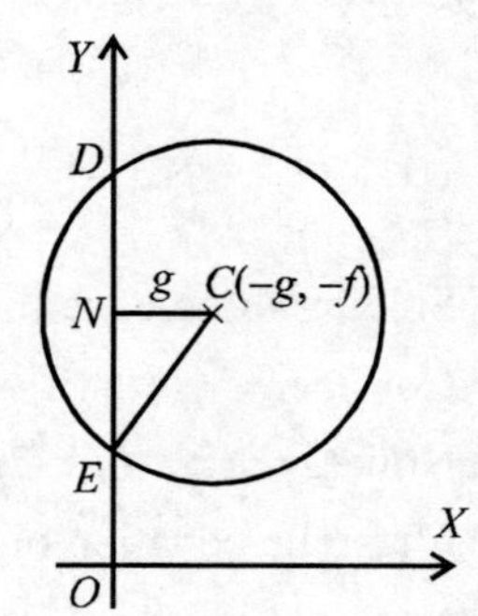

Clearly N in the Midpoint of DE.

Now, $\quad y\text{-intercept} = 2EN$

$\Rightarrow \quad y\text{-intercept} = 2\sqrt{CE^2 - CN^2}$

$\Rightarrow \quad y\text{-intercept} = 2\sqrt{g^2 + f^2 - c - g^2} \quad (\because CE = \text{radius}, CN = g)$

$\Rightarrow$ **y-intercept** $= 2\sqrt{f^2 - c}$

Example 15. Find the length of the chord of the circle $x^2 + y^2 - 6x + 4y + 5 = 0$ intercepted by x–axis.

Solution : The equation of the circle is $x^2 + y^2 - 6x + 4y + 5 = 0$

$$x - \text{intercept} = 2\sqrt{g^2 - c} = 2\sqrt{9 - 5} = 2\sqrt{4} = 4 \qquad (\because g = 3, c = 5)$$

Example 16. Find the length of the chord of the circle $x^2 + y^2 + 3x - 2 = 0$ intercepted by y–axis.

Solution : The equation of the circle is

$$x^2 + y^2 + 3x - 2 = 0$$

$$y - \text{intercept} = 2\sqrt{f^2 - c} = 2\sqrt{0 + 2} = \mathbf{2\sqrt{2}} \qquad (\because f = 0, c = -2)$$

Example 17. Find the equation of the circle passing through the origin and making positive intercepts 3 and 5 on the coordinate axes.

Solution : Let the equation of the circle be $x^2 + y^2 + 2gx + 2fy + c = 0$

The circle passes through the origin $\Rightarrow$ $\mathbf{c = 0}$. By data x – intercept = 3 and y – intercept = 5

$\Rightarrow \quad 2\sqrt{g^2 - c} = 3 \quad$ and $\quad 2\sqrt{f^2 - c} = 5$

$\Rightarrow \quad 2\sqrt{g^2} = 3 \quad$ and $\quad 2\sqrt{f^2} = 5 \qquad (\because c = 0)$

$\Rightarrow \quad -2g = 3 \quad$ and $\quad -2f = 5$

(We take –ve sign, as the centre must lie in the first quadrant)

Thus the equation of the circle is $\mathbf{x^2 + y^2 - 3x - 5y = 0}$

Example 18. Show that the length of the chord intercepted by a line with the given circle is given by $2\sqrt{r^2 - p^2}$, where r is the radius of the circle and p is the length of the perpendicular from the centre onto the line.

Solution : Let the given line cut the given circle at A and B. Clearly the length of the chord is AB. Let C be the centre of the circle.

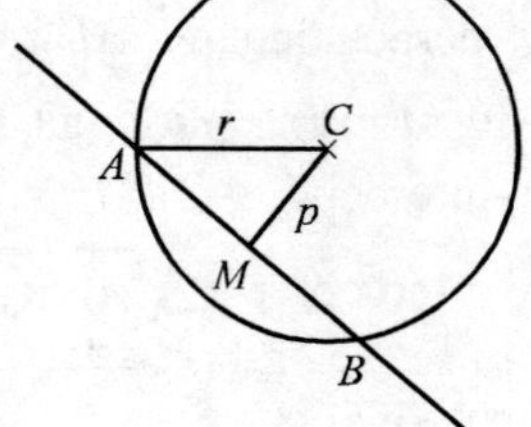

Draw CM perpendicular to the line AB.

Now, $\quad AB = 2AM$

$\Rightarrow \quad AB = 2\sqrt{CA^2 - CM^2}$

$$\Rightarrow \qquad AB = 2\sqrt{r^2 - p^2} \quad (\because CA = r, CM = p)$$

Thus, **the length of the chord = $2\sqrt{r^2 - p^2}$.**

Example 19. Find the length of the chord intercepted by the circle $x^2 + y^2 + 4x + 6y - 12 = 0$ and $x + 4y - 6 = 0$.

Solution : The equation of the circle is $x^2 + y^2 + 4x + 6y - 12 = 0$

Equation of the line is, $x + 4y - 6 = 0$

Now, centre C and radius r of the circle are

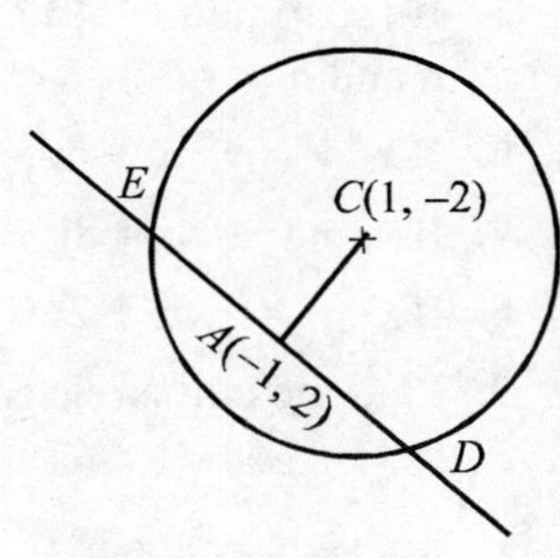

$$C = (-2, -3), \quad r = \sqrt{4 + 9 + 12} = 5$$

p = length of perpendicular from C onto the line.

$$p = \left|\frac{-2 + 4(-3) - 6}{\sqrt{1 + 16}}\right|^{*} = \left|\frac{-2 - 12 - 6}{\sqrt{17}}\right| = \frac{20}{\sqrt{17}}$$

Now, length of the chord $= 2\sqrt{r^2 - p^2} = 2\sqrt{25 - \frac{400}{17}} = 2\sqrt{\frac{425 - 400}{17}} = 2\sqrt{\frac{25}{17}} = \mathbf{\frac{10}{\sqrt{17}}}$

Example 20. Find the equation of the chord of the circle $x^2 + y^2 - 2x + 4y - 17 = 0$ bisected at $(-1, 2)$.

Solution : The equation of the circle is $x^2 + y^2 - 2x + 4y - 17 = 0$

The centre $C = (1, -2)$. The mid point of the chord $A = (-1, 2)$.

The chord is perpendicular to AC.

$$\therefore \quad \text{(Slope of the chord) (Slope of } AC) = -1 \quad \Rightarrow \quad \text{slope of the chord} = \frac{-1}{\left(\frac{2 + 2}{-1 - 1}\right)} = \frac{1}{2}$$

The chord ED passes through $(-1, 2)$. Thus the equation of the chord ED is

$$y - 2 = \frac{1}{2}(x + 1) \Rightarrow 2y - 4 = x + 1 \Rightarrow \mathbf{x - 2y + 5 = 0}$$

Exercise

I.

1. Find the centre and the radius of the circles

(a) $x^2 + y^2 - 4x - y - 5 = 0$

(b) $2x^2 + 2y^2 - 8x - 12y - 3 = 0$

(c) $3x^2 + 3y^2 - 6x - 12y - 2 = 0$

(d) $x^2 + y^2 + 8x + 4y - 5 = 0$

(e) $4x^2 + 4y^2 - 6x - 9 = 0$

(f) $x^2 + y^2 - 2x \cos\alpha - 2y \sin\alpha - 1 = 0$

(g) $(x - 2)(x - 4) + (y - 1)(y + 3) = 0$.

---

* Length of the perpendicular from (x_1, y_1) onto the line $ax + by + c = 0$ is given by $p = \frac{ax_1 + by_1 + c}{\sqrt{a^2 + b^2}}$

2. If the radius of the circle $x^2 + y^2 + 2x + 4y + p = 0$ is 5 units find p.
3. If the radius of the circle $x^2 + y^2 + 4x - 2y - k = 0$ is 4 units find k.
4. Find the other end of the diameter, if one end of the diameter of the circle

   (a) $x^2 + y^2 = 25$ is $(5, 0)$ **(b)** $x^2 + y^2 + 6x - 8y - 15 = 0$ is $(3, 2)$

   (c) $x^2 + y^2 + 4x - 6y - 12 = 0$ is $(-5, -1)$ **(d)** $x^2 + y^2 - 6x + 2y - 31 = 0$ is $(7, 4)$.
5. If (a, b) and $(-5, 1)$ are the end points of a diameter of the circle $x^2 + y^2 + 4x - 4y - 2 = 0$, find a and b.
6. If $x^2 + y^2 + ax + by - 3 = 0$ represents a circle with centre at $(1, -3)$, find a and b.
7. If $x^2 + y^2 - 4x - 8y + k = 0$ represents a point circle find k.
8. If $x^2 + y^2 + ax + 2y + 5 = 0$ represents a point circle find a.
9. Find the unit circle concentric with the circle $x^2 + y^2 - 8x + 4y = 8$.

II.

1. Find the equation of the circle whose centre is same as the centre of the circle $x^2 + y^2 + 6x + 2y + 1 = 0$, and passing through the point $(-2, -1)$.
2. Find the equation of the circle whose centre is same as the centre of the circle $x^2 + y^2 - 6x + 4y + 9 = 0$, and passing through the point $(-2, 3)$.
3. Find the equation of the circle whose centre is $(-2, 3)$ and passing through the centre of the circle $x^2 + y^2 - 6x + 4y + 9 = 0$.
4. Find the equation of the circle passing through the centre of the circle $x^2 + y^2 - 2x - 4y - 20 = 0$, and centre at $(4, -2)$.
5. Find the equation of the circle two of whose diameters are $x + y = 3$ and $2x + y = 2$, and passing through the centre of the circle $x^2 + y^2 - 4x + 2y - 1 = 0$.
6. Find the equation of the circle concentric with the centre of the circle $x^2 + y^2 - 2x + 2y - 1 = 0$, and double its area.
7. Find the equation of the circle concentric with the centre of the circle $3x^2 + 3y^2 - 6x + 9y - 2 = 0$, and having $\frac{2}{3}$ of its area.
8. Find the equation of the diameter of the circle $x^2 + y^2 + 6x - 2y = 6$ which when produced passes through the point $(1, -2)$.
9. Find the equation of the diameter of the circle $2x^2 + 2y^2 + 3x - 5y - 1 = 0$, which when produced passes through the point $(-1, 2)$.
10. Write one of the diameters of the circle $x^2 + y^2 - 6x + 2y = 0$.
11. Write one of the diameters of the circle $x^2 + y^2 - 2y + 1 = 0$.
12. Show that the line $4x - y = 17$ passes through the centre of the circle $x^2 + y^2 - 8x + 2y = 0$.
13. Find the length of the chord of the circle $x^2 + y^2 - 6x + 15y - 16 = 0$ intercepted by the x – axis.
14. Find the length of the chord of the circle $x^2 + y^2 + 3x - y - 6 = 0$ intercepted by the y – axis.
15. Find the lengths of the chords of the circle $x^2 + y^2 - 6x - 4y - 12 = 0$ on the coordinate axes.

III.

1. Find the equation of the circle

(a) passing through the origin, having its centre on the x – axis and radius 2 units.

(b) passing through (2, 3), having its centre on the x – axis and radius 5 units.

(c) passing through the points (5, 1), (3, 4) and has its centre on the x – axis.

(d) passing through the points (1, 2) and (2, 1) and has its centre on the y – axis.

(e) passing through the points (0, 5) and (6, 1) and has its centre on the line $12x + 5y = 25$.

(f) passing through the points (1, –4) and (5, 2) and has its centre on the line $x - 2y + 9 = 0$.

(g) passing through the points (0, –3) and (0, 5) and whose centre lies on $x - 2y + 5 = 0$.

(h) passing through (1, 1) and (2, 2) and having radius 1.

2. Find the equation of the circle passing through the points

(a) (0, 2), (3, 0), (3, 2) **(b)** (1, 1), (–2, 2), (–6, 0)

(c) (1, 1), (5, –5), (6, –4) **(d)** (1, 0), (3, 0), (0, 2)

(e) (5, 7), (6, 6), (2, –2) **(b)** (p, q), $(p, 0)$, $(0, q)$

(g) (0, 1), (2, 3), (–2, 5) **(b)** (0, 0), $(a, 0)$, $(0, b)$

3. Show that the following points are concyclic

(a) (0, 0), (1, 1), (5, –5), (6, –4) **(b)** (2, –4), (3, –1), (3, –3), (0, 0)

(c) (1, 0), (2, –7), (8, 1), (9, –6)

4. Find the equations of the circles whose radius is 5 and which passes through the points on x – axis at distances 3 from the origin.

5. A circle has radius 3 units and its centre lies on the line $y = x - 1$. Find the equation of the circle if it passes through (7, 3).

6. Find the equation of the circle which cuts intercepts of the lengths 'a' and 'b' on axes and passing through the origin.

7. Find the equation of the circle passing through the origin and making positive x-intercept 'a' units and negative y-intercepts 'b' units.

8. Find the length of the chord intercepted by the

(a) circle $x^2 + y^2 - 8x - 6y = 0$ and the line $x - 7y - 8 = 0$.

(b) circle $x^2 + y^2 = 9$ and the line $x + 2y = 3$.

(c) circle $x^2 + y^2 - 6x - 2y + 5 = 0$ and the line $x - y + 1 = 0$.

9. Find the equation of the chord of the circle

(a) $x^2 + y^2 = r^2$ bisected at (a, b). **(b)** $x^2 + y^2 + 2gx + 2fy + c = 0$ bisected at (a, b).

(c) $x^2 + y^2 + 4x - 6y - 9 = 0$ bisected at (0, 1). **(d)** $x^2 + y^2 - 4x - 2y - 20 = 0$ bisected at (2, 3).

10. Find the points of intersection of the circle

(a) $x^2 + y^2 = 9$ and the line $x + 2y = 3$.

(b) $x^2 + y^2 - 6x - 2y + 5 = 0$ and the line $x - y + 1 = 0$

(c) $x^2 + y^2 + 4x + 6y - 12 = 0$ and the line $x + 4y - 6 = 0$

Also find the length of the chord.

Answers

I. 1. (a) $\left(2, \frac{1}{2}\right), \frac{\sqrt{37}}{2}$ **(b)** $(2, 3), \sqrt{\frac{29}{2}}$ **(c)** $(1, 2), \sqrt{\frac{17}{3}}$ **(d)** $(-4, -2), 2$ **(e)** $\left(\frac{3}{4}, 0\right), \frac{3\sqrt{5}}{4}$

(f) $(\cos\alpha, \sin\alpha), \sqrt{2}$ **(g)** $(3, -1), \sqrt{5}$ **2.** $p = -20$ **3.** $k = 11$

4. (a) $(-5, 0)$ **(b)** $(-9, 6)$ **(c)** $(1, 7)$ **(d)** $(-1, -6)$

5. $a = 1, b = 3$ **6.** $a = -2, b = 6$ **7.** $k = 20$ **8.** $a = \pm 4$

9. $x^2 + y^2 - 8x + 4y + 19 = 0.$

II. 1. $x^2 + y^2 + 6x + 2y + 9 = 0$ **2.** $x^2 + y^2 - 6x + 4y - 37 = 0$

3. $x^2 + y^2 + 4x - 6y - 37 = 0$ **4.** $x^2 + y^2 - 8x + 4y - 5 = 0$

5. $x^2 + y^2 + 2x - 8y - 17 = 0$ **6.** $x^2 + y^2 - 2x + 2y - 4 = 0$

7. $36x^2 + 36y^2 - 72x + 108y + 23 = 0$ **8.** $3x + 4y + 5 = 0$

9. $3x + y + 1 = 0$

10. $y + 1 = 0,$ **11.** $y = x + 1$ **13.** 10 units **14.** 5 units **15.** $2\sqrt{21}$, 8

III.

1. (a) $x^2 + y^2 \pm 4x = 0$ **(b)** $\begin{cases} x^2 + y^2 - 12x + 11 = 0 \\ x^2 + y^2 + 4x - 21 = 0 \end{cases}$

(c) $2x^2 + 2y^2 - x - 47 = 0$ **(d)** $x^2 + y^2 - 5 = 0$

(e) $3x^2 + 3y^2 - 10x - 6y - 45 = 0$ **(f)** $x^2 + y^2 + 6x - 6y - 47 = 0$

(g) $x^2 + y^2 + 6x - 2y - 15 = 0$ **(h)** $\begin{cases} x^2 + y^2 - 2x - 4y + 4 = 0 \\ x^2 + y^2 - 4x - 2y + 4 = 0 \end{cases}$

2. (a) $x^2 + y^2 - 3x - 2y = 0$ **(b)** $x^2 + y^2 + 4x + 6y - 12 = 0$

(c) $x^2 + y^2 - 6x + 4y = 0$ **(d)** $2x^2 + 2y^2 - 8x - 7y + 6 = 0$

(e) $x^2 + y^2 - 4x - 6y - 12 = 0$ **(f)** $x^2 + y^2 - px - qy = 0$

(g) $3x^2 + 3y^2 + 2x - 20y + 17 = 0$ **(h)** $x^2 + y^2 - ax - by = 0$

4. $x^2 + y^2 - 8y - 9 = 0, x^2 + y^2 + 8y - 9 = 0$ **5.** $x^2 + y^2 - 14x - 12y + 76 = 0, x^2 + y^2 - 8x - 6y + 16 = 0$

6. $x^2 + y^2 \pm ax \pm by = 0$ **7.** $x^2 + y^2 - ax + by = 0$

8. (a) $5\sqrt{2}$ **(b)** $\frac{12\sqrt{5}}{5}$ **(c)** $\sqrt{2}$

9. (a) $ax + by = a^2 + b^2$ **(b)** $(a + g)x + (b + f)y = (a^2 + b^2 + ag + bf)$

(c) $x - y + 1 = 0$ **(d)** $y - 3 = 0$

10. (a) $(3, 0), \left(-\frac{9}{5}, \frac{12}{5}\right); \frac{12}{\sqrt{5}}$ **(b)** $(2, 3), (1, 2); \sqrt{2}$ **(c)** $(-2, 2), \left(\frac{6}{17}, \frac{24}{17}\right); \frac{10}{\sqrt{17}}$

Chapter 5

Conic Sections

5.1 Introduction

In this chapter we shall study the curves called **Parabola, Ellipse** and **Hyperbola**, which are known as **conic sections** or simply **conics**. The curves known as conics were named after their discovery as the sections of a plane with a right circular cone.

The surface generated by a line L moving about a circle and passing through a fixed point V, which lies on the line L' perpendicular to the plane of the circle and passing through the centre of the circle, is called a **right circular cone**. The fixed point V is called the **vertex** of the cone. The vertex separates the cone in to two parts called **nappes**. The line L is called **generator**. The circle about which the generator moves is called the **base** of the cone. The line L' through V and perpendicular to the base is called the **axis** of the cone. A portion of a right circular cone of two nappes is shown in the Fig. 1.

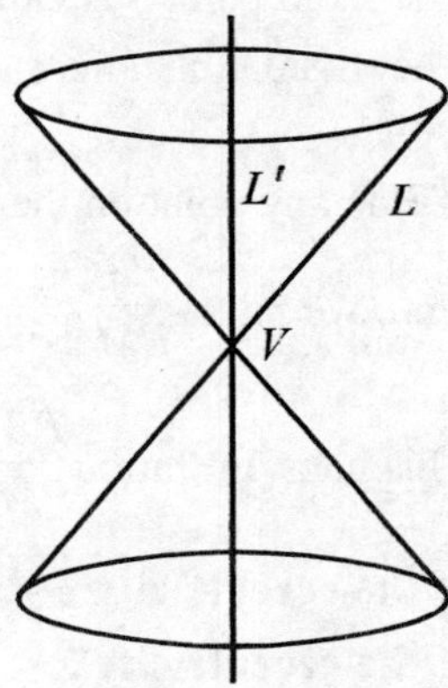

Fig. 1

Greek mathematician, **Applonius** (260-170 BC) found that the section of a right circular cone of two nappes, by a plane (not passing through the vertex of the cone), turned out to be three important curves latter known as **parabola, ellipse** and **hyperbola**, these together called **conic sections** or **conics**.

The common points of the cone and the plane is called a section of the cone by a plane.

The section of the cone by a plane, which is parallel to one of the generator is called a **parabola**. (Fig. 2(a)).

The section of the cone by a plane which is not parallel to any generator, is called an **ellipse** (Fig. 2(b)).

The section of the cone by a plane parallel to the axis and intersecting both the nappes, is called **hyperbola** (Fig. 2(c)).

Now, we shall define a conic section analytically, as a locus of the point which moves under definite condition. The particular cases of this definition will give us the three different conics – **parabola**, **ellipse** and **hyperbola**.

(a) Parabola

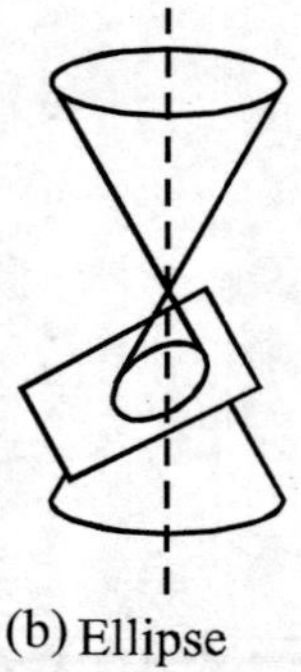
(b) Ellipse

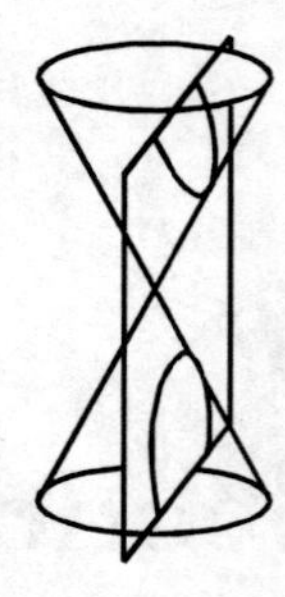
(c) Hyperbola

Fig. 2

5.2 Conic Sections

Definition : A conic section is the locus of the point which moves such that the ratio of its distance from the fixed point S to its distance from the fixed line l, is always a fixed positive constant.

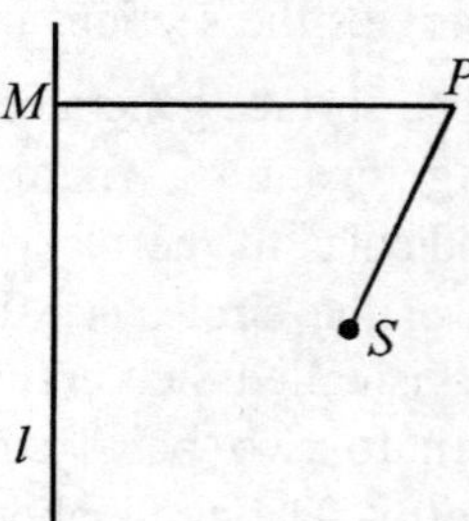

The fixed point S is called the **focus** of the conic section.

The fixed line l is called the **directrix** of the conic section.

If P is any point on the locus (conic section), then by the definition, we have $\frac{PS}{PM}$ = a fixed constant (> 0).

This constant ratio $\frac{PS}{PM}$ is called the **eccentricity** of the conic section and it is denoted by e.

If the eccentricity $e = 1$, the conic section is called a parabola.

If the eccentricity $e < 1$, the conic section is called a ellipse.

If the eccentricity $e > 1$, the conic section is called a hyperbola.

We shall see each one of the conic section separately and study few properties and results associated to them.

5.3 Parabola

Parabola is a conic section whose eccentricity is 1.

That is Parabola is the locus of the point which moves such that its distance from the fixed point (called the focus) is equal to its distance from the fixed line (called the directrix).

We shall derive the equation of the parabola in its standard form.

Theorem : The equation of a parabola with proper choice of coordinate axes is $y^2 = 4ax$.

Proof : Let S be the focus and the line l be the directrix. Draw SZ, through S and perpendicular to the directrix. Let O be the midpoint of the line segment SZ.

Let $SZ = 2a \quad (a > 0) \Rightarrow OZ = OS = a$

We shall choose O as the origin and the line $ZOSX$ as x-axis. The line YOY' through O, perpendicular to the x-axis will be y-axis.

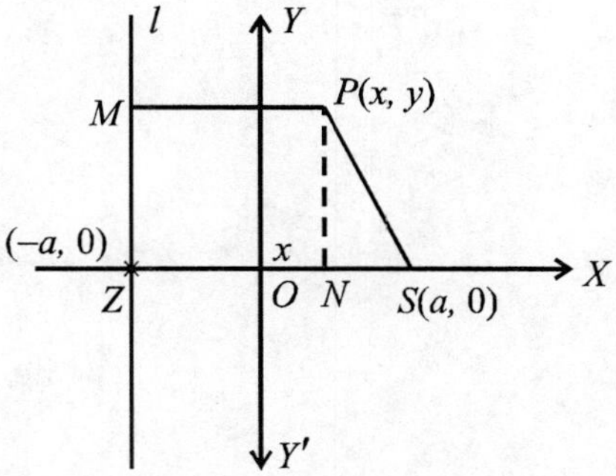

With this choice of coordinate axes, we have,

$$\boldsymbol{S = (a, 0)} \quad \text{and} \quad \boldsymbol{Z = (-a, 0)}$$

The equation of the **directrix l is $x = -a$.**

Let $P(x, y)$ be any point on the parabola. Then by the definition of parabola we have

Distance of P from S = Distance of P from l

$$\Rightarrow \quad |SP| = |PM|$$

$$\Rightarrow \quad SP^2 = ZN^2 \qquad (\because PM = ZN)$$

$$\Rightarrow \quad SP^2 = (OZ + ON)^2$$

$$\Rightarrow \quad (x-a)^2 + y^2 = (a+x)^2 \qquad (\because |OZ| = a, ON = x)$$

$$\Rightarrow \quad x^2 - 2ax + a^2 + y^2 = a^2 + 2ax + x^2$$

$$\Rightarrow \quad \boldsymbol{y^2 = 4ax}$$

which is the equation of the parabola.

Note : The equation $y^2 = 4ax$ is also called **standard form** of the equation of the parabola.

5.4 Shape of the Parabola $y^2 = 4ax$

We shall note few observations from the equation $y^2 = 4ax$, which will help us to trace the curve parabola.

1. If y is replaced by $-y$ in the equation, the equation remains same, i.e., $(-y)^2 = 4ax \Rightarrow y^2 = 4ax$. This shows that if (x, y) is any point on the curve $y^2 = 4ax$, then $(x, -y)$ is also a point on the curve. Thus, the curve is symmetric about the x-axis, i.e., the shape of the curve above the x-axis is the mirror image (about the x-axis) of the shape of the curve below the x-axis.

2. If $x < 0$, then y^2 will be a negative quantity (note that $a > 0$) and therefore $y^2 = 4ax$ will have no real solution for y. This shows that no part of the curve lies to the left side of the x-axis.

3. If $y = 0$, the only value of x we get is zero. Thus the curve cuts the x-axis at the origin (0, 0).

4. If $x = 0$, we get $y^2 = 0$, which gives $y = 0$ (twice). Thus, the curve cuts the y-axis at the origin and further the y-axis meets the curve only at the origin. That is, y-axis is a tangent to the curve at the origin.

5. For any point (x, y) on the parabola, we have

$$y^2 = 4ax \quad \Rightarrow \quad y = \pm 2\sqrt{ax}$$

i.e.,

$$y = 2\sqrt{ax} \quad \text{and} \quad y = -2\sqrt{ax}$$

This shows that, as x increases from 0 to ∞, y also increases from 0 to ∞ or y increases from 0 to $-\infty$. Thus, the two branches of the parabola, laying on opposite sides of the x-axis, will extend to infinity towards the positive directions of the x-axis.

From the above discussion and by plotting few points, whose coordinates satisfy $y^2 = 4ax$, it is found the shape of the parabola is as shown in the following figure.

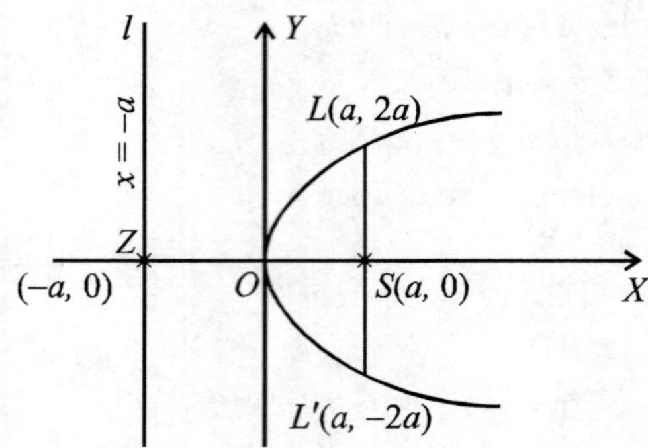

The origin O is called the **vertex** of the parabola, $y^2 = 4ax$, it is also denoted by V (0, 0).

The line ZSX is called the **axis of the parabola**, its equation is $y = 0$.

The focus $S = (a, 0)$ and the equation of the directrix is $x = -a$ i.e., $x + a = 0$.

The y-axis is called the **tangent at the vertex** – its equation is $x = 0$.

Note : The distance between the vertex and the focus is a, the distance between, the vertex and the directrix is a, and the distance between the directrix and the focus is $2a$.

Definition : The chord passing through the focus and perpendicular to the axis of the parabola is called the latus rectum of the parabola.

In the figure LSL' is the latus rectum. The points L and L' on the parabola are called **ends of latus rectum**, the length LL' is called the **length of the latus rectum**.

Clearly, the x-coordinates of L and L' is a because $OS = a$. To find the corresponding y-coordinates, we shall put $x = a$ in $y^2 = 4ax$, we get $y = \pm 2a$. Thus, y-coordinates of L and L' respectively $2a$ and $-2a$.

$$\therefore \qquad \boldsymbol{L = (a, 2a)} \quad \text{and} \quad \boldsymbol{L' = (a, -2a)}$$

Consider, $$|L\,L'| = \sqrt{(a-a)^2 + (2a+2a)^2} = \sqrt{(4a)^2} = 4a$$

Thus, the length of the latus rectum, $LL' = 4a$.

Any chord passing through the focus is called a **focal chord**.

5.5 Other Forms of Parabolas

1. Consider the equation $\boldsymbol{y^2 = 4ax}$

Supposing, we change x to $-x$, we get the equation

$$\boldsymbol{y^2 = -4ax} \quad (a > 0)$$

Clearly, this parabola, is the reflection of the parabola

$$y^2 = 4ax$$

about the y-axis. The shape of the parabola is as shown in Fig. 1.

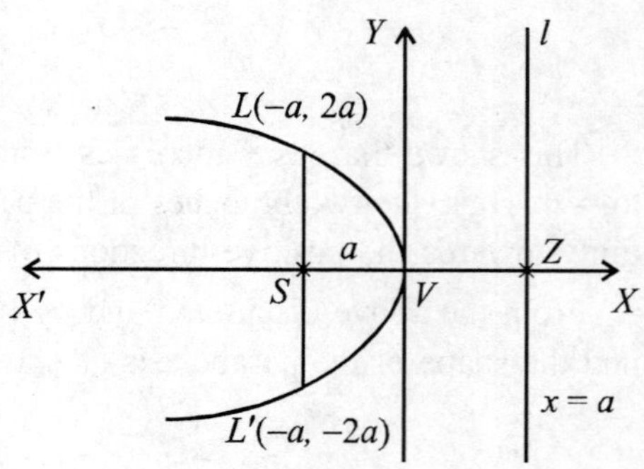

Fig. 1

For this parabola, we have,

Vertex $V = (0, 0)$, Focus $S = (-a, 0)$, Directrix : $x = a$,

Axis : $y = 0$. Tangent at the vertex : $x = 0$

The ends of latus rectum : $L = (-a, 2a)$, $L' = (-a, -2a)$

2. Consider the equation $y^2 = 4ax$, $(a > 0)$

Supposing, we change x and y, then we get the equation

$$x^2 = 4ay \quad (a > 0)$$

Clearly, this parabola will be a vertical parabola, opening upwards. The shape of the parabola is as shown in the Fig. 2.

For this parabola, we have

Vertex $V = (0, 0)$, Focus $S = (0, a)$, Directrix : $y = -a$,

Axis : $x = 0$, Tangent at the vertex : $y = 0$

The ends of latus rectum : $L = (2a, a)$, $L' = (-2a, a)$

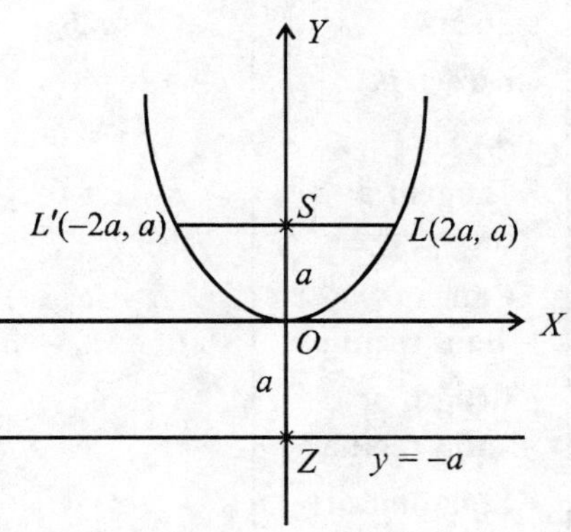

Fig. 2

3. Consider the equation $x^2 = 4ay$, $(a > 0)$

Supposing, we change y and $-y$, then we get the equation

$$x^2 = -4ay$$

Clearly, this parabola, will be a reflection of the parabola $x^2 = 4ay$ about the x-axis. The shape of the parabola is as shown in Fig. 3.

For this parabola, we have

Vertex $V = (0, 0)$, Focus $S = (0, -a)$, Directrix : $y = a$,

Axis : $x = 0$, Tangent at the vertex : $y = 0$

The ends of latus rectum : $L = (2a, -a)$, $L' = (-2a, -a)$

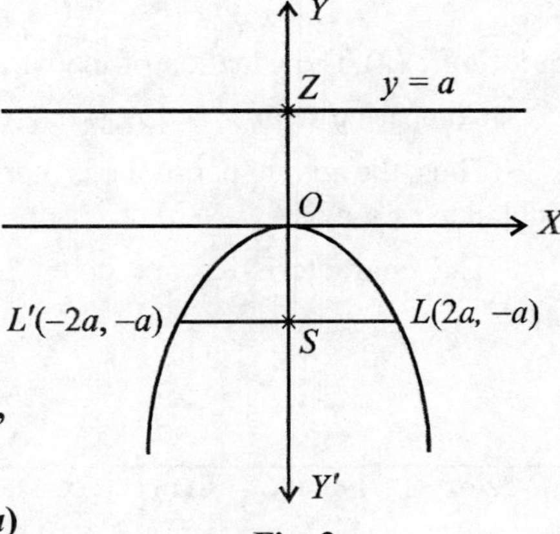

Fig. 3

Note : 1. For all the parabolas, mentioned above the length of the latus rectum is $4a$.

2.(a) The equation $y^2 = 4ax$, $(a > 0)$ represents a horizontal parabola, opening to the right side of y-axis.

(b) The equation $y^2 = -4ax$, $(a > 0)$ represents a horizontal parabola, opening to the left side of y-axis.

(c) The equation $x^2 = 4ay$, $(a > 0)$ represents a vertical parabola, opening upwards (i.e., above x-axis)

(d) The equation $x^2 = -4ay$, $(a > 0)$ represents a vertical parabola, opening downwards (i.e., below x-axis).

3. The vertex, focus, directrix, axis, tangent at the vertex, ends of latus rectum, length of the latus rectum, are known as **characteristics** of the parabola.

The following table gives the characteristics of all the forms of parabola together.

	$y^2 = 4ax$	$y^2 = -4ax$	$x^2 = 4ay$	$x^2 = -4ay$
Vertex	(0, 0)	(0, 0)	(0, 0)	(0, 0)
Focus	$(a, 0)$	$(-a, 0)$	$(0, a)$	$(0, -a)$
Directrix	$x = -a$	$x = a$	$y = -a$	$y = a$
Axis	$y = 0$	$y = 0$	$x = 0$	$x = 0$
Tangent at the vertex	$x = 0$	$x = 0$	$y = 0$	$y = 0$
Ends of latus rectum	$L = (a, 2a)$ $L' = (a, -2a)$	$L = (-a, 2a)$ $L' = (-a, -2a)$	$L = (2a, a)$ $L' = (-2a, a)$	$L = (2a, -a)$ $L' = (-2a, -a)$
Length of latus rectum	$4a$	$4a$	$4a$	$4a$
Equation of latus rectum	$x = a$	$x = -a$	$y = a$	$y = -a$

Example 1. Write the characteristics of the parabolas:

(a) $y^2 = 8x$ (b) $x^2 = 16y$ (c) $y^2 = -16x$ (d) $x^2 = -8y$.

Solution : (a) The equation of the parabola is $y^2 = 8x$.

Comparing with $y^2 = 4ax$, we have, $4a = 8 \Rightarrow$ **$a = 2$.**

Thus the given parabola is horizontal, turning right side of y-axis.

The characteristics are given in the following table.

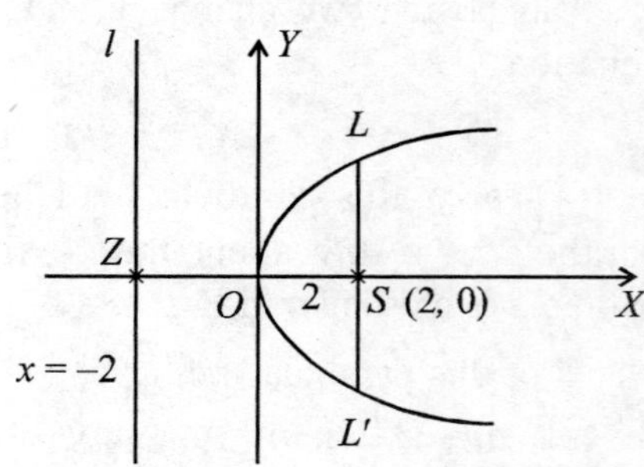

Vertex	Focus	Directrix	Axis	Tangent at vertex	Ends of latus rectum	Length of latus rectum
(0, 0)	$(a, 0)$	$x = -a$	$y = 0$	$x = 0$	$(a, 2a), (a, -2a)$	$4a$
(0, 0)	(2, 0)	$x = -2$	$y = 0$	$x = 0$	(2, 4), (2, –4)	8

Equation of latus rectum is $x = a$; $x = 2$.

(b) The equation of the parabola is $x^2 = 16y$.

Comparing with $x^2 = 4ay$, we have, $4a = 16 \Rightarrow$ **$a = 4$.**

Thus the given parabola is vertical turning upwards.

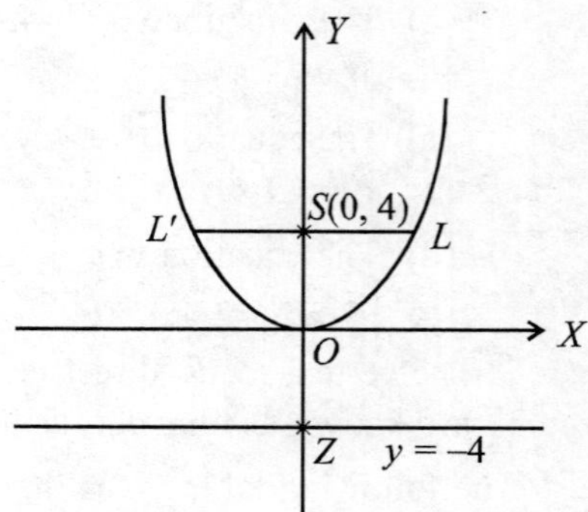

The characteristics are

Vertex	Focus	Directrix	Axis	Tangent at vertex	Ends of latus rectum	Length of latus rectum
(0, 0)	**(0, *a*)**	$y = -a$	$x = 0$	$y = 0$	**(2*a*, *a*), (–2*a*, *a*)**	**4*a***
(0, 0)	**(0, 4)**	$y = -4$	$x = 0$	$y = 0$	**(8, 4), (–8, 4)**	**16**

Equation of latus rectum is $y = a$; i.e., $y = 4$.

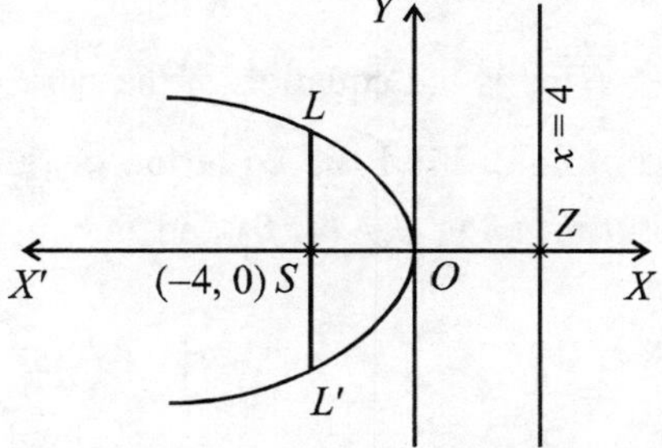

(c) The equation of the parabola is $y^2 = -16x$.

Comparing with $y^2 = -4ax$,

we have, $4a = 16 \Rightarrow$ $\mathbf{a = 4}$.

Thus, the given parabola is horizontal, turning left side of y-axis.

The characteristics are

Vertex	Focus	Directrix	Axis	Tangent at vertex	Ends of latus rectum	Length of latus rectum
(0, 0)	**(–*a*, 0)**	$x = a$	$y = 0$	$x = 0$	**(–*a*, 2*a*), (–*a*, –2*a*)**	**4*a***
(0, 0)	**(–4, 0)**	$x = 4$	$y = 0$	$x = 0$	**(–4, 8), (–4, –8)**	**16**

Equation of latus rectum is $\mathbf{x = -a}$; i.e., $\mathbf{x = -4}$.

(d) The equation of the parabola is $x^2 = -8y$.

Comparing with $x^2 = -4ay$,

We have, $4a = 8 \Rightarrow$ $\mathbf{a = 2}$.

Thus, the given parabola is vertical turning downwards.

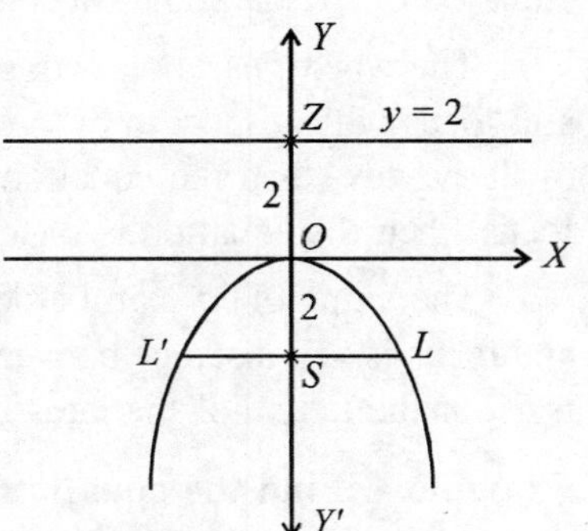

The characteristics are given in the following table.

Vertex	Focus	Directrix	Axis	Tangent at vertex	Ends of latus rectum	Length of latus rectum
(0, 0)	**(0, –*a*)**	$y = a$	$x = 0$	$y = 0$	**(2*a*, –*a*), (–2*a*, –*a*)**	**4*a***
(0, 0)	**(0, –2)**	$y = 2$	$x = 0$	$y = 0$	**(4, –2), (–4, –2)**	**8**

Equation of latus rectum is $\mathbf{y = -a}$; i.e., $\mathbf{y = -2}$.

Example 2. Find the equation of the parabola whose focus is (1, –1) and directrix is $x + y + 7 = 0$.

Solution : Let $S = (1, -1)$ and $P(x, y)$ be any point on the parabola. By the definition of parabola, we have,

$$\Rightarrow \quad PS = \text{distance of } P \text{ from } x + y + 7 = 0$$

$$\Rightarrow \quad \sqrt{(x-1)^2 + (y+1)^2} = \left|\frac{x+y+7}{\sqrt{1+1}}\right|$$

$$\Rightarrow \quad (x-1)^2 + (y+1)^2 = \frac{1}{2}(x+y+7)^2$$

$$\Rightarrow \quad 2(x^2 - 2x + 1 + y^2 + 2y + 1) = x^2 + y^2 + 49 + 2xy + 14y + 14x$$

$$\Rightarrow \quad \mathbf{x^2 + y^2 - 2xy - 18x - 10y - 45 = 0}$$

This is the equation of the parabola.

Example 3. Find the equation of the parabola whose focus is (4, 0) and directrix is $x = -4$.

Solution : Let $S = (4, 0)$ and $P(x, y)$ be any point on the parabola. By the definition of parabola, we have,

$$\Rightarrow \quad PS = \text{distance of } P \text{ from } x + 4 = 0$$

$$\Rightarrow \quad \sqrt{(x-4)^2 + y^2} = \left|\frac{x+4}{\sqrt{1+0}}\right|$$

$$\Rightarrow \quad (x-4)^2 + y^2 = (x+4)^2 \Rightarrow x^2 - 8x + 16 + y^2 = x^2 + 8x + 16 \Rightarrow \mathbf{y^2 = 16x}$$

This is the equation of the parabola.

Note : To determine the equation of a parabola completely, in general, we require two informations.

First is which one of the four type of the parabola is required. This gives us the form of the equation of the parabola. Secondly, the value of 'a'. These will determine the equation completely.

The type of parabola required, can be made out by observing *relative positions* of vertex, directrix and focus with respect to each other, from the given data. For example, if the focus is to the right side of the vertex, then the equation will be of the form $y^2 = 4ax$. If the directrix is above the vertex (or focus) then the equation is of the form $x^2 = -4ay$ etc.

The value of 'a' can be determined from the data by noting that the distance between the vertex and focus is 'a', distance between vertex and directrix is 'a' or distance between the directrix and focus is $2a$ or the length of the latus rectum is $4a$ etc.

Example 4. Find the equation of the parabola, given that

(a) its vertex is (0, 0) and focus is (3, 0)

(b) its vertex is (0, 0) and directrix is $y = -3$.

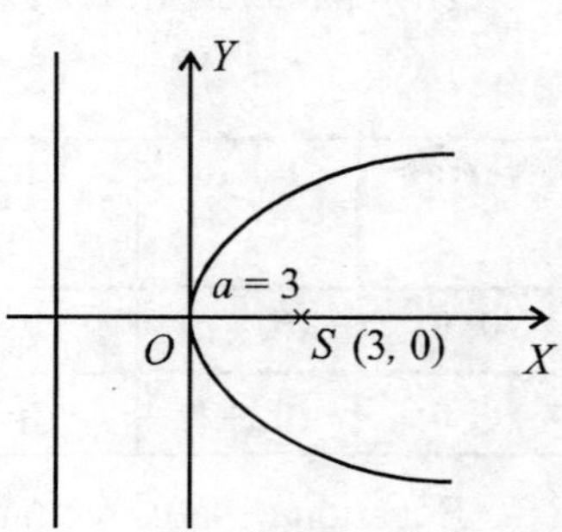

Solution : (a) By data $V = (0, 0)$, $S = (3, 0)$

Clearly, the focus is to right side of the vertex. Thus, the equation of the parabola is of the form

$$y^2 = 4ax$$

Now, $a = VS = 3$. Thus the equation of the parabola is

$$\mathbf{y^2 = 12x}$$

(b) By data $V = (0, 0)$ and the directrix is $y = -3$.

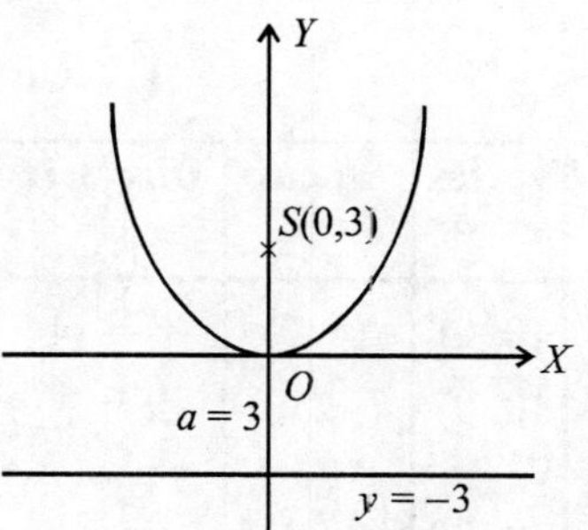

Clearly, directrix is a line parallel to x-axis and it is below the origin i.e., vertex. Thus, the parabola is a vertical parabola turning upwards.

$\therefore$ its equation is of the form $x^2 = 4ay$.

Now, a = distance between vertex and the directrix $y = -3$.

i.e., $a = 3$. Thus the equation is $\mathbf{x^2 = 12y}$

Example 5. Find the equation of the parabola given its vertex is (0, 0), axis is x-axis and it passes through (–1, 2).

Solution : By data vertex = (0, 0) and the axis is x-axis. Thus, the equation of the parabola is of one of the forms

$$y^2 = 4ax \quad \text{or} \quad y^2 = -4ax$$

Again, by data the parabola passes through the point (– 1, 2) which is in the second quadrant. Thus, the required equation is $y^2 = -4ax$ (because no point in second quadrant lies on $y^2 = 4ax$).

Now, (– 1, 2) lies on $y^2 = -4ax \Rightarrow 4 = 4a \Rightarrow a = 1$. Thus, the equation is $\mathbf{y^2 = -4x}$

Exercise

I.

1. Write the characteristics of the following parabolas

(a) $y^2 = 16x$ **(b)** $y^2 = -8x$ **(c)** $3x^2 = -8y$ **(d)** $x^2 = 8y$

(e) $y^2 = 8x$ **(f)** $y^2 + 4x = 0$ **(g)** $3x^2 + 4y = 0$ **(h)** $x^2 + 16y = 0$

2. If the length of the latus rectum of $y^2 = 8kx$ is 4, find k.

3. If the length of the latus rectum of $x^2 = 4ky$ is 8, find the value of k.

II.

1. Find the equation of the parabola given that its

(a) vertex is (0, 0) and focus is (4, 0)

(b) vertex is (0, 0) and directrix is $y = -3$

(c) focus is (1, 0) and directrix is $x = -1$

(d) focus is (– 4, 0) and directrix is $x = 4$

(e) focus is (0, – 3) and directrix is $y = 3$

(f) focus is (0, 6) and vertex is (0, 0)

(g) vertex is (0, 0), axis is y-axis and passes thro' $\left(\frac{1}{2}, 2\right)$

(h) vertex is (0, 0), axis is y-axis and passes thro' (– 1, – 3)

2. Find the equation of the parabola given that

(a) its focus is (– 1, 1) and directrix is $x + y + 1 = 0$

(b) its focus is (– 2, – 4) and directrix is $y = 1$

(c) its focus is (– 2, – 1) and directrix is $x = -4$.

Answers

I. 1.

	Vertex	Focus	Directrix	Axis	Tangent at vertex	Ends of latus rectum	L.L.R.	Eqn. to of latus rectum
(a)	(0, 0)	(4, 0)	$x = -4$	$y = 0$	$x = 0$	(4, 8), (4, –8)	16	$x = 4$
(b)	(0, 0)	(–2, 0)	$x = 2$	$y = 0$	$x = 0$	(–2, 4), (–2, –4)	8	$x = -2$
(c)	(0, 0)	$\left(0, -\frac{2}{3}\right)$	$y = \frac{2}{3}$	$x = 0$	$y = 0$	$\left(\frac{4}{3}, -\frac{2}{3}\right), \left(-\frac{4}{3}, -\frac{2}{3}\right)$	$\frac{8}{3}$	$y = -\frac{2}{3}$
(d)	(0, 0)	(0, 2)	$y = -2$	$x = 0$	$y = 0$	(4, 2), (–4, 2)	8	$y = 2$
(e)	(0, 0)	(2, 0)	$x = -2$	$y = 0$	$x = 0$	(2, 4), (2, –4)	8	$x = 2$
(f)	(0, 0)	(–1, 0)	$x = 1$	$y = 0$	$x = 0$	(–1, 2), (–1, –2)	4	$x = -1$
(g)	(0, 0)	$\left(0, -\frac{1}{3}\right)$	$y = \frac{1}{3}$	$x = 0$	$y = 0$	$\left(\frac{2}{3}, -\frac{1}{3}\right), \left(-\frac{2}{3}, -\frac{1}{3}\right)$	$\frac{4}{3}$	$y = -\frac{1}{3}$
(h)	(0, 0)	(0, –4)	$y = 4$	$x = 0$	$y = 0$	(8, –4), (–8, –4)	16	$y = -4$

2. (a) $k = \frac{1}{2}$ **3.** $k = 2$.

II. **1. (a)** $y^2 = 16x$ **(b)** $x^2 = 12y$ **(c)** $y^2 = 4x$ **(d)** $y^2 = -16x$

(e) $x^2 = -12y$ **(f)** $x^2 = 24y$ **(g)** $8x^2 - y = 0$ **(h)** $3x^2 + y = 0$

2. (a) $x^2 - 2xy + y^2 + 2x - 6y + 3 = 0$ **(b)** $x^2 + 4x + 10y + 19 = 0$

(c) $y^2 - 4x + 2y - 11 = 0$

5.6 Ellipse

Ellipse is a conic section whose eccentricity is less than one.

The formal definition of the ellipse is given below.

Definition : Ellipse is the locus of the point which moves such that the ratio of its distance from the fixed point (called focus) to its distance from the fixed line (called directrix) is always a fixed constant which is less than 1.

Theorem : The equation of the ellipse (with proper choice of coordinate axes) in its standard form is

$$\frac{x^2}{a^2} + \frac{y^2}{b^2} = 1$$

Proof : Let S be the focus and the line l be the directrix of the ellipse. Let e (< 1) be the eccentricity of the ellipse. Draw SZ perpendicular to the directrix.

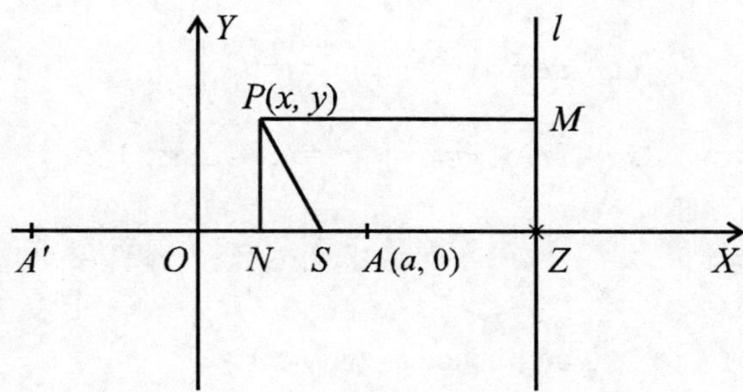

Divide the line segment SZ internally at A and externally at A' in the ratio $e : 1$.

Thus, we have, $\quad \dfrac{SA}{AZ} = e \quad$ and $\quad \dfrac{SA'}{A'Z} = e$

$\Rightarrow \quad \boldsymbol{SA = e \cdot AZ} \quad$ and $\quad \boldsymbol{SA' = e \cdot A'Z}$

Clearly, A and A' are points on the ellipse.

Let O be the midpoint of AA' and $\boldsymbol{AA' = 2a} \;\Rightarrow\; \boldsymbol{OA = OA' = a}$

Choose O as the origin, the line $A'OAZ$ as x-axis and the line through O and perpendicular to x-axis as y-axis.

Clearly, $A = (a, 0)$ and $A' = (-a, 0)$. Now, consider,

$$SA + SA' = e \cdot AZ + e \cdot A'Z \quad \Rightarrow \quad AA' = e\,(AZ + A'Z)$$

$$\Rightarrow \quad 2a = e\,(OZ - OA + OA' + OZ)$$

$$\Rightarrow \quad 2a = e\,(2OZ) \qquad (\because OA = OA')$$

$$\Rightarrow \quad \boldsymbol{OZ = \frac{a}{e}}$$

Thus, **the equation of the directrix is $\boldsymbol{x = \dfrac{a}{e}}$.**

Again consider, $\quad SA' - SA = e \cdot A'Z - e \cdot AZ$

$$\Rightarrow \quad (OA' + OS - OA + OS) = e\,(A'Z - AZ)$$

$$\Rightarrow \quad 2OS = e \cdot AA' \qquad (\because OA' = OA)$$

$$\Rightarrow \quad 2OS = e\,(2a) \qquad (\because AA' = 2a)$$

$$\Rightarrow \quad \boldsymbol{OS = ae}$$

Thus, **the coordinates of focus $\boldsymbol{S}$ is given by $\boldsymbol{S = (ae, 0)}$.**

Let $P(x, y)$ be any point on the ellipse. PM be its distance from the directrix and PS be its distance from the focus S.

By the definition of ellipse, we have

$$\frac{PS}{PM} = e \quad \Rightarrow \quad PS = e \cdot (PM)$$

$$\Rightarrow \quad PS^2 = e^2 \cdot (PM)^2$$

$$\Rightarrow \quad (x - ae)^2 + y^2 = e^2 \cdot (NZ)^2$$

$$\Rightarrow \quad (x - ae)^2 + y^2 = e^2 \cdot (OZ - ON)^2$$

$$\Rightarrow \quad (x - ae)^2 + y^2 = e^2 \cdot \left(\frac{a}{e} - x\right)^2 \qquad \left(\because OZ = \frac{a}{e},\ ON = x\right)$$

$$\Rightarrow \quad x^2 - \cancel{2aex} + ae^2 + y^2 = a^2 - \cancel{2aex} + x^2e^2$$

$$\Rightarrow \quad x^2(1 - e^2) + y^2 = a^2(1 - e^2) \quad \Rightarrow \quad \frac{x^2}{a^2} + \frac{y^2}{a^2(1 - e^2)} = 1$$

Since $e < 1$, we have $1 - e^2 > 0$ and hence $a^2(1 - e^2) > 0$.

Putting, $a^2(1 - e^2) = b^2$, we get $\boldsymbol{\frac{x^2}{a^2} + \frac{y^2}{b^2} = 1}$ which is the required equation.

5.7 Shape and the Characteristics of the Ellipse $\frac{x^2}{a^2} + \frac{y^2}{b^2} = 1$

Consider the equation of the ellipse $\frac{x^2}{a^2} + \frac{y^2}{b^2} = 1$

The following observations will help us to trace the curve ellipse.

1. Put $x = 0$ in the equation $\frac{x^2}{a^2} + \frac{y^2}{b^2} = 1$, we get $y^2 = b^2 \Rightarrow y = \pm b$

Thus, the curve cuts the y-axis at $\boldsymbol{B = (0, b)}$ and $\boldsymbol{B' = (0, -b)}$

Put $y = 0$ in the equation $\frac{x^2}{a^2} + \frac{y^2}{b^2} = 1$, we get, $x^2 = a^2 \Rightarrow x = \pm a$

Thus, the curve cuts the x-axis at $\boldsymbol{A = (a, 0)}$ and $\boldsymbol{A' = (-a, 0)}$.

2. The equation $\frac{x^2}{a^2} + \frac{y^2}{b^2} = 1$, does not alter, if y is changed to $-y$. Thus, the curve is symmetric about x-axis. That is if (x, y) is any point on the curve then $(x, -y)$ is also a point on the curve.

Again, the equation does not alter, if x is changed to $-x$. Thus, the curve is symmetric about y-axis. That is if (x, y) is any point on the curve then $(-x, y)$ is also a point on the curve.

3. We have, $b^2 = a^2(1 - e^2)$. Now, $e < 1 \Rightarrow e^2 < 1$.

Thus, we have, $0 < 1 - e^2 < 1 \quad \Rightarrow \quad a^2(1 - e^2) < a^2 \quad \Rightarrow \quad b^2 < a^2 \quad \Rightarrow \quad \boldsymbol{b < a}$

Hence, the points $(0, b)$ and $(0, -b)$ are nearer to the origin than the points $(a, 0)$ and $(-a, 0)$.

4. Consider the equation

$$\frac{x^2}{a^2} + \frac{y^2}{b^2} = 1 \quad \Rightarrow \quad \frac{y^2}{b^2} = 1 - \frac{x^2}{a^2} \quad \Rightarrow \quad 1 - \frac{x^2}{a^2} \geq 0 \qquad \left(\because \frac{y^2}{b^2} \geq 0\right)$$

$$\Rightarrow \quad -\frac{x^2}{a^2} \geq -1 \quad \Rightarrow \quad \frac{x^2}{a^2} \leq 1 \quad \Rightarrow \quad x^2 \leq a^2$$

Similarly, $y^2 \leq b^2$. Thus, we have $-a \leq x \leq a$ and $-b \leq y \leq b$.

Hence, the curve entirely lies inside the rectangle bounded by the lines $x = \pm a$ and $y = \pm b$.

The shape of the curve ellipse $\frac{x^2}{a^2} + \frac{y^2}{b^2} = 1$, $a > b$ is as shown in the following figure.

Let S' be a point on the negative side of the origin, along the x-axis, such that $OS = OS' = ae$ and Z' be the point such that $CZ' = CZ = \frac{a}{e}$ and l' is the line through Z' and perpendicular to x-axis. It can be show that, equation of the locus of the point which moves such that, the ratio of its distance from S' to its distance from the line l' is e (< 1), is again $\frac{x^2}{a^2} + \frac{y^2}{b^2} = 1$. Thus, the ellipse $\frac{x^2}{a^2} + \frac{y^2}{b^2} = 1$ has two foci and two directrices. The foci S and S' are respectively denoted by S_1 and S_2 (see fig. below) and the directrices l and l' are denoted by l_1 and l_2 (see fig. below).

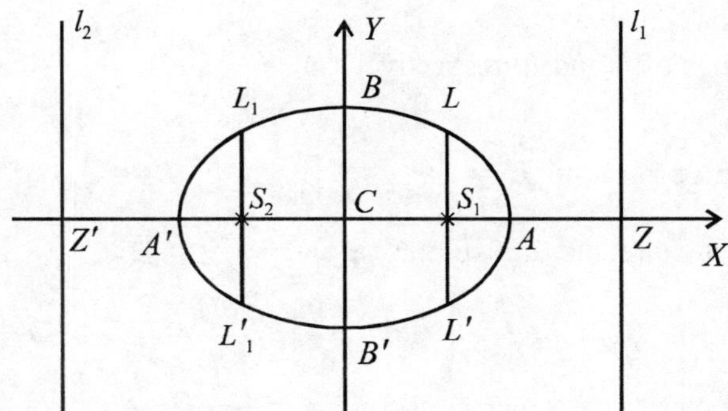

The following are the characteristics of the ellipse

$$\frac{x^2}{a^2} + \frac{y^2}{b^2} = 1. \quad a > b$$

1. The point $C = (0, 0)$ is called the **centre** of the ellipse.
2. The line AA' is called the **Major Axis** of the ellipse. The points $A = (a, 0)$ and $A' = (-a, 0)$ are called the **ends of Major Axis.** These are also called the **vertices** of the ellipse.

   The length $AA' = 2a$ is called the **length of Major Axis.**
3. The line BB' is called **Minor Axis** of the ellipse. The points $B = (0, b)$ and $B' = (0, -b)$ are called the **ends of Minor Axis.**

   The length $BB' = 2b$ is called the **length of Minor Axis.**
4. The **foci** S_1 and S_2 are $S_1 = (ae, 0)$ and $S_2 = (-ae, 0)$.

   The length S_1S_2 is called **distance between the foci** = $2ae$.
5. The directrices l_1 and l_2 are given by $x = \frac{a}{e}$ **and** $x = -\frac{a}{e}$,

   The **distance between the directrices** = $\frac{2a}{e}$
6. **Latus rectum**

   The line through the focus S_1, perpendicular to major axis, is called the **latus rectum** at S_1.

   The points L and L' where the latus rectum meets the ellipse are called the **ends of the latus rectum**.

   Clearly, the x-coordinates of L and L' are the x-coordinates of S_1, i.e., ae.

Now putting $x = ae$ in $\frac{x^2}{a^2} + \frac{y^2}{b^2} = 1$. We get,

$$\frac{a^2e^2}{a^2} + \frac{y^2}{b^2} = 1 \Rightarrow \frac{y^2}{b^2} = 1 - e^2 \Rightarrow y^2 = b^2(1 - e^2)$$

$$\Rightarrow y^2 = b^2\left(\frac{b^2}{a^2}\right) \qquad (\because a^2(1 - e^2) = b^2)$$

$$\Rightarrow y = \pm\frac{b^2}{a}$$

Thus, the y-coordinates of L and L' respectively are $\frac{b^2}{a}$ and $-\frac{b^2}{a}$. Thus, we have

Ends of latus rectum $L = \left(ae, \frac{b^2}{a}\right), \quad L' = \left(ae, -\frac{b^2}{a}\right)$

Similarly, the ends of the latus rectum drawn at S_2 are

$$L_1 = \left(-ae, \frac{b^2}{a}\right), \quad L'_1 = \left(-ae, -\frac{b^2}{a}\right)$$

Also, **the length of latus rectum = LL' (or $L_1L'_1$) =** $\frac{2b^2}{a}$

7. The eccentricity e of the ellipse is given by

$$b^2 = a^2(1 - e^2) \Rightarrow a^2e^2 = a^2 - b^2 \Rightarrow e^2 = \frac{a^2 - b^2}{a^2}$$

5.8 Another Form of Ellipse

Consider the equation $\frac{x^2}{a^2} + \frac{y^2}{b^2} = 1$, with $a > b$

As we have seen, for this ellipse, the major axis is along the x-axis and therefore the foci are on x-axis. This ellipse is called "horizontal ellipse".

Supposing in the above equation, $b > a$. Even then the equation represents an ellipse.

That is the equation $\frac{x^2}{a^2} + \frac{y^2}{b^2} = 1$, with $b > a$

represents an ellipse. For this ellipse the major axis will be along y-axis and therefore the foci are also along the y-axis. This ellipse is called "vertical ellipse" (see fig). The characteristics are given below.

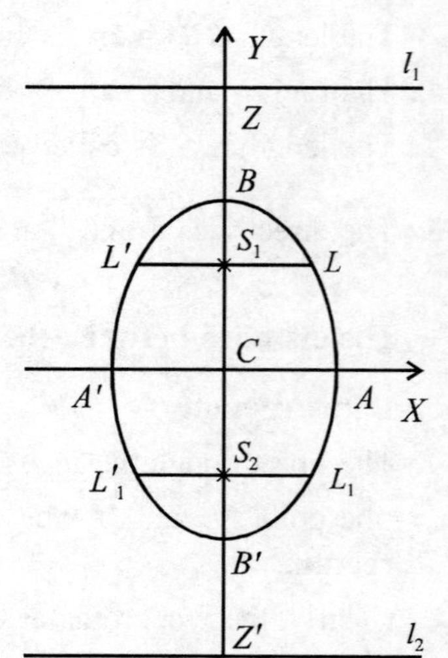

1. Centre $C = (0, 0)$.

2. Major Axis of the ellipse.

Ends: $\boldsymbol{B = (0, b)}$ and $\boldsymbol{B' = (0, -b)}$. Length: $\boldsymbol{BB' = 2b}$

3. Minor Axis

Ends: $\boldsymbol{A = (a, 0)}$ and $\boldsymbol{A' = (-a, 0)}$. Length: $\boldsymbol{AA' = 2a}$

4. Foci: $S_1 = (0, be)$, $S_2 = (0, -be)$.

5. Distance between the foci = $2be$.

6. Directrices: $y = \frac{b}{e}$ and $x = -\frac{b}{e}$.

Distance between directrix = $\frac{2b}{e}$

7. Ends of latus rectum $L = \left(\frac{a^2}{b}, be\right)$, $L' = \left(-\frac{a^2}{b}, be\right)$; $L_1 = \left(\frac{a^2}{b}, -be\right)$, $L' = \left(-\frac{a^2}{b}, -be\right)$

8. Length of latus rectum = $\frac{2a^2}{b}$.

9. Eccentricity is given by $e^2 = \frac{b^2 - a^2}{b^2}$.

Note : To write the characteristics of the ellipse $\frac{x^2}{a^2} + \frac{y^2}{b^2} = 1$, with $b > a$, interchange the x and y coordinates in the corresponding characteristics of the ellipse $\frac{x^2}{a^2} + \frac{y^2}{b^2} = 1$ with $a < b$, and change 'a' to 'b' and 'b' to 'a'.

The following table gives the characteristics of both the forms of the ellipses together.

Ellipse / Characteristics	$\frac{x^2}{a^2} + \frac{y^2}{b^2} = 1,\ a > b$	$\frac{x^2}{a^2} + \frac{y^2}{b^2} = 1,\ b > a$
Centre	$(0, 0)$	$(0, 0)$
Major Axis : Ends	$A = (a, 0)$, $A' = (-a, 0)$	$B = (0, b)$, $B' = (0, -b)$
Length	$2a$	$2b$
Minor Axis : Ends	$B = (0, b)$, $B' = (0, -b)$	$A = (a, 0)$, $A' = (-a, 0)$
Length	$2b$	$2a$
Foci	$S_1 = (ae, 0)$, $S_2 = (-ae, 0)$	$S_1 = (0, be)$, $S_2 = (0, -be)$
Distance between foci	$2ae$	$2be$
Directrices :	$x = \frac{a}{e}$, $x = -\frac{a}{e}$	$y = \frac{b}{e}$, $y = -\frac{b}{e}$
Distance between directrices	$\frac{2a}{e}$	$\frac{2b}{e}$
Ends of latus rectum	$L = \left(ae, \frac{b^2}{a}\right)$ $L' = \left(ae, -\frac{b^2}{a}\right)$	$L = \left(\frac{a^2}{b}, be\right)$ $L' = \left(-\frac{a^2}{b}, be\right)$
	$L_1 = \left(-ae, \frac{b^2}{a}\right)$ $L_1' = \left(-ae, -\frac{b^2}{a}\right)$	$L_1 = \left(\frac{a^2}{b}, -be\right)$ $L_1' = \left(-\frac{a^2}{b}, -be\right)$
Length of latus rectum	$\frac{2b^2}{a}$	$\frac{2a^2}{b}$
Eccentricity	$e^2 = \frac{a^2 - b^2}{a^2}$	$e^2 = \frac{b^2 - a^2}{b^2}$
Equations of latus rectum	$x = ae$, $x = -ae$	$y = be$, $y = -be$

Example 1. Find the characteristics of the following ellipses.

$$\textbf{(a)}\ \frac{x^2}{25}+\frac{y^2}{9}=1 \qquad \textbf{(b)}\ \frac{x^2}{36}+\frac{y^2}{64}=1$$

Solution : (a) The equation of the ellipse is $\frac{x^2}{25}+\frac{y^2}{9}=1$

Here, $a^2 = 25,\ b^2 = 9 \Rightarrow \boldsymbol{a = 5,\ b = 3}$, also $\boldsymbol{a > b}$.

Now, the eccentricity e is given by

$$e^2=\frac{a^2-b^2}{a^2} \Rightarrow e^2=\frac{25-9}{25} \Rightarrow e^2=\frac{16}{25} \Rightarrow \boldsymbol{e=\frac{4}{5}}$$

The characteristics are given below.

1. Centre = (0, 0)

2. Major Axis: Ends: $A = (a, 0) = \mathbf{(5, 0)}$, $A' = (-a, 0) = \mathbf{(-5, 0)}$; Length: $\mathbf{2a = 10}$

3. Minor Axis: Ends: $B = (0, b) = \mathbf{(0, 3)}$, $B' = (0, -b) = \mathbf{(0, -3)}$; Length: $\mathbf{2b = 6}$

4. Foci: $S_1 = (ae, 0) = \mathbf{(4, 0)}$, $S_2 = (-ae, 0) = \mathbf{(-4, 0)}$. Distance between the foci $= S_1S_2 = \mathbf{2ae = 8}$

5. Directrices:

$$\left[\begin{array}{l} x=\frac{a}{e} \Rightarrow x=\frac{5}{(4/5)} \Rightarrow \boldsymbol{x=\frac{25}{4}} \\ x=-\frac{a}{e} \Rightarrow x=-\frac{5}{(4/5)} \Rightarrow \boldsymbol{x=-\frac{25}{4}} \end{array}\right.$$

Distance between directrices: $\frac{2a}{e}=2\left(\frac{25}{4}\right)=\mathbf{\frac{25}{2}}$

6. Ends of latus rectums:

$$\left[\begin{array}{lll} L=\left(ae, \frac{b^2}{a}\right) & \Rightarrow & \boldsymbol{L=\left(4, \frac{9}{5}\right)} \\ L'=\left(ae, -\frac{b^2}{a}\right) & \Rightarrow & \boldsymbol{L'=\left(4, -\frac{9}{5}\right)} \\ L_1=\left(-ae, \frac{b^2}{a}\right) & \Rightarrow & \boldsymbol{L_1=\left(-4, \frac{9}{5}\right)} \\ L_1'=\left(-ae, -\frac{b^2}{a}\right) & \Rightarrow & \boldsymbol{L_1'=\left(-4, -\frac{9}{5}\right)} \end{array}\right.$$

7. Length of latus rectum: $LL' = \frac{2b^2}{a} = \mathbf{\frac{18}{5}}$

8. Equations of latus rectums:

$$\left[\begin{array}{lll} x = ae & \Rightarrow & \boldsymbol{x = 4} \\ x = -ae & \Rightarrow & \boldsymbol{x = -4} \end{array}\right.$$

(b) The equation of the ellipse is $\frac{x^2}{36}+\frac{y^2}{64}=1$

Here, $a^2 = 36,\ \ b^2 = 64 \ \Rightarrow\ \boldsymbol{a = 6,\ \ b = 8,}$ also $\boldsymbol{b > a}$.

Now, the eccentricity e is given by

$$e^2=\frac{b^2-a^2}{a^2} \Rightarrow e^2=\frac{64-36}{64} \Rightarrow e^2=\frac{28}{64} \Rightarrow \boldsymbol{e=\frac{\sqrt{7}}{4}}$$

The characteristics are given below.

1. Centre = (0, 0)

2. Major Axis: $\left[\begin{array}{l}\text{Ends: } B=(0,b)=\mathbf{(0,8)},\ B'=(0,-b)=\mathbf{(0-8)}\\ \text{Length: } BB'=2b=\mathbf{16}\end{array}\right.$

3. Minor Axis: $\left[\begin{array}{l}\text{Ends: } A=(a,0)=\mathbf{(6,0)},\ A'=(-a,0)=\mathbf{(-6,0)}\\ \text{Length: } AA'=2a=\mathbf{12}\end{array}\right.$

4. Foci: $S_1=(0,be)=\mathbf{(0,2\sqrt{7})}$, $S_2=(0,-be)=\mathbf{(0,-2\sqrt{7})}$

Distance between the foci = $S_1S_2 = \mathbf{2be = 4\sqrt{7}}$

5. Directrices: $\left[\begin{array}{l} y=\frac{b}{e} \Rightarrow y=\frac{8}{(\sqrt{7}/4)} \Rightarrow \boldsymbol{y=\frac{32}{\sqrt{7}}}\\ y=-\frac{b}{e} \Rightarrow y=-\frac{8}{(\sqrt{7}/4)} \Rightarrow \boldsymbol{y=-\frac{32}{\sqrt{7}}}\end{array}\right.$

6. Ends of latus rectums: $\left[\begin{array}{lll} L=\left(\frac{a^2}{b}, be\right) & \Rightarrow & \boldsymbol{L=\left(\frac{9}{2}, 2\sqrt{7}\right)}\\ L'=\left(-\frac{a^2}{b}, be\right) & \Rightarrow & \boldsymbol{L'=\left(-\frac{9}{2}, 2\sqrt{7}\right)}\\ L_1=\left(\frac{a^2}{b}, -be\right) & \Rightarrow & \boldsymbol{L_1=\left(\frac{9}{2}, -2\sqrt{7}\right)}\\ L_1'=\left(-\frac{a^2}{b}, -be\right) & \Rightarrow & \boldsymbol{L_1'=\left(-\frac{9}{2}, -2\sqrt{7}\right)}\end{array}\right.$

7. Length of latus rectum: $LL'=\frac{2a^2}{b} \Rightarrow \boldsymbol{LL'=9}$

8. Equations of latus rectums: $\left[\begin{array}{l} y=be \Rightarrow \boldsymbol{x=2\sqrt{7}}\\ y=-be \Rightarrow \boldsymbol{x=-2\sqrt{7}}\end{array}\right.$

Example 2. If the length of the latus rectum of an ellipse is one half of its minor axis, find the eccentricity of the ellipse.

Solution : By data $LL'=\frac{1}{2}(2b)$. But $LL'=\frac{2b^2}{a}$

$\therefore$ We have, $\dfrac{2b^2}{a} = \dfrac{1}{2}(2b) \Rightarrow 2b^2 = ab \Rightarrow \mathbf{2b = a.}$

Now, the eccentricity is given by

$$e^2 = \frac{a^2 - b^2}{a^2} \Rightarrow e^2 = \frac{4b^2 - b^2}{4b^2} \Rightarrow e^2 = \frac{3}{4} \Rightarrow \boldsymbol{e = \frac{\sqrt{3}}{2}} \qquad (\because a = 2b)$$

Example 3. If the distance between the directrices of the ellipse $\dfrac{x^2}{a^2} + \dfrac{y^2}{b^2} = 1$ ($a > b$) is three times the distance between the foci, find the eccentricity.

Solution : By data, we have

distance between the directrices = 3 (distance between the foci)

$$\Rightarrow \frac{2a}{e} = 3(2ae) \Rightarrow 3e^2 = 1 \Rightarrow e^2 = \frac{1}{3} \Rightarrow \boldsymbol{e = \frac{1}{\sqrt{3}}}$$

Example 4. If the major axis is double the minor axis of an ellipse find the eccentricity.

Solution : By data, we have,

Length of major axis = 2 (length of minor axis) $\Rightarrow 2a = 2(2b) \Rightarrow \mathbf{a = 2b.}$

Now, the eccentricity e is given by

$$e^2 = \frac{a^2 - b^2}{a^2} \Rightarrow e^2 = \frac{4b^2 - b^2}{4b^2} \Rightarrow e^2 = \frac{3}{4} \Rightarrow \boldsymbol{e = \frac{\sqrt{3}}{2}}$$

Example 5. If the distance between the foci of the ellipse $\dfrac{x^2}{a^2} + \dfrac{y^2}{b^2} = 1$ ($b > a$) is 10 and $e = \dfrac{1}{2}$, find the length of the major axis.

Solution : By data, we have the equation $\dfrac{x^2}{a^2} + \dfrac{y^2}{b^2} = 1$ with $b > a$. Again by data we have

$$\text{distance between the foci} = 10 \Rightarrow 2be = 10 \Rightarrow be = 5$$

$$\Rightarrow b\left(\frac{1}{2}\right) = 5 \qquad \left(\because e = \frac{1}{2}\right)$$

$$\Rightarrow \mathbf{b = 10}$$

Thus, the length of the major axis is given by $\mathbf{2b = 20}$.

Example 6. Find the equation of the ellipse in the form $\dfrac{x^2}{a^2} + \dfrac{y^2}{b^2} = 1$ ($a > b$), if

(a) one of its foci is (3, 0) and eccentricity is $\dfrac{3}{5}$.

(b) the distance between the foci is 3 and $e = \dfrac{1}{2}$

(c) the length of the latus rectum is 5 and $e = \dfrac{1}{2}$.

Solution : (a) By data one of the foci is (3, 0) and $e = \frac{3}{5}$.

$\Rightarrow \quad (ae, 0) = 3, 0 \Rightarrow ae = 3 \Rightarrow a\left(\frac{3}{5}\right) = 3 \Rightarrow \mathbf{a = 5.}$

Now, consider, $b^2 = a^2(1 - e^2) \Rightarrow b^2 = 25\left(1 - \frac{9}{25}\right) \Rightarrow \mathbf{b^2 = 16}$

Thus, the equation of the ellipse is $\mathbf{\frac{x^2}{25} + \frac{y^2}{16} = 1.}$

(b) By data, distance between foci = 3 and $e = \frac{1}{2} \Rightarrow 2ae = 3 \Rightarrow 2a\left(\frac{1}{2}\right) = 3 \Rightarrow \mathbf{a = 3}$

Consider, $b^2 = a^2(1 - e^2) \Rightarrow b^2 = 9\left(1 - \frac{1}{4}\right) \Rightarrow b^2 = \frac{27}{4}$

Thus, the equation of the ellipse is $\mathbf{\frac{x^2}{9} + \frac{y^2}{(27/4)} = 1.}$

(c) By data, $e = \frac{1}{2}$ and the length of latus rectum = 5 $\Rightarrow \frac{2b^2}{a} = 5 \Rightarrow 2b^2 = 5a \Rightarrow \mathbf{b^2 = \frac{5}{2}a}$

Consider, $b^2 = a^2(1 - e^2) \Rightarrow \frac{5}{2}a = a^2\left(1 - \frac{1}{4}\right) \Rightarrow 5 = 2a\left(\frac{3}{4}\right) \Rightarrow \mathbf{a = \frac{10}{3}} \Rightarrow \mathbf{b^2 = \frac{25}{3}}$

Thus, the equation of the ellipse is $\mathbf{\frac{x^2}{(10/3)} + \frac{y^2}{(25/3)} = 1.}$

Example 7. Two ends of the major axis of an ellipse $\frac{x^2}{a^2} + \frac{y^2}{b^2} = 1$ $(a > b)$ are (5, 0) and (–5, 0). If one of the foci lies on 3x – 5y – 9 = 0, find the equation of the ellipse.

Solution : By data the ends of major axis are : (5, 0) and (–5, 0) $\Rightarrow$ **$a = 5$**

Again, by data the foci $(ae, 0)$ lies on $3x - 5y - 9 = 0$.

$\Rightarrow \quad 3(ae) - 9 = 0 \Rightarrow 15e - 9 = 0 \Rightarrow e = \mathbf{\frac{3}{5}} \qquad (\because a = 5)$

Now, consider, $b^2 = a^2(1 - e^2) \Rightarrow b^2 = 25\left(1 - \frac{9}{25}\right) \Rightarrow \mathbf{b^2 = 16}$

Thus, the equation of the ellipse is $\mathbf{\frac{x^2}{25} + \frac{y^2}{16} = 1.}$

Example 8. Find the equation of an ellipse, in the form $\frac{x^2}{a^2} + \frac{y^2}{b^2} = 1$ $(a > b)$, given that the length of the latus rectum is 8 and the distance between the foci is $2\sqrt{21}$.

Solution : By data, length of latus rectum = 8 $\Rightarrow$ $\frac{2b^2}{a} = 8 \Rightarrow \boldsymbol{b^2 = 4a}$

Again by data distance between the foci = $2\sqrt{21}$.

$\Rightarrow$ $2ae = 2\sqrt{21}$ $\Rightarrow$ $\boldsymbol{ae = \sqrt{21}}$

Consider, $b^2 = a^2(1 - e^2)$ $\Rightarrow$ $b^2 = a^2 - a^2e^2$

$\Rightarrow$ $4a = a^2 - 21$ $(\because ae = \sqrt{21})$

$\Rightarrow$ $a^2 - 4a - 21 = 0$

$\Rightarrow$ $(a - 7)(a + 3) = 0$ $\Rightarrow$ $a = 7$ or $a = -3$

Rejecting the negative value, we have, $\boldsymbol{a = 7 \Rightarrow b^2 = 28}$.

Thus, the equation of the ellipse is $\boldsymbol{\frac{x^2}{49} + \frac{y^2}{28} = 1}$.

Example 9. Find the equation of an ellipse, in its standard form passing through the points (–2, 2) and (3, 1).

Solution : Let the equation of the ellipse be $\frac{x^2}{a^2} + \frac{y^2}{b^2} = 1$.

By data this passes through the points (–2, 2) and (3, 1) $\Rightarrow$ $\frac{4}{a^2} + \frac{4}{b^2} = 1$ and $\frac{9}{a^2} + \frac{1}{b^2} = 1$

Let $\frac{1}{a^2} = p$ and $\frac{1}{b^2} = q$, we have, $4p + 4q = 1$ and $9p + q = 1$

Solving these equations we get, $\boldsymbol{p = \frac{3}{32}}$ and $\boldsymbol{q = \frac{5}{32}}$. Thus, $\boldsymbol{a^2 = \frac{32}{3}}$ and $\boldsymbol{b^2 = \frac{32}{5}}$

Hence, the equation of the ellipse is $\boldsymbol{\frac{x^2}{\left(32/3\right)} + \frac{y^2}{\left(32/5\right)} = 1}$.

Example 10. Find the equation of an ellipse whose focus is (2, –3), directrix is $\boldsymbol{x + y - 3 = 0}$ and $\boldsymbol{e = \frac{2}{\sqrt{5}}}$.

Solution : We have, focus $S = (2, -3)$, $e = \frac{2}{\sqrt{5}}$, directrix : $x + y - 3 = 0$

Let $P(x, y)$ be any point on the ellipse. Then by the definition of ellipse, we have

$$\frac{PS}{PM} = \frac{2}{\sqrt{5}}, \text{ where } PM = \text{distance of } P \text{ from } x + y - 3 = 0$$

$\Rightarrow$ $\sqrt{5}\,PS = 2 \cdot PM$ $\Rightarrow$ $5 \cdot (PS)^2 = 4 \cdot (PM)^2$

$\Rightarrow$ $5[(x-2)^2 + (y+3)^2] = {}^{2}4\,\frac{(x + y - 3)^2}{2}$

$\Rightarrow \quad 5x^2 - 20x + 20 + 5y^2 + 30y + 45 = 2x^2 + 2y^2 + 18 + 4xy - 12y - 12x$

$\Rightarrow \quad \mathbf{3x^2 - 4xy + 3y^2 - 8x + 42y + 47 = 0}$

Exercise

I.

1. Find the characteristics of the following ellipse

(a) $\frac{x^2}{25} + \frac{y^2}{16} = 1$ **(b)** $\frac{x^2}{16} + \frac{y^2}{9} = 1$ **(c)** $\frac{x^2}{9} + \frac{y^2}{4} = 1$

(d) $\frac{x^2}{36} + \frac{y^2}{49} = 1$ **(e)** $9x^2 + 4y^2 = 144$ **(f)** $\frac{x^2}{4} + \frac{y^2}{9} = 1$

2. Find the eccentricity of the ellipse $\frac{x^2}{a^2} + \frac{y^2}{b^2} = 1$, $a > b$, if

(a) its latus rectum is equal to length of the semi-major axis.

(b) the distance between the directrices is three times the distance between the foci.

(c) the axes are in the ratio 5 : 3.

(d) its latus rectum is equal to half of its major axis.

(e) its major axis is 4 times long as the minor axis

(f) the distance between the foci is 50 and the length of semi major axis is 150.

(g) its latus rectum is one half of its minor axis.

3. If the distance between the directrices of the ellipse $\frac{x^2}{a^2} + \frac{y^2}{b^2} = 1$, $(b > a)$ is three times the distance between the foci, find the eccentricity.

4. If the length of the major axis of the ellipse $\frac{x^2}{a^2} + \frac{y^2}{b^2} = 1$, $(b > a)$ is 8 and the distance between directrices is 16, find the eccentricity.

5. If the length of major axis of the ellipse $\frac{x^2}{a^2} + \frac{y^2}{b^2} = 1$, $(b > a)$ is 12 and the distance between the foci is 6, find the eccentricity.

6. A focus of an ellipse in its standard form is (3, 0) and its eccentricity is $\frac{3}{5}$. Find its major axis and the length of latus rectum.

7. If the distance between the foci of an ellipse is 10 and $e = \frac{1}{2}$, find the length of major axis.

8. If the distance between the foci is 4 and the eccentricity is $\frac{1}{2}$, find the square of the semi minor axis.

9. If the length of latus rectum is 8 and eccentricity is $\frac{1}{\sqrt{2}}$, find the square of the semiminor axis.

II.

1. Find the equation of the ellipse in the form $\frac{x^2}{a^2} + \frac{y^2}{b^2} = 1$, with $a > b$, if

(a) the length of latus rectum is 5 and $e = \frac{2}{3}$

(b) the distance between the directrices is 5 and the distance between the foci is 4.

(c) the length of the major axis is 10 and the distance between foci is 8.

(d) the distance between the foci is 4 and $e = \frac{1}{3}$

(e) the distance between the foci is 4 and $e = \frac{1}{4}$

(f) the distance between the directrices is 32 and $e = \frac{1}{2}$

(g) the distance between the directrices is $\frac{32}{3}$ and $e = \frac{3}{4}$

(h) the length of the latus rectum is $\frac{5}{2}$ and $e = \frac{\sqrt{3}}{2}$

(i) the distance between the foci is 4 and the distance between the directrices is 16.

(j) the distance between the foci is $2\sqrt{5}$ and the length of the latus rectum is $\frac{8}{3}$.

(k) the major axis is three times minor axis and the length of latus rectum is 2.

(*l*) the length of the major axis is 8 and the distance between the directrices is 16.

(m) the length of the latus rectum is 10 and the length of minor axis is equal to the distance between the foci.

(n) the length of the minor axis is 10 and the distance between the foci $5\sqrt{2}$.

2. Find the equation of the ellipse in the form $\frac{x^2}{a^2} + \frac{y^2}{b^2} = 1$, with $a > b$, if

(a) its foci are $(\pm 4, 0)$ and vertices are $(\pm 5, 0)$ (i.e., ends of major axis)

(b) a focus is $(3, 0)$ and $e = \frac{3}{5}$

(c) its vertices are $(\pm 7, 0)$ and $e = \frac{6}{7}$

(d) it passes through the points $(2, 2)$ and $(3, 1)$

(e) it passes through the points $(1, 4)$ and $(-6, 1)$

(f) its foci are $(\pm 9, 0)$ and $e = \frac{3}{4}$

(g) its foci are $(\pm 5, 0)$ and one of its directrices is $x = \frac{36}{5}$

(h) its $e = \frac{4}{5}$ and passing through $\left(\frac{5\sqrt{3}}{2}, \frac{3}{2}\right)$

3. Find the equation of the ellipse in the form $\frac{x^2}{a^2} + \frac{y^2}{b^2} = 1$, with $b > a$, if

(a) the vertices are $(0, \pm 6)$ and foci are $(0, \pm 4)$

(b) the foci are $(0, \pm 8)$ and $e = \frac{2}{3}$

(c) the vertices are $(0, \pm 12)$ and $e = \frac{2}{3}$ **(d)** its eccentricity is $\frac{4}{5}$ and passing through $(3\sqrt{2}, 5\sqrt{2})$

(e) its eccentricity is $\frac{3}{4}$ and passing through $(6, 4)$

4. Find the equation of the ellipse whose focus, directrix and eccentricity are given by

(a) focus = (1, 1), directrix : $4x + 3y - 1 = 0$, $e = \frac{5}{6}$

(b) focus = (0, 0), directrix : $3x + 4y - 1 = 0$, $e = \frac{5}{7}$

(c) focus = (– 1, 1), directrix : $x - y + 3 = 0$, $e = \frac{1}{2}$

Answers

I.

1. Read the answers in the following order.
centre, ends of major axis, length of major axis, ends of minor axis, length of minor axis, foci, distance between foci, equations of directrices, distance between directrices, ends of latus rectums, length of latus rectum, equations of latus rectums, eccentricity.

(a) (0, 0), $(\pm 5, 0)$, 10, $(0, \pm 4)$, 8, $(\pm 3, 0)$, 6, $x = \pm \frac{25}{3}$, $\frac{50}{3}$, $\left(3, \pm\frac{16}{5}\right)$ and $\left(-3, \pm\frac{16}{5}\right)$, $\frac{32}{5}$, $x = \pm 3$, $\frac{3}{5}$

(b) (0, 0), $(\pm 4, 0)$, 8, $(0, \pm 3)$, 6, $(\pm\sqrt{7}, 0)$, $2\sqrt{7}$, $x = \pm\frac{16}{\sqrt{7}}$, $\frac{32}{\sqrt{7}}$, $\left(\sqrt{7}, \pm\frac{9}{4}\right)$ and $\left(-\sqrt{7}, \pm\frac{9}{4}\right)$, $\frac{9}{2}$, $x = \pm\sqrt{7}$, $\frac{\sqrt{7}}{4}$

(c) (0, 0), $(\pm 3, 0)$, 6, $(0, \pm 2)$, 4, $(\pm\sqrt{5}, 0)$, $2\sqrt{5}$, $x = \pm\frac{9}{\sqrt{5}}$, $\frac{18}{\sqrt{5}}$, $\left(\sqrt{5}, \pm\frac{4}{3}\right)$ and $\left(-\sqrt{5}, \pm\frac{4}{3}\right)$, $\frac{8}{3}$, $x = \pm\sqrt{5}$, $\frac{\sqrt{5}}{3}$

(d) (0, 0), $(0, \pm 7)$, 14, $(\pm 6, 0)$, 12, $(0, \pm\sqrt{13})$, $2\sqrt{13}$, $y = \pm\frac{49}{\sqrt{13}}$, $\frac{98}{\sqrt{13}}$, $\left(\pm\frac{36}{7}, \sqrt{13}\right)$ and $\left(\pm\frac{36}{7}, -\sqrt{13}\right)$, $\frac{72}{7}$, $y = \pm\sqrt{13}$, $\frac{\sqrt{13}}{7}$

(e) (0, 0), $(0, \pm 6)$, 12, $(\pm 4, 0)$, 8, $(0, \pm 2\sqrt{5})$, $4\sqrt{5}$, $y = \pm\frac{18}{\sqrt{5}}$, $\frac{36}{\sqrt{5}}$, $\left(\pm\frac{8}{3}, 2\sqrt{5}\right)$ and $\left(\pm\frac{8}{3}, -2\sqrt{5}\right)$, $\frac{16}{3}$, $y = \pm 2\sqrt{5}$, $\frac{\sqrt{5}}{3}$

(f) $(0, 0)$, $(0, \pm 3)$, 6, $(\pm 2, 0)$, 4, $(0, \pm\sqrt{5})$, $2\sqrt{5}$, $y = \pm\dfrac{9}{\sqrt{5}}$, $\dfrac{18}{\sqrt{5}}$, $\left(\pm\dfrac{4}{3}, \sqrt{5}\right)$ and $\left(\pm\dfrac{4}{3}, -\sqrt{5}\right)$, $\dfrac{8}{3}$, $y = \pm\sqrt{5}$, $\dfrac{\sqrt{5}}{3}$

2. **(a)** $\dfrac{1}{\sqrt{2}}$ **(b)** $\dfrac{1}{\sqrt{3}}$ **(c)** $\dfrac{4}{5}$ **(d)** $\dfrac{1}{\sqrt{2}}$ **(e)** $\dfrac{\sqrt{15}}{4}$ **(f)** $\dfrac{1}{6}$ **(g)** $\dfrac{\sqrt{3}}{2}$

3. $\dfrac{1}{\sqrt{3}}$ **4.** $\dfrac{1}{2}$ **5.** $\dfrac{1}{2}$ **6.** 10 and $\dfrac{32}{5}$ **7.** 20 **8.** $2\sqrt{3}$ **9.** 32

II.

1. **(a)** $\dfrac{x^2}{(81/4)} + \dfrac{y^2}{(45/4)} = 1$ **(b)** $\dfrac{x^2}{5} + \dfrac{y^2}{1} = 1$ **(c)** $\dfrac{x^2}{25} + \dfrac{y^2}{9} = 1$ **(d)** $\dfrac{x^2}{36} + \dfrac{y^2}{32} = 1$

(e) $\dfrac{x^2}{64} + \dfrac{y^2}{60} = 1$ **(f)** $\dfrac{x^2}{64} + \dfrac{y^2}{48} = 1$ **(g)** $\dfrac{x^2}{16} + \dfrac{y^2}{7} = 1$ **(h)** $\dfrac{x^2}{25} + \dfrac{y^2}{\left(25/4\right)} = 1$

(i) $\dfrac{x^2}{16} + \dfrac{y^2}{12} = 1$ **(j)** $\dfrac{x^2}{9} + \dfrac{y^2}{4} = 1$ **(k)** $\dfrac{x^2}{81} + \dfrac{y^2}{9} = 1$ **(l)** $\dfrac{x^2}{16} + \dfrac{y^2}{12} = 1$

(m) $\dfrac{x^2}{100} + \dfrac{y^2}{50} = 1$ **(n)** $\dfrac{x^2}{(75/2)} + \dfrac{y^2}{25} = 1$

2. **(a)** $\dfrac{x^2}{25} + \dfrac{y^2}{9} = 1$ **(b)** $\dfrac{x^2}{25} + \dfrac{y^2}{16} = 1$ **(c)** $\dfrac{x^2}{49} + \dfrac{y^2}{13} = 1$ **(d)** $\dfrac{x^2}{(32/2)} + \dfrac{y^2}{(32/5)} = 1$

(e) $\dfrac{x^2}{(115/3)} + \dfrac{y^2}{(115/7)} = 1$ **(f)** $\dfrac{x^2}{144} + \dfrac{y^2}{63} = 1$ **(g)** $\dfrac{x^2}{36} + \dfrac{y^2}{11} = 1$ **(h)** $\dfrac{x^2}{25} + \dfrac{y^2}{9} = 1$

(i) $\dfrac{x^2}{25} + \dfrac{y^2}{9} = 1$

3. **(a)** $\dfrac{x^2}{20} + \dfrac{y^2}{36} = 1$ **(b)** $\dfrac{x^2}{80} + \dfrac{y^2}{144} = 1$ **(c)** $\dfrac{x^2}{80} + \dfrac{y^2}{144} = 1$

(d) $\dfrac{x^2}{36} + \dfrac{y^2}{100} = 1$ **(e)** $\dfrac{x^2}{43} + \dfrac{y^2}{688} = 1$

4. **(a)** $20x^2 - 24xy + 27y^2 - 64x - 66y + 71 = 0$

(b) $40x^2 - 24xy + 33y^2 + 6x + 8y - 1 = 0$

(c) $7x^2 + 2xy + 7y^2 + 10x - 10y + 7 = 0$

5.9 Hyperbola

Hyperbola is a conic section whose eccentricity is greater than one.

The formal definition of the hyperbola is given below.

Definition : Hyperbola is the locus of the point which moves such that the ratio of its distance from the fixed point (called focus) to its distance from the fixed line (called directrix) is always a fixed constant which is greater than 1.

Theorem : The equation of the Hyperbola (with proper choice of coordinate axes) in its standard form is $\frac{x^2}{a^2} - \frac{y^2}{b^2} = 1$.

Proof : Let S be the focus and the line l be the directrix of the hyperbola. Let e (> 1) be the eccentricity of the hyperbola. Draw SZ perpendicular to the directrix l.

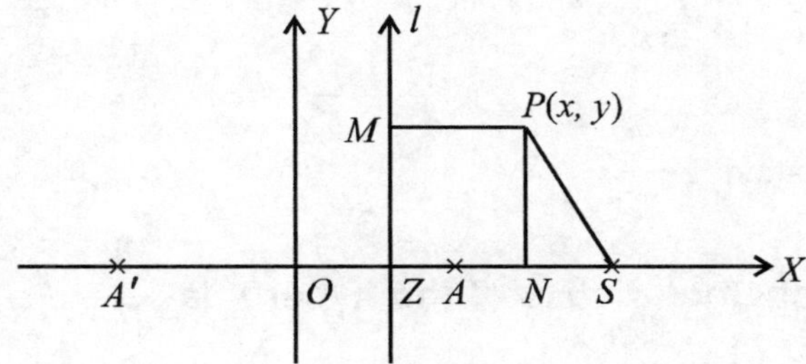

Divide the line segment SZ internally at A and externally at A' in the ratio $e : 1$.

Thus we have, $\frac{SA}{AZ} = e$ and $\frac{SA'}{A'Z} = e$

$\Rightarrow$ $\boldsymbol{SA = e \cdot AZ}$ and $\boldsymbol{SA' = e \cdot A'Z}$

Clearly, A and A' are the points on the hyperbola.

Let O be the mid point of AA' and $\boldsymbol{AA' = 2a} \Rightarrow \boldsymbol{OA = OA' = a}$. Choose O as the origin, the line $A'OZA$ as x-axis and the line OY, through O perpendicular to x-axis as y-axis.

Clearly, $\boldsymbol{A = (a, 0)}$ and $\boldsymbol{A' = (-a, 0)}$.

Consider, $SA = e \cdot AZ \Rightarrow OS - OA = e\,(OA - OZ)$ .... (1)

$SA' = e \cdot A'Z \Rightarrow OS + OA' = e\,(OA' + OZ)$ .... (2)

Consider (1) + (2), we get

$$2OS = e\,(OA + OA') \quad (\because OA = OA')$$

$$\Rightarrow \quad 2OS = 2ae \quad (\because OA = OA' = a)$$

$$\Rightarrow \quad \boldsymbol{OS = ae \Rightarrow S = (ae, 0)}$$

Again consider (2) – (1), we get $\Rightarrow 2a = 2e \cdot (OZ) \Rightarrow \boldsymbol{OZ = \frac{a}{e}}$ $(\because OA = OA' = a)$

Thus the equation of the directrix is $\boldsymbol{x = \frac{a}{e}}$

Let $P(x, y)$ be any point on the hyperbola. PM be its distance from the directrix and PS be its distance from the focus S.

By the definition of hyperbola, we have,

$$\frac{PS}{PM} = e \quad \Rightarrow \quad PS = e \cdot PM$$

$$\Rightarrow \quad PS^2 = e^2 \cdot (PM)^2$$

$$\Rightarrow \quad (x - ae)^2 + y^2 = e^2\,(NZ)^2$$

$$\Rightarrow \quad (x - ae)^2 + y^2 = e^2\,(ON - OZ)^2$$

$$\Rightarrow \quad (x - ae)^2 + y^2 = e^2\left(x - \frac{a}{e}\right)^2 \qquad \left(\because OZ = \frac{a}{e},\ ON = x\right)$$

$$\Rightarrow \quad x^2 - \cancel{2aex} + a^2e^2 + y^2 = e^2x^2 - \cancel{2aex} + a^2$$

$$\Rightarrow \quad x^2\,(e^2 - 1) - y^2 = a^2\,(e^2 - 1) \quad \Rightarrow \quad \frac{x^2}{a^2} - \frac{y^2}{a^2\left(e^2 - 1\right)} = 1 \qquad (\because e^2 - 1 > 0)$$

Since $e > 1$, we have $e^2 - 1 > 1$, and hence $a^2\,(e^2 - 1) > 0$. Putting, $b^2 = a^2\,(e^2 - 1)$,

we get, $$\boldsymbol{\frac{x^2}{a^2} - \frac{y^2}{b^2} = 1}$$

which is the required equation.

5.10 Shape and the characteristics of the hyperbola $\frac{x^2}{a^2} - \frac{y^2}{b^2} = 1$

Consider the equation of the hyperbola $\dfrac{x^2}{a^2} - \dfrac{y^2}{b^2} = 1$

The following observation will help to trace the curve hyperbola:

1. The equation satisfies the coordinates of the points $(a, 0)$ and $(-a, 0)$. Thus the curve hyperbola cuts the x-axis at $A\,(a, 0)$ and $A'(-a, 0)$.
2. By putting $x = 0$ in the equation we get
$$-\frac{y^2}{b^2} = 1 \quad \Rightarrow \quad y^2 = -b^2 \quad \Rightarrow \quad y = \pm\, ib$$
Thus, the curve cuts the y-axis at imaginary points.
3. Clearly the curve is symmetric about both the coordinate axes.
4. Consider the equation $\dfrac{x^2}{a^2} - \dfrac{y^2}{b^2} = 1 \;\Rightarrow\; \dfrac{y^2}{b^2} = \dfrac{x^2}{a^2} - 1$

   As $\dfrac{y^2}{b^2} > 0$, we have, $\dfrac{x^2}{a^2} - 1 > 0 \;\Rightarrow\; x^2 \geq a^2 \;\Rightarrow\; \boldsymbol{x \geq a}$ **or** $\boldsymbol{x \leq -a}$

   Thus no point of the curve lies between $x = a$ and $x = -a$.
5. Again rewriting the equation as $\dfrac{x^2}{a^2} = 1 + \dfrac{y^2}{b^2}$, we have, $\dfrac{x}{a} = \pm\sqrt{\dfrac{y^2 + b^2}{b^2}}$

This show as y increases, the variable x also increases. Also as $y \to \infty$, x also tends to ∞. That is the hyperbola extends to infinity on both the sides of x and y axes. That is the curve hyperbola consists of two symmetrical branches each extending to infinity in two directions.

Further as in the case of ellipse, in the case of hyperbola also, we can show that it has two foci S_1 and S_2 and two directrices l_1 and l_2.

The shape of the curve is shown in the following figure.

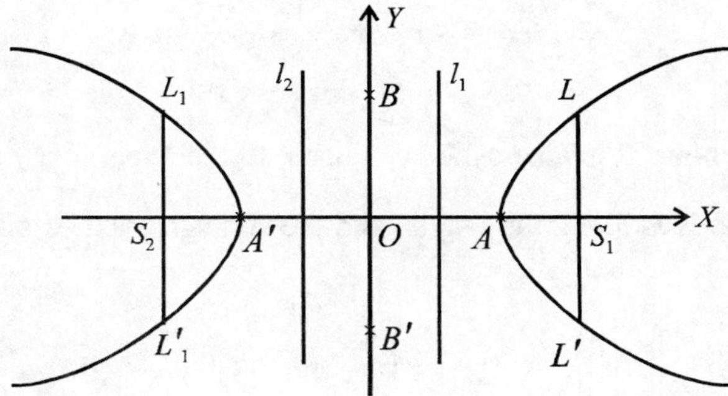

Note : Since $e > 1$, $b^2 = a^2 (e^2 - 1)$ implies, $b > a$, $a = b$ or $b < a$. That is as in the case of ellipse, b need not be less than a (or greater than a).

Following are the characteristics of the hyperbola $\frac{x^2}{a^2} - \frac{y^2}{b^2} = 1$.

1. The point $C = (0, 0)$ is called the **centre** of the ellipse.
2. The line AA' is called the **transverse axis.**

   $A = (a, 0)$ and $A' = (-a, 0)$ are called the **ends of transverse axis**

   $AA' = 2a$ is called **length of transverse axis.**
3. Along the y-axis, take two point B and B' such that $OB = OB' = b$. The line BB' is called **conjugate axis.**

   $B = (0, b)$ and $B' = (0, -b)$ are called the **ends of conjugate axis.**

   $BB' = 2b$ is called **length of conjugate axis.**
4. The foci S_1 and S_2 are $S_1 = (ae, 0)$, $S_2 = (-ae, 0)$. S_1S_2 = **distance between the foci = $2ae$**
5. The equations of directrices are $x = \frac{a}{e}$, $x = -\frac{a}{e}$. **Distance between the directrices** $= \frac{2a}{e}$
6. **Latus rectum.**

   As in the case of ellipse, there are two latus recta.

   The ends of latus rectra are given by

$$L = \left(ae, \frac{b^2}{a}\right), \; L' = \left(ae, -\frac{b^2}{a}\right), \; L_1 = \left(-ae, \frac{b^2}{a}\right), \; L_1' = \left(-ae, -\frac{b^2}{a}\right)$$

The length of latus rectum $= LL' = L_1L_1' = \frac{2b^2}{a}$

7. The eccentricity e of the hyperbola is given by

$$b^2 = a^2(e^2 - 1) \Rightarrow a^2 e^2 = a^2 + b^2 \Rightarrow e^2 = \frac{a^2 + b^2}{a^2}$$

5.11 Conjugate Hyperbola and Rectangular Hyperbola

The hyperbola represented by the equation

$$\frac{x^2}{a^2} - \frac{y^2}{b^2} = -1 \quad \text{or} \quad \frac{y^2}{b^2} - \frac{x^2}{a^2} = 1$$

is called **conjugate hyperbola**. It is also called conjugate to the hyperbola $\frac{x^2}{a^2} - \frac{y^2}{b^2} = 1$. Its centre is the origin, the transverse axis is along the y-axis, and the conjugate axis along the x-axis (see fig.).

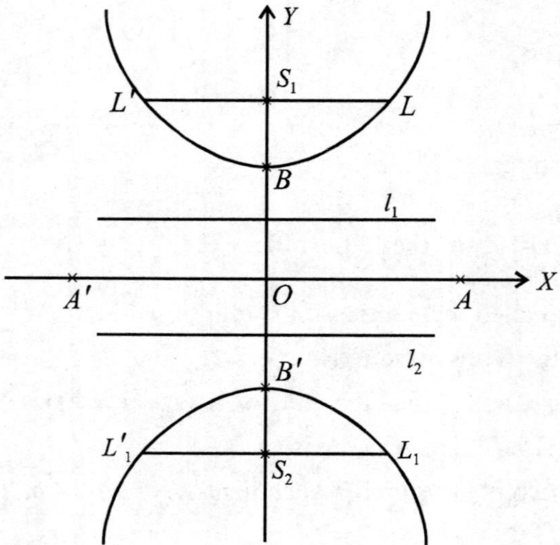

To write the characteristics of the conjugate hyperbola, interchange the x and y coordinates in the corresponding characteristics of the hyperbola $\frac{x^2}{a^2} - \frac{y^2}{b^2} = 1$ and change a to b and b to a.

The following table gives the characteristics of both the forms of the hyperbolae together.

Characteristics \ Hyperbola	$\frac{x^2}{a^2}-\frac{y^2}{b^2}=1$	$\frac{x^2}{a^2}-\frac{y^2}{b^2}=-1$
Centre	**(0, 0)**	**(0, 0)**
Transverse axis:		
Ends:	$A=(a,0), A'=(-a,0)$	$B=(0,b), B'=(0,-b)$
Length:	$2a$	$2b$
Conjugate axis:		
Ends:	$B=(0,b), B'=(0,-b)$	$A=(a,0), A'=(-a,0)$
Length:	$2b$	$2a$
Foci:	$(\pm ae, 0)$	$(0, \pm be)$
Distance between foci:	$2ae$	$2be$
Directrices:	$x=\pm\frac{a}{e}$	$y=\pm\frac{b}{e}$
Distance between directrices	$\frac{2a}{e}$	$\frac{2b}{e}$
Ends of latus rectum	$\left(ae, \pm\frac{b^2}{a}\right)\left(-ae, \pm\frac{b^2}{a}\right)$	$\left(\pm\frac{a^2}{b}, be\right)\left(\pm\frac{a^2}{b}, -be\right)$
Length of latus rectum	$\frac{2b^2}{a}$	$\frac{2a^2}{b}$
Eccentricity	$e^2=\frac{a^2+b^2}{a^2}$	$e^2=\frac{a^2+b^2}{b^2}$
Equations of latus rectum	$x=\pm ae$	$y=\pm be$
Relation between a, b & e.	$b^2=a^2(e^2-1)$	$a^2=b^2(e^2-1)$

Definition : The hyperbola in which the length of the transverse axis is equal to the length of the conjugate axis is called a rectangular hyperbola.

The equation of rectangular hyperbola is given by

$$\frac{x^2}{a^2}-\frac{y^2}{b^2}=1 \quad\Rightarrow\quad x^2-y^2=a^2$$

Theorem : For a rectangular hyperbola $x^2-y^2=a^2$ the eccentricity is $\sqrt{2}$.

Proof : The equation of rectangular hyperbola is, $x^2-y^2=a^2$

Now, we have $b^2=a^2(e^2-1) \Rightarrow a^2=a^2(e^2-1)$ $(\because b=a)$

$$\Rightarrow e^2=2 \Rightarrow e=\sqrt{2}$$

Example 1. Write the characteristics of the following hyperbolae.

$$\frac{x^2}{25}-\frac{y^2}{16}=1$$

Solution : The equation of the hyperbola is

$$\frac{x^2}{25}-\frac{y^2}{16}=1 \quad\Rightarrow\quad a^2=25,\ b^2=16 \quad\Rightarrow\quad a=5, \quad b=4$$

Now the eccentricity e is given by

$$\Rightarrow \qquad e^2=\frac{a^2+b^2}{a^2} \quad\Rightarrow e^2=\frac{25+16}{25} \quad\Rightarrow e^2=\frac{41}{25} \quad\Rightarrow e=\frac{\sqrt{41}}{5}$$

The characteristics are

1. Centre $C=(0, 0)$
2. Transverse axis: Ends, $A=(a, 0)=\mathbf{(5, 0)}$, $A'=(-a, 0)=\mathbf{(-5, 0)}$; Length $=2a=\mathbf{10}$.
3. Conjugate axis: Ends. $B=(0, b)=\mathbf{(0, 4)}$, $B'=(0, -b)=\mathbf{(0, -4)}$; Length $=2b=\mathbf{8}$
4. Foci: $S_1=(ae, 0)=\left(\sqrt{41}, 0\right)$, $S_2=(-ae, 0)=\left(-\sqrt{41}, 0\right)$; Distance between foci $=2ae=2\sqrt{41}$
5. Directrices: $x=\frac{a}{e} \Rightarrow x=\frac{25}{\sqrt{41}}$, $x=-\frac{a}{e} \Rightarrow x=-\frac{25}{\sqrt{41}}$

   Distance between directrices $=\frac{2a}{e}=\frac{50}{\sqrt{41}}$
6. Ends of latus rectums

$$L=\left(ae, \frac{b^2}{a}\right)=\left(\sqrt{41}, \frac{16}{5}\right), \qquad L'=\left(ae, -\frac{b^2}{a}\right)=\left(\sqrt{41}, -\frac{16}{5}\right)$$

$$L_1=\left(-ae, \frac{b^2}{a}\right)=\left(-\sqrt{41}, \frac{16}{5}\right), \qquad L_1'=\left(-ae, -\frac{b^2}{a}\right)=\left(-\sqrt{41}, -\frac{16}{5}\right)$$

   Length of the latus rectum $=\frac{2b^2}{a}=\frac{32}{5}$
7. Equations of latus rectums: $x=ae \Rightarrow x=\sqrt{41}$; $x=-ae \Rightarrow x=-\sqrt{41}$

Example 2. If the distance between the foci of a hyperbola is 120 and the length of transverse axis is 40, find the eccentricity.

Solution : By data, $2ae=120$ and $2a=40 \quad \therefore \ e=3$

Example 3. If for a hyperbola the distance between the foci is 12 and the distance between the directrices is $\frac{32}{3}$ find the length of the latus rectum.

Solution : By data, $2ae=12$ and $\frac{2a}{e}=\frac{32}{3}$

Consider, $(2ae)\left(\frac{2a}{e}\right)=(12)\left(\frac{32}{3}\right) \Rightarrow 4a^2=128 \Rightarrow a^2=32$

Also we have, $b^2 = a^2(e^2-1) \Rightarrow b^2 = a^2e^2 - a^2 \Rightarrow b^2 = 36 - 32 \Rightarrow \boldsymbol{b^2 = 4}$

$$\text{Length of latus rectum} = \frac{2b^2}{a} = \frac{2\times 4}{\sqrt{32}} = \frac{8}{4\sqrt{2}} = \sqrt{2}$$

Example 4. Find the equation of the hyperbola, in the form $\frac{x^2}{a^2} - \frac{y^2}{b^2} = 1$, if the distance between the foci is $2\sqrt{5}$ and the length of the latus rectum is $\frac{4\sqrt{3}}{3}$.

Solution : By data $2ae = 2\sqrt{5} \Rightarrow \boldsymbol{ae = \sqrt{5}}$

Again by data $\frac{2b^2}{a} = \frac{4\sqrt{3}}{3} \Rightarrow b^2 = \frac{2\sqrt{3}}{3}a$

Consider, $b^2 = a^2(e^2-1) \Rightarrow b^2 = a^2e^2 - a^2$

$$\Rightarrow \frac{2\sqrt{3}}{3}a = 5 - a^2$$

$$\Rightarrow 3a^2 + 2\sqrt{3}\,a - 15 = 0$$

$$\Rightarrow 3a^2 + 5\sqrt{3}\,a - 3\sqrt{3}\,a - 15 = 0$$

$$\Rightarrow \sqrt{3}\,a(\sqrt{3}\,a + 5) - 3(\sqrt{3}\,a + 5) = 0$$

$$\Rightarrow (\sqrt{3}\,a + 5)(\sqrt{3}\,a - 3) = 0$$

$$\Rightarrow a = -\frac{5}{\sqrt{3}} \text{ or } a = \sqrt{3}$$

-45

$5\sqrt{3}$ $\quad -3\sqrt{3}$

Rejecting the negative value, we have $a = \sqrt{3} \Rightarrow \boldsymbol{a^2 = 3} \Rightarrow \boldsymbol{b^2 = 2}$.

Thus the equation of the hyperbola is $\frac{x^2}{3} - \frac{y^2}{2} = 1$.

Exercise

I.

1. Write the characteristics of the hyperbola

   (a) $\frac{x^2}{25} - \frac{y^2}{9} = 1$ **(b)** $\frac{x^2}{16} - \frac{y^2}{25} = 1$ **(c)** $\frac{x^2}{9} - \frac{y^2}{27} = 1$

   (d) $\frac{x^2}{25} - \frac{y^2}{9} = -1$ **(e)** $\frac{y^2}{4} - \frac{x^2}{16} = 1$ **(f)** $x^2 - y^2 = 4$

2. The length of the semi conjugate axis of a hyperbola is 4 and the point (3, 3) lies on it. Find the eccentricity and length of latus rectum.
3. If one of the foci of the hyperbola is $(8\sqrt{2}, 0)$ and the eccentricity is $2\sqrt{2}$, find the length of transverse axis.
4. If the distance between the foci is 4 times the distance between the directrices, find e.
5. If the distance between the foci is 120 and the length of the transverse axis is 40, find e.
6. Find the eccentricity of the hyperbola, if the length of the latus rectum is $\frac{1}{3}$ of its transverse axis.

II.

1. Find the equation of the hyperbola in the form $\frac{x^2}{a^2} - \frac{y^2}{b^2} = 1$, if

   (a) the distance between the foci is 10 and length of conjugate axis is 8.

   (b) the distance between the foci is 32 and $e = \sqrt{8}$

   (c) the distance between the foci is 12 and $e = 2$

   (d) the length of transverse axis is 16 and $e = \frac{5}{4}$

   (e) the distance between the directrices is $\frac{8}{3}$ and $e = \frac{3}{2}$

   (f) $e = \frac{5}{4}$ and the foci are $(\pm 15, 0)$

   (g) the distance between the directrices is 20 and $e = 2$

   (h) the length of conjugate axis is 12 and latus rectum is $\frac{27}{4}$

   (i) the distance between the foci is 24 and the distance between the directrices is 6.

   (j) the distance between foci 10 and the length of latus rectum is $\frac{9}{2}$

   (k) $e = \sqrt{2}$ and the length of latus rectum is $8\sqrt{2}$

   (*l*) the distance between directrices is $\frac{32}{5}$ and the length of conjugate axis is 6.

   (m) the length of conjugate axis is 6 and the distance between the foci is 8.

   (n) the length of the conjugate axis is 7 and passes through $(3, -2)$.

2. Find the equation of the hyperbola in the form $\frac{x^2}{a^2} - \frac{y^2}{b^2} = 1$, if

   (a) it passes through the points $(3, -2)$ and $(2, -1)$

   (b) it passes through the points $(6, -1)$ and $(-8, 2\sqrt{2})$

   (c) it passes through the points $(2, -1)$ and $e = 2$

3. The foci of a hyperbola coincides with the foci of the ellipse $\frac{x^2}{25} + \frac{y^2}{9} = 1$, find the equation of the hyperbola if its eccentricity is 2.

4. Find the equation of the hyperbola whose foci lie at the vertices of the ellipse $\frac{x^2}{100} + \frac{y^2}{64} = 1$ and whose directrices passes through the foci of the ellipse.

Answers

I.

1. **Read the answers in the following order.**
   eccentricity, centre, ends of transverse axis, length of transverse axis, ends of conjugate axis, length of conjugate axis, foci, distance between foci, equations of directrices, distance between directrices, ends of latus rectums, length of latus rectum, equations of latus rectums.

(a) $e = \frac{\sqrt{34}}{5}$; (0, 0); (5, 0), (– 5, 0); 10; (0, 3), (0, – 3); 6; $(\pm\sqrt{34}, 0)$; $2\sqrt{34}$; $x = \pm\frac{25}{\sqrt{34}}$; $\frac{50}{\sqrt{34}}$; $\left(\pm\sqrt{34}, \pm\frac{9}{5}\right)$; $\frac{18}{5}$; $x = \pm\sqrt{34}$

(b) $e = \frac{\sqrt{41}}{4}$; (0, 0); (± 4, 0); 8; (0, ± 5); 10; $(\pm\sqrt{41}, 0)$; $2\sqrt{41}$; $x = \pm\frac{16}{\sqrt{41}}$; $\frac{32}{\sqrt{41}}$; $\left(\pm\sqrt{41}, \pm\frac{25}{4}\right)$; $\frac{25}{2}$; $x = \pm\sqrt{41}$

(c) $e = 2$; (0, 0); (± 3, 0); 6; $(0, \pm 3\sqrt{3})$; $6\sqrt{3}$; (±6, 0); 12; $x = \pm\frac{3}{2}$; 3; (± 6, ± 9); 18; $x = \pm 6$

(d) $e = \frac{\sqrt{34}}{3}$; (0, 0); (0, ± 3); 6; (± 5, 0); 10; $(0, \pm\sqrt{34})$; $2\sqrt{34}$; $y = \pm\frac{9}{\sqrt{34}}$; $\frac{18}{\sqrt{34}}$; $\left(\pm\frac{25}{3}, \pm\sqrt{34}\right)$; $\frac{50}{3}$; $y = \pm\sqrt{34}$

(e) $e = \sqrt{5}$; (0, 0); (0, ± 2); 4; (± 4, 0); 8; $(0, \pm 2\sqrt{5})$; $4\sqrt{5}$; $y = \pm\frac{2}{\sqrt{5}}$; $\frac{4}{\sqrt{5}}$; $(\pm 8, \pm 2\sqrt{5})$; 16; $y = \pm 2\sqrt{5}$

(f) $e = \sqrt{2}$; (0, 0); (± 2, 0); 4; (0, ± 2); 4; $(\pm 2\sqrt{2}, 0)$; $4\sqrt{2}$; $x = \pm\frac{2}{\sqrt{2}}$; $2\sqrt{2}$; $(\pm 2\sqrt{2}, \pm 2)$; 4; $x = \pm 2\sqrt{2}$

2. $e = \frac{\sqrt{34}}{3}, \frac{40}{3}$ **4.** 8 **5.** $e = 2$ **6.** $e = 3$ **7.** $e = \frac{2}{\sqrt{3}}$

II.

1. (a) $\frac{x^2}{9} - \frac{y^2}{16} = 1$ **(b)** $\frac{x^2}{32} - \frac{y^2}{224} = 1$ **(c)** $\frac{x^2}{9} - \frac{y^2}{27} = 1$ **(d)** $\frac{x^2}{64} - \frac{y^2}{36} = 1$

(e) $\frac{x^2}{4} - \frac{y^2}{5} = 1$ **(f)** $\frac{x^2}{144} - \frac{y^2}{81} = 1$ **(g)** $\frac{x^2}{400} - \frac{y^2}{1200} = 1$ **(h)** $\frac{9x^2}{1024} - \frac{y^2}{36} = 1$

(i) $\frac{x^2}{36} - \frac{y^2}{108} = 1$ **(j)** $\frac{x^2}{16} - \frac{y^2}{9} = 1$ **(k)** $\frac{x^2}{32} - \frac{y^2}{32} = 1$ **(l)** $\frac{x^2}{16} - \frac{y^2}{9} = 1$

(m) $\frac{x^2}{7} - \frac{y^2}{9} = 1$ **(n)** $\frac{65x^2}{441} - \frac{4y^2}{49} = 1$

2. (a) $\frac{3x^2}{7} - \frac{5y^2}{7} = 1$ **(b)** $\frac{x^2}{32} - \frac{y^2}{8} = 1$ **(c)** $\frac{3x^2}{11} - \frac{y^2}{11} = 1$

3. $\frac{x^2}{4} - \frac{y^2}{12} = 1$ **5.** $\frac{x^2}{60} - \frac{y^2}{40} = 1$

UNIT IV & V

Differential Calculus

Chapter 1

Differential Calculus

1.1 Introduction

The readers are already familiar with the concept of limit and continuity of a function, which are the basic requirements for the study of differential calculus. The subject differential calculus was first introduced independently by two mathematicians, Sir Isac Newton and Leibnitz during the same period – i.e., the year 1684. In the present chapter we shall introduce the concept of derivative of a function, and process of differentiation of different form of functions.

Before we go over to the concept of differentiation, we shall mention the results related to the limit of a function and continuity of a function.

1.2 Limit and Continuity of a Function

Let $y = f(x)$ be a real valued function of a real variable. We say the limit of the function $y = f(x)$ is 'l' as x tends to a, if the value of the function $f(x)$ is "nearer" and "nearer" to the value l, whenever the value of the variable x is "nearer" and "nearer" to the value 'a'. This is denoted by $\lim_{x\to a} f(x) = l$.

The readers are familiar with the various methods of evaluating the limit of a function under different circumstances. During the process of evaluating the limit, we use the result of certain standard limits, we list it below for the benefit of the readers:

1. If n is any rational number, $\lim_{x\to a} \dfrac{x^n - a^n}{x - a} = na^{n-1}$

2. If θ is measured in radians, then **(a)** $\lim_{\theta\to 0} \dfrac{\sin\theta}{\theta} = 1$ **(b)** $\lim_{\theta\to 0} \dfrac{\tan\theta}{\theta} = 1.$

3. $\lim_{n\to\infty} \left(1 + \dfrac{1}{n}\right)^n = e$ **4.** $\lim_{x\to 0} \dfrac{a^x - 1}{x} = \log_e a \quad (a > 0)$

5. $\lim_{x\to 0} \left(\dfrac{e^x - 1}{x}\right) = 1$ **6.** $\lim_{x\to 0} \dfrac{\log(1 + x)}{x} = 1$

A function $y = f(x)$ is said to be continuous at $x = a$ if

(i) $f(x)$ is defined at $x = a$ **(ii)** $\lim_{x\to a} f(x)$ exists **(iii)** $\lim_{x\to a} f(x) = f(a)$.

Geometrically, a function $y = f(x)$ is said to continuous at $x = a$, if the graph of the function has no break or a jump at the point $x = a$.

If $f(x)$ and $g(x)$ are two continuous functions, at $x = a$. Then,

(i) $f(x) \pm g(x)$ is continuous at $x = a$.

(ii) $f(x) \cdot g(x)$ is continuous at $x = a$.

(iii) $\frac{f(x)}{g(x)}$ is continuous at $x = a$, provided $g(a) \neq 0$.

Following are few basic continuous functions:

(i) A constant function, $y = f(x) = k$, is a continuous function.

(ii) A polynomial function $f(x) = a_0 x^n + a_1 x^{n-1} + ... + a_n$ is a continuous function for $x \in R$.

(iii) $y = \sin x$, $y = \cos x$, $y = e^x$ are continuous functions for all $x \in R$.

(iv) $y = \tan x$ is continuous for all x, except at $x = (2n - 1)\frac{\pi}{2}$, where n is an integer.

1.3 Derivative of a Function

In this section we shall introduce the basic concept of differential calculus, called differentiation.

Let $y = f(x)$ be a function of real variable, defined over an interval $I \subset R$. Let the variable x be changed from x to $x + \delta x$.

Let the corressponding change in y be from y to $y + \delta y$. That is $y + \delta y$, is the value of the function at $x = x + \delta x$.

$\therefore$ $$y + \delta y = f(x + \delta x); \quad \text{Also,} \quad y = f(x)$$

On subtraction, we have, $$\delta y = f(x + \delta x) - f(x)$$

This relation gives us the change in the dependent variable, due to change in the independent variable from x to $x + \delta x$.

Consider, $$\frac{\delta y}{\delta x} = \frac{f(x + \delta x) - f(x)}{\delta x}$$

This ratio gives us average rate of change of y w.r.t x, per unit.

Taking the limit as $\delta x \to 0$ on both sides, then we have. $\lim_{\delta x \to 0} \frac{\delta y}{\delta x} = \lim_{\delta x \to 0} \frac{f(x + \delta x) - f(x)}{\delta x}$

If this limit exists and finte, then we say the function $y = f(x)$ is differentiable at x and the limit is called the **differential coefficient** of or **derivative of** or **rate of change of $y = f(x)$ w.r.t x.** This is denoted by $\frac{dy}{dx}$, $\frac{d}{dx}[f(x)]$ or $f'(x)$.

Thus we have the following definitation.

Definition : Let $y = f(x)$ be a function defined in the internal $I \subset R$. Then the derivative or differential coefficient of $y = f(x)$, w.r.t x is denoted and defined by

$$f'(x) = \frac{dy}{dx} = \lim_{\delta x \to 0} \frac{\delta y}{\delta x} = \lim_{\delta x \to 0} \frac{f(x + \delta x) - f(x)}{\delta x}$$

where, δy is the change in the dependent variable, corresponding to a change δx in independent variable.

If the increment δx of x is denoted by h, then the above formula takes the form

$$f'(x) = \frac{dy}{dx} = \lim_{h \to 0} \frac{f(x+h) - f(x)}{h}$$

Supposing, we put $x = a$, then we have, $f'(a) = \lim_{h \to 0} \frac{f(a+h) - f(a)}{h}$

This gives us the derivative (or differential coefficient) of the function $f(x)$ at $x = a$. This is also denoted by $\left(\frac{dy}{dx}\right)_{x=a}$

Definition : The function $y = f(x)$ is said to be differentiable at $x = a$, if

$$\lim_{h \to 0} \frac{f(a+h) - f(a)}{h}$$

exists and it is denoted by $f'(a)$ or $\left(\frac{dy}{dx}\right)_{x=a}$

Note : If we put $a + h = x$ in the above expresion, then $h \to 0$ changes to $x \to a$ and we have

$$f'(a) = \lim_{x \to a} \frac{f(x) - f(a)}{x - a}$$

This is another form of expressing the derivative of $f(x)$ at $x = a$.

We know that $\lim_{x \to a} f(x)$, exists, if both left hand limit and right hand limit of $f(x)$ as $x \to a$ exists and are equal.

i.e., $\lim_{x \to a} f(x)$ exists $\Rightarrow$ $\lim_{x \to a^-} f(x)$ and $\lim_{x \to a^+} f(x)$ exists and are equal.

Thus, a function $y = f(x)$ is said to be differentiable at $x = a$, if both left hand limit and right hand limit of $\frac{f(x) - f(a)}{x - a}$ as $x \to a$, exists and are equal.

The left hand limit and right hand limit of $\frac{f(x) - f(a)}{x - a}$ as $x \to a$ are respectively called **left hand derivative** and **right hand derivative** of $f(x)$ at $x = a$ and are denoted by $L\,f'(a)$ and $R\,f'(a)$. The formal definitions are given below.

Definition (Left hand derivative) : The left hand derivative of a function $y = f(x)$ at $x = a$ is denoted and defined by $L\,\{f'(a)\} = \lim_{x \to a^-} \frac{f(x) - f(a)}{x - a}$ provided the limit exists.

Definition (Right hand derivative) : The right hand derivative of a function $y = f(x)$ at $x = a$ is denoted and defined by $R\,\{f'(a)\} = \lim_{x \to a^+} \frac{f(x) - f(a)}{x - a}$ provided the limit exists.

A function $y = f(x)$ is said to be differentiable at $x = a$, if $L\,\{f'(a)\}$ and $R\,\{f'(a)\}$ exists and equal.

This is denoted by $f'(a)$ or $\left(\frac{dy}{dx}\right)_{x=a}$

If for a function $y = f(x)$, any one of $L\ \{f'(a)\}$ or $R\ \{f'(a)\}$ or both does not exist, or even they exists but not equal, then we say $f(x)$ is **not differentiable** at $x = a$.

1.4 Differentiability and Continuity

The following theorem gives us the relation between differentiability and continuity of a function.

Theorem : If $y = f(x)$ is differentiable at $x = a$, then $f(x)$ is continuous at $x = a$.

Proof : By data $y = f(x)$ is differentiable at $x = a$. Thus $\lim\limits_{x \to a} \dfrac{f(x) - f(a)}{x - a}$ exists and it is denoted by $f'(a)$.

Consider, $$f(x) - f(a) = \frac{f(x) - f(a)}{x - a} \cdot (x - a)$$

Taking the limit both sides as $x \to a$ we get

$$\lim_{x \to a} \{f(x) - f(a)\} = \lim_{x \to a} \left\{\frac{f(x) - f(a)}{x - a}\right\} \cdot \lim_{x \to a} (x - a)$$

$$\Rightarrow \quad \lim_{x \to a} \{f(x)\} - f(a) = f'(a) \cdot 0 \qquad (\because f(a) \text{ is a constant})$$

$$\Rightarrow \quad \lim_{x \to a} \{f(x)\} - f(a) = 0 \Rightarrow \lim_{x \to a} f(x) = f(a) \Rightarrow \mathbf{f(x) \text{ is continuous at } x = a.}$$

Note : Converse of the above theorem is not true. That is if a function $y = f(x)$ is continuous at $x = a$, then it need not be differentiable at $x = a$. For example, consider the function $y = f(x) = |x|$.

This function is defined by $$f(x) = \begin{cases} x & x > 0 \\ 0 & x = 0 \\ -x & x < 0 \end{cases}$$

We shall show that $f(x)$ is continuous at $x = 0$ but not differentiable at $x = 0$.

Consider, $$\lim_{x \to 0^-} f(x) = \lim_{x \to 0} (-x) = 0 \qquad (\because f(x) = -x, \text{ if } x < 0)$$

Again, $$\lim_{x \to 0^+} f(x) = \lim_{x \to 0} x = 0 \qquad (\because f(x) = x, \text{ if } x > 0)$$

Thus, $$\lim_{x \to 0} f(x) = 0 = f(0) \Rightarrow \mathbf{f(x) \text{ is continuous at } x = 0.}$$

Now consider, $$L\ \{f'(0)\} = \lim_{x \to 0^-} \frac{f(x) - f(0)}{x - 0} = \lim_{x \to 0} \frac{-x}{x} \qquad (\because f(x) = -x, \text{ when } x < 0, f(0) = 0)$$
$$= \lim_{x \to 0} (-1) = -1 \Rightarrow L\ [f'(0)] = -1.$$

Again consider, $$R\ \{f'(0)\} = \lim_{x \to 0^+} \frac{f(x) - f(0)}{x - 0} = \lim_{x \to 0} \frac{x}{x} \qquad (\because f(x) = x, x > 0, f(0) = 0)$$
$$= \lim_{x \to 0} = 1 \Rightarrow R\ \{f'(0)\} = 1.$$

Thus $L\ \{f'(0)\} \neq R\ \{f'(0)\}$. Hence $f(x) = |x|$ is not differentiable at $x = 0$.

Note : 1. From the above theorem we have the following results.

(i) Differentiability $\Rightarrow$ continuity

(ii) Continuity does not imply differentiability

(iii) not continuous $\Rightarrow$ not differentiable.

2. We have seen, for a function $y = f(x)$, its derivative (if it exists) is denoted by $\frac{dy}{dx}$; this is only a notation to indicate the derivative of y w.r.t x. Further $\frac{d}{dx}$ is a notation for an operator when it operates on a function $y = f(x)$, it gives us its differential coefficient – for this reason $\frac{d}{dx}$ is also called **differential operator.**

3. The process of finding the derivative of a function is called **differentiation.**

1.5 Differentiation from the First Principles

In this section, we shall consider the derivatives of certain basic functions, using the definition of differentiation. Obtaining the derivative of a function using the definition, is called **"differentiation from the first principles."**

1. To differentiate a constant function from the first principles

Let $y = f(x) = k$, $\forall\, x$ where k is a constant, be a constant function. By definition, we have

$$\frac{dy}{dx} = \lim_{h\to 0} \frac{f(x+h) - f(x)}{h} \Rightarrow \frac{dy}{dx} = \lim_{h\to 0} \frac{k-k}{h} \qquad (\because f(x+h) = k)$$

$$\Rightarrow \frac{dy}{dx} = \lim_{h\to 0} 0 = 0 \qquad (\because h \neq 0)$$

Thus the derivative of a constant function is zero

i.e., $$y = f(x) = k \Rightarrow \frac{dy}{dx} = 0.$$

2. To differentiate x^n from the first principles.

Let $y = f(x) = x^n$. By definition we have

$$\frac{dy}{dx} = \lim_{h\to 0} \frac{f(x+h) - f(x)}{h} \Rightarrow \frac{dy}{dx} = \lim_{h\to 0} \frac{(x+h)^n - x^n}{h} \qquad (\because f(x) = x^n)$$

$$\Rightarrow \frac{dy}{dx} = \lim_{\substack{h\to 0\\(x+h)\to x}} \frac{(x+h)^n - x^n}{(x+h) - x} \qquad (\because \text{ as } h \to 0, (x+h) \to x)$$

$$\Rightarrow \frac{dy}{dx} = n \cdot x^{n-1} \qquad \left(\because \lim_{x\to a} \frac{x^n - a^n}{x - a} = na^{n-1}\right)$$

Thus the derivative of x^n w.r.t x is nx^{n-1}.

i.e., $$y = x^n \Rightarrow \frac{dy}{dx} = nx^{n-1}$$

Examples :

(i) $y = x^7$, $\frac{dy}{dx} = 7x^6$. (ii) $y = x^{3/4}$, $\frac{dy}{dx} = \frac{3}{4} x^{\frac{3}{4}-1} = \frac{3}{4} x^{-1/4}$

(iii) $y = \frac{1}{x^5} \Rightarrow y = x^{-5} \Rightarrow \frac{dy}{dx} = -5 \cdot x^{-5-1} \Rightarrow \frac{dy}{dx} = -5x^{-6} \Rightarrow \frac{dy}{dx} = \frac{-5}{x^6}$.

Note : Consider the function $y = \frac{1}{x^n}$, then

$$y = \frac{1}{x^n} \Rightarrow y = x^{-n} \Rightarrow \frac{dy}{dx} = -n \cdot x^{-n-1} \Rightarrow \frac{dy}{dx} = \frac{-n}{x^{n+1}}.$$

Thus, **if $y = \frac{1}{x^n}$ then $\frac{dy}{dx} = \frac{-n}{x^{n+1}}$**

Using this formula we can write the derivatives of the function of the form $\frac{1}{x^n}$ directly.

Examples :

(i) $y = \frac{1}{x^3} \Rightarrow \frac{dy}{dx} = \frac{-3}{x^4}$ (ii) $y = \frac{1}{x^{4/5}} \Rightarrow \frac{dy}{dx} = \frac{-4}{5x^{9/5}}$

In particular, if $y = \frac{1}{x}$, then $\frac{dy}{dx} = -\frac{1}{x^2}$ and if $y = \sqrt{x}$, then $\frac{dy}{dx} = \frac{1}{2\sqrt{x}}$.

3. To differentiate sin x w.r.t x from the first principles

Let $y = f(x) = \sin x$, where x is measured in radians.

By definition, we have,

$$\frac{dy}{dx} = \lim_{h\to 0} \frac{f(x+h) - f(x)}{h} \Rightarrow \frac{dy}{dx} = \lim_{h\to 0} \frac{\sin(x+h) - \sin x}{h} \quad (\because f(x) = \sin x)$$

$$\Rightarrow \frac{dy}{dx} = \lim_{h\to 0} \frac{2\cos\left(\frac{x+h+x}{2}\right) \cdot \sin\left(\frac{x+h-x}{2}\right)}{h}$$

$$\left[\text{Using } \sin C - \sin D = 2\cos\frac{C+D}{2} \cdot \sin\frac{C-D}{2}\right]$$

$$\Rightarrow \frac{dy}{dx} = \lim_{h\to 0} \frac{2\cos\left(\frac{2x+h}{2}\right) \cdot \sin\left(\frac{h}{2}\right)}{h}$$

$$\Rightarrow \frac{dy}{dx} = \lim_{h\to 0} 2\cos\left(\frac{2x+h}{2}\right) \cdot \frac{\sin\left(\frac{h}{2}\right)}{\left(\frac{h}{2}\right)} \quad \text{(Note this step)}$$

$$\Rightarrow \frac{dy}{dx} = \lim_{h\to 0} \cos\left(\frac{2x+0}{2}\right) \cdot 1 \quad \left(\because \lim_{\theta\to 0} \frac{\sin\theta}{\theta} = 1\right)$$

$$\Rightarrow \frac{dy}{dx} = \cos x$$

Thus, the derivative of sin x w.r.t x is cos x. i.e., $y = \sin x \Rightarrow \frac{dy}{dx} = \cos x$.

4. To differentiate cos x w.r.t x from the first principles

Let $y = f(x) = \cos x$, where x is measured in radians. By definition, we have

$$\frac{dy}{dx} = \lim_{h\to 0} \frac{f(x+h)-f(x)}{h} \Rightarrow \frac{dy}{dx} = \lim_{h\to 0} \frac{\cos(x+h)-\cos x}{h} \qquad (\because f(x) = \cos x)$$

$$\Rightarrow \frac{dy}{dx} = \lim_{h\to 0} \frac{-2\sin\left(\frac{x+h+x}{2}\right)\cdot \sin\left(\frac{x+h-x}{2}\right)}{h}$$

$$\left[\because \cos C - \cos D = -2\sin\left(\frac{C+D}{2}\right)\cdot \sin\left(\frac{C-D}{2}\right)\right]$$

$$\Rightarrow \frac{dy}{dx} = \lim_{h\to 0} -\sin\left(\frac{2x+h}{2}\right)\cdot \frac{\sin(h/2)}{(h/2)} \qquad \text{(note this step)}$$

$$\Rightarrow \frac{dy}{dx} = -\sin\left(\frac{2x+0}{2}\right)\cdot 1 \qquad \left(\because \lim_{\theta\to 0}\frac{\sin\theta}{\theta} = 1\right)$$

$$\Rightarrow \frac{dy}{dx} = -\sin x.$$

Thus, the derivative of $\cos x$ w.r.t x is $-\sin x$. i.e., $\mathbf{y = \cos x \Rightarrow \frac{dy}{dx} = -\sin x}$

5. To differentiate tan *x* w.r.t *x* from the first principles

Let $y = f(x) = \tan x$, where x is measured in radians. By definition, we have,

$$\frac{dy}{dx} = \lim_{h\to 0} \frac{f(x+h)-f(x)}{h} \Rightarrow \frac{dy}{dx} = \lim_{h\to 0} \frac{\tan(x+h)-\tan x}{h} \qquad (\because f(x) = \tan x)$$

$$\Rightarrow \frac{dy}{dx} = \lim_{h\to 0} \frac{1}{h}\left[\frac{\sin(x+h)}{\cos(x+h)} - \frac{\sin x}{\cos x}\right]$$

$$\Rightarrow \frac{dy}{dx} = \lim_{h\to 0} \frac{1}{h}\left[\frac{\sin(x+h)\cdot\cos x - \cos(x+h)\cdot\sin x}{\cos(x+h)\cdot\cos x}\right]$$

$$\Rightarrow \frac{dy}{dx} = \lim_{h\to 0} \frac{1}{h}\cdot\frac{\sin(x+h-x)}{\cos(x+h)\cdot\cos x}$$

$$\Rightarrow \frac{dy}{dx} = \lim_{h\to 0} \frac{1}{\cos(x+h)\cdot\cos x}\left(\frac{\sin h}{h}\right)$$

$$\Rightarrow \frac{dy}{dx} = \frac{1}{\cos(x+0)\cdot\cos x}\cdot 1 \qquad \left(\because \lim_{h\to 0}\frac{\sin h}{h} = 1\right)$$

$$\Rightarrow \frac{dy}{dx} = \frac{1}{\cos x\cdot\cos x} = \sec^2 x.$$

Thus, the derivative of $\tan x$ w.r.t. x is $\sec^2 x$. i.e., $\mathbf{y = \tan x \Rightarrow \frac{dy}{dx} = \sec^2 x}$.

6. To differentiate cosec x w.r.t x from the first principles

Let $y = f(x) = \text{cosec } x$, where x is measured in radians.

By definition, we have,

$$\frac{dy}{dx} = \lim_{h\to 0}\frac{f(x+h)-f(x)}{h} \Rightarrow \frac{dy}{dx} = \lim_{h\to 0}\frac{\text{cosec }(x+h)-\text{cosec } x}{h} \qquad (\because f(x) = \text{cosec } x)$$

$$\Rightarrow \frac{dy}{dx} = \lim_{h\to 0}\frac{1}{h}\left[\frac{1}{\sin(x+h)} - \frac{1}{\sin x}\right]$$

$$\Rightarrow \frac{dy}{dx} = \lim_{h\to 0}\frac{1}{h}\left[\frac{\sin x - \sin(x+h)}{\sin(x+h)\cdot\sin x}\right]$$

$$\Rightarrow \frac{dy}{dx} = \lim_{h\to 0}\frac{1}{h}\frac{2\cos\left(\frac{x+x+h}{2}\right)\sin\left(\frac{x-x-h}{2}\right)}{\sin(x+h)\cdot\sin x}$$

$$\Rightarrow \frac{dy}{dx} = \lim_{h\to 0}\frac{-1}{h}\frac{2\cos\left(\frac{2x+h}{2}\right)\sin\left(\frac{h}{2}\right)}{\sin(x+h)\cdot\sin x} \qquad (\because \sin(-\theta) = -\sin\theta)$$

$$\Rightarrow \frac{dy}{dx} = \lim_{h\to 0} - \frac{2\cos\left(\frac{2x+h}{2}\right)}{\sin(x+h)\cdot\sin x}\cdot\left(\frac{\sin\frac{h}{2}}{\frac{h}{2}}\right) \qquad \text{(note this step)}$$

$$\Rightarrow \frac{dy}{dx} = -\frac{\cos\left(\frac{2x+0}{2}\right)}{\sin(x+0)\cdot\sin x}\cdot 1 \qquad \left(\because \lim_{\theta\to 0}\frac{\sin\theta}{\theta} = 1\right)$$

$$\Rightarrow \frac{dy}{dx} = -\frac{\cos x}{\sin x\cdot\sin x} = -\frac{1}{\sin x}\cdot\frac{\cos x}{\sin x} = \mathbf{-\text{cosec } x\cdot\cot x.}$$

Thus the derivative of cosec x w.r.t x is $-\text{cosec } x\cdot\cot x$.

i.e., $\quad \mathbf{y = \text{cosec } x \Rightarrow \frac{dy}{dx} = -\text{cosec } x\cdot\cot x.}$

7. To differentiate sec x w.r.t x from the first principles

Let $y = f(x) = \sec x$, where x is measured in radians. By definition, we have,

$$\frac{dy}{dx} = \lim_{h\to 0}\frac{f(x+h)-f(x)}{h} \Rightarrow \frac{dy}{dx} = \lim_{h\to 0}\frac{\sec(x+h)-\sec x}{h} \qquad (\because f(x) = \sec x)$$

$$\Rightarrow \frac{dy}{dx} = \lim_{h\to 0}\frac{1}{h}\left[\frac{1}{\cos(x+h)} - \frac{1}{\cos x}\right]$$

$$\Rightarrow \frac{dy}{dx} = \lim_{h\to 0}\frac{1}{h}\left[\frac{\cos x - \cos(x+h)}{\cos(x+h)\cdot\cos x}\right]$$

$$\Rightarrow \frac{dy}{dx} = \lim_{h\to 0} \frac{1}{h}\left[\frac{-\sin\left(\frac{x+x+h}{2}\right)\cdot \sin\left(\frac{\not x-\not x-h}{2}\right)}{\cos(x+h)\cdot\cos x}\right]$$

$$\Rightarrow \frac{dy}{dx} = \lim_{h\to 0} \frac{1}{h}\,\frac{\sin\left(\frac{2x+h}{2}\right)\cdot \sin\left(\frac{h}{2}\right)}{\cos(x+h)\cdot\cos x} \quad (\because \sin(-\theta) = -\sin\theta)$$

$$\Rightarrow \frac{dy}{dx} = \lim_{h\to 0} \frac{\sin\left(\frac{2x+h}{2}\right)}{\cos(x+h)\cdot\cos x}\cdot\left(\frac{\sin\frac{h}{2}}{\frac{h}{2}}\right) \quad \text{(note this step)}$$

$$\Rightarrow \frac{dy}{dx} = \frac{\sin\left(\frac{2x+0}{2}\right)}{\cos(x+0)\cdot\cos x}\cdot 1 \quad \left(\because \lim_{\theta\to 0}\frac{\sin\theta}{\theta} = 1\right)$$

$$\Rightarrow \frac{dy}{dx} = \frac{\sin x}{\cos x\cdot\cos x} = \frac{1}{\cos x}\cdot\frac{\sin x}{\cos x} = \mathbf{\sec x\cdot\tan x.}$$

Thus the derivative of sec x w.r.t x is sec $x\cdot$ tan x. i.e., $y = \sec x \Rightarrow \frac{dy}{dx} = \sec x\cdot\tan x.$

8. To differentiate cot *x* w.r.t *x* from the first principles

Let $y = f(x) = \cot x$, where x is measured in radians.

By definition, we have,

$$\frac{dy}{dx} = \lim_{h\to 0}\frac{f(x+h)-f(x)}{h} \Rightarrow \frac{dy}{dx} = \lim_{h\to 0}\frac{\cot(x+h)-\cot x}{h} \quad (\because f(x) = \cot x)$$

$$\Rightarrow \frac{dy}{dx} = \lim_{h\to 0}\frac{1}{h}\left[\frac{\cos(x+h)}{\sin(x+h)} - \frac{\cos x}{\sin x}\right]$$

$$\Rightarrow \frac{dy}{dx} = \lim_{h\to 0}\frac{1}{h}\left[\frac{\sin x\cdot\cos(x+h) - \cos x\sin(x+h)}{\sin(x+h)\cdot\sin x}\right]$$

$$\Rightarrow \frac{dy}{dx} = \lim_{h\to 0}\frac{1}{h}\left[\frac{\sin(\not x-\not x-h)}{\sin(x+h)\cdot\sin x}\right]$$

$$\Rightarrow \frac{dy}{dx} = \lim_{h\to 0} -\frac{1}{\sin(x+h)\cdot\sin x}\cdot\left(\frac{\sin h}{h}\right) \quad (\because \sin(-\theta) = -\sin\theta)$$

$$\Rightarrow \frac{dy}{dx} = \frac{1}{\sin(x+0)\ \sin x}\cdot 1 = \frac{1}{\sin^2 x} = \mathbf{-\operatorname{cosec}^2 x.}$$

Thus the derivative of cot x w.r.t x is $-\operatorname{cosec}^2 x$. i.e., $y = \cot x \Rightarrow \frac{dy}{dx} = -\operatorname{cosec}^2 x$

9. To differentiate e^x w.r.t x from the first principles

Let $y = f(x) = e^x$. By definition,

$$\frac{dy}{dx} = \lim_{h\to 0} \frac{f(x+h) - f(x)}{h} \Rightarrow \frac{dy}{dx} = \lim_{h\to 0} \frac{e^{x+h} - e^x}{h} \qquad (\because f(x) = e^x)$$

$$\Rightarrow \frac{dy}{dx} = \lim_{h\to 0} e^x \left(\frac{e^h - 1}{h}\right)$$

$$\Rightarrow \frac{dy}{dx} = e^x \; 1 = e^x \qquad \left(\because \lim_{h\to 0} \frac{e^h - 1}{h} = 1\right)$$

Thus the derivative of e^x w.r.t x is e^x. i.e., $y = e^x \Rightarrow \frac{dy}{dx} = e^x$.

10. To differentiate $\log x$, $(x > 0)$ w.r.t x from the first principles

Let $y = f(x) = \log x$ (base e is understood).

By definition,

$$\frac{dy}{dx} = \lim_{h\to 0} \frac{f(x+h) - f(x)}{h} \Rightarrow \frac{dy}{dx} = \lim_{h\to 0} \frac{\log(x+h) - \log x}{h} \qquad (\because f(x) = \log x)$$

$$\Rightarrow \frac{dy}{dx} = \lim_{h\to 0} \frac{1}{h} \log\left(\frac{x+h}{x}\right)$$

$$\Rightarrow \frac{dy}{dx} = \lim_{h\to 0} \frac{1}{h} \log\left(1 + \frac{h}{x}\right)$$

$$\Rightarrow \frac{dy}{dx} = \lim_{h\to 0} \frac{\log\left(1 + \frac{h}{x}\right)}{\left(\frac{h}{x}\right)} \cdot \frac{1}{x} \qquad \textbf{(note this step)}$$

$$\Rightarrow \frac{dy}{dx} = 1 \cdot \frac{1}{x} \qquad \left(\because \lim_{h\to 0} \frac{\log(1+x)}{x} = 1\right)$$

Thus the derivative of $\log x$ w.r.t x is $\frac{1}{x}$. i.e., $y = \log x \Rightarrow \frac{dy}{dx} = \frac{1}{x}$.

11. To differentiate a^x, where $(a > 0)$ and $a \neq 1$, w.r.t x from the first principles.

Let $y = f(x) = a^x$. By definition,

$$\frac{dy}{dx} = \lim_{h\to 0} \frac{f(x+h) - f(x)}{h} \Rightarrow \frac{dy}{dx} = \lim_{h\to 0} \frac{a^{x+h} - a^x}{h} \qquad (\because f(x) = a^x)$$

$$\Rightarrow \frac{dy}{dx} = \lim_{h\to 0} a^x \cdot \left(\frac{a^h - 1}{h}\right)$$

$$\Rightarrow \frac{dy}{dx} = a^x \cdot \lim_{h\to 0} \left(\frac{a^h - 1}{h}\right)$$

$$\Rightarrow \quad \frac{dy}{dx} = a^x \log_e a \qquad \left(\because \lim_{h\to 0} \frac{a^h - 1}{h} = \log_e a\right)$$

Thus the derivative of a^x w.r.t x is $a^x \log_e a$. i.e., $y = a^x, (a \neq 1, a > 0)$ then $\frac{dy}{dx} = \log_e a$.

We list below the derivatives of the basic functions which we have seen above.

$y = f(x)$	$\frac{dy}{dx} = f'(x)$	$y = f(x)$	$\frac{dy}{dx} = f'(x)$
1. $y = k$	$\frac{dy}{dx} = 0$	2. $y = x^n$	$\frac{dy}{dx} = nx^{n-1}$
3. $\frac{1}{x^n}$	$\frac{-n}{x^{n+1}}$	4. $\frac{1}{x}$	$-\frac{1}{x^2}$
5. $\sqrt{x}$	$\frac{1}{2\sqrt{x}}$	6. $\sin x$	$\cos x$
7. $\cos x$	$-\sin x$	8. $\tan x$	$\sec^2 x$
9. $\operatorname{cosec} x$	$-\operatorname{cosec} x \cdot \cot x$	10. $\sec x$	$\sec x \cdot \tan x$
11. $\cot x$	$-\operatorname{cosec}^2 x$	12. e^x	e^x
13. $\log x, (x > 0)$	$\frac{1}{x}$	14. $a^x \; (a > 0, a \neq 1)$	$a^x \cdot \log a$.

1.6 Algebra of Derivatives

In this section we shall see the rule of differentiating, scalar multiple of a function, sum of two or more functions, product of two or more functions and the quotient of two functions.

1. Derivative of a scalar multiple of a function

Let $y = k \cdot u$, where k is a constant and u is a differentiable a function of x.

Then, $$\frac{dy}{dx} = k \cdot \frac{du}{dx}$$

Proof : Let the variable x be changed from x to $x + \delta x$ and let the corresponding change in y be from y to $y + \delta y$, and in u be u to $u + \delta u$.

$$\therefore \quad y + \delta y = k(u + \delta u) \quad \Rightarrow \quad y = k \cdot u.$$

On subtraction, we get

$$\delta y = k(u + \delta u) - ku \Rightarrow \delta y = k \cdot \delta u \quad \Rightarrow \quad \frac{\delta y}{\delta x} = k \cdot \frac{\delta u}{\delta x}$$

$$\Rightarrow \frac{dy}{dx} = \lim_{\delta x\to 0} \frac{\delta y}{\delta x} = k \cdot \lim_{\delta x\to 0} \frac{\delta u}{\delta x} \Rightarrow \frac{dy}{dx} = k \cdot \frac{du}{dx} \quad \left(\because \lim_{\delta x\to 0} \frac{\delta u}{\delta x} = \frac{du}{dx}\right)$$

Hence the result.

Illustrations :

(i) $y = 4x^4 \quad \Rightarrow \quad \frac{dy}{dx} = 4 \cdot \frac{d}{dx}\left(x^4\right) = 4 \cdot 4x^3 = 16x^3$.

(ii) $y = 5 \tan x \Rightarrow \frac{dy}{dx} = 5 \cdot \frac{d}{dx}(\tan x) = 5 \cdot \sec^2 x.$

(iii) $y = \frac{3}{2} e^x \Rightarrow \frac{dy}{dx} = \frac{3}{2} \cdot e^x$. (iv) $y = \frac{5}{3} x^{3/2} \Rightarrow \frac{dy}{dx} = \frac{5}{3} \cdot \frac{3}{2} x^{1/2} = \frac{5}{2} x^{1/2}$.

2. Derivative of sum of two functions

Let $y = u + v$, where u and v are differentiable a function of x, then $\frac{dy}{dx} = \frac{du}{dx} + \frac{dv}{dx}$.

Proof : Let the variable x be changed from x to $x + \delta x$. Let the corresponding change in u,v and y be from u to $u + \delta u$, v to $v + \delta v$, and y to $y + \delta y$, respectively.

$\therefore \quad y + \delta y = u + \delta u + v + \delta v \quad$ Also, $\quad v = v + v.$

On subtraction, we have,

$$\delta y = (u + \delta u) + (v + \delta v) - (u + v) \Rightarrow \delta y = \delta u + \delta v \Rightarrow \frac{\delta y}{\delta x} = \frac{\delta u}{\delta x} + \frac{\delta v}{\delta x}$$

Taking the limit both sides as $\delta x \to 0$, we get

$$\lim_{\delta x \to 0} \frac{\delta y}{\delta x} = \lim_{\delta x \to 0} \frac{\delta u}{\delta x} + \lim_{\delta x \to 0} \frac{\delta v}{\delta x} \Rightarrow \frac{dy}{dx} = \frac{du}{dx} + \frac{dv}{dx}.$$

Note : 1. If $y = u - v$ and u and v are functions of x, then, $\frac{dy}{dx} = \frac{du}{dx} - \frac{dv}{dx}$

2. The generalised result of the above, is given by,

If, $\quad y = u_1 + u_2 + u_3 + \ldots\ldots + u_n.$

Where $u_1, u_2, u_3 \ldots\ldots u_n$ are finite numbers of functions of x, then

$$\frac{dy}{dx} = \frac{du_1}{dx} + \frac{du_2}{dx} + \frac{du_3}{dx} + \ldots\ldots + \frac{du_n}{dx}$$

Illustrations :

(i) $y = 3x^2 + 4 \sin x \Rightarrow \frac{dy}{dx} = 6x + 4 \cos x.$

(ii) $y = x^4 + 3 \tan x + 3e^x + \log x + 7 \Rightarrow \frac{dy}{dx} = 4x^3 + 3 \sec^2 x + 3e^x + \frac{1}{x} \quad \left(\because \frac{d}{dx}(7) = 0 \right).$

3. Derivative of the product of two functions (product-rule)

Let $y = u \cdot v$, where u and v are differentiable functions of x, then $\frac{dy}{dx} = u \cdot \frac{dv}{dx} + v \cdot \frac{du}{dx}$

Taking u as first function and v as second function, we can state the above rule as :

Derivative of the product = (first function) × (derivative of the second) + (second function) × (derivative of the first)

Proof : Let the variable x be changed from x to $x + \delta x$. Let the corresponding change in u,v and y be from u to $u + \delta u$, v to $v + \delta v$, and y to $y + \delta y$. Then we have,

$$y + \delta y = (u + \delta u)\ (v + \delta v) \Rightarrow y + \delta y = uv + u \cdot \delta v + v \cdot \delta u + \delta u \cdot \delta v. \quad \text{Also,} \quad y = uv.$$

On subtraction, we have,

$$\delta y = u \cdot \delta v + v \cdot \delta u + \delta u \cdot \delta v \ \Rightarrow\ \frac{\delta y}{\delta x} = u \cdot \frac{\delta v}{\delta x} + v \cdot \frac{\delta u}{\delta x} + \frac{\delta u}{\delta x} \cdot \delta v.$$

Taking the limit both sides, we get,

$$\lim_{\delta x \to 0}\left(\frac{\delta y}{\delta x}\right) = u \cdot \lim_{\delta x \to 0}\frac{\delta v}{\delta x} + v \cdot \lim_{\delta x \to 0}\frac{\delta u}{\delta x} + \lim_{\delta x \to 0}\left(\frac{\delta u}{\delta x} \cdot \delta v\right)$$

$$\Rightarrow \quad \frac{dy}{dx} = u \cdot \frac{dv}{dx} + v \cdot \frac{du}{dx} + \frac{du}{dx} \cdot 0 \qquad (\because \delta v \to 0 \text{ as } \delta x \to 0)$$

$$\Rightarrow \quad \frac{dy}{dx} = u \cdot \frac{dv}{dx} + v \cdot \frac{du}{dx}.$$

which is the required result.

Note : 1. Consider, the product rule for $y = u \cdot v$.

$$\frac{dy}{dx} = u \cdot \frac{dv}{dx} + v\frac{du}{dx} \quad \Rightarrow \quad \frac{1}{y}\frac{dy}{dx} = \frac{u}{y} \cdot \frac{dv}{dx} + \frac{v}{y}\frac{du}{dx}$$

$$\Rightarrow \quad \frac{1}{y} \cdot \frac{dy}{dx} = \frac{u}{u \cdot v} \cdot \frac{dv}{dx} + \frac{v}{u \cdot v}\frac{du}{dx} \qquad (\because y = u\,v)$$

$$\Rightarrow \quad \frac{1}{y}\frac{dy}{dx} = \frac{1}{v} \cdot \frac{dv}{dx} + \frac{1}{u} \cdot \frac{du}{dx}$$

2. The product rule can be extended to three or more functions. If u, v, w are functions of x and $y = u \cdot v \cdot w$, then,

$$\frac{dy}{dx} = uv \cdot \frac{dw}{dx} + u \cdot \frac{dv}{dx} \cdot w + \frac{du}{dx} \cdot vw. \quad \text{Also,} \quad \frac{1}{y}\frac{dy}{dx} = \frac{1}{w}\frac{dw}{dx} + \frac{1}{v}\frac{dv}{dx} + \frac{1}{u}\frac{du}{dx}$$

Illustrations :

(i) $y = x^3 \cdot \cos x \ \Rightarrow\ \dfrac{dy}{dx} = x^3 \cdot \dfrac{d}{dx}(\cos x) + \cos x \cdot \dfrac{d}{dx}\left(x^3\right)$

$$\Rightarrow \frac{dy}{dx} = -x^3 \sin x + \cos x \cdot 3x^2 \ \Rightarrow\ \frac{dy}{dx} = -x^3 \sin x + 3x^2 \cdot \cos x.$$

(ii) $y = x^4 \cdot \log x \ \Rightarrow\ \dfrac{dy}{dx} = x^4 \cdot \dfrac{d}{dx}(\log x) + \dfrac{d}{dx}\left(x^4\right) \log x$

$$\Rightarrow \frac{dy}{dx} = x^4 \cdot \frac{1}{x} + 4x^3 \cdot \log x \ \Rightarrow\ \frac{dy}{dx} = x^3(1 + 4\log x).$$

(iii) $y = x^2 \cdot \sin x \cdot e^x \ \Rightarrow\ \dfrac{dy}{dx} = x^2 \cdot \sin x \cdot \dfrac{d}{dx}\left(e^x\right) + x^2 \cdot \dfrac{d}{dx}(\sin x) \cdot e^x + \dfrac{d}{dx}\left(x^2\right) \cdot \sin x \cdot e^x.$

$$\Rightarrow \frac{dy}{dx} = x^2 \cdot \sin x \cdot e^x + x^2 \cdot \cos x \cdot e^x + 2x \cdot \sin x \cdot e^x.$$

4. Derivative of quotient of two functions (Quotient rule)

Let $y = \dfrac{u}{v}$, where u and v are differentiable functions of x, then $\dfrac{dy}{dx} = \dfrac{v \cdot \dfrac{du}{dx} - u\dfrac{dv}{dx}}{v^2}$.

Proof : Let the variable x be changed from x to $x + \delta x$. Let the corresponding change in u, v and y be from u to $u + \delta u$, v to $v + \delta v$, and y to $y + \delta y$. Then we have

$$y + \delta y = \frac{u + \delta u}{u + \delta v}, \quad \text{Also,} \quad y = \frac{u}{v}.$$

On subtraction, we have,

$$\delta y = \frac{u + \delta u}{v + \delta v} - \frac{u}{v} \Rightarrow \delta y = \frac{v(u + \delta u) - u(v + \delta v)}{v(v + \delta v)} \Rightarrow \delta y = \frac{uv + v\delta u - uv - u\delta v}{v(v + \delta v)}$$

$$\Rightarrow \frac{\delta y}{\delta x} = \frac{v \cdot \dfrac{\delta u}{\delta x} - u \cdot \dfrac{\delta v}{\delta x}}{v(v + \delta v)}.$$

Taking the limit both sides as $\delta x \to 0$, we get

$$\lim_{\delta x \to 0} \frac{\delta y}{\delta x} = \frac{v \cdot \lim\limits_{\delta x \to 0}\left(\dfrac{\delta u}{\delta x}\right) - u \cdot \lim\limits_{\delta x \to 0} \dfrac{\delta v}{\delta x}}{\lim\limits_{\delta x \to 0}\left[v(v + \delta v)\right]} \Rightarrow \frac{dy}{dx} = \frac{v \cdot \dfrac{du}{dx} - u\dfrac{dv}{dx}}{v^2} \quad (\because \delta v \to 0 \text{ as } \delta x \to 0)$$

Which is the required result.

The above result can be stated as follows.

Derivative of quotient of two functions:

$$\frac{(\text{denominator})\,(\text{derivative of numerator}) - (\text{numerator})(\text{derivative of denominator})}{(\text{denominator})^2}$$

Note : Taking $u = 1$, in the quotient rule, we have if $y = \dfrac{1}{v}$, then

$$\frac{dy}{dx} = \frac{d}{dx}\left(\frac{1}{v}\right) = \frac{v \cdot \dfrac{d}{dx}(1) - 1 \cdot \dfrac{dv}{dx}}{v^2} \Rightarrow \frac{d}{dx}\left(\frac{1}{v}\right) = -\frac{1}{v^2} \cdot \frac{dv}{dx} \quad \left(\because \frac{d}{dx}(1) = 0\right)$$

Illustrations :

(i) $$y = \frac{\sin x}{x^2} \Rightarrow \frac{dy}{dx} = \frac{x^2 \cdot \dfrac{d}{dx}(\sin x) - \sin x \cdot \dfrac{d}{dx}\left(x^2\right)}{\left(x^2\right)^2}$$

$$\Rightarrow \frac{dy}{dx} = \frac{x^2 \cos x - \sin x \cdot 2x}{x^4} \Rightarrow \frac{dy}{dx} = \frac{x \cos x - 2 \sin x}{x^3}$$

(ii) $$y = \frac{x^2 + 1}{x - 1} \Rightarrow \frac{dy}{dx} = \frac{(x-1)(2x) - \left(x^2 + 1\right) \cdot 1}{(x-1)^2} \Rightarrow \frac{dy}{dx} = \frac{2x^2 - 2x - x^2 - 1}{(x-1)^2} = \frac{x^2 - 2x - 1}{(x-1)^2}$$

(iii) $y = \frac{e^x}{e^x + 1} \Rightarrow \frac{dy}{dx} = \frac{(e^x + 1)\cdot e^x - e^x(e^x + 0)}{(e^x + 1)^2} \Rightarrow \frac{dy}{dx} = \frac{e^{2x} + e^x - e^{2x}}{(e^x + 1)^2} = \frac{e^x}{(e^x + 1)^2}$

The readers are adviced to follow the following:

1. Commit to memory the derivatives of basic functions.
2. Practice to apply the rules of differentiating the sum, product and quotient of functions.

Example 1. Differentiate the following w.r.t. x

(a) $3x^3 - 4x^2 + 2x + 1$ **(b)** $4x^3 - 9 - 5x^2$ **(c)** $e^x + 3\cos x$

(d) $x^{3/4} - 4\sqrt{x}$ **(e)** $\frac{1}{x^{4/3}} - \frac{3}{x^{3/2}}$ **(f)** $\log x + 3\tan x + 7$.

Solution : (a) Let $y = 3x^3 - 4x^2 + 2x + 1 \Rightarrow \frac{dy}{dx} = 3\cdot\frac{d}{dx}(x^3) - 4\frac{d}{dx}(x^2) + 2\frac{d}{dx}(x) + \frac{d}{dx}(1)$

$$\Rightarrow \frac{dy}{dx} = 3\cdot(3x^2) - 4(2x) + 2\cdot 1 + 0 = \mathbf{9x^2 - 8x + 2.}$$

(b) Let $y = 4x^3 - 9 - 5x^2$

$$\Rightarrow \frac{dy}{dx} = 4\frac{d}{dx}(x^3) - \frac{d}{dx}(9) - 5\frac{d}{dx}(x^2) \Rightarrow \frac{dy}{dx} = 4(3x^2) - 0 - 5(2x) = \mathbf{12x^2 - 10x.}$$

(c) Let $y = e^x + 3\cos x$

$$\Rightarrow \frac{dy}{dx} = \frac{d}{dx}(e^x) + 3\frac{d}{dx}(\cos x) \Rightarrow \frac{dy}{dx} = e^x + 3(-\sin x) = \mathbf{e^x - 3\sin x}$$

(d) Let $y = x^{3/4} - 4\sqrt{x} \Rightarrow \frac{dy}{dx} = \frac{3}{4}x^{\frac{3}{4}-1} - 4\cdot\frac{1}{2}x^{\frac{1}{2}-1}$ $\left(\because \sqrt{x} = x^{\frac{1}{2}}\right)$

$$\Rightarrow \frac{dy}{dx} = \frac{3}{4}x^{-1/4} - 2x^{-\frac{1}{2}} \Rightarrow \frac{dy}{dx} = \mathbf{\frac{3}{4x^{1/4}} - \frac{2}{x^{1/2}}}$$

(e) Let $y = \frac{1}{x^{4/3}} - \frac{3}{x^{3/2}} \Rightarrow \frac{dy}{dx} = \frac{-(4/3)}{x^{(4/3+1)}} - \frac{3\cdot(-3/2)}{x^{(3/2+1)}}$ $\left(\because \frac{d}{dx}\left(\frac{1}{x^n}\right) = -\frac{n}{x^{n+1}}\right)$

$$\Rightarrow \frac{dy}{dx} = -\frac{4}{3x^{7/3}} + \frac{9}{2x^{5/2}}$$

Students, should practice to write the derivatives of $\frac{1}{x^n}$ directly :

$$\frac{d}{dx}\left(\frac{1}{x^{10}}\right) = \frac{-10}{x^9}, \quad \frac{d}{dx}\left(\frac{1}{x^{7/3}}\right) = -\frac{7}{3x^{10/3}} \quad \left(\frac{1}{n} + 1 = \frac{n+1}{n}\right)$$

$$\frac{d}{dx}\left(\frac{1}{x^{2/3}}\right) = -\frac{2}{3x^{5/3}} \quad \frac{d}{dx}\left(\frac{1}{x^{3/4}}\right) = -\frac{3}{4x^{7/4}}.$$

(f) Let $y = \log x + 3\tan x + 7 \Rightarrow \frac{dy}{dx} = \frac{1}{x} + 3\sec^2 x + 0 \Rightarrow \frac{dy}{dx} = \mathbf{\frac{1}{x} + 3\sec^2 x}.$

Example 2. Differentiate the following w.r.t. x,

(a) $(1-x^2)\tan x$ (b) $\sec x\tan x$ (c) $(1-2\tan x)\ (5+4\cos x)$ (d) $x^9\cdot 9^x$

(e) $\sec x\cdot(\log x-3x^2)$ (f) $\sqrt{x}\ (\cot x+3)$ (g) $(3e^x-\log x)\operatorname{cosec} x$ (h) $\cot x\cdot(\sqrt{3}-4e^x)$

Solution : (a) Let $y=(1-x^2)\cdot\tan x$

$$\Rightarrow \frac{dy}{dx}=(1-x^2)\cdot\frac{d}{dx}(\tan x)+\frac{d}{dx}(1-x^2)\cdot\tan x \qquad \text{(Note this step)}$$

$$\Rightarrow \mathbf{\frac{dy}{dx}=(1-x^2)\cdot\sec^2 x-2x\cdot\tan x}$$

(b) Let, $y=\sec x\cdot\tan x \Rightarrow \frac{dy}{dx}=\sec x\cdot\frac{d}{dx}(\tan x)+\frac{d}{dx}(\sec x)\cdot\tan x$

$$\Rightarrow \frac{dy}{dx}=\sec x\cdot\sec^2 x+\sec x\cdot\tan x\cdot\tan x$$

$$\Rightarrow \mathbf{\frac{dy}{dx}=\sec x(\sec^2 x+\tan^2 x)}$$

(c) Let, $y=(1-2\tan x)\cdot(5+4\cos x)$

$$\Rightarrow \frac{dy}{dx}=(1-2\tan x)\frac{d}{dx}(5+4\cos x)+\left[\frac{d}{dx}(1-2\tan x)\right]\cdot(5+4\cos x)$$

$$\Rightarrow \frac{dy}{dx}=(1-2\tan x)(-4\sin x)+(-2\sec^2 x)(5+4\cos x)$$

$$\Rightarrow \mathbf{\frac{dy}{dx}=-4\sin x+8\tan x\sin x-10\sec^2 x-8\sec x}$$

(d) Let $y=x^9\cdot 9^x$

$$\Rightarrow \frac{dy}{dx}=x^9\cdot\frac{d}{dx}(9^x)+\frac{d}{dx}(x^9)\cdot 9^x \Rightarrow \frac{dy}{dx}=x^9\cdot 9^x\cdot\log 9+9^x\cdot 9x^8 \Rightarrow \mathbf{\frac{dy}{dx}=9^x\cdot x^8[x\log 9+9]}$$

(e) Let $y=\sec x\cdot(\log x-3x^2) \Rightarrow \frac{dy}{dx}=\sec x\cdot\left(\frac{1}{x}-6x\right)+(\log x-3x^2)\cdot\sec x\cdot\tan x$

$$\Rightarrow \mathbf{\frac{dy}{dx}=\sec x\left[\frac{1-6x^2}{x}+\left(\log x-3x^2\right)\tan x\right]}$$

Note that, we have avoided the second step that we have written for other problems. Students are adviced to write differential coefficients directly, during the process of applying the product rule. Further differentiate the function wherever it is, by keeping once the first function as it is, and again keeping the second function as it is.

(f) Let $y=\sqrt{x}\cdot(\cot x+3) \Rightarrow \frac{dy}{dx}=\sqrt{x}\cdot(-\operatorname{cosec}^2 x)+\frac{1}{2\sqrt{x}}\cdot(\cot x+3)$

$$\Rightarrow \mathbf{\frac{dy}{dx}=-\sqrt{x}\operatorname{cosec}^2 x+\frac{\cot x+3}{2\sqrt{x}}}$$

(g) Let $y=(3e^x-\log x)\cdot\operatorname{cosec} x \Rightarrow \frac{dy}{dx}=(3e^x-\log x)\cdot(-\operatorname{cosec} x\cdot\cot x)+\left(3e^x-\frac{1}{x}\right)\cdot\operatorname{cosec} x$

$$\Rightarrow \mathbf{\frac{dy}{dx}=\operatorname{cosec} x\left[\frac{3xe^x-1}{x}-\cot x\cdot\left(3e^x-\log x\right)\right]}$$

(h) Let $y = \cot x \cdot (\sqrt{3} - 4e^x)$ $\Rightarrow$ $\frac{dy}{dx} = \cot x \cdot (-4e^x) + (-\text{cosec}^2 x) \cdot (\sqrt{3} - 4e^x)$

$$\Rightarrow \quad \frac{dy}{dx} = -4e^x \cot x - \text{cosec}^2 x \cdot (\sqrt{3} - 4e^x)$$

Example 3. Differentiate the following w.r.t. x,

(a) $\frac{x^2 - 3x + 5}{x^2 + 3x + 5}$ (b) $\frac{\log x}{1 + \log x}$ (c) $\frac{x + \sin x}{x + \cos x}$ (d) $\frac{\log x}{1 + \sqrt{x}}$

(e) $\frac{3^x - 2^x}{x + 1}$ (f) $\frac{\sec x - 1}{\sec x + 1}$ (g) $\frac{\cos x + \sin x}{\cos x - \sin x}$ (h) $\frac{e^x - 1}{e^x + 1}$

Solution : (a) Let $y = \frac{x^2 - 3x + 5}{x^2 + 3x + 5}$

$$\Rightarrow \quad \frac{dy}{dx} = \frac{(x^2 + 3x + 5)\frac{d}{dx}(x^2 - 3x + 5) - (x^2 - 3x + 5)\frac{d}{dx}(x^2 + 3x + 5)}{(x^2 + 3x + 5)^2}$$

$$\Rightarrow \quad \frac{dy}{dx} = \frac{(x^2 + 3x + 5)(2x - 3) - (x^2 - 3x + 5)(2x + 3)}{(x^2 + 3x + 5)^2}$$

$$\Rightarrow \quad \frac{dy}{dx} = \frac{2x(\cancel{x^2} + 3x + \cancel{5} - \cancel{x^2} + 3x - \cancel{5}) + 3(-x^2 - \cancel{3x} - 5 - x^2 + \cancel{3x} - 5)}{(x^2 + 3x + 5)^2}$$

$$\Rightarrow \quad \frac{dy}{dx} = \frac{2x(6x) - 3(2x^2 + 10)}{(x^2 + 3x + 5)^2} = \frac{6x^2 - 30}{(x^2 + 3x + 5)^2}$$

(b) Let $y = \frac{\log x}{1 + \log x}$

$$\Rightarrow \quad \frac{dy}{dx} = \frac{(1 + \log x)\left(\frac{1}{x}\right) - \log x \cdot \left(\frac{1}{x}\right)}{(1 + \log x)^2} \Rightarrow \frac{dy}{dx} = \frac{\frac{1}{x} + \frac{1}{x}\log x - \frac{1}{x}\log x}{(1 + \log x)^2} = \frac{1}{x(1 + \log x)^2}$$

Note that in the above problem, the functions are differentiated directly during the process of applying the quotient rule. Students are adviced to practice in this way.

(c) Let $y = \frac{x + \sin x}{x + \cos x}$ $\Rightarrow$ $\frac{dy}{dx} = \frac{(x + \cos x)(1 + \cos x) - (x + \sin x)(1 - \sin x)}{(x + \cos x)^2}$

$$\Rightarrow \quad \frac{dy}{dx} = \frac{\cancel{x} + \cos x + x\cos x + \cos^2 x - \cancel{x} - \sin x + x\sin x + \sin^2 x}{(x + \cos x)^2}$$

$$\Rightarrow \quad \frac{dy}{dx} = \frac{\cos x - \sin x + x(\cos x + \sin x) + 1}{(x + \cos x)^2}$$

(d) Let $y = \dfrac{\log x}{1+\sqrt{x}} \Rightarrow \dfrac{dy}{dx} = \dfrac{(1+\sqrt{x})\left(\frac{1}{x}\right) - \log x\left(\frac{1}{2\sqrt{x}}\right)}{(1+\sqrt{x})^2}$ $\left(\because \dfrac{d}{dx}\sqrt{x} = \dfrac{1}{2\sqrt{x}}\right)$

$$\Rightarrow \frac{dy}{dx} = \frac{2(1+\sqrt{x}) - \sqrt{x}\log x}{2x(1+\sqrt{x})^2}$$

(e) Let $y = \dfrac{3^x - 2^x}{x+1} \Rightarrow \dfrac{dy}{dx} = \dfrac{(x+1)\left[3^x \log 3 - 2^x \log 2\right] - (3^x - 2^x)\cdot 1}{(x+1)^2}$

(f) Let $y = \dfrac{\sec x - 1}{\sec x + 1} \Rightarrow \dfrac{dy}{dx} = \dfrac{(\sec x + 1)(\sec x \tan x) - (\sec x - 1)(\sec x \tan x)}{(\sec x + 1)^2}$

$$\Rightarrow \frac{dy}{dx} = \frac{\cancel{\sec^2 x \tan x} + \sec x \tan x - \cancel{\sec^2 x \tan x} + \sec x \tan x}{(\sec x + 1)^2}$$

$$\Rightarrow \frac{dy}{dx} = \frac{2\sec x \tan x}{(\sec x + 1)^2}$$

(g) Let $y = \dfrac{\cos x + \sin x}{\cos x - \sin x} \Rightarrow \dfrac{dy}{dx} = \dfrac{(\cos x - \sin x)(-\sin x + \cos x) - (\cos x + \sin x)(-\sin x - \cos x)}{(\cos x - \sin x)^2}$

$$\Rightarrow \frac{dy}{dx} = \frac{(\cos x - \sin x)^2 + (\cos x + \sin x)^2}{(\cos x - \sin x)^2}$$

$$\Rightarrow \frac{dy}{dx} = \frac{2(\cos^2 x + \sin^2 x)}{(\cos x - \sin x)^2} \qquad (\because (a-b)^2 + (a+b)^2 = 2(a^2 + b^2))$$

$$\Rightarrow \frac{dy}{dx} = \frac{2}{(\cos x - \sin x)^2} \qquad (\because \cos^2 x + \sin^2 x = 1)$$

(h) Let $y = \dfrac{e^x - 1}{e^x + 1} \Rightarrow \dfrac{dy}{dx} = \dfrac{(e^x + 1)(e^x) - (e^x - 1)(e^x)}{(e^x + 1)^2}$

$$\Rightarrow \frac{dy}{dx} = \frac{e^{2x} + e^x - e^{2x} + e^x}{(e^x + 1)^2} = \frac{2e^x}{(e^x + 1)^2}$$

Example 4. Find $\dfrac{dy}{dx}$, if (a) $y = x^2 \cdot e^x \cdot (\cos x - 4)$ (b) $y = \sin x \cdot \log x \cdot (e^x + 3)$

Solution : (a) Let $y = x^2 \cdot e^x \cdot (\cos x - 4)$

$$\Rightarrow \frac{dy}{dx} = x^2 \cdot e^x \cdot \frac{d}{dx}(\cos x - 4) + x^2 \cdot \frac{d}{dx}(e^x)\cdot(\cos x - 4) + \frac{d}{dx}(x^2)\cdot e^x \cdot (\cos x - 4)$$

$$\Rightarrow \frac{dy}{dx} = -x^2 \cdot e^x \cdot \sin x + x^2 \cdot e^x \cdot (\cos x - 4) + 2x\, e^x \cdot (\cos x - 4)$$

$$\Rightarrow \frac{dy}{dx} = x\, e^x\, [x(\cos x - 4) - x \sin x + 2\cdot(\cos x - 4)]$$

(b) Let $y = \sin x \cdot \log x \cdot (e^x + 3)$

$$\Rightarrow \quad \frac{dy}{dx} = \sin x \cdot \log x \cdot (e^x) + \sin x \cdot \frac{1}{x} \cdot (e^x + 3) + \cos x \cdot \log x\, (e^x + 3)$$

$$\Rightarrow \quad \frac{dy}{dx} = e^x \cdot \sin x \cdot \log x + \frac{\sin x}{x}(e^x + 3) + \cos x \cdot \log x \cdot (e^x + 3)$$

Exercise

1. Differentiate the following w.r.t. x

(a) $\sqrt[3]{x^5}$ **(b)** $\frac{4}{\sqrt[4]{x^5}}$ **(c)** $\frac{1}{x^7}$ **(d)** $x\sqrt{x}$ **(e)** $4x^3 - 2x^2 - 7x + 1$

(f) $x^2 + 3\sqrt{x}$ **(g)** $\frac{4x^2 - 3x}{x}$ **(h)** $\frac{2x^2 + 3x + 4}{\sqrt{x}}$ **(i)** $\left(\sqrt{x} + \frac{1}{\sqrt{x}}\right)^2$ **(j)** $\left(x - \frac{1}{x}\right)^3$

(k) $5 \tan x - 3e^x + 4 \log x + 5^x$ **(l)** $5e^x - \log x + 3 \sec x$ **(m)** $4 \cot x - 4\sqrt{x} + 7$

(n) $\operatorname{cosec} x + 3 \log x + 2\sqrt{x}$ **(o)** $5e^x - \log x - \sqrt[3]{x^2} + 3 \cdot 4^x$

2. Differentiate the following w.r.t. x

(a) $(1 + 3x^2)(1 + 5x)$ **(b)** $(x^2 + 1)(x^3 - 2x + 1)$ **(c)** $\sqrt{x}\left(x^2 + \frac{1}{x}\right)$ **(d)** $e^x \sin x$

(e) $\tan x \cdot e^x$ **(f)** $\sqrt{x} \cdot \operatorname{cosec} x$ **(g)** $(4 \sin x + 3)\, e^x$ **(h)** $x^2 (\operatorname{cosec} x - \cot x)$

(i) $(x^2 - 2x + 1)(\sec x + 4)$ **(j)** $(x^2 + 3x) \log x$ **(k)** $(\sqrt{x} + 1)(3^x + 4)$

(l) $(4^x + 3)(e^x + 7)$ **(m)** $2^x (\cos x - 1)$ **(n)** $e^x (\operatorname{cosec} x - 7)$ **(o)** $\left(\frac{1}{2}\right)^x (3e^x + \log x)$

(p) $x\, e^x \cdot \sin x$ **(q)** $(2x - 1)(2x^2 - 4)(4x^2 - 3)$

3. Differentiate the following w.r.t. x

(a) $\frac{\sin x}{x^2}$ **(b)** $\frac{\sqrt{x}}{2x + 1}$ **(c)** $\frac{1 + \sin x}{e^x}$ **(d)** $\frac{4x^2 - 2x + 1}{2x^2 + 4x - 1}$ **(e)** $\frac{\sin x}{1 + \tan x}$ **(f)** $\frac{1 - \sin x}{1 + \sin x}$

(g) $\frac{e^x + \sin x}{1 - \log x}$ **(h)** $\frac{\sqrt{a} + \sqrt{x}}{\sqrt{a} - \sqrt{x}}$ **(i)** $\frac{x \tan x}{\sec x + \tan x}$ **(j)** $\frac{e^x (x + 1)}{x - 1}$ **(k)** $\frac{2^x \cot x}{\sqrt{x}}$ **(l)** $\frac{\sec x - 1}{\sec x + 1}$

Answers

1. (a) $\frac{5}{3} x^{2/3}$ **(b)** $-\frac{5}{x^{9/4}}$ **(c)** $-\frac{7}{x^8}$ **(d)** $\frac{3}{2} x^{1/2}$ **(e)** $12x^2 - 4x - 7$ **(f)** $2x + \frac{3}{2\sqrt{x}}$

(g) 4 **(h)** $3\sqrt{x} + \frac{3}{2\sqrt{x}} - \frac{2}{x^{3/2}}$ **(i)** $1 - \frac{1}{x^2}$ **(j)** $3x^2 + \frac{3}{x^4} - \frac{3}{x^2} - 3$

(k) $5 \sec^2 x - 3e^x + \frac{4}{x} + 5^x \log 5$ **(l)** $5e^x - \frac{1}{x} + 3 \sec x \tan x$ **(m)** $-4 \operatorname{cosec}^2 x - \frac{2}{\sqrt{x}}$

(n) $-\operatorname{cosec} x \cot x + \frac{3}{x} + \frac{1}{\sqrt{x}}$ **(o)** $5e^x - \frac{1}{x} - \frac{2}{3} x^{-1/3} + 3 \cdot 4^x \log 4$

2. (a) $60x^3 + 9x^2 + 5$ (b) $5x^4 - 3x^2 + 2x - 2$ (c) $\frac{5}{2} x^{3/2} - \frac{1}{2x^{3/2}}$ (d) $e^x (\cos x + \sin x)$

(e) $e^x (\tan x + \sec^2 x)$. (f) $\frac{\operatorname{cosec} x}{2\sqrt{x}} (1 - 2x \cdot \cot x)$ (g) $(4 \sin x + 3 + 4 \cos x) e^x$

(h) $(\operatorname{cosec} x - \cot x)(x^2 \operatorname{cosec} x + 2x)$ (i) $(x^2 - 2x + 1)(\sec x \tan x) + (2x - 2)(\sec x + 4)$

(j) $(x + 3) + (2x + 3) \log x$ (k) $(\sqrt{x} + 1)\, 3^x \log 3 + \frac{1}{2\sqrt{x}} (3^x + 4)$

(l) $(1 + \log 4)\, 4^x e^x + 3e^x + 7\, 4^x \log 4$ (m) $2^x (\log 2 \cos x - \sin x) - 2^x \log 2$

(n) $e^x [\operatorname{cosec} x - 7 - \operatorname{cosec} x \cot x]$ (o) $\left(\frac{1}{2}\right)^x \left[\left(1 + \log \frac{1}{2}\right) 3e^x + \frac{1}{x} + \log x \cdot \log \frac{1}{2}\right]$

(p) $e^x [x \cos x + x \sin x + \sin x]$ (q) $(2x - 1)(16x^3 - 32x) + (4x^2 - 3)(12x^2 - 4x - 8)$

(a) $\frac{x \cos x - 2 \sin x}{x^3}$ (b) $\frac{1 - 2x}{2\sqrt{x}\,(2x + 1)^2}$ (c) $\frac{\cos x - \sin x - 1}{e^x}$ (d) $\frac{20x^2 - 12x - 2}{\left(2x^2 + 4x - 1\right)^2}$

(e) $\frac{\cos x + \sin x - \sin x \sec^2 x}{(1 + \tan x)^2}$ (f) $\frac{-2 \cos x}{(1 + \sin x)^2}$

(g) $\frac{x(1 - \log x)\left(e^x + \cos x\right) + e^x + \sin x}{x \cdot (1 - \log x)^2}$ (h) $\frac{\sqrt{a}}{\sqrt{x}\left(\sqrt{a} - \sqrt{x}\right)^2}$

(i) $\frac{x \sec^2 x - x \sec x \tan x + \tan x}{(\sec x + \tan x)}$ (j) $\frac{e^x \left(x^2 - 3\right)}{(x - 1)^2}$

(k) $\frac{2x\left[2^x\left(\log 2 \cot x - \operatorname{cosec}^2 x\right)\right] - 2^x \cot x}{x}$ (l) $\frac{2 \sec x \tan x}{(\sec x + 1)^2}$

1.7 Differentiation of Composite Functions – Chain Rule

In this section we shall discuss the most powerful rule of differential calculus, known as the **chain rule**. This rule deals with the method of differentiating the function which is a composition of two or more basic functions.

For example, consider the function $y = \sin(e^x)$. This is a composition of two functions – one is trigonometric function and the other is exponential function. By putting $f(x) = \sin x$, and $g(x) = e^x$, we can write $y = \sin(e^x)$ as $y = f(g(x))$; i.e., $y = f \circ g(x)$. The chain rule is one which gives us the method of differentiating such functions.

Theorem (Chain Rule) : Let $y = f(u)$ and $u = g(x)$ be two differentiable functions of u and x respectively such that the composite function $y = f(g(x))$ is defined. Then $\frac{dy}{dx} = \frac{dy}{du} \cdot \frac{du}{dx}$.

Proof : Let the variable x be changed from x to $x + \delta x$ (and $\delta x \neq 0$). Let the corresponding change in u be from u to $u + \delta u$, and in turn the change in y be from y to $y + \delta y$. We have,

$$y = f(u), \quad u = g(x), \quad \text{and} \quad y = f(g(x))$$

$$\therefore \quad y + \delta y = f(u + \delta u), \quad u + \delta u = g(x + \delta x) \quad \text{and hence} \quad y + \delta y = f[g(x + \delta x)].$$

Now, by definition, we have,

$$\frac{dy}{dx} = \lim_{\delta x \to 0} \frac{f[g(x+\delta x) - f[g(x)]]}{\delta x} = \lim_{\delta x \to 0} \left[\frac{f(u+\delta u) - f(u)}{\delta u} \cdot \frac{\delta u}{\delta x}\right] \quad (\because g(x+\delta x) = u + \delta u)$$

$$= \lim_{\delta x \to 0} \left[\frac{f(u+\delta u) - f(u)}{\delta u} \cdot \frac{(u+\delta u) - u}{\delta x}\right]$$

$$= \lim_{\delta x \to 0} \left[\frac{f(u+\delta u) - f(u)}{\delta u} \cdot \frac{g(x+\delta x) - g(x)}{\delta x}\right]$$

$$(\because u + \delta u = g(x + \delta x))$$

We have, $\delta u \to 0$ as $\delta x \to 0$, thus, $\dfrac{dy}{dx} = \lim\limits_{\delta u \to 0} \dfrac{f(u+\delta u) - f(u)}{\delta u} \cdot \lim\limits_{\delta x \to 0} \dfrac{g(x+\delta x) - g(x)}{\delta x}$

$$\Rightarrow \frac{dy}{dx} = \frac{dy}{du} \cdot \frac{du}{dx} \quad (\because y = f(u) \text{ and } u = g(x))$$

which is the required result.

Note : The chain rule can be extended to the functions which are composition of more than two function. If, $y = f(u)$, $u = g(v)$, $v = h(x)$ then $\dfrac{dy}{dx} = \dfrac{dy}{du} \cdot \dfrac{du}{dv} \cdot \dfrac{dv}{dx}$

Example 1. Differentiate the following w.r.t. x

(a) $\sin^3 x$ (b) $\sqrt{\tan x}$ (c) $\cos(\log x)$ (d) $e^{\sec x}$ (e) $\log(\text{cosec } x)$

Solution : (a) Let $y = \sin^3 x = (\sin x)^3$ and $u = \sin x$

$$\Rightarrow \quad y = u^3 \text{ and } u = \sin x \Rightarrow \frac{dy}{du} = 3u^2 \text{ and } \frac{du}{dx} = \cos x$$

Now, $\dfrac{dy}{dx} = \dfrac{dy}{du} \cdot \dfrac{du}{dx} \Rightarrow \dfrac{dy}{dx} = 3u^2 \cdot \cos x \Rightarrow \dfrac{dy}{dx} = 3\sin^2 x \cdot \cos x \quad (\because u = \sin x)$

(b) Let, $y = \sqrt{\tan x}$ and $u = \tan x \Rightarrow y = \sqrt{u}$ and $u = \tan x \Rightarrow \dfrac{dy}{du} = \dfrac{1}{2\sqrt{u}}$ and $\dfrac{du}{dx} = \sec^2 x$

Now, $\dfrac{dy}{dx} = \dfrac{dy}{du} \cdot \dfrac{du}{dx} \Rightarrow \dfrac{dy}{dx} = \dfrac{1}{2\sqrt{u}} \cdot \sec^2 x \Rightarrow \dfrac{dy}{dx} = \dfrac{1}{2\sqrt{\tan x}} \sec^2 x \quad (\because u = \tan x)$

(c) Let, $y = \cos(\log x)$ and $u = \log x \Rightarrow y = \cos u$ and $u = \log x \Rightarrow \dfrac{dy}{du} = -\sin u$ and $\dfrac{du}{dx} = \dfrac{1}{x}$

Now, $\dfrac{dy}{dx} = \dfrac{dy}{du} \cdot \dfrac{du}{dx} \Rightarrow \dfrac{dy}{dx} = -\sin u \cdot \dfrac{1}{x} \Rightarrow \dfrac{dy}{dx} = -\sin(\log x) \cdot \dfrac{1}{x} \quad (\because u = \log x)$

(d) Let, $y = e^{\sec x}$ and $u = \sec x \Rightarrow y = e^u$ and $u = \sec x \Rightarrow \dfrac{dy}{du} = e^u$ and $\dfrac{du}{dx} = \sec x \cdot \tan x$

Now, $\dfrac{dy}{dx} = \dfrac{dy}{du} \cdot \dfrac{du}{dx} \Rightarrow \dfrac{dy}{dx} = e^u \cdot \sec x \cdot \tan x \Rightarrow \dfrac{dy}{dx} = e^{\sec x} \cdot \sec x \cdot \tan x \quad (\because u = \sec x)$

(c) Let, $y = \log(\text{cosec } x)$ and $u = \text{cosec } x \Rightarrow y = \log u$ and $u = \text{cosec } x$

$$\Rightarrow \frac{dy}{du} = \frac{1}{u} \text{ and } \frac{du}{dx} = -\text{cosec } x \cdot \cot x$$

Now, $\frac{dy}{dx} = \frac{dy}{du} \cdot \frac{du}{dx} \Rightarrow \frac{dy}{dx} = \frac{1}{u} \cdot (-\text{cosec } x \cdot \cot x) \Rightarrow \frac{dy}{dx} = -\frac{1}{\text{cosec } x} \cdot (\text{cosec } x \cdot \cot x)$

$$\Rightarrow \frac{dy}{dx} = -\cot x \quad (\because u = \text{cosec } x)$$

Example 2. Differentiate the following w.r.t. x

(a) $e^{\cos x^2}$ **(b)** $\log(\tan \sqrt{x})$ **(c)** $(\sin^{-1} \sqrt{x})^3$ **(d)** $\tan^{-1} \sqrt{\log x}$

Solution :

(a) Let $y = e^{\cos x^2}$, $u = \cos x^2$, $v = x^2$

$\Rightarrow y = e^u$, $u = \cos v$, $v = x^2$

$\Rightarrow \frac{dy}{du} = e^u$, $\frac{du}{dv} = -\sin v$, $\frac{dv}{dx} = 2x$

We have, $\frac{dy}{dx} = \frac{dy}{du} \cdot \frac{du}{dv} \cdot \frac{dx}{dx}$

$$\Rightarrow \frac{dy}{dx} = e^u \cdot (-\sin v) \cdot 2x \Rightarrow \frac{dy}{dx} = e^{\cos x^2} \cdot (-\sin x^2) \cdot 2x \Rightarrow \frac{dy}{dx} = -e^{\cos x^2} \cdot \sin x^2 \cdot 2x$$

(b) Let $y = \log(\tan \sqrt{x})$, $u = \tan \sqrt{x}$, $v = \sqrt{x}$

$\Rightarrow y = \log u$, $u = \tan v$, $v = \sqrt{x}$

$\Rightarrow \frac{dy}{du} = \frac{1}{u}$, $\frac{du}{dv} = \sec^2 v$, $\frac{dv}{dx} = \frac{1}{2\sqrt{x}}$

We have, $\frac{dy}{dx} = \frac{dy}{du} \cdot \frac{du}{dv} \cdot \frac{dv}{dx} \Rightarrow \frac{dy}{dx} = \frac{1}{u} \cdot \sec^2 v \cdot \frac{1}{2\sqrt{x}} \Rightarrow \frac{dy}{dx} = \frac{1}{\tan\sqrt{x}} \cdot \sec^2 \sqrt{x} \cdot \frac{1}{2\sqrt{x}}$

(c) Let $y = (\sin^{-1} \sqrt{x})^3$, $u = \sin^{-1} \sqrt{x}$, $v = \sqrt{x}$

$\Rightarrow y = u^3$, $u = \sin^{-1} v$, $v = \sqrt{x}$

$\Rightarrow \frac{dy}{du} = 3u^2$, $\frac{du}{dv} = \frac{1}{\sqrt{1-v^2}}$, $\frac{dv}{dx} = \frac{1}{2\sqrt{x}}$

We have, $\frac{dy}{dx} = \frac{dy}{du} \cdot \frac{du}{dv} \cdot \frac{dv}{dx} \Rightarrow \frac{dy}{dx} = 3u^2 \cdot \frac{1}{\sqrt{1-v^2}} \cdot \frac{1}{2\sqrt{x}} \Rightarrow \frac{dy}{dx} = 3(\sin^{-1} \sqrt{x})^2 \cdot \frac{1}{\sqrt{1-x}} \cdot \frac{1}{2\sqrt{x}}$

(d) Let, $y = \tan^{-1} \sqrt{\log x}$, $u = \sqrt{\log x}$, $v = \log x$

$\Rightarrow y = \tan^{-1} u$, $u = \sqrt{v}$, $v = \log x$

$\Rightarrow \frac{dy}{du} = \frac{1}{1+u^2}$, $\frac{du}{dv} = \frac{1}{2\sqrt{v}}$, $\frac{dv}{dx} = \frac{1}{x}$

We have, $\frac{dy}{dx} = \frac{dy}{du} \cdot \frac{du}{dv} \cdot \frac{dv}{dx} \Rightarrow \frac{dy}{dx} = \frac{1}{1+u^2} \cdot \frac{1}{2\sqrt{v}} \cdot \frac{1}{x} \Rightarrow \frac{dy}{dx} = \frac{1}{1+\log x} \cdot \frac{1}{2\sqrt{\log x}} \cdot \frac{1}{x}$

During the process of differentiating the given composite function, by applying the chain rule, we have to make suitable substitutions and bring the function into several basic functions. Applying this process for every problem not only consumes time but also, in a way, tedious.

In order to apply the chain rule directly we shall consider particular cases of the chain rule related to the basic functions, which we have seen already.

Consider the function $y = x^n$. We know that $\frac{dy}{dx} = nx^{n-1}$. Supposing x is replaced by some expression in x say $f(x)$. Then we get the composite function $y = [f(x)]^n$. In this case to find $\frac{dy}{dx}$, we put $f(x) = u$ and apply the chain rule.

$$y = u^n \text{ and } u = f(x) \Rightarrow \frac{dy}{du} = nu^{n-1} \text{ and } \frac{du}{dx} = f'(x)$$

Now, $\frac{dy}{dx} = \frac{dy}{du} \cdot \frac{du}{dx} \Rightarrow \frac{dy}{dx} = nu^{n-1} \cdot f'(x) \Rightarrow \frac{dy}{dx} = n[f(x)]^{n-1} \cdot f'(x)$

$\therefore \quad y = [f(x)]^n \Rightarrow \frac{dy}{dx} = n[f(x)]^{n-1} \cdot f'(x)$

Thus, to differentiate $[f(x)]^n$, differentiate it treating $f(x)$ as a single variable and write $n[f(x)]^{n-1}$, next differentiate $f(x)$ w.r.t. x, and take the product.

In other words, first differentiate $[f(x)]^n$ w.r.t. $f(x)$ (which is $n[f(x)]^{n-1}$) and then write the derivative of $f(x)$ w.r.t. x.

Illustration : Consider, $y = \sin^3 x = (\sin x)^3$

$\Rightarrow \quad \frac{dy}{dx} = 3(\sin x)^2 \cdot \frac{d}{dx}(\sin x)$ (where $f(x) = \sin x$)

$\Rightarrow \quad \frac{dy}{dx} = 3\sin^2 x \cdot \cos x$

In a similar way the corresponding chain rule to each basic function, when the variable x, is replaced by $f(x)$, can be established. The corresponding rule for each function is mentioned below, the readers are adviced to understand the principle behind it and practice.

1. $y = \sqrt{f(x)} \Rightarrow \frac{dy}{dx} = \frac{1}{2\sqrt{f(x)}} \cdot f'(x)$
2. $y = \frac{1}{f(x)} \Rightarrow \frac{dy}{dx} = -\frac{1}{[f(x)]^2} \cdot f'(x)$
3. $y = \frac{1}{[f(x)]^n} \Rightarrow \frac{dy}{dx} = \frac{-n}{[f(x)]^{n+1}} \cdot f'(x)$
4. $y = \sin[f(x)] \Rightarrow \frac{dy}{dx} = \cos[f(x)] \cdot f'(x)$
5. $y = \cos[f(x)] \Rightarrow \frac{dy}{dx} = -\sin[f(x)] \cdot f'(x)$
6. $y = \tan[f(x)] \Rightarrow \frac{dy}{dx} = \sec^2[f(x)] \cdot f'(x)$
7. $y = \operatorname{cosec}[f(x)] \Rightarrow \frac{dy}{dx} = -\operatorname{cosec}[f(x)] \cdot \cot[f(x)] \cdot f'(x)$

8. $y = \sec[f(x)] \Rightarrow \frac{dy}{dx} = \sec[f(x)] \cdot \tan[f(x)] \cdot f'(x)$

9. $y = \cot[f(x)] \Rightarrow \frac{dy}{dx} = -\text{cosec}^2[f(x)] \cdot f'(x)$

10. $y = e^{f(x)} \Rightarrow \frac{dy}{dx} = e^{f(x)} \cdot f'(x)$ 11. $y = \log[f(x)] \Rightarrow \frac{dy}{dx} = \frac{1}{f(x)} \cdot f'(x)$

11. $y = a^{f(x)} \Rightarrow \frac{dy}{dx} = a^{f(x)} \cdot \log a \cdot f'(x)$

We shall consider some functions which are in examples 1 and 2, and see how we can write directly the derivatives.

Example 3. Differentiate the following w.r.t. x

(a) $\sin^3 x$ (b) $e^{\sec x}$ (c) $\log(\sqrt{\tan x})$

Solution :

(a) Let $y = \sin^3 x = (\sin x)^3 \Rightarrow \frac{dy}{dx} = 3(\sin x)^2 \cdot \frac{d}{dx}(\sin x) \Rightarrow \frac{dy}{dx} = 3\sin^2 x \cdot \cos x$

(b) Let $y = e^{\sec x} \Rightarrow \frac{dy}{dx} = e^{\sec x} \cdot \frac{d}{dx}(\sec x) \Rightarrow \frac{dy}{dx} = e^{\sec x} \cdot \sec x \cdot \tan x$

(c) Let $y = \log(\tan\sqrt{x}) \Rightarrow \frac{dy}{dx} = \frac{1}{\tan\sqrt{x}} \cdot \frac{d}{dx}(\tan\sqrt{x})$

$\Rightarrow \frac{dy}{dx} = \frac{1}{\tan\sqrt{x}} \cdot \sec^2\sqrt{x}\ \frac{d}{dx}(\sqrt{x}) \Rightarrow \frac{dy}{dx} = \frac{1}{\tan\sqrt{x}} \cdot \sec^2\sqrt{x} \cdot \frac{1}{2\sqrt{x}}$

Example 4. Differentiate the following w.r.t. x

(a) $(3x^2 - 7x + 4)^{5/2}$ (b) $\frac{1}{4x-5}$ (c) $\text{cosec}(2\sqrt{\cos x})$ (d) $\sqrt{\tan\sqrt{x}}$

(e) $\tan(\log\sin x)$ (f) $e^{\sin x^2}$ (g) $\sec(\log\sqrt{4x-1})$ (h) $\cos^3(\log\sec x)$

Solution :

(a) Let $y = (3x^2 - 7x + 4)^{5/2}$

$\Rightarrow \frac{dy}{dx} = \frac{5}{2}(3x^2 - 7x + 4)^{\frac{5}{2}-1} \cdot \frac{d}{dx}(3x^2 - 7x + 4) \Rightarrow \frac{dy}{dx} = \frac{5}{2}(3x^2 - 7x + 4)^{\frac{3}{2}} \cdot (6x - 7)$

(b) Let, $y = \frac{1}{4x-5} \Rightarrow \frac{dy}{dx} = \frac{-1}{(4x-5)^2} \cdot \frac{d}{dx}(4x-5)$ $\left(\because \frac{d}{dx}\left(\frac{1}{f(x)}\right) = \frac{-1}{[f(x)]^2} \cdot f'(x)\right)$

$\Rightarrow \frac{dy}{dx} = \frac{-1}{(4x-5)^2}\ 4 = \frac{-4}{(4x-5)^2}$

(c) Let, $y = \text{cosec}(2\sqrt{\cos x})$

$\Rightarrow \frac{dy}{dx} = -\text{cosec}(2\sqrt{\cos x}) \cdot \cot(2\sqrt{\cos x}) \cdot \frac{d}{dx}(2\sqrt{\cos x})$

$$\Rightarrow \quad \frac{dy}{dx} = -\operatorname{cosec}(2\sqrt{\cos x}) \cdot \cot(2\sqrt{\cos x}) \cdot 2 \cdot \frac{1}{2\sqrt{\cos x}} \cdot \frac{d}{dx}(\cos x)$$

$$\Rightarrow \frac{dy}{dx} = -\frac{\cos\left(2\sqrt{\cos x}\right) \cdot \cot\left(2\sqrt{\cos x}\right)}{\sqrt{\cos x}} \cdot (-\sin x) \Rightarrow \frac{dy}{dx} = \cos\left(2\sqrt{\cos x}\right) \cdot \cot\left(2\sqrt{\cos x}\right) \frac{\sin x}{\sqrt{\cos x}}$$

(d) Let $y = \sqrt{\tan \sqrt{x}}$ $\Rightarrow \frac{dy}{dx} = \frac{1}{2\sqrt{\tan \sqrt{x}}} \cdot \frac{d}{dx}\left(\tan \sqrt{x}\right)$

$$\Rightarrow \frac{dy}{dx} = \frac{1}{2\sqrt{\tan \sqrt{x}}} \cdot \sec^2 \sqrt{x}\ \frac{d}{dx}(\sqrt{x})$$

$$\Rightarrow \frac{dy}{dx} = \frac{1}{2\sqrt{\tan \sqrt{x}}} \cdot \sec^2 \sqrt{x} \cdot \frac{1}{2\sqrt{x}} \Rightarrow \frac{dy}{dx} = \frac{\sec^2 \sqrt{x}}{4\sqrt{x}\ \sqrt{\tan \sqrt{x}}}$$

(e) Let $y = \tan(\log \sin x) \Rightarrow \frac{dy}{dx} = \sec^2(\log \sin x) \cdot \frac{d}{dx}(\log \sin x)$

$$\Rightarrow \frac{dy}{dx} = \sec^2(\log \sin x) \cdot \frac{1}{\sin x} \cdot \frac{d}{dx}(\sin x)$$

$$\Rightarrow \frac{dy}{dx} = \sec^2(\log \sin x)\ \frac{1}{\sin x} \cdot \cos x \Rightarrow \frac{dy}{dx} = \cot x \cdot \sec^2(\log \sin x)$$

(f) Let $y = e^{\sin x^2} \Rightarrow \frac{dy}{dx} = e^{\sin x^2} \cdot \frac{d}{dx}(\sin x^2) \Rightarrow \frac{dy}{dx} = e^{\sin x^2} \cdot \cos x^2 \cdot \frac{d}{dx}(x^2)$

$$\Rightarrow \frac{dy}{dx} = e^{\sin x^2} \cdot \cos x^2 \cdot 2x$$

$$\Rightarrow \frac{dy}{dx} = 2x \cdot \cos x^2 \cdot e^{\sin x^2}$$

(g) Let $y = \sec(\log \sqrt{4x-1})$

$$\Rightarrow \frac{dy}{dx} = \sec(\log \sqrt{4x-1}) \cdot \tan(\log \sqrt{4x-1}) \cdot \frac{d}{dx}\left(\log \sqrt{4x-1}\right)$$

$$\Rightarrow \frac{dy}{dx} = \sec(\log \sqrt{4x-1}) \cdot \tan(\log \sqrt{4x-1}) \cdot \frac{1}{\sqrt{4x-1}} \cdot \frac{d}{dx}(\sqrt{4x-1})$$

$$\Rightarrow \frac{dy}{dx} = \sec(\log \sqrt{4x-1}) \cdot \tan(\log \sqrt{4x-1}) \cdot \frac{1}{\sqrt{4x-1}} \cdot \frac{1}{2\sqrt{4x-1}} \cdot \frac{d}{dx}(4x-1)$$

$$\Rightarrow \frac{dy}{dx} = \sec(\log \sqrt{4x-1}) \cdot \tan(\log \sqrt{4x-1}) \cdot \frac{2}{(4x-1)}$$

(h) Let $y = \cos^3(\log \sec x) = [\cos(\log \sec x)]^3$

$$\Rightarrow \frac{dy}{dx} = 3[\cos(\log \sec x)]^2 \cdot \frac{d}{dx}[\cos(\log \sec x)]$$

$$\Rightarrow \frac{dy}{dx} = 3\cos^2(\log \sec x) \cdot [-\sin(\log \sec x)] \cdot \frac{d}{dx}(\log \sec x)$$

$$\Rightarrow \quad \frac{dy}{dx} = 3\cos^2(\log\sec x)\cdot[-\sin(\log\sec x)]\,\frac{1}{\sec x}\cdot(\sec x\cdot\tan x)$$

$$\Rightarrow \quad \frac{dy}{dx} = -3\cos^2(\log\sec x)\sin(\log\sec x)\cdot\tan x$$

The readers are adviced to recognise the basic functions in succession, that appears in the given composite function and practice to write the derivative directly.

For example, consider the function, $y = \log\left(\sqrt{\sin e^x}\right)$

Here, the basic functions that are involved in succession are log function, root function, sine function and exponential function.

Now, $$\frac{dy}{dx} = \frac{1}{\sqrt{\sin e^x}}\cdot\frac{1}{2\sqrt{\sin e^x}}\cdot\cos e^x\cdot e^x$$

Observe that there are as many factors as the number of basic functions composed to form the composite function. Each one corresponds to the derivative of corresponding basic function in the order they are composed.

Example 5. Differentiate the following w.r.t. x

(a) $\sin^2(\cos 3x)$ **(b)** $\cos(2^{\log x})$ **(c)** $\log(x^2 - 2\tan x)$

(d) $\log\left(x + \sqrt{1+x^2}\right)$ **(e)** $e^{\sqrt{\cos x}}$ **(f)** $\cot\left(x^2 + \frac{1}{x^2}\right)$

Solution : (a) Let $y = \sin^2(\cos 3x) = [\sin(\cos 3x)]^2$

$$\Rightarrow \frac{dy}{dx} = 2\sin(\cos 3x)\cdot\cos(\cos 3x)\cdot(-\sin 3x)\cdot 3 \Rightarrow \frac{dy}{dx} = -6\sin(\cos 3x)\cdot\cos(\cos 3x)\cdot(\sin 3x)$$

(b) Let $y = \cos(2^{\log x}) \Rightarrow \dfrac{dy}{dx} = -\sin(2^{\log x})\cdot(2^{\log x}\cdot\log 2)\cdot\dfrac{1}{x}$

(c) Let $y = \log(x^2 - 2\tan x) \Rightarrow \dfrac{dy}{dx} = \dfrac{1}{x^2 - 2\tan x}\cdot\left(2x - 2\sec^2 x\right) = \dfrac{2\left(x - \sec^2 x\right)}{x^2 - 2\tan x}$

(d) Let $y = \log\left(x + \sqrt{1+x^2}\right)$

$$\Rightarrow \frac{dy}{dx} = \frac{1}{x+\sqrt{1+x^2}}\left[1 + \frac{1}{2\sqrt{1+x^2}}(2x)\right] \Rightarrow \frac{dy}{dx} = \frac{1}{x+\sqrt{1+x^2}}\left[\frac{\sqrt{1+x^2}+x}{\sqrt{1+x^2}}\right] = \frac{1}{\sqrt{1+x^2}}$$

(e) Let $y = e^{\sqrt{\cos x}} \Rightarrow \dfrac{dy}{dx} = e^{\sqrt{\cos x}}\cdot\dfrac{1}{2\sqrt{\cos x}}\cdot(-\sin x) \Rightarrow \dfrac{dy}{dx} = -\dfrac{\sin x}{2\sqrt{\cos x}}\cdot e^{\sqrt{\cos x}}$

(f) Let $y = \cot\left(x^2 + \frac{1}{x^2}\right) \Rightarrow \dfrac{dy}{dx} = -\operatorname{cosec}^2\left(x^2 + \dfrac{1}{x^2}\right)\cdot\left(2x - \dfrac{2}{x^3}\right) \quad \left(\because \dfrac{d}{dx}\left(\dfrac{1}{x^n}\right) = \dfrac{-n}{x^{n+1}}\right)$

$$\Rightarrow \frac{dy}{dx} = -2\left(x - \frac{1}{x^3}\right)\cdot\operatorname{cosec}^2\left(x^2 + \frac{1}{x^2}\right)$$

Example 6. Differentiate the following w.r.t. x

(a) $x^3 \tan x^2$ (b) $e^{\sin x} \cdot \sin(e^x)$ (c) $\tan(x^3 \cdot e^x)$ (d) $\sin^2 x \cdot \cos^3 x$

Solution :

(a) Let $y = x^3 \cdot \tan x^2 \Rightarrow \frac{dy}{dx} = x^3 \cdot \frac{d}{dx}(\tan x^2) + \frac{d}{dx}(x^3) \cdot \tan x^2$

$$\Rightarrow \frac{dy}{dx} = x^3 \cdot \sec^2 x^2 \cdot 2x + 3x^2 \cdot \tan x^2 \Rightarrow \frac{dy}{dx} = x^2 [2x^2 \sec^2 x^2 + 3 \tan x^2]$$

(b) Let $y = e^{\sin x} \cdot \sin(e^x)$

$$\Rightarrow \frac{dy}{dx} = e^{\sin x} \cdot \frac{d}{dx}(\sin e^x) + \frac{d}{dx}(e^{\sin x}) \cdot \sin(e^x) \Rightarrow \frac{dy}{dx} = e^{\sin x} \cdot (\cos e^x \cdot e^x) + e^{\sin x} \cdot \cos x \cdot \sin(e^x)$$

$$\Rightarrow \frac{dy}{dx} = e^{\sin x} [e^x \cdot \cos e^x + \cos x \cdot \sin(e^x)]$$

(c) Let $y = \tan^{-1}(x^3 \cdot e^x) \Rightarrow \frac{dy}{dx} = \frac{1}{1 + \left(x^3 \cdot e^x\right)^2} \cdot \frac{d}{dx}\left(x^3 \cdot e^x\right)$

$$\Rightarrow \frac{dy}{dx} = \frac{1}{1 + \left(x^6 \cdot e^{2x}\right)} \cdot \left[x^3 \cdot e^x + 3x^2 \cdot e^x\right] \Rightarrow \frac{dy}{dx} = \frac{x^2 \cdot e^x (x+3)}{1 + x^6 \cdot e^{2x}}$$

(d) Let $y = \sin^2 x \cdot \cos^3 x = (\sin x)^2 \cdot (\cos x)^3$

$$\Rightarrow \frac{dy}{dx} = (\sin x)^2 \cdot 3(\cos x)^2 \cdot (-\sin x) + 2 \sin x \cdot \cos x \cdot (\cos x)^3$$

$$\Rightarrow \frac{dy}{dx} = -3\cos^2 x \cdot \sin^3 x + 2 \sin x \cdot \cos^4 x \Rightarrow \frac{dy}{dx} = \cos^2 x \cdot \sin x\,(2\cos^2 x - 3\sin^2 x)$$

Exercise

I. Differentiate the following w.r.t. x

1. $\sin 4x$ 2. $\cos 6x$ 3. $\tan 2x$ 4. $\sec 5x$ 5. $\cos x^3$ 6. $\sin^3 \sqrt{x}$ 7. $\tan^3 x$

8. e^{x^4} 9. $(3 + 7x)^5$ 10. $\frac{1}{(x+4)^2}$ 11. $\frac{1}{\sqrt{x^2 - 2}}$ 12. $\sqrt{\sin x^3}$ 13. $[\log(\cos x)]^2$

14. $\log \sqrt{\tan x}$ 15. $\sqrt{\sec x + \tan x}$ 16. $\sqrt{5x^2 + 3\log x}$ 17. $\sec\left(x + \frac{1}{x}\right)$

18. $\cot\left(x^2 + \frac{1}{x^2}\right)$ 19. $3^{(x^2 - 4x)}$ 20. $7^{\sin\sqrt{x}}$ 21. $\sqrt{\cot\sqrt{x}}$

22. $\log\left(\sec\frac{x}{2} + \tan\frac{x}{2}\right)$ 23. $\log(\sin\sqrt{x})$ 24. $\sqrt{\log(\log \tan x)}$

25. $\sin\left(e^{x^2}\right)$ 26. $\cot^{-1}(6x^2)$ 27. $(\cot^{-1} x)^2$ 28. $\tan^{-1}(x\sqrt{x})$

II. Differentiate the following w.r.t. x

1. $\cos 3x \sin 5x$ 2. $\sin x \sin 2x$ 3. $e^x \cdot \log\sqrt{x}$ 4. $e^{2x} \cdot \sin 3x$ 5. $e^{4x} \cdot \cos 3x$

6. $e^{-4x} \cdot \cot 4x$ 7. $e^{2x} \cdot \sin 3x$ 8. $(5 + 2x)\sqrt{2x - 1}$ 9. $x^4 \cdot \sin^4 x$ 10. $\log(e^{3x} \cdot \sqrt{x})$

11. $\log(2x+3)\cdot\left(e^{2x^2}\right)$ **12.** $-\text{cosec } 2x\cdot\sec^2 4x$ **13.** $e^{\sin x}\cdot\tan\sqrt{x}$ **14.** $\cos^2 x\cdot\log(\tan x)$

15. $\cos^5 x\cdot\cos(x^5)$ **16.** $3^{x^2}\ \log x$

III. Differentiate the following w.r.t. x

1. $\dfrac{\log(\sin x)}{\sin(\log x)}$ **2.** $\sqrt{\dfrac{\sin x}{x}}$ **3.** $\dfrac{x}{\sqrt{x^2-1}}$ **4.** $\sqrt{\dfrac{1-x}{1+x}}$ **5.** $\dfrac{x}{\sqrt{2x-1}}$ **6.** $\sqrt{\dfrac{1+\sin x}{1-\sin x}}$

7. $\dfrac{\sin^3 2x}{1+e^{-x}}$ **8.** $\dfrac{e^{\sin x}}{\sqrt{\log x}}$ **9.** $\dfrac{\cos^3 x}{e^x+1}$ **10.** $\sqrt{\dfrac{1+e^x}{1-e^x}}$ **11.** $\log\left(\dfrac{1+\sin x}{1-\sin x}\right)$

12. $\log\left(\sqrt{x}+\dfrac{1}{\sqrt{x}}\right)$ **13.** $\dfrac{x\ \text{cosec}^{-1}x}{\sqrt{x^2-1}}$

Answers

I. **1.** $4\cos 4x$ **2.** $-6\sin 6x$ **3.** $2\sec^2 2x$ **4.** $5\sec 5x\cdot\tan 5x$ **5.** $-3x^2\sin x^3$

6. $3\sin^2\sqrt{x}\cdot\cos\sqrt{x}\cdot\dfrac{1}{2\sqrt{x}}$ **7.** $3\tan^2 x\cdot\sec^2 x$ **8.** $4x^3\cdot e^{x^4}$ **9.** $35(3+7x)^4$

10. $\dfrac{-2}{(x+4)^3}$ **11.** $\dfrac{-x}{(x^2-2)^{3/2}}$ **12.** $\dfrac{3x^2\cdot\cos x^3}{2\sqrt{\sin x^3}}$ **13.** $-2\tan x\cdot\log(\cos x)$ **14.** $\dfrac{\sec^2 x}{2\tan x}$

15. $\dfrac{1}{2}\sec x\sqrt{\sec x+\tan x}$ **16.** $\dfrac{10x^2+3}{2x\sqrt{5x^2+3\log x}}$ **17.** $\cos\left(x+\dfrac{1}{x}\right)\cdot\left(1-\dfrac{1}{x^2}\right)$

18. $-2\left(\dfrac{x^4+1}{x^3}\right)\text{cosec}^2\left(x^2-\dfrac{1}{x^2}\right)$ **19.** $(\log 3)(2x-4)\cdot 3^{(x^2-4x)}$ **20.** $\dfrac{(\log 7)\cos\sqrt{x}\cdot 7^{\sin\sqrt{x}}}{2\sqrt{x}}$

21. $\dfrac{-\text{cosec}^2\sqrt{x}}{2\sqrt{x}\sqrt{\cot\sqrt{x}}}$ **22.** $\dfrac{1}{2}\sec\dfrac{x}{2}$ **23.** $\dfrac{\cot\sqrt{x}}{2\sqrt{x}}$

24. $\dfrac{\sec^2 x}{2\tan x\cdot\log(\tan x)\sqrt{\log(\log\tan x)}}$ **25.** $2x\,e^{x^2}\cdot\cos\left(e^{x^2}\right)$ **26.** $\dfrac{-12x^2}{1+36x^4}$

27. $\dfrac{-2\cos^{-1}x}{\sqrt{1-x^2}}$ **28.** $\dfrac{3\sqrt{x}}{2(1+x^3)}$

II. **1.** $5\cos 3x\cos 5x-3\sin 3x\sin 5x$ or $\dfrac{1}{2}(8\cos 8x+2\cos 4x)$

2. $\dfrac{1}{2}(-\sin x+3\sin 3x)$ **3.** $\dfrac{e^x}{2}\left(\dfrac{1}{x}+\log x\right)$ **4.** $e^{2x}(3\cos 3x+2\sin 3x)$

5. $e^{4x}[4\cos 3x-3\sin 3x]$ **6.** $-e^{-4x}[\text{cosec}^2 x+4\cot x]$ **7.** $e^{2x}(3\cos 3x+2\sin 3x)$

8. $\dfrac{6x+3}{\sqrt{2x-1}}$ **9.** $(4x^3\sin^3 x)[x\cos x+\sin x]$ **10.** $\dfrac{1+6x}{2x}$

11. $2e^{2x^2}\left[2x\log(2x+3)+\dfrac{1}{2x+3}\right]$ **12.** $(4\tan 4x-\cot 2x)\,2\sec^2 4x\cdot\operatorname{cosec} 2x$

13. $e^{\sin x}\left[\dfrac{\sec^2\sqrt{x}}{2\sqrt{x}}+\cos x\cdot\tan\sqrt{x}\right]$ **14.** $\cot x-\sin 2x\cdot\log\tan x$

15. $-5\cos^4 x\,[x^4\cos x\cdot\sin x^5+\sin x\cos x^5]$ **16.** $\dfrac{3^{x^2}}{x}[1+2x^2\log 3\cdot\log x]$

III. 1. $\dfrac{x\sin(\log x)\cdot\cot x-\log(\sin x)\cdot\cos(\log x)}{x\left[\sin^2(\log x)\right]}$ **2.** $2\sqrt{\dfrac{x}{\sin x}}\left[\dfrac{x\cos x-\sin x}{x^2}\right]$

3. $\dfrac{-1}{(x^2-1)^{3/2}}$ **4.** $-\sqrt{\dfrac{1+x}{1-x}}\cdot\dfrac{1}{(1+x)^2}$ **5.** $\dfrac{x-1}{(2x-1)^{3/2}}$ **6.** $\sqrt{\dfrac{1-\sin x}{1+\sin x}}\cdot\left[\dfrac{\cos x}{(1-\sin x)^2}\right]$

7. $\dfrac{6\left(1+e^{-x}\right)\sin^2 2x\cdot\cos 2x+e^{-x}\cdot\sin^3 2x}{\left(1+e^{-x}\right)^2}$ **8.** $\dfrac{(2x\log x\cdot\cos x-1)\,e^{\sin x}}{2x\,(\log x)^{3/2}}$

9. $\dfrac{-\cos^2 x\left[3\left(e^x+1\right)\sin x+e^x\cos x\right]}{\left(e^x+1\right)^2}$ **10.** $\dfrac{e^x}{\left(1-e^x\right)^2}\sqrt{\dfrac{1-e^x}{1+e^x}}$ **11.** $2\sec x$

12. $\dfrac{x-1}{2x\,(x+1)}$ **13.** $\dfrac{-\left[\sqrt{x^2-1}+\operatorname{cosec}^{-1}x\right]}{\left(x^2-1\right)\sqrt{x^2-1}}$

1.8 Logarithmic Differentiation

Supposing we come across a function of the form $y=u^v$, where u and v are functions of x and we have to find $\dfrac{dy}{dx}$. In such case we take logarithm both sides and then differentiate. This technique of differentiation is known as **"logarithmic differentiation"**. While differentiating, after taking logarithm, we apply chain rule to differentiate $\log y$. i.e., $\dfrac{d}{dx}(\log y)=\dfrac{1}{y}\dfrac{dy}{dx}$.

This method of differentiation is also useful to differentiate a function which is a product of two or more functions and quotient of two functions in which numerator and/or denominator are/is product of two or more functions.

The technique of logarithmic differentiation is made more clear in the following examples.

Example 1. Differentiate the following w.r.t. x

(a) x^x **(b)** $(\sin x)^{\tan x}$ **(c)** $(\sin x)^{\log x}+(\log x)^{\sin x}$

Solution : (a) Let $y=x^x$. Taking log both sides we get, $\log y=x\log x$

Differentiating both sides w.r.t. x, we get,

$$\frac{1}{y}\frac{dy}{dx}=x\cdot\frac{1}{x}+1\cdot\log x \Rightarrow \frac{1}{y}\frac{dy}{dx}=1+\log x \Rightarrow \frac{dy}{dx}=y\,(1+\log x)$$

$$\Rightarrow \frac{dy}{dx}=x^x(1+\log x) \qquad (\because\ y=x^x)$$

(b) Let $y = (\sin x)^{\tan x}$. Taking log both sides we get, $\log y = \tan x \cdot \log (\sin x)$

Differentiating both sides w.r.t. x, we get,

$$\frac{1}{y}\frac{dy}{dx} = \tan x \cdot \frac{1}{\sin x}(\cos x) + \sec^2 x \cdot \log (\sin x)$$

$$\Rightarrow \quad \frac{dy}{dx} = y\,[\tan x \cdot \cot x + \sec^2 x \cdot \log (\sin x)]$$

$$\Rightarrow \quad \frac{dy}{dx} = (\sin x)^{\tan x}\,[1 + \sec^2 x \cdot \log (\sin x)] \qquad (\because\ y = (\sin x)^{\tan x})$$

(c) Let $y = (\sin x)^{\log x} + (\log x)^{\sin x}$.

Here, we can not take log both sides directly, because the r.h.s is a sum of two functions and further $\log (a + b) \neq \log a + \log b$. In this case we have to take each term of r.h.s separately and differentiate; the final derivative will be sum of these two differential coefficients.

Let $u = (\sin x)^{\log x}$ and $v = (\log x)^{\sin x}$

Now, $u = (\sin x)^{\log x} \Rightarrow \log u = \log x \cdot \log (\sin x)$

Differentiating both sides w.r.t. x, we get

$$\frac{1}{u}\frac{du}{dx} = \log x \cdot \frac{1}{\sin x} \cdot \cos x + \frac{1}{x} \cdot \log (\sin x)$$

$$\Rightarrow \quad \frac{du}{dx} = u\left[\log x \cdot \cot x + \frac{\log (\sin x)}{x}\right] \Rightarrow \frac{du}{dx} = (\sin x)^{\log x}\left[\log x \cdot \cot x + \frac{\log (\sin x)}{x}\right]$$

Now, $v = (\log x)^{\sin x} \Rightarrow \log v = \sin x \cdot \log (\log x)$

Differentiating both sides w.r.t. x, we get,

$$\frac{1}{v}\frac{dv}{dx} = \sin x \cdot \frac{1}{\log x} \cdot \frac{1}{x} + \cos x \cdot \log (\log x)$$

$$\Rightarrow \quad \frac{du}{dx} = v\left[\frac{\sin x}{x \log x} + \cos x \cdot \log (\log x)\right] \Rightarrow \frac{dv}{dx} = (\log x)^{\sin x}\left[\frac{\sin x}{x \log x} + \cos x \cdot \log (\log x)\right]$$

Now, $$y = u + v \Rightarrow \frac{dy}{dx} = \frac{du}{dx} + \frac{dv}{dx}$$

Note : Consider the function $y = u^v$, where u and v are functions of x. Taking log both sides we get

$$\log y = v \cdot \log u$$

Differentiating both sides w.r.t. x, we get,

$$\frac{1}{y}\frac{dy}{dx} = v \cdot \frac{1}{u}\frac{du}{dx} + \log u \cdot \frac{dv}{dx} \Rightarrow \frac{dy}{dx} = y\left[\frac{v}{u}\frac{du}{dx} + \log u \cdot \frac{dv}{dx}\right]$$

$$\Rightarrow \frac{dy}{dx} = u^v\left[\frac{v}{u}\frac{du}{dx} + \log u \cdot \frac{dv}{dx}\right]$$

This result can be stated as below.

If $y = u^v$, **then** $$\frac{dy}{dx} = u^v\left[\frac{\textbf{power fun.}}{\textbf{base fun.}} \cdot \frac{d}{dx}[\textbf{base fun.}] + \log (\textbf{base fun.}) \cdot \frac{d}{dx}[\textbf{power fun.}]\right]$$

Example 2. If $y = (\tan x)^{\sin x}$ find $\frac{dy}{dx}$

Solution :

Let $y = (\tan x)^{\sin x}$ $\Rightarrow \frac{dy}{dx} = (\tan x)^{\sin x}\left[\frac{\sin x}{\tan x}\cdot\frac{d}{dx}(\tan x) + \log(\tan x)\cdot\frac{d}{dx}(\sin x)\right]$

$$\Rightarrow \frac{dy}{dx} = (\tan x)^{\sin x}\left[\frac{\sin x}{\tan x}\cdot\sec^2 x + \log(\tan x)\cdot\cos x\right]$$

$$\Rightarrow \frac{dy}{dx} = (\tan x)^{\sin x}\,[\sec x + \cos x\cdot\log(\tan x)]$$

Example 3. Differentiate the following w.r.t. x, $x\cdot e^x\cdot\sin x^2$.

Solution : Let $y = x\cdot e^x\cdot\sin x^2$. Taking log both sides we get,

$\log y = \log x + \log e^x + \log(\sin x^2) \Rightarrow \log y = \log x + x + \log(\sin x^2)$ $(\because \log e = 1)$

Differentiating both sides w.r.t. x, we get,

$$\frac{1}{y}\frac{dy}{dx} = \frac{1}{x} + 1 + \frac{1}{\sin x^2}\cdot\left(\cos x^2\right)\cdot 2x$$

$$\Rightarrow \frac{dy}{dx} = y\left[\frac{1}{x} + 1 + \frac{2x\cos x^2}{\sin x^2}\right] \Rightarrow \frac{dy}{dx} = (x\cdot e^x\cdot\sin x^2)\left[\frac{1}{x} + 1 + 2x\cot x^2\right]$$

Exercise

I. Differentiate the following w.r.t. x,

1. $x^{\sqrt{x}}$ 2. $x^{\sin x}$ 3. $(\tan x)^{1/x}$ 4. $(\sin x)^{\log x}$ 5. $\sin(x^x)$ 6. $(\sin x)^x$
7. $(\log x)^x$ 8. $(\tan x)^{\log x}$ 9. $[\tan^{-1}(x^2+1)]^{\cos x}$ 10. $x^{5+\log x}$ 11. $x^{(\sin x - \cos x)}$
12. $(x^2+a^2)^x$ 13. $(x\log x)^x$ 14. $\left(1+\frac{1}{x}\right)^2$ 15. $(x^x)^x$ 16. $x^{(x^x)}$
17. $x^{\sin x}\cdot(\sin x)^x$ 18. $x^{\cos x}\cdot(\cos x)^x$ 19. $(e^x+e^{-x})^{2x}$

II. Differentiate the following w.r.t. x,

1. $x^3\cdot e^{2x}\cdot\sec^2 x$ 2. $\frac{x\cos^{-1}x}{\sqrt{1-x^2}}$ 3. $\tan^3(x^2)\sqrt{\sin x^3}$ 4. $\frac{\log(1+\tan x)}{(1+x^2)\sqrt{1+x}}$

Answers

I. 1. $\frac{x^{\sqrt{x}}}{2\sqrt{x}}(2+\log x)$ 2. $x^{\sin x}\left[\frac{\sin x}{x} + \cos x\cdot\log x\right]$ 3. $(\tan x)^{1/x}\left[\frac{\sec^2 x}{x\tan x} - \frac{\log(\tan x)}{x^2}\right]$

4. $(\sin x)^{\log x}\left[\log x\cdot\cot x + \frac{1}{x}\log(\sin x)\right]$ 5. $\cos(x^x)\,x^x(1+\log x)$

6. $(\sin x)^x[x\cot x + \log(\sin x)]$ 7. $(\log x)^x\left[\frac{1}{\log x} + \log(\log x)\right]$

8. $(\tan x)^{\log x}\left[\frac{\log x \sec^2 x}{\tan x} + \frac{1}{x}\log(\tan x)\right]$

9. $y\left[\frac{2x\cos x}{\tan^{-1}(x^2+1)}\right] - \sin x \log[\tan^{-1}(x^2+1)]$ where $y = [\tan^{-1}(x^2+1)]^{\cos x}$

10. $x^{(5+\log x)}\left[\frac{1}{x}(5+2\log x)\right]$

11. $y\left[\frac{\sin x - \cos x}{x} + (\cos x + \sin x)\log x\right]$, where $y = x^{(\sin x - \cos x)}$

12. $(x^2+a^2)^x\left[\frac{2x^2}{x^2+a^2} + \log(x^2+a^2)\right]$

13. $(x\log x)^x\left[1 + \frac{1}{\log x} + \log(x\log x)\right]$

14. $\left(1+\frac{1}{x}\right)^x\left[\log\left(1+\frac{1}{x}\right) - \frac{1}{x+1}\right]$

15. $(x^x)^x\,[x + 2x\log x]$

16. $x^{(x^x)}\left[x^x\left\{\frac{1}{x} + \log x + (\log x)^2\right\}\right]$

17. $y\left[\frac{\sin x}{x} + \cos x \log x + x\cot x + \log(\sin x)\right]$ where $y = x^{\sin x}\cdot(\sin x)^x$

18. $y\left[\frac{\cos x}{x} - \sin x\log x - x\tan x + \log(\cos x)\right]$ where $y = x^{\cos x}\cdot(\cos x)^x$

19. $y\left[\frac{e^x - e^{-x}}{e^x + e^{-x}}\cdot 2x + \log(e^x+e^{-x})^2\right]$ where $y = (e^x + e^{-x})^{2x}$

II. 1. $(x^3\cdot e^{2x}\cdot\sec^2 x)\left(\frac{3}{x} + 2 + 2\tan x\right)$

2. $\frac{x\cos^{-1}x}{\sqrt{1-x^2}}\left[\frac{1}{x} - \frac{1}{\sqrt{1-x^2}\cos^{-1}x} - \frac{x}{1-x^2}\right]$

3. $y\left[\frac{6x\sec^2 x^2}{\tan x^2} + \frac{3x^2\cos x^3}{\sin x^3}\right]$

4. $y\left[\frac{\sec^2 x}{(1+\tan x)\log(1+\tan x)} - \frac{2x}{1+x^2} - \frac{1}{2(1+x)}\right]$

1.9 Successive Differentiation

Let $y = f(x)$ be a differentiable function, then $\frac{dy}{dx} = f'(x)$ exists and is called **first order derivative** of the function $y = f(x)$ w.r.t x. The first order derivative of $y = f(x)$ is also denoted by y', y_1 or Dy. That is y_1 denotes the derivative y w.r.t. x.

Now, if $\frac{dy}{dx} = f'(x)$ is also a differentiable function, then its derivative exists – i.e., $\frac{d}{dx}\left(\frac{dy}{dx}\right)$ exists. This derivative is called the **second order derivative** of $y = f(x)$ w.r.t. x.

The second order derivative of $y = f(x)$ is denoted by

$$\frac{d}{dx}\left(\frac{dy}{dx}\right) = \frac{d^2y}{dx^2},\quad f''(x) \text{ or } y'' \text{ or } y_2$$

i.e, if $y = f(x)$, then y_1 is the first order derivative and y_2 is the second derivative of y. Also the **derivative of y_1 w.r.t. x is y_2.**

Similarly, if the derivative of $\frac{d^2y}{dx^2}$ exists then $\frac{d}{dx}\left(\frac{d^2y}{dx^2}\right)$ it is called **third order derivative** of $y = f(x)$ w.r.t. x, and is denoted by $\frac{d^3y}{dx^3}, y_3$ or $f'''(x)$.

In general the process of differentiating the same function again and again is called **successive differentiation**, further $\frac{d^ny}{dx^n}, y_n$ or $f^{(n)}(x)$ denotes the nth order derivative (if it exists) of the function $y = f(x)$.

In the present book we consider upto and including second ordered derivatives of a given function.

Example 1. Find the second derivative of the following functions w.r.t x

(a) $x^3 + 3x^2 + \frac{4}{x}$ (b) $\log(3x - 4)$ (c) $4x + 3e^{2x}$ (d) $\frac{\log x}{x}$

Solution : (a) Let, $y = x^3 + 3x^2 + \frac{4}{x}$

$$\Rightarrow \frac{dy}{dx} = 3x^2 + 6x - \frac{4}{x^2} \quad \Rightarrow \frac{d^2y}{dx^2} = 6x + 6 + \frac{8}{x^3}$$

(b) Let, $y = \log(3x - 4) \Rightarrow \frac{dy}{dx} = \frac{3}{3x - 4}$

$$\Rightarrow \frac{d^2y}{dx^2} = 3 \cdot \left[\frac{-1}{(3x-4)^2}\right](3) \quad \Rightarrow \frac{d^2y}{dx^2} = \frac{-9}{(3x-4)^2}$$

(c) Let, $y = 4x + 3e^{2x} \Rightarrow \frac{dy}{dx} = 4^x \cdot \log 4 + 3 \cdot 2e^{2x}$

$$\Rightarrow \frac{d^2y}{dx^2} = \log 4 \cdot 4^x \cdot \log 4 + 6 \cdot 2e^{2x} \Rightarrow \frac{d^2y}{dx^2} = 4^x \cdot (\log 4)^2 + 12\, e^{2x}$$

(d) Let $y = \frac{\log x}{x} \Rightarrow \frac{dy}{dx} = \frac{x(1/x) - \log x \cdot 1}{x^2}$

$$\Rightarrow \frac{dy}{dx} = \frac{1 - \log x}{x^2}$$

$$\Rightarrow \frac{d^2y}{dx^2} = \frac{x^2(-1/x) - (1 - \log x) \cdot 2x}{x^4}$$

$$\Rightarrow \frac{d^2y}{dx^2} = \frac{-x - 2x + 2x \log x}{x^4} = \frac{x(2 \log x - 3)}{x^4} = \frac{2 \log x - 3}{x^3}$$

Example 2. Find $\frac{d^2y}{dx^2}$, if (a) $x = at^2,\ y = 2at$ (b) $x = ct,\ y = \frac{c}{t}$

Solution : (a) We have, $x = at^2,\ y = 2at$

$$x = at^2 \Rightarrow \frac{dx}{dt} = 2at,\quad y = 2at \Rightarrow \frac{dy}{dt} = 2a$$

Now, $\frac{dy}{dx} = \frac{dy/dt}{dx/dt} \Rightarrow \frac{dy}{dx} = \frac{2a}{2at} = \frac{1}{t}$

Now, $\frac{dy}{dx} = \frac{1}{t} \Rightarrow \frac{d^2y}{dx^2} = -\frac{1}{t^2}\cdot\frac{dt}{dx}$ (note this step)

$$\Rightarrow \frac{d^2y}{dx^2} = -\frac{1}{t^2}\left(\frac{1}{2at}\right) \qquad \left(\because \frac{dx}{dt} = 2at\right)$$

$$\Rightarrow \frac{d^2y}{dx^2} = -\frac{1}{2at^3}$$

(b) $x = ct \Rightarrow \frac{dx}{dt} = c,\quad y = \frac{c}{t} \Rightarrow \frac{dy}{dt} = -\frac{c}{t^2}$

Now, $\frac{dy}{dx} = \frac{dy/dt}{dx/dt} \Rightarrow \frac{dy}{dx} = \frac{-c/t^2}{c} = -\frac{1}{t^2}$

$$\frac{dy}{dx} = -\frac{1}{t^2} \Rightarrow \frac{d^2y}{dx^2} = -\left(\frac{-2}{t^3}\right)\frac{dt}{dx}$$

$$\Rightarrow \frac{d^2y}{dx^2} = \frac{2}{t^3}\cdot\frac{1}{c} \qquad \left(\because \frac{dx}{dt} = c\right)$$

$$\Rightarrow \frac{d^2y}{dx^2} = \frac{2}{ct^3}$$

Example 3. If $xy + 6y = 2x$ show that $\frac{d^2y}{dx^2} = \frac{-24}{(x+6)^3}$.

Solution : We have, $xy + 6y = 2x$

$\Rightarrow y(x+6) = 2x$

$\Rightarrow y = \frac{2x}{x+6}$

$$\Rightarrow \frac{dy}{dx} = \frac{(x+6)\cdot 2 - 2x(1)}{(x+6)^2} = \frac{12}{(x+6)^2}$$

$$\Rightarrow \frac{d^2y}{dx^2} = 12\left[\frac{-2}{(x+6)^3}\right] = \frac{-24}{(x+6)^3}$$

Example 4. **If $2x^2 + 3xy - y^2 = 1$ show that $\dfrac{d^2y}{dx^2} = \dfrac{-34}{(2y-3x)^3}$**

Solution : We have, $2x^2 + 3xy - y^2 = 1$

Differentiating both sides w.r.t x, we get

$$4x + 3\left(x\frac{dy}{dx} + y\right) - 2y\frac{dy}{dx} = 0$$

$$\Rightarrow \quad \frac{dy}{dx}(3x - 2y) = -(4x + 3y)$$

$$\Rightarrow \quad \frac{dy}{dx} = \frac{4x + 3y}{2y - 3x}$$

$$\Rightarrow \quad \frac{d^2y}{dx^2} = \frac{(2y-3x)\left(4 + 3\dfrac{dy}{dx}\right) - (4x+3y)\left(2\dfrac{dy}{dx} - 3\right)}{(2y-3x)^2}$$

$$\Rightarrow \quad \frac{d^2y}{dx^2} = \frac{(8y - 12x + 12x + 9y) + \dfrac{dy}{dx}(6y - 9x - 8x - 6y)}{(2y-3x)^2}$$

$$\Rightarrow \quad \frac{d^2y}{dx^2} = \frac{17y - 17x\left(\dfrac{4x+3y}{2y-3x}\right)}{(2y-3x)^2} \qquad \left(\because \frac{dy}{dx} = \frac{4x+3y}{2y-3x}\right)$$

$$\Rightarrow \quad \frac{d^2y}{dx^2} = \frac{34y^2 - 51xy - 68x^2 - 51xy}{(2y-3x)^3}$$

$$\Rightarrow \quad \frac{d^2y}{dx^2} = \frac{-34\left(2x^2 + 3xy - y^2\right)}{(2y-3x)^3}$$

$$\Rightarrow \quad \frac{d^2y}{dx^2} = \frac{-34}{(2y-3x)^3} \qquad (\because 2x^2 + 3xy - y^2 = 1)$$

Example 5. **If $x^2 - xy + y^2 = a^2$ show that $\dfrac{d^2y}{dx^2} = \dfrac{6a^2}{(x-2y)^3}$**

Solution : We have,

$$x^2 - xy + y^2 = a^2$$

Differentiate w.r.t x both sides, we get

$$2x - \left(x\frac{dy}{dx} + y\right) + 2y\frac{dy}{dx} = 0$$

$$\Rightarrow \quad \frac{dy}{dx}(2y - x) = y - 2x$$

$$\Rightarrow \quad \frac{dy}{dx} = \frac{2x - y}{x - 2y} \qquad \text{(note)}$$

$$\Rightarrow \quad \frac{d^2y}{dx^2} = \frac{(x-2y)\left(2-\frac{dy}{dx}\right)-(2x-y)\left(1-2\frac{dy}{dx}\right)}{(x-2y)^2}$$

$$\Rightarrow \quad \frac{d^2y}{dx^2} = \frac{(2x-4y-2x+y)+\frac{dy}{dx}(-x+2y+4x-2y)}{(x-2y)^2}$$

$$\Rightarrow \quad \frac{d^2y}{dx^2} = \frac{-3y+\left(\frac{2x-y}{x-2y}\right)(3x)}{(x-2y)^2} \qquad \left(\because \frac{dy}{dx} = \frac{2x-y}{x-2y}\right)$$

$$\Rightarrow \quad \frac{d^2y}{dx^2} = \frac{-3xy+6y^2+6x^2-3xy}{(x-2y)^3} \Rightarrow \frac{d^2y}{dx^2} = \frac{6(x^2-xy+y^2)}{(x-2y)^3}$$

$$\Rightarrow \quad \frac{d^2y}{dx^2} = \frac{6a^2}{(x-2y)^3} \qquad (\because x^2-xy+y^2=a^2)$$

Example 6. If $ax^2 + 2hxy + by^2 = 1$ show that $\frac{d^2y}{dx^2} = \frac{h^2-ab}{(hx+by)^3}$.

Solution : We have, $ax^2 + 2hxy + by^2 = 1$

Differentiate w.r.t x both sides, we get

$$2ax + 2h\left(x\frac{dy}{dx}+y\right) + 2by\frac{dy}{dx} = 0$$

$$\Rightarrow \quad \frac{dy}{dx}(2hx+2by) = -2ax-2hy$$

$$\Rightarrow \quad \frac{dy}{dx} = -\left(\frac{ax+hy}{hx+by}\right)$$

$$\Rightarrow \quad \frac{d^2y}{dx^2} = -\left[\frac{(hx+by)\left(a+h\frac{dy}{dx}\right)-(ax+hy)\left(h+b\frac{dy}{dx}\right)}{(hx+by)^2}\right]$$

$$\Rightarrow \quad \frac{d^2y}{dx^2} = -\left[\frac{(ahx+aby-ahx-h^2y)+\frac{dy}{dx}(h^2x+hby-abx-hby)}{(hx+by)^2}\right]$$

$$\Rightarrow \quad \frac{d^2y}{dx^2} = -\left[\frac{(ab-h^2)y-\left(\frac{ax+hy}{hx+by}\right)(h^2-ab)x}{(hx+by)^2}\right]$$

$$\Rightarrow \quad \frac{d^2y}{dx^2} = -\left[\frac{(ab-h^2)\{hxy+by^2+ax^2+hxy\}}{(hx+by)^3}\right]$$

$$\Rightarrow \quad \frac{d^2y}{dx^2} = \frac{(h^2-ab)(ax^2+2hxy+by^2)}{(hx+by)^3}$$

$$\Rightarrow \quad \frac{d^2y}{dx^2} = \frac{h^2-ab}{(hx+by)^3} \qquad (\because ax^2+2hxy+by^2=1)$$

Example 7. If $y = \left[x+\sqrt{a^2+x^2}\right]^n$ show that $(a^2+x^2)y_2 + xy_1 - n^2y = 0$.

Solution : We have, $\quad y = \left[x+\sqrt{a^2+x^2}\right]^n \qquad \text{.... (1)}$

$$\Rightarrow \quad y_1 = n\left[x+\sqrt{a^2+x^2}\right]^{n-1}\frac{d}{dx}\left[x+\sqrt{a^2+x^2}\right]$$

$$\Rightarrow \quad y_1 = n\left[x+\sqrt{a^2+x^2}\right]^{n-1}\left[1+\frac{2x}{2\sqrt{a^2+x^2}}\right]$$

$$\Rightarrow \quad y_1 = n\left[x+\sqrt{a^2+x^2}\right]^{n-1}\left[\frac{\sqrt{a^2+x^2}+x}{\sqrt{a^2+x^2}}\right]$$

$$\Rightarrow \quad \sqrt{a^2+x^2}\cdot y_1 = n\left[x+\sqrt{a^2+x^2}\right]^n$$

$$\Rightarrow \quad \sqrt{a^2+x^2}\cdot y_1 = ny \qquad \text{(from (1))}$$

Squaring both sides we get

$$\Rightarrow \quad (a^2+x^2)\,y_1^2 = n^2y^2$$

Differentiating both sides w.r.t x we get

$$\Rightarrow \quad (a^2+x^2)\,2y_1\cdot y_2 + 2x\cdot y_1^2 = n^2\cdot 2y\cdot y_1$$

Dividing throughout by $2y_1$, we get

$$\Rightarrow \quad (a^2+x^2)\,y_2 + xy_1 - n^2y = 0$$

Example 8. If $y = \log\left(x+\sqrt{x^2+1}\right)$ show that $(x^2+1)\,y_2 + xy_1 = 0$.

Solution : We have, $y = \log\left(x+\sqrt{x^2+1}\right)$

$$\Rightarrow \quad y_1 = \frac{1}{x+\sqrt{x^2+1}}\left[1+\frac{2x}{2\sqrt{x^2+1}}\right]$$

$$\Rightarrow \quad y_1 = \frac{1}{x+\sqrt{x^2+1}}\left[\frac{\sqrt{x^2+1}+x}{\sqrt{x^2+1}}\right]$$

$$\Rightarrow \qquad y_1 = \frac{1}{\sqrt{x^2+1}}$$

$$\Rightarrow \qquad \sqrt{x^2+1}\cdot y_1 = 1 \quad \Rightarrow \quad (x^2+1)\,y_1^2 = 1$$

Differentiating both sides w.r.t x, we get

$$(x^2+1)\cdot 2y_1\cdot y_2 + 2xy_1^2 = 0$$

$$\Rightarrow \qquad (x^2+1)\,y_2 + xy_1 = 0 \qquad \text{(dividing throughout by } 2y_1)$$

Example 9. If $y = e^x \cdot \log x$ show that $xy_2 - (2x-1)\,y_1 + (x-1)\,y = 0$.

Solution : We have, $y = e^x \cdot \log x$

$$\Rightarrow \qquad \frac{dy}{dx} = e^x\cdot\frac{1}{x} + e^x\cdot\log x$$

$$\Rightarrow \qquad y_1 = \frac{e^x}{x} + y \qquad (\because\ e^x\cdot\log x = y)$$

$$\Rightarrow \qquad xy_1 = e^x + yx$$

Differentiating both sides w.r.t x we get

$$xy_2 + y_1\cdot 1 = e^x + y_1x + y$$

$$\Rightarrow \qquad xy_2 + y_1 = xy_1 - yx + xy_1 + y \qquad (\because\ e^x = xy_1 - yx)$$

$$\Rightarrow \qquad xy_2 + y_1 - 2xy_1 + xy - y = 0 \ \Rightarrow\ xy_2 - (2x-1)\,y_1 + (x-1)\,y = 0$$

Exercise

I. Find $\frac{d^2y}{dx^2}$, if

(a) $x = \log(4x-3)$, (b) $y = e^{3x+2}$ (c) $x = x^3\cdot\log x$

(d) $y = \log x + a^x$ (e) $y = x^m\cdot\log mx$ (f) $y = e^{ax}\cdot\log x$

(g) $y = 7^{(x^2+2)}$ (h) $y = 3x^3 + 4x^2 + 7$ (i) $y = \sqrt{2x+3}$

II. Find $\frac{d^2y}{dx^2}$, if

1. (a) $x = \log(1+t),\ y = \frac{1}{1+t}$ (b) $x = \log t,\ y = \frac{1}{t}$ (c) $x = \left(t - \frac{1}{t}\right),\ y = \left(t + \frac{1}{t}\right)$

2. If $x = \frac{e^\theta + e^{-\theta}}{2}$ and $y = \frac{e^\theta - e^{-\theta}}{2}$ show that $\frac{d^2y}{dx^2} = \frac{y^2 - x^2}{y^3}$.

3. If $x = \frac{2t}{1+t^2}$ and $y = \frac{1-t^2}{1+t^2}$ show that $y^3\frac{d^2y}{dx^2} + (x^2+y^2) = 0$.

4. If $x = \frac{t^2}{1-t^2}$ and $y = \frac{t^2}{1+t^2}$ show that $\frac{d^2y}{dx^2} = \frac{2y^2(y-x)}{x^4}$

III.

1. If $xy + 4y = 3x$ show that $\frac{d^2y}{dx^2} = \frac{-24}{(x+4)^3}$.

2. If $x^2 + xy + y^2 = a^2$ show that $\frac{d^2y}{dx^2} = -\frac{6a^2}{(x+2y)^3}$.

3. If $y^2 + 2y = x^2$ show that $y_2 = \dfrac{1}{(1+y)^3}$. **4.** If $x^2 + 2xy + 3y^2 = 1$ show that $y_2 = \dfrac{-2}{(x+3y)^3}$.

5. If $y = 2 + \log x$ show that $xy_2 + y_1 = 0$.

6. If $y = x + \sqrt{x^2+1}$ show that $(x^2 - 1)y_2 + xy_1 - y = 0$.

7. If $y = \left(x + \sqrt{x^2+1}\right)^m$ show that $(x^2 + 1)y_2 + xy_1 - m^2y = 0$.

8. If $y = (x^2 - 1)^n$ show that $(x^2 - 1)y_2 + 2x(1-n)\,y_1 - 2ny = 0$.

9. If $xy = ax^2 + \dfrac{b}{x}$ show that $x^2y_2 + 2(xy_1 - y) = 0$.

10. If $y^{\frac{1}{m}} + y^{-\frac{1}{m}} = 2x$ show that $(x^2 - 1)y_2 + xy_1 - m^2y = 0$.

11. If $y = (a^2 + x^2)^6$ show that $(x^2 + a^2)y_2 - 10\,xy_1 - 12y = 0$.

12. If $y = ax^{n+1} + \dfrac{b}{x^n}$ show that $x^2y_2 - n(n+1)y = 0$.

13. If $y = A\left(x + \sqrt{x^2-1}\right)^m + B\left(x - \sqrt{x^2-1}\right)^m$ show that $(x^2 - 1)y_2 + xy_1 - m^2y = 0$.

14. If $y = A\,e^{\sqrt{1+x}} + Be^{-\sqrt{1+x}}$ show that $4(1+x)y_2 + 2y_1 - y = 0$.

15. If $y = (A + Bt)\,e^{-nt}$ show that $y_2 + 2ny_1 + n^2y = 0$.

16. If $y = be^x + ce^{2x}$ show that $\dfrac{d^2y}{dx^2} - 3\dfrac{dy}{dx} + 2y = 0$.

Answers

I. (a) $\dfrac{-16}{(4x-3)^2}$ (b) $9y$ (c) $x(5 + 6\log x)$ (d) $-\dfrac{1}{x^2} + (\log a)^2 \cdot a^x$

(e) $\dfrac{1}{2}\,[m(m-1)\log mx + 2m - 1]\,x^{m-2}$ (f) $e^{ax}\left[\dfrac{2a}{x} - \dfrac{1}{x^2} + a^2\log x\right]$

(g) $(2\log 7)\,7^{x^2+2}\,[2x^2\log 7 + 1]$ (h) $18x + 8$ (i) $-\dfrac{1}{(2x+3)^{3/2}}$

II.1. (a) $\dfrac{1}{1+t}$ (b) $\dfrac{1}{t}$ (c) $\dfrac{4t^3}{a\left(1+t^2\right)^3}$

1.10 Maxima and Minima

In this section we shall discuss the function for its attainment of maximum or minimum at a point. To this end we have the following definitions.

Definition : A function $f(x)$ is said to have a maximum at a point $x = a$, if the value of the function at $x = a$ (i.e., $f(a)$), is greater than any other value that $f(x)$ can have for $a - h < x < a + h$, where h is a small positive real number. The value $f(a)$ is called the maximum value.

That is, $f(x)$ is said to have maximum value at $x = a$ if we can find a positive real number h such that $f(x) < f(a)$ for all x such that $a - h < x < a + h$ (except at a).

Thus, it follows that the function $f(x)$ will have maximum at $x = a$, if $f(x)$ is an increasing function for $x < a$ and a decreasing function for $x > a$ (Fig (i)).

Definition : A function $f(x)$ is said to have a minimum at a point $x = a$, if the value of the function at $x = a$ (i.e., $f(a)$) is less than any other value that $f(x)$ can have for $a - h < x < a + h$ where h is a small positive real number. The value $f(a)$ is called the minimum alue.

That is $f(x)$ is said to have minimum value at $x = a$ if we can find a positive real number h such that $f(x) > f(a)$ for all x such that $a - h < x < a + h$ (except at a).

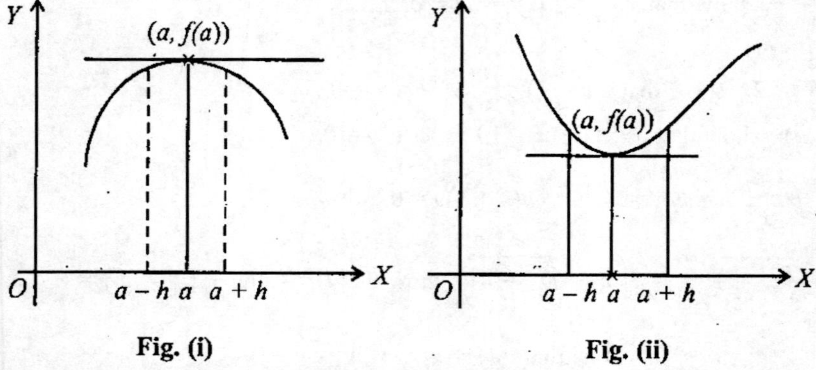

Fig. (i) **Fig. (ii)**

Thus it follows that the function $f(x)$ will have minimum at $x = a$, if $f(x)$ is a decreasing function for $x < a$ and an increasing function for $x > a$ (Fig. (ii)).

Note : The maximum and minimum values of a function $f(x)$ are not necessarily the greatest and least values of the function. Actually it is the greatest and the least values of the function relative to some interval sufficiently small and containing 'a'. For this reason the maximum and minimum values defined above are known as **local** (or **relative**) **maximum** and **local** (or **relative**) **minimum** of the function $f(x)$. A function may have several maxima and several minima. Further at some point maximum value of the function may be less than the minimum value at some other point. In fact in the fig. (iii) the function $f(x)$ attains maximum at $x = a$ and minimum at $x = b$, and the value of $f(x)$ at $x = b$ is great than the value at $x = a$.

In what follows we discuss only local maximum and/or local minimum.

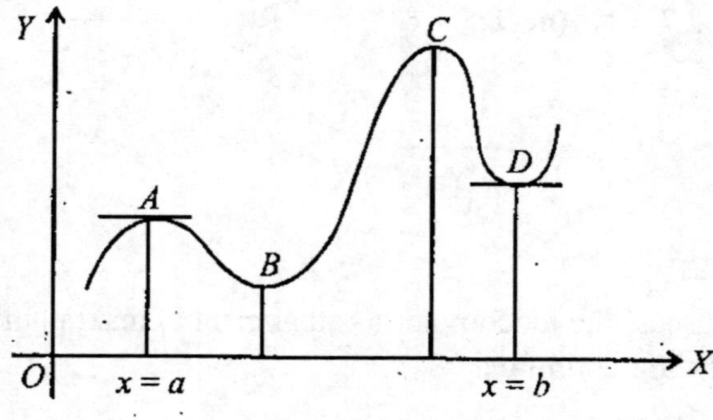

Fig. (iii)

The following theorem gives us a necessary condition for maxima or minima.

Theorem : If $f(x)$ be a differentiable function, defined for $x_1 < x < x_2$, has a relative maximum or minimum at $x = a$ $(x_1 < a < x_2)$ then $f'(a) = 0$.

Proof : Let $f(x)$ has a maximum at $x = a$.

$\Rightarrow$ $f(x)$ is an increasing function for $x < a$ and decreasing for $x > a$

$\Rightarrow$ $f'(x) > 0$ for $a - h < x < a$ and $f'(x) < 0$ for $a < x < a + h$ for some positive real number h.

$\Rightarrow$ $f'(x)$ changes from +ve to –ve as x passes through the value 'a'. But $f'(x)$ cannot change the sign without passing the value zero, which must be attained at $x = a$. Thus $f'(a) = 0$.

In a similar way it can be shown $f'(a) = 0$, if $f'(x)$ has a minimum at $x = a$.

Hence the proof.

Note : 1. The points where $f'(x) = 0$ are called **stationary points** or **critical points** or **turning points.** Thus, at stationary points the tangents drawn to curves are parallel to x-axis.

2. The points where $f'(x)$ is undefined (does not exist) but $f(x)$ is defined are also known as critical points – for example, one can discuss and show that $f(x) = |x|$ has minimum at $x = 0$ but $f'(0)$ does not exist.

3. The condition that $f'(x) = 0$ for maxima or minima is only necessary condition and not sufficient. For example, consider $f(x) = x^3$. For this curve $f'(x) = 3x^2$ and $f'(0) = 0$. But the function attains neither maxima nor minima at $x = 0$ (see Fig. (iv)).

Y X O

Fig. (iv)

The following theorem gives us a sufficient condition for a maximum and minimum.

Theorem : Let $f(x)$ be a differentiable function defined for $x_1 < x < x_2$ and at $x = a$ $(x_1 < a < x_2)$ let $f'(a) = 0$ and $f''(a)$ exists and is non zero, then $f(x)$ attains maximum if $f''(a) < 0$ and attains minimum if $f''(a) > 0$.

Proof : Let $f(x)$ has a maximum at $x = a$. Then $f(x)$ will be an increasing function for $x < a$ and decreasing function for $x > a$. That is $f'(x)$ changes sign from +ve to –ve at $x = a$. i.e., $f'(x)$ as a function of x, continuously decreases from +ve to –ve in neighbourhood of a.

Thus the its derivative, i.e., $\frac{d}{dx}(f'(x)) = f''(x)$ will be negative at $x = a$.

Thus, $f(x)$ attains maximum at $x = a$ if $f''(a) < 0$.

Similarly, it can be shown that $f(x)$ attains minimum at $x = a$ if $f''(a) > 0$. Hence the proof.

Thus, we have the following conditions for maxima and minima called **second derivative test.**

(i) If $f'(a) = 0$ and $f''(a) < 0$ then $f(x)$ has maximum at $x = a$, and $f(a)$ is the maximum value.

(ii) If $f'(a) = 0$ and $f''(a) > 0$ then $f(x)$ has minimum at $x = a$, and $f(a)$ is the minimum value.

Note : 1. The second derivative tests are only sufficient conditions, but not necessary. For example, for $f(x) = x^6, f'(x) = 6x^5$ and $f''(x) = 30x^4$. Now, $f''(0) = 0$. Thus, the second derivative test fails. However, $f'(a) < 0$ for $x < 0$ and $f'(x) > 0$ for $x > 0$. That is, $f'(x)$ change sign from –ve to +ve as x passes through 0. Thus, $f(x)$ attains maximum at $x = 0$.

2. Supposing for a function $f(x)$, we have $f'(a) = 0$ and $f''(a) = 0$, further $f'(x)$ does not change the sign as x passes through 'a', then $f(x)$ attains neither maximum nor minimum at $x = a$. Such a point is called **a point of inflection** (see Fig. (iv)).

Working rule to discuss maxima and minima for $f(x)$.

1. **Find $f'(x)$ and solve $f'(x) = 0$ for turining points. Let these points be $x = a, b, c, \ldots$**
2. **Find $f''(x)$ and its value at $x = a, b, c, \ldots$ separately**
3. **If $f''(a) < 0$, then decide that $f(x)$ attains maximum at $x = a$, and find the corresponding maximum value $f(a)$.**

   If $f''(a) > 0$, then decide that $f(x)$ attains minimum at $x = a$ and find the corresponding minimum value $f(a)$.

Similarly, for the points $x = a, b, c, \ldots$

If $f''(a) = 0$, then find the signs of $f'(x)$ for $x = a - h$ and for $x = a + h$, where h is a small +ve real number. If $f(a - h)$ is +ve and $f(a + h)$ is –ve then decide $f(x)$ attains maximum at $x = a$. If $f(a - h)$ is –ve and $f(a + h)$ is +ve then decide $f(x)$ attains minimum (this test is called "first derivative test").

If $f(a - h)$ and $f(a + h)$ both +ve or both –ve then decide $f(x)$ has neither maximum nor minimum. i.e., a is a point of inflection.

To decide a point of inflection we can also use the following test.

If $f'(a) = 0$, $f''(a) = 0$ but $f'''(a) \neq 0$, then $x = a$ is a point of inflection.

The following results are useful.

(i) The function $f(x) = \dfrac{k}{g(x)}$, where k is a non zero constant is maximum or minimum according as $g(x)$ is minimum or maximum.

(ii) The function $f(x) = \log(g(x))$ is maximum or minimum according as $g(x)$ is maximum or minimum

(iii) The functions $f(x) = k \cdot g(x)$ and $[g(x)]^k$ where k is a non zero constant are maximum or minimum according as $g(x)$ is maximum or minimum.

Example 1. Find the maxima and minima of the function

$$f(x) = 3x^3 - 9x^2 - 27x + 30.$$

Solution : Let, $f(x) = 3x^3 - 9x^2 - 27x + 30$

$\Rightarrow \quad f'(x) = 9x^2 - 18x - 27 \Rightarrow f'(x) = 9(x^2 - 2x - 3) \Rightarrow f'(x) = 9(x - 3)(x + 1)$

For maxima or minima, $f'(x) = 0 \Rightarrow (x - 3)(x + 1) = 0 \Rightarrow$ $\mathbf{x = 3}$ or $\mathbf{x = -1}$.

That is $f(x)$ attains maxima or minima at $x = 3$ and $x = -1$. That is $x = 3$ and $x = -1$ are stationary points or turning points of $f(x)$.

To decide whether the function attains maximum or minimum, at these points, we shall consider

$$f''(x) = 18x - 18$$

Now at $x = 3$, we have, $f''(3) = 54 - 18 = 36 > 0$

$\Rightarrow$ **$f(x)$ attains minimum at $x = 3$ and minimum value is**

$$f(3) = 3(27) - 9(9) - 27(3) + 30 \Rightarrow f(3) = \mathbf{-51}.$$

Again at $x = -1$, we have $f''(-1) = -18 - 18 = -36 < 0$

$\Rightarrow$ **$f(x)$ attains maximum at $x = -1$,** and **maximum value is**

$$f(-1) = 3(-1) - 9(1) - 27(-1) + 30 \Rightarrow \mathbf{f(-1) = 45.}$$

Example 2. Investigate the maximum and minimum values of the function $f(x) = x^5 - 5x^4 + 5x^3 - 1$.

Solution : Let; $f(x) = x^5 - 5x^4 + 5x^3 - 1$

$\Rightarrow$ $f'(x) = 5x^4 - 20x^3 + 15x^2 \Rightarrow f'(x) = 5x^2(x^2 - 4x + 3) \Rightarrow f'(x) = 5x^2(x-3)(x-1)$

For maxima or minima, $f'(x) = 0 \Rightarrow$ **$x = 0$, $x = 3$** or **$x = 1$**

Consider, $f''(x) = 20x^3 - 60x^2 + 30x = 10x(2x^2 - 6x + 3)$

Now at $x = 1$, $f''(1) = 10(2 - 6 + 3) = -10 < 0$.

$\Rightarrow$ **$f(x)$ attains maximum at $x = 1$;** max. value $= \mathbf{f(1) = 1 - 5 + 5 - 1 = 0}$.

Now at $x = 3$, $f''(3) = 30(18 - 18 + 3) = 90 > 0$

$\Rightarrow$ **$f(x)$ attains minimum at $x = 3$;** minimum value $= f(3) = 3^5 - 5 \cdot 3^4 + 5 \cdot 3^3 - 1 = \mathbf{28}$.

Now at $x = 0$, $f''(0) = 0$.

Consider, $f'''(x) = 60x - 120x + 30$ and $f'''(0) = 30 \neq 0$.*

Thus $f(x)$ attains neither maximum nor minimum at $x = 0$. That is, **$x = 0$ a point of inflection.**

Example 3. Find the points of maxima and minima, if any, for the following functions. Find the maximum and minimum values as the case may be

(a) $\dfrac{\log x}{x}$ $(x \neq 0)$ **(b)** $a \sin x + b \cos x$ **(c)** $2\cos x + x;\ 0 < x < \pi$

Solution : (a) Let $f(x) = \dfrac{\log x}{x} \Rightarrow f'(x) = \dfrac{x\left(\frac{1}{x}\right) - (\log x) \cdot 1}{x^2} \Rightarrow f'(x) = \dfrac{1 - \log x}{x^2}$

For maxima or minima $f'(x) = 0$.

$$\Rightarrow \frac{1 - \log x}{x^2} = 0 \Rightarrow 1 - \log x = 0 \Rightarrow \log_e x = 1 \Rightarrow \mathbf{x = e.}$$

Consider, $f''(x) = \dfrac{x^2\left(-\frac{1}{x}\right) - (1 - \log x)(2x)}{x^4} \Rightarrow f''(x) = \dfrac{-x - 2x + 2x\log x}{x^4} = \dfrac{2\log x - 3}{x^3}$

Now, $f''(e) = \dfrac{2\log e - 3}{e^3} \Rightarrow f''(e) = -\left(\dfrac{1}{e^3}\right) < 0$ $(\because \log e = 1$ and $e > 0)$

$\Rightarrow$ **$f(x)$ attains maximum at $x = e$.** Maximum value $= f(e) = \dfrac{\log e}{e} = \dfrac{1}{e}$

(b) Let, $f(x) = a\sin x + b\cos x \Rightarrow f'(x) = a\cos x - b\sin x$.

For maxima or minima, $f'(x) = 0 \Rightarrow a\cos x - b\sin x = 0$

Now, $a\cos x - b\sin x = 0 \Rightarrow b\sin x = a\cos x$

$$\Rightarrow \frac{\sin x}{a} = \frac{\cos x}{b} = \pm\frac{\sqrt{\sin^2 x + \cos^2 x}}{\sqrt{a^2 + b^2}} = \pm\frac{1}{\sqrt{a^2 + b^2}}$$

---

* One can also verify that $f'(x) > 0$ for $x < 0$ and also for $x > 0$. Infact, $f'(-.2) = .768$ and $f'(.2) = .448$ – both are +ve.

Thus we have two cases.

(i) $\sin x = \dfrac{a}{\sqrt{a^2+b^2}}$ and $\cos x = \dfrac{b}{\sqrt{a^2+b^2}}$ **(ii)** $\sin x = \dfrac{-a}{\sqrt{a^2+b^2}}$ and $\cos x = \dfrac{-b}{\sqrt{a^2+b^2}}$

Consider $f''(x) = -a \sin x - b \cos x = -(a \sin x + b \sin x)$

Now, for the set of values (i), we have $f''(x) < 0 \Rightarrow$ **$f(x)$ attains maximum for the values in (i).**

$$\text{Maximum value} = \frac{a^2}{\sqrt{a^2+b^2}} + \frac{b^2}{\sqrt{a^2+b^2}} = \frac{a^2+b^2}{\sqrt{a^2+b^2}} = \sqrt{a^2+b^2}$$

Again for the set of values (ii), we have $f''(x) > 0 \Rightarrow$ **$f(x)$ attains minimum for the values in (ii)**

$$\text{Minimum value} = \frac{-a^2}{\sqrt{a^2+b^2}} - \frac{b^2}{\sqrt{a^2+b^2}} = \frac{-(a^2+b^2)}{\sqrt{a^2+b^2}} = -\sqrt{a^2+b^2}$$

Second method.

Consider, $f(x) = a \cos x + b \sin x$

$$\Rightarrow \quad f(x) = \sqrt{a^2+b^2}\left(\frac{a}{\sqrt{a^2+b^2}}\cos x + \frac{b}{\sqrt{a^2+b^2}}\sin x\right)$$

$$\Rightarrow \quad f(x) = \sqrt{a^2+b^2}\,(\cos x \cdot \cos \alpha + \sin x \cdot \sin \alpha)$$

Where, $\cos \alpha = \dfrac{a}{\sqrt{a^2+b^2}}$ and $\sin \alpha = \dfrac{b}{\sqrt{a^2+b^2}}$

$$\Rightarrow \quad f(x) = \sqrt{a^2+b^2}\,\cos(x-\alpha)$$

We know that $-1 \le \cos \theta \le 1$ for all values of θ. $\therefore$ Maximum value of $\cos(x-\alpha)$ is 1.

Thus, the **maximum value of $f(x)$ is $\sqrt{a^2+b^2}$** $(\because \cos(x-\alpha) = 1)$

Again minimum value of $\cos(x-\alpha)$ is -1. Thus, the **minimum value of $f(x)$ is $-\sqrt{a^2+b^2}$.**

(c) Let, $f(x) = 2\cos x + x \Rightarrow f'(x) = -2\sin x + 1.$

For maxima or minima, $f'(x) = 0 \Rightarrow -2\sin x + 1 = 0 \Rightarrow \sin x = \dfrac{1}{2}$

Now, $\sin x = \dfrac{1}{2} \Rightarrow x = \dfrac{\pi}{6}, \dfrac{5\pi}{6}$ for $0 < x < \pi$. Now, $f''(x) = -2\cos x$

For $x = \dfrac{\pi}{6}$, we have, $f''\left(\dfrac{\pi}{6}\right) = -2\cos\dfrac{\pi}{6} \Rightarrow f''\left(\dfrac{\pi}{6}\right) = -2\left(\dfrac{\sqrt{3}}{2}\right) = -\sqrt{3} < 0.$

$\Rightarrow$ **$f(x)$ attains maximum at $x = \dfrac{\pi}{6}$; Maximum value** $= f\left(\dfrac{\pi}{6}\right) = 2\cos\dfrac{\pi}{6} + \dfrac{\pi}{6} = \sqrt{3} + \dfrac{\pi}{6}.$

Again at $x = \dfrac{5\pi}{6}$, we have $f''\left(\dfrac{5\pi}{6}\right) = -2\cos\dfrac{5\pi}{6} = -2\left(-\dfrac{\sqrt{3}}{2}\right) = \sqrt{3} > 0$ $\left(\because \cos\dfrac{5\pi}{6} = -\dfrac{\sqrt{3}}{2}\right)$

$\Rightarrow$ **$f(x)$ attains minimum at $x = \frac{5\pi}{6}$; Minimum value** $= f\left(\frac{5\pi}{6}\right) = 2\cos\frac{5\pi}{6} + \frac{5\pi}{6} = -\sqrt{3} + \frac{5\pi}{6}$.

Example 3. Show that x^x is minimum at $x = \frac{1}{e}$.

Solution : Let, $y = x^x \Rightarrow \log y = x \log x$

Differentiating both sides w.r.t x, we get

$$\frac{1}{y}\frac{dy}{dx} = x \cdot \frac{1}{x} + \log x \Rightarrow \frac{dy}{dx} = y(1 + \log x)$$

For maxima or minima, $\frac{dy}{dx} = 0 \Rightarrow y(1 + \log x) = 0$

$$\Rightarrow 1 + \log x = 0 \Rightarrow \log_e x = -1 \Rightarrow x = e^{-1}.$$

Now, $\frac{dy}{dx} = y(1 + \log x) \Rightarrow \frac{d^2y}{dx^2} = y \cdot \left(\frac{1}{x}\right) + (1 + \log x)\frac{dy}{dx}$

$$\Rightarrow \frac{d^2y}{dx^2} = \frac{y}{x} + (1 + \log x)\frac{dy}{dx}$$

Now, $\left(\frac{d^2y}{dx^2}\right)_{x=\frac{1}{e}} = \left(\frac{1}{e}\right)^{1/e} \cdot e + [1 + (-1)]\left(\frac{dy}{dx}\right)_{x=\frac{1}{e}}$ $\quad\left(\because \log\frac{1}{e} = -1\right)$

$$= \left(\frac{1}{e}\right)^{1/e} \cdot e > 0 \qquad \left(\because e > 0, \left(\frac{1}{e}\right)^{1/e} > 0\right)$$

$\Rightarrow$ **$y = x^x$ attains minimum at $x = \frac{1}{e}$,** and **minimum value** $= \left(\frac{1}{e}\right)^{1/e}$.

Example 4. Divide 16 into two parts such that their product will be maximum.

Solution : Let one part be x then the other part will be $16 - x$.

Consider, $f(x) = x(16 - x)$ (i.e., the product of the two parts)

We shall discuss for what value of x, $f(x)$ will be maximum

Now, $f'(x) = 16 - 2x$, $\quad (\because f(x) = 16x - x^2)$

For maxima or minimum, $f'(x) = 0 \Rightarrow 16 - 2x = 0 \Rightarrow x = 8$.

Now, $f''(x) = -2$ always negative. Thus $f(x)$ attains maximum at $x = 8$. Thus, the required numbers are **8 and 8.**

Example 5. A line AB of length 8 cm is divided into two parts AP and PB by a point P. Find the position of P if $AP^2 + BP^2$ is minimum.

Solution : $AP = x$ cm. Then $PB = AB - AP = (8 - x)$ cm

Consider, $AP^2 + PB^2 = x^2 + (8 - x)^2 = 2x^2 - 16x + 64$

Let, $f(x) = 2x^2 - 16x + 64 \Rightarrow f'(x) = 4x - 16$.

For maxima or minima, $f'(x) = 0 \Rightarrow x = 4$

Now, $f''(x) = 4$ is always positive. Thus, $f(x)$ attains minimum at $x = 4$.

Thus, $AP^2 + BP^2$ is minimum when $AP = PB = 4$ cm. Thus, **P is the mid point of AB.**

Example 6. Find the rectangle of greatest perimeter which can be inscribed in a circle of radius 'a' units.

Solution : Let $ABCD$ be the rectangle inscribed in the circle of radius 'a' units.

Let $AB = 2x$ and $BC = 2y$. From the figure we have,

$OE = y$, $EB = x$ and $OB = a$ $\Rightarrow$ $x^2 + y^2 = a^2$

$$\Rightarrow \quad y^2 = a^2 - x^2 \Rightarrow y = \sqrt{a^2 - x^2} \quad .... (1)$$

Now, the perimeter of $ABCD$ is given by

$$P = 4(x + y) \Rightarrow \frac{dP}{dx} = 4\left(1 + \frac{-2x}{2\sqrt{a^2 - x^2}}\right) = 4\left(1 - \frac{x}{\sqrt{a^2 - x^2}}\right)$$

$$\Rightarrow \frac{dP}{dx} = 4\left(\frac{\sqrt{a^2 - x^2} - x}{\sqrt{a^2 - x^2}}\right)$$

For maxima or minima $\frac{dP}{dx} = 0 \Rightarrow \sqrt{a^2 - x^2} - x = 0$

$$\Rightarrow \sqrt{a^2 - x^2} = x \Rightarrow 2x^2 = a^2 \Rightarrow x = \frac{a}{\sqrt{2}} \quad (\because x > 0)$$

Now, $$\frac{dP}{dx} = 4\left(1 - \frac{x}{\sqrt{a^2 - x^2}}\right)$$

$$\Rightarrow \frac{d^2P}{dx^2} = -4\left[\frac{\sqrt{a^2 - x^2} \cdot 1 - x \cdot \left(\frac{-2x}{2\sqrt{a^2 - x^2}}\right)}{a^2 - x^2}\right]$$

$$\Rightarrow \frac{d^2P}{dx^2} = -4\left[\frac{a^2 - x^2 + x^2}{(a^2 - x^2)\sqrt{a^2 - x^2}}\right] \Rightarrow \frac{d^2P}{dx^2} = -\frac{4a^2}{(a^2 - x^2)^{3/2}}$$

Now, at $x = \frac{a}{\sqrt{2}}$, $\frac{d^2P}{dx^2} = \frac{-4a^2}{\left(a^2 - \frac{a^2}{2}\right)^{3/2}} = -\text{ve} \quad \left(\because a^2 - \frac{a^2}{2} > 0\right)$

Thus, **P attains maximum at $x = \frac{a}{\sqrt{2}}$**

Now, (1) $\Rightarrow$ $y = \sqrt{a^2 - \frac{a^2}{2}}$ $\Rightarrow$ $y = \frac{a}{\sqrt{2}}$.

Thus, $AB = 2x = \sqrt{2}\,a$ and $BC = 2y = \sqrt{2}\,a$ $\Rightarrow$ $AB = BC$.

Hence, the rectangle of greatest perimeter which can be inscribed in a given circle is a square.

Example 7. Two sides of a triangle are given. Find the angle between them such the area shall be maximum.

Solution : Let the lengths of the given sides be 'a' and 'b' and let θ be the angle between them.

Then the area of the triangle is given by $A = \frac{1}{2}\,ab\sin\theta$ $\Rightarrow$ $\frac{dA}{d\theta} = \frac{1}{2}\,ab\cos\theta$

For maxima or minima, $\frac{dA}{d\theta} = 0 \Rightarrow \frac{1}{2}\,ab\cos\theta = 0 \Rightarrow \theta = \frac{\pi}{2}$

Now, $\frac{d^2A}{d\theta^2} = -\frac{1}{2}\,ab\sin\theta \Rightarrow \left(\frac{d^2A}{d\theta^2}\right)_{\theta=\frac{\pi}{2}} = -\frac{1}{2}\,ab < 0$

Thus, A attains maximum at $\theta = \frac{\pi}{2}$. Thus, the area will be maximum when $\theta = \frac{\pi}{2}$ i.e., when the triangle is a right angled triangle.

Alternative method.

Area is given by $A = \frac{1}{2}\,ab\sin\theta$.

A will be maximum when $\sin\theta$ is maximum, and it will be maximum when $\theta = \frac{\pi}{2}$. Thus, the triangle must be a right angled triangle.

Example 8. What is the largest size of rectangle that can be inscribed in a semicircle of radius 'a' units so that two vertices lie on the diameter.

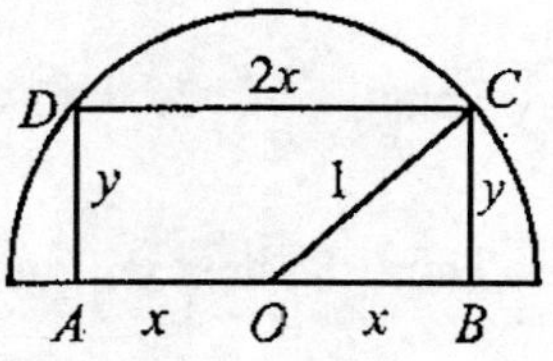

Solution : Let $ABCD$ be the rectangle inscribed in the given semicircle of radius 1 unit.

Let AB = length = $2x$ and AD = breadth = y

From the figure we have, $OB = x$, $BC = y$.

Thus, we have, $x^2 + y^2 = 1 \Rightarrow y^2 = 1 - x^2 \Rightarrow y = \sqrt{1 - x^2}$.

The area A of the rectangle is given by

$$A = 2x \cdot y \Rightarrow A = 2x \cdot \sqrt{1 - x^2} \qquad \left(\because y = \sqrt{1 - x^2}\right)$$

$$\Rightarrow A = \sqrt{4x^2\left(1 - x^2\right)} = \sqrt{4x^2 - 4x^4}$$

Let, $z = 4x^2 - 4x^4$. The maxima or minima of A is same as that of z.

Now, $z = 4x^2 - 4x^4 \Rightarrow \frac{dz}{dx} = 8\,(x - 2x^3)$

For maxima or minima, $\frac{dz}{dx} = 0 \Rightarrow x(1 - 2x^2) = 0 \Rightarrow x = \frac{1}{\sqrt{2}}$ $(\because x \neq 0)$

Again, $\frac{d^2z}{dx^2} = 8(1 - 6x) \Rightarrow \left(\frac{d^2z}{dx^2}\right)_{x = \frac{1}{\sqrt{2}}} = 8(1 - 3\sqrt{2}) < 0$

Example 9. A window in the form of a rectangle, surmounted by a semicircle. If the perimeter be 30 meters, find the dimensions so that the greatest possible amount of light may be admitted.

Solution : Let x be the radius of the semicircle. Then we have, $AB = DC = 2x$. Let $AD = BC = y$

By data perimeter = 30 metres i.e., $AD + DC + BC + \text{arc } APB = 30 \Rightarrow y + 2x + y + \frac{1}{2}(2\pi x)$

$\Rightarrow$ $2x + 2y + \pi x = 30 \Rightarrow \mathbf{2y = 30 - 2x - \pi x}$ .... (1)

The maximum amount of light will be admitted when the area of the figure is maximum. Let A be the area of the figure. Then

$$A = \text{Area } ABCD + \frac{1}{2}(\text{area of the circle})$$

$\Rightarrow$ $A = 2xy + \frac{1}{2}\pi x^2 \Rightarrow A = x(30 - \pi x - 2x) + \frac{1}{2}\pi x^2$ (from (1))

$\Rightarrow A = 30x - \pi x^2 - 2x^2 + \frac{1}{2}\pi x^2$

$\Rightarrow A = 30x - 2x^2 - \frac{1}{2}\pi x^2 \Rightarrow \frac{dA}{dx} = 30 - 4x - \pi x$

For maxima or minima, $\frac{dA}{dx} = 0 \Rightarrow 30 - 4x - \pi x = 0 \Rightarrow x = \frac{30}{\pi + 4}$

Now, $\frac{d^2A}{dx^2} = -4 - \pi = -(4 + \pi)$ which is always –ve. Thus, A attains maximum when $x = \frac{30}{\pi + 4}$.

Thus, $2y = 30 - \pi x - 2x \Rightarrow 2y = 30 - \frac{30\pi}{\pi + 4} - \frac{60}{\pi + 4} \Rightarrow 2y = \frac{60}{\pi + 4}$.

Thus, the area will be maximum, when $\mathbf{x = y = \frac{30}{\pi + 4}}$.

Example 10. The perimeter of a sector is l. Find the radius when the area of the sector is maximum.

Solution : Let r be the radius of the sector and θ be the angle subtended at the center. Thus, we have

Arc $AB = r\theta$. (θ measured in radians)

Now the perimeter l is given by

$$l = OA + AB + BO$$

$\Rightarrow$ $l = r + r\theta + r \Rightarrow l = 2r + r\theta \Rightarrow \theta = \frac{l - 2r}{}$ .... (1)

Now the area A of the sector is given by

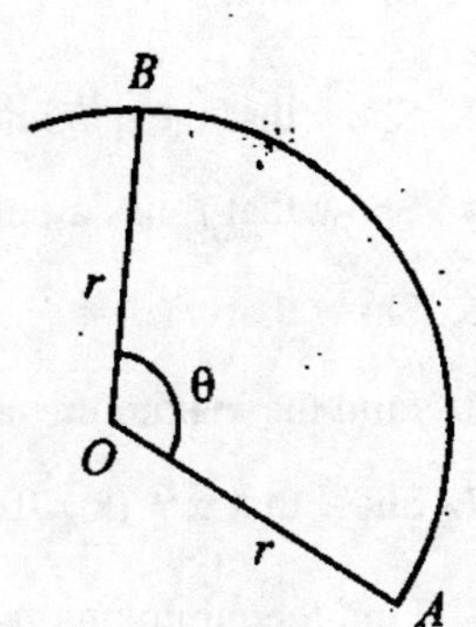

$$A = \frac{1}{2} r^2 \theta \Rightarrow A = \frac{1}{2} r^2 \left(\frac{l - 2r}{r}\right) \text{ (from (1))}$$

$$\Rightarrow A = \frac{1}{2} (lr - 2r^2) \Rightarrow \frac{dA}{dr} = \frac{1}{2} (l - 4r)$$

For maxima or minima, $\frac{dA}{dr} = 0 \Rightarrow \frac{1}{2} (l - 4r) = 0 \Rightarrow r = \frac{l}{4}$.

Now, $\frac{d^2 A}{dr^2} = \frac{1}{2} (-4) = -2 < 0$ always.

Thus, A attains maximum when $r = \frac{l}{4}$, i.e., when **radius** $= \frac{l}{4}$.

Example 11. The perimeter of a triangle is 10 cm. If one of the side is 4 cm, what are the other two sides for maximum area of the triangle?

Solution : The perimeter of a triangle is given by $2s = a + b + c$, where a, b and c are the lengths of the sides.

By data, $2s = 10 \Rightarrow s = 5$, let $a = 4$ cm $\therefore$ $b + c = 2s - a \Rightarrow \mathbf{b + c = 6}$

The area is given by

$$\Delta = \sqrt{s(s-a)(s-b)(s-c)} \Rightarrow \Delta = \sqrt{5(5-4)(s-b)(s-c)}$$

$$\Rightarrow \Delta = \sqrt{5(5-6+c)(5-c)} \quad (\because a = 6 - c \text{ and } s = 5)$$

$$\Rightarrow y = \Delta^2 = 5(c-1)(5-c) = 5(-c^2 + 6c - 5)$$

$$\Rightarrow \frac{dy}{dc} = 5(-2c + 6)$$

For maxima or minima $\frac{dy}{dc} = 0 \Rightarrow -2c + 6 = 0 \Rightarrow \mathbf{c = 3}$

Also, $\frac{d^2y}{dc^2} = -10 < 0$. Thus y attains maximum at $c = 3$.

That is area is maximum when $c = 3$ and $b = 6 - c \Rightarrow b = 3$.

Thus, the triangle must be an **isosceles triangle.**

Exercise

1. Determine the maximum and minimum values of the following functions.

(a) $9x^2 + 12x + 2$ (b) $2x^3 - 15x^2 + 36x + 10$ (c) $x^4 - 62x^2 + 120x + 9$

(d) $3x^5 - 25x^3 + 60x$ (e) $2x^3 - 21x^2 + 36x - 20$ (f) $\frac{(x-1)(x-6)}{x-10}$

(g) $\frac{x^2 + x + 1}{x^2 - x + 1}$ (h) $x\sqrt{1-x}, (x > 0)$ (i) $\sin x + \frac{1}{2}\cos 2x \ (0 < x < \pi)$

(j) $x^2 e^x$ (k) $12x^5 - 45x^4 + 40x^3 + 6$ (l) $\frac{4}{x^3 - 3x + 1}$

2. Show that $x^3 - 6x^2 + 12x - 3$ has neither a maximum nor a minimum at $x = 2$.

3. Show that the maximum value of $\left(\frac{1}{x}\right)^x$ is $e^{1/e}$

4. Show that f has a point of inflection at $x = 0$, where $f(x) = 4x^5 - 25x^4 + 40x^3 - 3$.

5. Show that $f(x) = \frac{2}{3}x^3 - 4x^2 + 10x + 2$ has neither a maximum nor a minimum.

6. Find the maximum value of $x \cdot e^{-x}$.

7. Show that $x^{1/x}$ $(x \neq 0)$ is maximum at $x = e$ and find the maximum value.

8. Find the minimum value of the function $x^2 + \frac{250}{x}$.

9. Find the maximum value of $3\cos x + \sqrt{3}\sin x, 0 < x < \pi$.

10. Show that the maximum value of $x \cdot \log x$ is $-\frac{1}{e}$.

11. Show that $x + \frac{1}{x}$ has maximum and minimum, and that maximum is less than minimum.

12. Divide the number 20 in to two parts such that their product is maximum.

13. Divide the number 84 into two parts such that the product of one part and the square of the other is maximum.

14. The product of two natural numbers is 36. Find the numbers if their sum is minimum.

15. Find two numbers whose sum is 12 and sum of whose cubes is minimum.

16. Find two positive numbers x and y such that $x + y = 2$ and x^3y is maximum.

17. Divide a given number in two parts such that the product of one part with the cube of the other is maximum.

18. Show that among the rectangles of given area the square has the least perimeter.

19. Show that among rectangles of given perimeter, the square has the greatest area.

20. Show that the rectangle of given perimeter which has the shortest diagonal is a square.

21. Show that the perimeter of a right angled triangle of a given hypotenuse, is maximum when the triangle is isosceles.

22. The sum of the hypotenuse and one side of right angled triangle is given. Show that the area of the triangle will be maximum when the angle between these sides is 60°.

23. A wire of given length is cut into two portions which are divided into the shapes of a circle and a square respectively. Show that the sum of the areas of the circle and the square will be least when the side of the square is equal to the diameter of the circle.

Answers

1. (a) $f\left(-\frac{2}{3}\right) = -2$, min. (b) $f(2) = 38$ max. $f(3) = 37$ min.

(c) $f(1) = 68$ max, $f(-6) = -1647$ min. $f(5) = -316$ min.

(d) $f(-2) = 92$ max. $f(2) = 16$ min. $f(-1) = -38$ min. $f(1) = 38$ max.

(e) $f(6) = -128$ min. $f(1) = -3$ min. (f) $f(16) = 25$ min. $f(4) = 1$ max

(g) $f(1)=3$ max. $f(-1)=\frac{1}{3}$ min. (h) $f\left(\frac{2}{3}\right)=\frac{2}{3\sqrt{3}}$ max. (i) $f\left(\frac{\pi}{2}\right)=\frac{1}{2}$ min. $f\left(\frac{\pi}{6}\right)=\frac{3}{4}$ max.

(j) $f(0)=0$ min. $f(-2)=\frac{4}{e^2}$ max. (k) $x=0$ point of inflection, $f(2)=-10$ min. $f(1)=13$ max.

(l) $f(1)=-4$ max. $f(-1)=\frac{4}{3}$ min. **6.** Max. value $\frac{1}{e}$ **7.** Max. value $e^{1/e}$ **8.** 75 **9.** $2\sqrt{3}$

12. 10 and 10 **13.** 56 and 28 **14.** 6 and 6 **15.** 6 and 6. **16.** $\frac{3}{2}$ and $\frac{1}{2}$ **17.** $\frac{3a}{4}, \frac{a}{4}$.

1.11 Partial Differentiation

We have so far dealt with functions of a single independent variable. We shall now consider the functions with more than one independent variables.

Definition : If for each pair of real values of the variable x and y a unique real number is associated to the variable z, then we say z is a function of two variables x and y. This we denote it by $z=f(x, y)$.

If $z=f(x, y)$ is a function of two variables, x and y are independent variables and z is the dependent variable.

Following are few examples of functions of two variables.

1. Area of rectangle is a function of two variables. i.e., length and the breadth.
2. Volume of a cylinder is given by $V=\pi r^2 h$. Thus V is a function of two variables r and h.
3. $z=(1-2xy+y^2)^{-1/2}$

In a similar manner, a function of three or more variables can be defined.

If u is a function of three variables, x, y and z then it is denoted by $u=f(x, y, z)$.

Consider a function $z=f(x, y)$ of two variables. The number 'l' is said to be the limit of the function $f(x, y)$ as $x \to a$ and $y \to b$, if the following condition holds.

$$\forall\, \varepsilon>0,\ \exists\, \delta>0 : |f(x, y)-l|<\varepsilon \text{ for } |x-a|<\delta \text{ and } |y-b|<\delta.$$

This we denoted by, $\lim\limits_{\substack{x\to a\\ y\to b}} f(x, y)=l.$

If this limit is equal to the value of the function $f(x, y)$ at $x=a, y=b$, then we say $f(x, y)$ is continuous at the point $x=a, y=b$.

Note : In what follows the functions of two or more variables are assumed to be continuous functions.

Partial Differentiation

In this section we shall see the process of differentiating the functions of two variables.

Consider a function $z=f(x, y)$ of two variables. Let this function be continuous in the domain of the definition.

Supposing one of the independent variables x or y, say y, is kept fixed and the variables x is allowed to vary, then z becomes a function of a single variable x alone. If this function possess the derivative then, this derivative is called the **partial derivative** of z w.r.t. x and is denoted by $\frac{\partial z}{\partial x}, f_x$ or $\frac{\partial f}{\partial x}$.

More explicitely,
$$\frac{\partial z}{\partial x} = \lim_{\delta x \to 0} \frac{f(x+\delta x, y) - f(x, y)}{\delta x}$$

provided the right hand limit exists.

Similarly the partial derivative of $z = f(x, y)$ w.r.t. y is also defined and is denoted by $\frac{\partial z}{\partial y}$, f_y or $\frac{\partial f}{\partial y}$.

These derivatives $\frac{\partial f}{\partial x}, \frac{\partial f}{\partial y}$ are called the **first order partial derivatives** of $z = f(x, y)$ w.r.t. x and w.r.t. y respectively.

Example 1. Find $\frac{\partial z}{\partial x}$ and $\frac{\partial z}{\partial y}$, if $z = 2x + 3xy + y^2 - 2$.

Solution :
$$z = x^2 + 3xy + y^2 - 2$$
$$\frac{\partial z}{\partial x} = 2x + 3y \text{ and } \frac{\partial z}{\partial y} = 3x + 2y$$

Example 2. If $z = 3\sin xy + 4\cos x - 1$, find $\frac{\partial z}{\partial x}$ and $\frac{\partial z}{\partial y}$.

Solution :
$$z = 3\sin xy + 4\cos x - 1$$
$$\frac{\partial z}{\partial x} = 3y\cos xy - 4\sin x \text{ and } \frac{\partial z}{\partial y} = 3x\cos xy$$

Example 3. If $z = e^{(x+2y)}$, find $\frac{\partial z}{\partial x}$ and $\frac{\partial z}{\partial y}$.

Solution :
$$z = e^{(x+2y)} \Rightarrow \frac{\partial z}{\partial x} = e^{x+2y} \text{ and } \frac{\partial z}{\partial y} = 2e^{x+2y}$$

Exercise

Find $\frac{\partial z}{\partial x}$ and $\frac{\partial z}{\partial y}$ for the following:

1. $z = 3x^2 - 2xy + 7y^2$
2. $z = 4xy - 3y^2$
3. $z = \sqrt{xy} - 2$
4. $z = \sin(x + 2y)$
5. $z = \cos(x \cdot y)$
6. $z = \tan(xy) + \cot x$
7. $z = \cos\sqrt{xy}$

Answers

1. $\dfrac{\partial z}{\partial x} = 6x - 2y, \qquad \dfrac{\partial z}{\partial y} = -2x + 14y$

2. $\dfrac{\partial z}{\partial x} = 4y, \qquad \dfrac{\partial z}{\partial y} = 4x - 6y$

3. $\dfrac{\partial z}{\partial x} = \dfrac{2y}{\sqrt{xy}}, \qquad \dfrac{\partial z}{\partial y} = \dfrac{2x}{\sqrt{xy}}$

4. $\dfrac{\partial z}{\partial x} = \cos(x + 2y) \qquad \dfrac{\partial z}{\partial y} = 2\cos(x + 2y)$

5. $\dfrac{\partial z}{\partial x} = y \sin(xy), \qquad \dfrac{\partial z}{\partial y} = -x \sin(xy)$

6. $\dfrac{\partial z}{\partial x} = y \sec^2(xy) - \text{cosec}^2 x, \qquad \dfrac{\partial z}{\partial y} = x \sec^2(xy)$

7. $\dfrac{\partial z}{\partial x} = -\cos\sqrt{xy} \cdot \left(\dfrac{y}{2\sqrt{xy}}\right), \qquad \dfrac{\partial z}{\partial y} = -\cos\sqrt{xy} \cdot \left(\dfrac{x}{2\sqrt{xy}}\right)$

UNIT VI

Integral Calculus

Chapter 1

Indefinite Integrals

1.1 Introduction

In this chapter we shall introduce the concept of integration of a function as an inverse process of differentiation and see techniques of integration. In fact the concept of integration has its origin in finding the areas of plane regions bounded by curves. However, later it was found it is the inverse process of differentiation and the subject is developed by finding different techniques of integration of different forms of functions. Further the term integration signifies "summation".

1.2 Indefinite Integral

Let $y = f(x)$ be a function such that $\frac{dy}{dx} = F(x)$. The function $F(x)$ is called the derivative or the differentiation coefficient of $f(x)$ w.r.t x. The function $f(x)$ itself is called **an integral, a primitive** or **an anti derivative** of $\boldsymbol{F(x)}$ **w.r.t** $\boldsymbol{x}$. This is denoted by $\int F(x)\,dx = f(x)$.

Thus, $$\int F(x)\,dx = f(x) \Leftrightarrow f'(x) = F(x)$$

We know that the derivative of a constant is zero.

$$\therefore \quad \frac{d}{dx}[f(x) + c] = \frac{d}{dx}[f(x)] = F(x) \quad \therefore \quad \boldsymbol{\int F(x)\,dx = f(x) + c}$$

The constant c is called a **constant of integration**. Further, as the constant c is arbitrary, the integral of a function is not unique. For this reason we say $f(x) + c$ is the **indefinite integral** of the function. The function $F(x)$ is called **integrand**.

Hence, we have,

$$\boldsymbol{\int F(x)\,dx = f(x) + c \Leftrightarrow f'(x) = F(x)}$$

The symbol $\int$ is called **integral sign**, it is an elongated S, which is the first letter of Summation.

Illustrations:

$$\frac{d}{dx}(x^2) = 2x \Leftrightarrow \int 2x\,dx = x^2 + c\,; \quad \frac{d}{dx}(\tan x) = \sec^2 x \Leftrightarrow \int \sec^2 x\,dx = \tan x + c.$$

1.3 Standard Integrals

We have defined integration as the inverse process of differentiation. Thus using the differential coefficients of basic elementary functions, which we have seen, we can prepare a list of integration of certain functions. This list is called the **list of elementary standard forms of integrals**.

List of Standard Integrals

1. $\int x^n\,dx = \dfrac{x^{n+1}}{n+1} + c\ (n \neq -1)$	$\because \dfrac{d}{dx}\left(\dfrac{x^{n+1}}{n+1}\right) = x^n$
2. $\int \dfrac{1}{x}\,dx = \log x + c$	$\because \dfrac{d}{dx}(\log x) = \dfrac{1}{x}$
3. $\int \dfrac{1}{x^n}\,dx = \dfrac{-1}{(n-1)\,x^{n-1}} + c\ (n \neq 1)$	$\because \dfrac{d}{dx}\left(\dfrac{-1}{(n-1)\,x^{n-1}}\right) = \dfrac{1}{x^n}$
4. $\int \dfrac{1}{x^2}\,dx = -\dfrac{1}{x} + c$	$\because \dfrac{d}{dx}\left(-\dfrac{1}{x}\right) = \dfrac{1}{x^2}$
5. $\int \dfrac{1}{\sqrt{x}}\,dx = 2\sqrt{x} + c$	$\because \dfrac{d}{dx}(2\sqrt{x}) = \dfrac{1}{\sqrt{x}}$
6. $\int e^x\,dx = e^x + c$	$\because \dfrac{d}{dx}(e^x) = e^x$
7. $\int a^x\,dx = \dfrac{a^x}{\log a} + c$	$\because \dfrac{d}{dx}(a^x) = \log a \cdot a^x$
8. $\int \sin x\,dx = -\cos x + c$	$\because \dfrac{d}{dx}(-\cos x) = \sin x$
9. $\int \cos x\,dx = \sin x$	$\because \dfrac{d}{dx}(\sin x) = \cos x$
10. $\int \sec^2 x\,dx = \tan x$	$\because \dfrac{d}{dx}(\tan x) = \sec^2 x$
11. $\int \operatorname{cosec}^2 x\,dx = -\cot x$	$\because \dfrac{d}{dx}(-\cot x) = \operatorname{cosec}^2 x$
12. $\int \sec x \cdot \tan x\,dx = \sec x$	$\because \dfrac{d}{dx}(\sec x) = \sec x \cdot \tan x$
13. $\int \operatorname{cosec} x \cdot \cot x\,dx = -\operatorname{cosec} x$	$\because \dfrac{d}{dx}(-\operatorname{cosec} x) = \operatorname{cosec} x \cdot \cot x$

Note : $\int k\,dx = kx + c$ where k is a constant. In particular $\int 1\,dx = x + c$.

The readers are advised to memories the above formulae for the reason, in evaluating the integral (in most of the cases), we bring the given integral to one of the standard forms given above and write the integral.

In the above set of standard integrals, if x is replaced by a linear expression $ax + b$, the same formula holds good with a factor $\frac{1}{a}$ on the right side.

i.e., if $\int f(x)\,dx = F(x) + c$ then $\int f(ax+b)\,dx = \frac{1}{a}F(ax+b) + c$.

Illustrations:

1. $\int e^{(3x+4)}\, dx = \frac{1}{3} e^{3x+4} + c$ **2.** $\int \sin 4x\, dx = -\frac{1}{4} \cos 4x + c$

3. $\int \frac{1}{2x-5}\, dx = \frac{1}{2} \log(2x-5) + c$ **4.** $\int \frac{1}{\sqrt{1-(4x)^2}}\, dx = \frac{1}{4} \sin^{-1} 4x + c$

Note : 1. From the definition of indefinite integral it follows that

(a) $\int \frac{d}{dx}[f(x)]\, dx = f(x) + c$ **(b)** $\frac{d}{dx} \int F(x)\, dx = \frac{d}{dx}[f(x) + c] = F(x)$.

Here $\frac{d}{dx}[f(x)] = F(x)$ and $\int F(x)\, dx = f(x) + c$.

2. If $f(x)$ is a differential function, then we usually write $\frac{d}{dx} f(x) = f'(x)$ .... (1)

We shall write this as $d[f(x)] = f'(x) \cdot dx$

We write (1) in this form only from the point of view of abbreviation. The symbol $d[f(x)]$ is called the "**differential of $f(x)$**" and dx is called the "**differential of x**".

With this notation our definition of integration can be expressed as

$$d f(x) = F(x)\, dx \quad \Leftrightarrow \quad \int F(x)\, dx = f(x)$$

For example,

$$dx^2 = 2x\, dx \quad \Leftrightarrow \quad \int 2x\, dx = x^2$$

$$d(\sin x) = \cos x \quad \Leftrightarrow \quad \int \cos x\, dx = \sin x$$

The differential of a function $f(x)$ is the product of the derivative of $f(x)$ w.r.t. x with the differential of x i.e., $d f(x) = f'(x) \cdot dx$

1.4 Two Basic Rules of Integration

1. If k is any non zero constant and $F(x)$ is a function, then $\int k \cdot F(x)\, dx = k \cdot \int F(x)\, dx$.

Proof : Let, $\int F(x)\, dx = f(x) \Rightarrow \frac{d}{dx}[f(x)] = F(x)$

Now, $\frac{d}{dx}[k \cdot f(x)] = k \cdot \frac{d}{dx}[f(x)] \Rightarrow \frac{d}{dx}[k \cdot f(x)] = k \cdot F(x)$

Thus, from the definition of integration, we have

$$\int k \cdot F(x)\, dx = k \cdot f(x) \Rightarrow \int \boldsymbol{k \cdot F(x)\, dx = k \cdot \int F(x)\, dx} \qquad \left(\because f(x) = \int F(x)\, dx\right)$$

2. If $F_1(x)$ and $F_2(x)$ are two functions, then, $\int [F_1(x) \pm F_2(x)]\, dx = \int F_1(x)\, dx \pm \int F_2(x)\, dx$

Proof : Consider,

$$\frac{d}{dx}\left[\int F_1(x)\, dx \pm \int F_2(x)\, dx\right] = \frac{d}{dx} \int F_1(x)\, dx \pm \frac{d}{dx} \int F_2(x)\, dx$$

$$\Rightarrow \quad \frac{d}{dx}\left[\int F_1(x)\, dx \pm \int F_2(x)\, dx\right] = F_1(x) \pm F_2(x) \qquad \text{(by Note (a))}$$

$$\Rightarrow \quad \int F_1(x) \pm F_2(x)\; dx = \int F_1(x)\, dx \pm \int F_2(x)\, dx \quad \text{(by the definition of Integration)}$$

Thus, the integration of the sum (or the difference) of two functions is the sum (or difference) of the integrals of the functions.

In general, we have

$$\int [k_1\, F_1(x) \pm k_2\, F_2(x) \pm k_3\, F_3(x) \pm \text{.....finite terms}]\, dx$$

$$= k_1 \int F_1(x)\, dx \pm k_2 \int F_2(x)\, dx \pm k_3 \int F_3(x)\, dx \pm \ldots\ldots$$

Example 1. Integrate the following w.r.t x

(a) $4x^2 - 2x + 1$ (b) $3 \cos x - 4$ (c) $3e^x + \dfrac{4}{x} + 7$ (d) $2 \sin x - 4 \sec^2 x + 1$.

Solution : Consider

(a) $$\int (4x^2 - 2x + 1)\, dx = \int 4x^2\, dx - \int 2x\, dx + \int 1 \cdot dx$$

$$= 4 \int x^2\, dx - 2 \int x\, dx + \int 1\, dx = 4\, \frac{x^3}{3} - 2 \cdot \frac{x^2}{2} + x + c$$

$$= \mathbf{\frac{4x^3}{3} - x^2 + x + c}.$$

(b) $$\int 3 \cos x - 4\, dx = 3 \int \cos x\, dx - 4 \int 1\, dx = \mathbf{3 \sin x - 4x + c}$$

(c) $$\int \left(3e^x + \frac{4}{x} + 7\right) dx = \int 3e^x\, dx + \int \frac{4}{x}\, dx + \int 7\, dx = 3 \int e^x\, dx + 4 \int \frac{1}{x}\, dx + 7 \int dx$$

$$= \mathbf{3e^x + 4 \log x + 7x + c}$$

(d) $$\int (2 \sin x - 4 \sec^2 x + 1)\, dx = 2 \int \sin x\, dx - 4 \int \sec^2 x\, dx + \int 1\, dx = \mathbf{-2 \cos x - 4 \tan x + x + c}$$

Example 2. Integrate the following w.r.t x

(a) $\dfrac{3x^2 + 7x + 12}{x^3}$ (b) $\left(x + \dfrac{1}{x}\right)\left(x^2 + \dfrac{1}{x^2}\right)$ (c) $\dfrac{4}{1 + x^2} + 9^x - 5 \operatorname{cosec}^2 x$.

Solution : Consider,

(a) $$\int \frac{3x^2 + 7x + 12}{x^3}\, dx = \int \left(\frac{3x^2}{x^3} + \frac{7x}{x^3} + \frac{12}{x^3}\right) dx$$

$$= 3 \int \frac{1}{x}\, dx + 7 \int \frac{1}{x^2}\, dx + 12 \int \frac{1}{x^3}\, dx$$

$$= 3 \log x - 7\, \frac{1}{x} + 12 \left(\frac{-1}{2x^2}\right) + c \qquad \left[\because \int \frac{1}{x^n}\, dx = \frac{-1}{(n-1)x^{n-1}}\right]$$

$$= \mathbf{3 \log x - \frac{7}{x} - \frac{6}{x^2} + c}.$$

(b) $$\int \left(x + \frac{1}{x}\right)\left(x^2 + \frac{1}{x^2}\right) dx = \int \left(x^3 + x + \frac{1}{x} + \frac{1}{x^3}\right) dx = \mathbf{\frac{x^4}{4} + \frac{x^2}{2} + \log x - \frac{1}{2x^2} + c}$$

(Integrating term by term)

(c) $\int\left(\frac{4}{1+x^2}+9^x-5\operatorname{cosec}^2 x\right)dx = 4\int\frac{1}{1+x^2}\,dx+\int 9^x\,dx-5\int\operatorname{cosec}^2 x\,dx$

$$= \mathbf{4\tan^{-1} x + \frac{9^x}{\log 9} + 5\cot x + c}.$$

Example 3. Integrate the following w.r.t x

(a) $(3x+2)^3$ **(b)** $\frac{5}{3x-7}$ **(c)** $\frac{1}{(2x+9)^{3/4}}$ **(d)** $4\sec 4x\tan 4x+\cos 2x$.

Solution : Consider,

(a) $\int(3x+2)^3\,dx = \frac{(3x+2)^4}{4}\cdot\left(\frac{1}{3}\right)+c = \mathbf{\frac{(3x+2)^4}{12}+c}$

(b) $\int\frac{5}{3x-7}\,dx = 5\int\frac{1}{3x-7}\,dx = \mathbf{\frac{5}{3}\log(3x-7)+k}$

(c) $\int\frac{1}{(2x+9)^{3/4}}\,dx = \frac{-1}{\left(\frac{3}{4}-1\right)(2x+9)^{(3/4-1)}}\cdot\frac{1}{2}+c = \mathbf{\frac{2}{(2x+9)^{-1/4}}+c}$

(d) $\int(4\sec 4x\cdot\tan 4x+\cos 2x)\,dx = 4\int\sec 4x\cdot\tan 4x\,dx+\int\cos 2x\,dx$

$$= 4\cdot(\sec 4x)\cdot\frac{1}{4}+(\sin 2x)\frac{1}{2}+c = \mathbf{\sec 4x+\frac{1}{2}\sin 2x+c}$$

Example 4. Integrate the following w.r.t x

(a) $\frac{2\cos 3x}{3\sin^2 3x}$ **(b)** $\frac{4\sin 2x}{5\cos^2 2x}$ **(c)** $\frac{x^2-1}{x^2+1}$

Solution : Consider,

(a) $\int\frac{2\cos 3x}{3\sin^2 3x}\,dx = \frac{2}{3}\int\frac{1}{\sin 3x}\cdot\frac{\cos 3x}{\sin 3x}\,dx = \frac{2}{3}\int\operatorname{cosec} 3x\cdot\cot 3x\,dx$

$$= -\frac{2}{3}\cdot\frac{1}{3}\operatorname{cosec} 3x+c = \mathbf{-\frac{2}{9}\operatorname{cosec} 3x+c}.$$

(b) $\int\frac{4\sin 2x}{5\cos^2 x}\,dx = \frac{4}{5}\int\frac{1}{\cos 2x}\cdot\frac{\sin 2x}{\cos 2x}\,dx = \frac{4}{5}\int\sec 2x\cdot\tan 2x\,dx$

$$= \frac{4}{5}\cdot\frac{1}{2}\text{ see } 2x+c = \mathbf{\frac{2}{5}\text{ see } 2x+c}.$$

(c) $\int\frac{x^2-1}{x^2+1}\,dx = \int\frac{(x^2+1)-2}{x^2+1}\,dx$ (note this step)

$$= \int\left(\frac{x^2+1}{x^2+1}-\frac{2}{x^2+1}\right)dx$$

$$= \int 1\,dx-2\int\frac{1}{x^2+1}\,dx = \mathbf{x-2\tan^{-1} x+c}$$

Example 5. Integrate the following w.r.t x

(a) $\sin^2 x$ **(b)** $\cos^3 x$ **(c)** $\tan^2 x$ **(d)** $\dfrac{1}{1+\cos x}$ **(e)** $\dfrac{1}{1-\sin x}$.

Solution : Consider,

(a) $\int \sin^2 x\, dx = \int \frac{1}{2}(1-\cos 2x)\, dx$ $(\because 2\sin^2 x = 1-\cos 2x)$

$= \frac{1}{2}\int (1-\cos 2x)\, dx = \frac{1}{2}\left(x - \frac{\sin 2x}{2}\right) + c$ (Integrating term by term)

(b) We have, $\cos 3x = 4\cos^3 x - 3\cos x \Rightarrow 4\cos^3 x = 3\cos x + \cos 3x$

$\therefore \int \cos^3 x\, dx = \frac{1}{4}\int (3\cos x + \cos 3x)\, dx = \frac{1}{4}\left(3\sin x + \frac{\sin 3x}{3}\right) + c.$

(c) $\int \tan^2 x\, dx = \int (\sec^2 x - 1)\, dx = \tan x - x + c. (\because 1+\tan^2 x = \sec^2 x)$

(d) $\int \frac{1}{1+\cos x}\, dx = \int \frac{1}{2\cos^2 \frac{x}{2}}\, dx$ $\left(\because 1+\cos x = 2\cos^2 \frac{x}{2}\right)$

$= \frac{1}{2}\int \sec^2 \frac{x}{2}\, dx = \frac{1}{2}\left(\frac{1}{(1/2)}\right) \tan \frac{x}{2} + c = \tan \frac{x}{2} + c$

Alternative method

$\int \frac{1}{1+\cos x}\, dx = \int \frac{1-\cos x}{1-\cos^2 x}\, dx = \int \frac{1-\cos x}{\sin^2 x}\, dx = \int \operatorname{cosec}^2 x\, dx - \int \operatorname{cosec} x \cdot \cot x\, dx$

$= -\cot x + \operatorname{cosec} x + c.$

(e) $\int \frac{1}{1-\sin x}\, dx = \int \frac{1+\sin x}{1-\sin^2 x}\, dx = \int \frac{1+\sin x}{\cos^2 x}\, dx = \int (\sec^2 x + \sec x \cdot \tan x)\, dx$

$= \tan x + \sec x + c.$

Example 6. Integrate the following w.r.t x

(a) $\sin 4x \cos 2x$ **(b)** $\sin 4x \sin 8x$ **(c)** $\dfrac{x^4+1}{x^2+1}$.

Solution : Consider,

(a) $\int \sin 4x \cdot \cos 2x\, dx = \frac{1}{2}\int [\sin(4x+2x) + \sin(4x-2x)]\, dx$

$[2\sin A \cdot \cos B = \sin(A+B) + \sin(A-B)$

$= \frac{1}{2}\int (\sin 6x + \sin 2x)\, dx$

$= \frac{1}{2}\left(-\frac{\cos 6x}{6}\right) + \frac{1}{2}\left(-\frac{\cos 2x}{2}\right) + c = -\frac{\cos 6x}{12} - \frac{\cos 2x}{4} + c.$

(b) Consider, $\sin 8x \sin 4x = \frac{1}{2}[\cos(8x-4x) - \cos(8x+4x)] = \frac{1}{2}[\cos 4x - \cos 12x]$

$\therefore \int \sin 8x \sin 4x\, dx = \frac{1}{2}\int (\cos 4x - \cos 12x)\, dx = \frac{1}{2}\left[\frac{\sin 4x}{4} - \frac{\sin 12x}{12}\right] + c.$

(c) Consider, $\dfrac{x^4+1}{x^2+1} = \dfrac{(x^2+1)^2 - 2x^2}{(x^2+1)} = (x^2+1) - \dfrac{2x^2}{x^2+1}$

$$= (x^2+1) - \frac{2(x^2+1)-2}{(x^2+1)} = (x^2+1) - 2 + \frac{2}{x^2+1}$$

$$\therefore \quad \int \frac{x^4+1}{x^2+1}\,dx = \int\left(x^2 - 1 + \frac{2}{x^2+1}\right)dx = \int x^2\,dx - \int dx + 2\int \frac{1}{1+x^2}\,dx$$

$$= \boldsymbol{\frac{x^3}{3} - x + 2\tan^{-1} x + c.}$$

Alternative method.

$$\int \frac{x^4+1}{x^2+1}\,dx = \int \frac{(x^4-1)+2}{x^2+1}\,dx \qquad \text{(Note this step)}$$

$$= \int \frac{(x^2-1)(x^2+1)+2}{(x^2+1)}\,dx = \int\left(x^2 - 1 + \frac{2}{1+x^2}\right)dx$$

$$= \boldsymbol{\frac{x^3}{3} - x + 2\tan^{-1} x + c.}$$

Exercise

1. Integrate the following w.r.t x

(a) $3x^2 - 4x + 1$ **(b)** $4x^{3/2} - 7\sqrt{x} + 1$ **(c)** $(3x-5)^{3/2}$ **(d)** $\sqrt{3x+7}$

(e) $\dfrac{1}{(3x-5)^3}$ **(f)** $\dfrac{1}{(4-2x)^{3/2}}$ **(g)** $\dfrac{x^3 - 4x^2 + 3x - 1}{\sqrt{x}}$ **(h)** $\left(x + \dfrac{1}{x}\right)^2$

(i) $\left(\sqrt{x} - \dfrac{1}{\sqrt{x}}\right)^2$ **(j)** $\dfrac{3x^3 - 4x^2 + 1}{\sqrt{x^3}}$ **(k)** $\dfrac{\sqrt{x} + \sqrt[3]{x^2}}{x}$.

2. Integrate the following w.r.t x

(a) $\sin(2-3x) + \sec 2x \cdot \tan 2x$ **(b)** $\operatorname{cosec} 4x \cdot \cot 4x + x^3$ **(c)** $\sec^2 3x - \operatorname{cosec}^2 4x + e^x$

(d) $\dfrac{\sin 4x}{\cos^2 4x}$ **(e)** $\dfrac{\cos 3x}{\sin^2 3x}$ **(f)** $4e^{3x} + 2e^{2x} - \dfrac{1}{x}$ **(g)** $\dfrac{3}{\sqrt{1-x^2}} + 4\sinh 3x$

(h) $\dfrac{3\cosh 2x}{\sinh^2 2x}$ **(i)** $\dfrac{4\sinh 3x}{\cosh^2 3x}$ **(j)** $\dfrac{4}{1+x^2} - 3e^{(3x+1)} + 4\cdot 5^{2x}$.

3. Integrate the following w.r.t x

(a) $\cos^2 x$ **(b)** $\sin^3 x$ **(c)** $\cot^2 x$ **(d)** $\sqrt{1+\cos x}$

(e) $\sqrt{1+\sin 2x}$ **(f)** $\dfrac{1}{1+\cos x}$ **(g)** $\dfrac{1+\cos x}{1-\cos x}$ **(h)** $\dfrac{1-\cos 2x}{1+\cos x}$

(i) $\dfrac{1}{\sin^2 x \cdot \cos^2 x}$ (j) $\dfrac{\sin^2 x - \cos^2 x}{\sin^2 x \cdot \cos^2 x}$ (k) $\dfrac{\sin x + \cos x}{\sqrt{1 + \sin 2x}}$ (l) $\dfrac{4 + 3 \sin x}{\cos^2 x}$

(m) $\sin^4 x$ (n) $\dfrac{\cos 2x + 2 \sin^2 x}{\cos^2 x}$ (o) $\sec^2 x \cdot \text{cosec}^2 x$ (p) $\dfrac{\sin x}{1 + \sin x}$

(q) $\dfrac{\sin^6 x + \cos^6 x}{\sin^2 x \cdot \cos^2 x}$.

4. Integrate the following w.r.t x

(a) $\dfrac{\text{cosec } x}{\text{cosec } x - \cot x}$ (b) $\dfrac{\sec x}{\sec x + \tan x}$ (c) $\dfrac{3e^{2x} + 3e^{4x}}{e^x + e^{-x}}$ (d) $\dfrac{x}{x + 4}$

(e) $\dfrac{2x^2}{x^2 + 1}$ (f) $\dfrac{x}{\sqrt{x + a}}$ (g) $\dfrac{x^2 - 1}{x^2 + 1}$ (h) $\dfrac{x^3}{2x + 1}$

(i) $\dfrac{1}{\sqrt{x + 1} - \sqrt{x - 2}}$.

5. Integrate the following w.r.t x

(a) $\sin 4x \cos 3x$ (b) $\cos 3x \cos 2x$ (c) $\cos 4x \sin x$ (d) $\sin 3x \sin x$.

Answers

Constant of inegration to be included.

1. (a) $x^3 - 2x^2 + 1 + c$ (b) $\dfrac{8}{5} x^{5/2} - \dfrac{14}{3} x^{3/2} + x + c$ (c) $\dfrac{2}{15}(3x - 5)^{5/2} + c$ (d) $\dfrac{2}{9}(3x + 7)^{3/2} + c$

(e) $\dfrac{1}{6(3x - 5)^2} + c$ (f) $\dfrac{1}{(4 - 2x)^{1/2}} + c$ (g) $\dfrac{2}{7} x^{7/2} - \dfrac{8}{5} x^{5/2} + 2x^{3/2} - 2\sqrt{x} + c$

(h) $\dfrac{x^3}{3} - \dfrac{1}{x} + 2x + k$ (i) $\dfrac{x^2}{2} + \log x - 2x + k$ (j) $\dfrac{6}{5} x^{5/2} - \dfrac{8}{3} x^{3/2} - \dfrac{2}{\sqrt{x}} + c$

(k) $2\sqrt{x} + \dfrac{3}{2} x^{2/3} + c$.

2. (a) $\dfrac{1}{3} \cos(2 - 3x) + \dfrac{1}{2} \sec 2x + k$ (b) $-\dfrac{1}{4} \text{cosec } 4x + \dfrac{x^4}{4} + k$

(c) $\dfrac{1}{3} \tan 3x + \dfrac{1}{4} \cot 4x + e^x + c$ (d) $\dfrac{1}{4} \sec 4x + c$ (e) $-\dfrac{1}{3} \text{cosec } 3x + c$

(f) $\dfrac{4}{3} e^{3x} + e^{2x} - \log x + c$ (g) $3 \sin^{-1} x + \dfrac{4}{3} \cosh 3x + c$ (h) $-\dfrac{3}{2} \text{cosech } 2x + k$

(i) $-\dfrac{4}{3} \text{sech } 3x + c$ (j) $4 \tan^{-1} x - e^{3x + 1} + 2 \cdot \dfrac{5^{2x}}{\log 5} + c$

3. (a) $\dfrac{1}{2}\left(x + \dfrac{\sin 2x}{2}\right) + c$ (b) $\dfrac{1}{4}\left(-3 \cos x + \dfrac{1}{3} \cos 3x\right) + c$ (c) $-\cot x - x + c$

(d) $2\sqrt{2} \sin \dfrac{x}{2} + c$ (e) $\sin x - \cos x + c$ (f) $\tan \dfrac{x}{2} + c$ (g) $-2 \cot \dfrac{x}{2} - x + c$

(h) $\tan x - x + c$ **(i)** $-2 \cot 2x + c$ **(j)** $\tan x + \cot x + c$ **(k)** $x + c$

(l) $4 \tan x + 3 \sec x + c$ **(m)** $\frac{1}{4}\left[x - \sin 2x + \frac{1}{2}x + \frac{\sin 4x}{8}\right] + k$ **(n)** $\tan x + c$

(o) $\tan x - \cot x + c$ **(p)** $\sec x - \tan x + x + c$ **(q)** $\tan x - \cot x - 3x + c.$

4. **(a)** $-\cot x - \text{cosec}\, x + c$ **(b)** $\tan x - \sec x + k$ **(c)** $3e^x + c$

(d) $x - 4 \log (x + 4) + k$ **(e)** $2x - 2 \tan^{-1} x + c$ **(f)** $\frac{2}{3}(x + a)^{3/2} - 2a\sqrt{x + a} + c$

(g) $x - 2 \tan^{-1} x + c$ **(h)** $\frac{x^3}{6} - \frac{x^2}{8} + \frac{x}{8} - \frac{1}{16}\log (2x + 1) + k$

(i) $\frac{2}{9}\left[(x + 1)^{3/2} + (x - 2)^{3/2}\right] + c.$

5. **(a)** $\frac{1}{2}\left(-\frac{\cos 7x}{7} - \cos x\right) + c$ **(b)** $\frac{1}{2}\left(\frac{\sin 5x}{5} + \sin x\right) + c$

(c) $\frac{1}{2}\left(-\frac{\cos 5x}{5} + \frac{\cos 3x}{3}\right) + c$ **(d)** $\frac{1}{2}\left(\frac{\sin 2x}{2} - \frac{\sin 4x}{4}\right) + c.$

1.5 Integration by Substitution

In the previous section, we have seen that during the process of integration, by some manipulation, we bring the given integral to one of the standard forms and finally we write the integral of the given function. In most of the cases we use this procedure. In fact there is no general method of integration as we have for differentiation (i.e., chain rule). The various procedures we adopt to integrate a function are called "techniques of integration". One such technique is called integration by the **method of substitution**.

In this method we reduce the given integral to one of the standard forms by changing the variable (say x) involved in the function to some other variable (say t) by means of substitution $x = g(t)$. During this process, the differential dx changes to $g'(t)\, dt$. For example consider

$$I = \int x \cdot e^{x^2}\, dx$$

Here we put $x^2 = t$. Correspondingly, $2x\, dx = dt \Rightarrow x\, dx = \frac{1}{2}\, dt.$

Thus, $I = \int e^{x^2} \cdot x\, dx = \frac{1}{2}\int e^t\, dt = \frac{1}{2} e^t + c$

Now, putting back $t = x^2$, we get

$$I = \int x \cdot e^{x^2}\, dx = \frac{1}{2} e^{x^2} + c$$

The most general **method of substitution** is illustrated below:

Supposing we know that $\int F(x)\, dx = f(x)$ .... (1)

and we have to evaluate $\int F(g(x)) \cdot g'(x)\, dx$ .... (2)

To this end we put, $g(x) = t \Rightarrow g'(x) \cdot dx = dt$

Thus, we have, $\int F(g(x)) \cdot g'(x)\, dx = \int F(t)\, dt = f(t) + c$ (from (1))

Now substituting back for t, we have, $\int F(g(x)) \cdot g'(x)\, dx = f(g(x)) + c$

We shall see three particular cases, which we come across most often.

1. $I_1 = \int [f(x)]^n f'(x)\, dx = \dfrac{[f(x)]^n}{n+1} + c \quad (n \neq -1)$

2. $I_2 = \int \dfrac{f'(x)}{f(x)}\, dx = \log [f(x)] + c$ 3. $I_3 = \int \dfrac{f'(x)}{\sqrt{f(x)}}\, dx = 2\sqrt{f(x)} + c.$

Each one can be proved by putting $f(x) = t \Rightarrow f'(x)\, dx = dt.$

Thus, we have,
$$I_1 = \int t^n\, dt = \frac{t^{n+1}}{n+1} + c = \frac{[f(x)]^{n+1}}{n+1} + c$$
$$I_2 = \int \frac{1}{t}\, dt = \log t + c = \log (f(x)) + c$$
$$I_3 = \int \frac{1}{\sqrt{t}}\, dt = 2\sqrt{t} + c = 2\sqrt{f(x)} + c.$$

In the formula $\int \dfrac{f'(x)}{f(x)}\, dx = \log [f(x)] + c$, we notice the function under the integral sign is of the form that the **numerator is the derivative of the denominator**. If that is the case then its integral is **log of the denominator.**

Using this fact, we evaluate the integrals of tan x, cot x, sec x and cosec x.

1. To evaluate $\int \tan x\, dx$

Consider, $\int \tan x\, dx = \int \dfrac{\sec x \tan x}{\sec x}\, dx$ (note)

$= \log (\sec x) + c \quad \left(\because \dfrac{d}{dx}(\sec x) = \sec x \cdot \tan x\right)$

Thus, $\int \tan x\, dx = \log (\sec x).$

2. To evaluate $\int \cot x\, dx$

Consider, $\int \cot x\, dx = \int \dfrac{\cos x}{\sin x}\, dx = \log (\sin x) + c \quad \left(\because \dfrac{d}{dx}(\sin x) = \cos x\right)$

Thus, $\int \cot x\, dx = \log (\sin x).$

3. To evaluate $\int \sec x\, dx$

Consider, $\int \sec x\, dx = \int \dfrac{\sec x (\sec x + \tan x)}{(\sec x + \tan x)}\, dx = \dfrac{\sec^2 x + \sec x \tan x}{\sec x + \tan x}\, dx$

Observe $\dfrac{d}{dx}(\sec x + \tan x) = \sec x \cdot \tan x + \sec^2 x$, which is the numerator.

Thus, $\int \sec x\, dx = \log (\sec x + \tan x) = \log \left[\tan \left(\dfrac{\pi}{4} + \dfrac{x}{2}\right)\right]$

4. To evaluate $\int \operatorname{cosec} x \, dx$

Consider, $\int \operatorname{cosec} x \, dx = \int \frac{\operatorname{cosec} x (\operatorname{cosec} x - \cot x)}{(\operatorname{cosec} x - \cot x)} dx = \int \frac{\operatorname{cosec}^2 x - \operatorname{cosec} x \cdot \cot x}{\operatorname{cosec} x - \cot x} dx$

Now, $\frac{d}{dx} (\operatorname{cosec} x - \cot x) = -\operatorname{cosec} x \cdot \cot x + \operatorname{cose}^2 x$

$\therefore \quad \int \operatorname{cosec} x \, dx = \log (\operatorname{cosec} x - \cot x) = \log \left(\tan \frac{x}{2} \right)$

Example 1. Integrate the following w.r.t x

(a) $\frac{x}{x^2 + 4}$ (b) $\frac{2x + 9}{x^2 + 9x + 10}$ (c) $\frac{\sec^2 x}{3 \tan x + 4}$ (d) $\frac{3e^x}{4e^x + 7}$.

Solution : Consider,

(a) $\int \frac{x}{x^2 + 4} dx = \frac{1}{2} \int \frac{2x}{x^2 + 4} dx$ (note this step)

$= \frac{1}{2} \log (x^2 + 4) + c$ $\left(\because \frac{d}{dx}(x^2 + 4) = 2x \right)$

(b) $\int \frac{2x + 9}{x^2 + 9x + 10} dx = \log (x^2 + 9x + 10) + c$ $\left(\because \frac{d}{dx}(x^2 + 9x + 10) = 2x + 9 \right)$

(c) $\int \frac{\sec^2 x}{3 \tan x + 4} dx = \frac{1}{3} \int \frac{3 \sec^2 x}{3 \tan x + 4} dx = \frac{1}{3} \log (3 \tan x + 4) + c$

$\left(\because \frac{d}{dx}(3 \tan x + 4) = 3 \sec^2 x \right)$

(d) $\int \frac{3e^x}{4e^x + 7} dx = \frac{3}{4} \int \frac{4e^x}{4e^x + 7} dx = \frac{3}{4} \log (4e^x + 7) + c$ **(note)**

Example 2. Integrate the following w.r.t x

(a) $(4x + 2) \sqrt{x^2 + x + 1}$ (b) $(4x^2 - 2x + 7)^{3/2} \cdot (4x - 1)$

(c) $\frac{\sin x}{(3 + 4 \cos x)^2}$ (d) $e^x \sqrt{3 e^x + 7}$.

Solution : Consider,

(a) $I = \int (4x + 2) \sqrt{x^2 + x + 1} \, dx = 2 \int \sqrt{x^2 + x + 1} \, (2x + 1) \, dx$

Put, $x^2 + x + 1 = t \Rightarrow (2x + 1) \, dx = dt$

$\therefore \quad I = 2 \int \sqrt{t} \cdot dt = 2 \cdot \frac{2}{3} \left(t^{3/2} \right) + c = \frac{4}{3} \left(x^2 + x + 1 \right)^{3/2} + c.$

Alternative method

$I = \int (4x + 2) \sqrt{x^2 + x + 1} \, dx = 2 \int \sqrt{x^2 + x + 1} \, (2x + 1) \, dx$

$$= 2\,\frac{\left(x^2+x+1\right)^{3/2}}{\left(3/2\right)} \qquad \left(\because \int [f(x)]^n\, f'(x)\,dx = \frac{[f(x)]^{n+1}}{n+1}\;k\right)$$

$$= \frac{4}{3}\left(x^2+x+1\right)^{3/2} + c.$$

(b) $\int \left(4x^2-2x+7\right)^{3/2}(4x-1)\,dx = \frac{1}{2}\int \left(4x^2-2x+7\right)^{3/2}(8x-2)\,dx$ (note this step)

$$= \frac{1}{2}\,\frac{\left(4x^2-2x+7\right)^{5/2}}{\left(5/2\right)} + c = \frac{1}{5}\left(4x^2-2x+7\right)^{5/2} + c.$$

Alternative method

Put, $4x^2-2x+7=t \;\Rightarrow\; (8x-2)\,dx = dt \;\Rightarrow\; 2(4x-1)\,dx = dt$

$\Rightarrow\; (4x-1)\,dx = \frac{1}{2}\,dt$

$$\therefore \quad \int \left(4x^2-2x+7\right)^{3/2}(4x-1)\,dx = \frac{1}{2}\int t^{3/2}\,dt = \frac{1}{2}\cdot\frac{2}{5}\,t^{5/2} + c = \frac{1}{5}\left(4x^2-2x+7\right)^{5/2} + c.$$

(c) $I = \int \frac{\sin x}{(3+4\cos x)^2}\,dx$; Put, $3+4\cos x = t \;\Rightarrow\; -4\sin x\,dx = dt$

$\Rightarrow\; \sin x\,dx = -\frac{1}{4}\,dt$

$$\therefore\; I = \left(-\frac{1}{4}\right)\int \frac{1}{t^2}\,dt = -\frac{1}{4}\left(-\frac{1}{t}\right) + c = \frac{1}{4(3+4\cos x)} + c.$$

(d) $I = \int e^x\sqrt{3e^x+7}\,dx = \frac{1}{3}\int \left(3e^x+7\right)^{1/2}\cdot 3e^x$ (note)

$$= \frac{1}{3}\,\frac{\left(3e^x+7\right)^{3/2}}{\left(3/2\right)} + c = \frac{2}{9}\left(3e^x+7\right)^{3/2} + c.$$

Example 3. Evaluate.

(a) $\int \frac{\sin^2(\log x)}{x}\,dx$ **(b)** $\int \sec^4 x\tan x\,dx$ **(c)** $\int \sec^4 x\tan x\,dx$ **(d)** $\int \cos^3 x\cdot\sin^4 x\,dx$.

Solution : Consider

(a) $I = \int \frac{\sin^2(\log x)}{x}\,dx = \int \sin^2(\log x)\cdot\frac{1}{x}\,dx$; Put, $\log x = t \;\Rightarrow\; \frac{1}{x}\,dx = dt$

$$\therefore\quad I = \int \sin^2 t\cdot dt = \frac{1}{2}\int (1-\cos 2t)\,dt = \frac{1}{2}\left(t - \frac{\sin 2t}{2}\right) + c = \frac{1}{2}\left(\log x - \frac{\sin(2\log x)}{2}\right) + c.$$

(b) $$I = \int \frac{\left(\tan^{-1}x\right)^3}{1+x^2}\,dx = \int \left(\tan^{-1}x\right)^3\cdot\frac{1}{1+x^2}\,dx = \frac{\left(\tan^{-1}x\right)^4}{4} + c.$$

(c) $I = \int \sec^4 x \tan x \, dx = \int \sec^3 x \cdot (\sec x \cdot \tan x) \, dx$

Put, $\sec x = t \Rightarrow \sec x \cdot \tan x \, dx = dt$

$\therefore \quad I = \int t^3 \cdot dt = \dfrac{t^4}{4} + c = \mathbf{\dfrac{\sec 4x}{4} + c}.$

(d) $I = \int \cos^3 x \cdot \sin^4 x \, dx = \int \sin^4 x \cdot \cos^2 x \cdot \cos x \, dx = \int \sin^4 x \left(1 - \sin^2 x\right) \cos x \, dx$

$= \int \left(\sin^4 x - \sin^6 x\right) \cos x \, dx$

Put, $\sin x = t \Rightarrow \cos x \, dx = dt$

$\therefore \quad I = \int \left(t^4 - t^6\right) dt = \dfrac{t^5}{5} - \dfrac{t^7}{7} + c = \mathbf{\dfrac{\sin^5 x}{5} - \dfrac{\sin^7 x}{7} + c}.$

Example 4. Evaluate

(a) $\int x \cdot \sin^3 \left(x^2\right) \cdot \cos \left(x^2\right) dx$ **(b)** $\int \dfrac{\tan \sqrt{x} \sec^2 \sqrt{x}}{\sqrt{x}} dx$ **(c)** $\int \dfrac{e^{\sin^{-1} x}}{\sqrt{1 - x^2}} dx$

(d) $\int e^x \cdot \sec \left(e^x\right) dx$ **(e)** $\int \dfrac{\cot (1 + \log x)}{x} dx$ **(f)** $\int \dfrac{(1 + x) e^x}{\sin^2 \left(x e^x\right)} dx$

(g) $\int \dfrac{\sec x \cdot \operatorname{cosec} x}{\log (\tan x)} dx$ **(h)** $\int \dfrac{1}{e^x + e^{-x}} dx$

Solution : Consider,

(a) $I = \int x \cdot \sin^3 \left(x^2\right) \cdot \cos \left(x^2\right) dx = \int \sin^3 \left(x^2\right) \cdot x \cos \left(x^2\right) dx$

Put, $\sin x^2 = t \Rightarrow 2x \cdot \cos x^2 \, dx = dt \Rightarrow x \cos (x^2) \, dx = \dfrac{1}{2} dt$

$\therefore \quad I = \dfrac{1}{2} \int t^3 \, dt \Rightarrow I = \dfrac{1}{2} \dfrac{t^4}{4} + c \Rightarrow \mathbf{I = \dfrac{1}{8} \sin^4 (x^2) + c}.$

(b) Consider $I = \int \dfrac{\tan \sqrt{x} \sec^2 \sqrt{x}}{\sqrt{x}} dx = \int \tan \sqrt{x} \cdot \left(\dfrac{\sec^2 \sqrt{x}}{\sqrt{x}}\right) dx$

Put, $\tan \sqrt{x} = t \Rightarrow \sec^2 \sqrt{x} \cdot \dfrac{1}{2 \sqrt{x}} dx = dt \Rightarrow \dfrac{\sec^2 \sqrt{x}}{\sqrt{x}} dx = 2 \, dt$

$\therefore \quad I = 2 \int t \, dt \Rightarrow I = 2 \cdot \dfrac{t^2}{2} + c \Rightarrow \mathbf{I = \tan^2 \sqrt{x} + c}.$

(c) $I = \int \dfrac{e^{\sin^{-1} x}}{\sqrt{1 - x^2}} dx = \int e^{\sin^{-1} x} \cdot \dfrac{1}{\sqrt{1 - x^2}} dx$; Put, $\sin^{-1} x = t \Rightarrow \dfrac{1}{\sqrt{1 - x^2}} dx = dt$

$\therefore \quad I = \int e^t \, dt \Rightarrow I = e^t + c = \mathbf{e^{\sin^{-1} x} + c}.$

(d) $I = \int e^x \cdot \sec \left(e^x\right) dx = \int \sec \left(e^x\right) \cdot e^x \, dx$; Put, $e^x = t \Rightarrow e^x \, dx = dt$

$\therefore \quad I = \int \sec t \, dt = \log (\sec t + \tan t) + c = \mathbf{\log (\sec e^x + \tan e^x) + c}.$

(e) $I = \int \frac{\cot(1+\log x)}{x}\,dx = \int \cot(1+\log x)\cdot\frac{1}{x}\,dx$; Put, $(1+\log x) = t \Rightarrow \frac{1}{x}\,dx = dt$

$\therefore$ $I = \int \cot t\,dt = \log(\sin t) + c = \mathbf{\log[\sin(1+\log x)) + c}.$

(f) $I = \int \frac{(1+x)\,e^x}{\sin^2(xe^x)}\,dx = \int \operatorname{cosec}^2(xe^x)\cdot(1+x)e^x\,dx$

Put, $x\cdot e^x = t \Rightarrow (xe^x + e^x)\,dx = dt \Rightarrow (1+x)\,e^x\,dx = dt$

$\therefore$ $I = \int \operatorname{cosec}^2 t\cdot dt = -\cot t + c = \mathbf{-\cot(xe^x) + c}.$

(g) $I = \int \frac{\sec x\cdot\operatorname{cosec} x}{\log(\tan x)}\,dx = \int \frac{1}{\log(\tan x)}\cdot\sec x\cdot\operatorname{cosec} x\,dx$

Put, $\log(\tan x) = t \Rightarrow \frac{1}{\tan x}\cdot\sec^2 x\,dx = dt \Rightarrow \frac{\cos x\cdot\sec^2 x}{\sin x}\,dx = dt$

$\Rightarrow (\operatorname{cosec} x\cdot\sec x)\,dx = dt$

$\therefore$ $I = \int \frac{1}{t}\,dt \Rightarrow I = \log t + c = \mathbf{\log[\log(\tan x)] + c}.$

(h) $I = \int \frac{1}{e^x + e^{-x}}\,dx = \int \frac{1}{e^{-x}(e^{2x}+1)}\,dx = \int \frac{e^x}{1+e^{2x}}\,dx$; Put, $e^x = t \Rightarrow e^x\,dx = dt$

$\therefore$ $I = \int \frac{1}{1+t^2}\,dt = \tan^{-1} t + c = \mathbf{\tan^{-1}(e^x) + c}.$

Example 5. Evaluate (a) $\int \tan^4 x\,dx$ **(b)** $\int \sin^5 x\,dx$ **(c)** $\int \frac{dx}{1+\cot x}$

Solution : Consider,

(a) $\int \tan^4 x\,dx = \int \tan^2 x\cdot\tan^2 x\,dx = \int \tan^2 x\left(\sec^2 x - 1\right)dx$

$$= \int \tan^2 x\cdot\sec^2 x\,dx - \int \tan^2 x\,dx$$

$$= \int (\tan x)^2\cdot\sec^2 x\,dx - \int\left(\sec^2 x - 1\right)dx$$

$$= \mathbf{\frac{(\tan x)^3}{3} - \tan x + x + c}.$$

$$\left(\because \int [f(x)]^n\, f'(x)\,dx = \frac{[f(x)]^{n+1}}{n+1}\right)$$

(b) $I = \int \sin^5 x\,dx = \int \sin^4 x\cdot\sin x\,dx = \int\left(1-\cos^2 x\right)^2\cdot\sin x\,dx$

Put, $\cos x = t \Rightarrow -\sin x\,dx = dt \Rightarrow \sin x\,dx = -dt$

$I = -\int\left(1-t^2\right)^2 dt = -\int\left(1 - 2t^2 + t^4\right)dt = -\left(t - \frac{2t^3}{3} + \frac{t^5}{5}\right) + c$

$$= \mathbf{-\left(\cos x - \frac{2\cos^3 x}{3} + \frac{\cos^5 x}{5}\right) + c}.$$

(c) $I = \int \frac{dx}{1 + \cot x} = \int \frac{\sin x}{\sin x + \cos x}\, dx$ $\left(\text{writing } \cot x = \frac{\cos x}{\sin x}\right)$

$$= \frac{1}{2} \int \frac{2 \sin x}{\sin x + \cos x}\, dx = \frac{1}{2} \int \frac{(\sin x + \cos x) + (\sin x - \cos x)}{(\sin x + \cos x)}\, dx \quad \text{(Note)}$$

$$= \frac{1}{2} \int 1\, dx - \frac{1}{2} \int \frac{\cos x - \sin x}{\sin x + \cos x}\, dx$$

$$= \mathbf{\frac{1}{2} x - \frac{1}{2} \log (\sin x + \cos x) + c}$$

$$\left(\because \int \frac{f'(x)}{f(x)}\, dx = \log f(x)\right)$$

Exercise

I. Integrate the following w.r.t x

(a) $\frac{x + 1}{x^2 + 2x + 4}$ **(b)** $\frac{x + 3}{x^2 + 6x + 7}$ **(c)** $\frac{4x + 2}{\sqrt{x^2 + x + 1}}$ **(d)** $\sqrt{x^2 + x + 1} \cdot (2x + 1)$

(e) $(2x - 3) \cdot \sqrt{x^2 - 3x + 2}$ **(f)** $\frac{x}{\sqrt{4 - x^2}}$ **(g)** $x^3 \sqrt{3 + 5x^4}$

(h) $\frac{2 \cos x - 3 \sin x}{6 \cos x + 4 \sin x}$ **(i)** $\frac{\left(\tan^{-1} x\right)^4}{1 + x^2}$.

II. Integrate the following w.r.t x

(a) $\frac{\sqrt{4 + \log x}}{x}$ **(b)** $\frac{1}{x \sqrt{1 + \log x}}$ **(c)** $\frac{\sqrt{2 + \log x}}{x}$

(d) $\frac{\sin^2 (\log x)}{x}$ **(e)** $\frac{\sec^2 (\log x)}{x}$ **(f)** $\frac{1}{x \sin^2 (1 + \log x)}$.

III. Integrate the following w.r.t x

(a) $x \sin (x^2 + 1)$ **(b)** $x^2 \sec x^3$ **(c)** $e^x \operatorname{cosec}^2 (e^x)$ **(d)** $x^2 e^{x^3} \cos \left(e^{x^3}\right)$

(e) $\frac{\cos \sqrt{x}}{\sqrt{x}}$ **(f)** $\frac{\sin \sqrt{x}}{\sqrt{x}}$ **(g)** $\frac{\sin \left(2 \tan^{-1} x\right)}{1 + x^2}$ **(h)** $\frac{x^3 \cdot \sin \left(\tan^{-1} x^4\right)}{1 + x^8}$

(i) $\frac{1 + \tan x}{x + \log \sec x}$ **(j)** $\sec x \cdot \log (\sec x + \tan x)$ **(k)** $\cos 3x \cdot \sqrt{2 + \sin 3x}$

(l) $\sin^3 x \cos^2 x$.

IV. Integrate the following w.r.t x

(a) $\frac{1}{1 + e^x}$ **(b)** $\frac{1}{e^x - 1}$ **(c)** $\frac{4}{3 + 7e^x}$ **(d)** $\frac{1}{e^x + e^{-x}}$ **(e)** $\frac{1}{\left(e^x + e^{-x}\right)^2}$

(f) $\frac{e^{2x}}{e^x + 1}$ **(g)** $\frac{e^x + e^{-x}}{e^x - e^{-x}}$ **(h)** $\frac{1}{\sqrt{2e^x - 1}}$ **(i)** $\frac{e^{2x} + 1}{e^{2x} - 1}$ **(j)** $\sqrt{e^x - 1}$.

V. Integrate the following w.r.t x

(a) $\dfrac{10x^9 + 10^x \log 10}{10^x + x^{10}}$ (b) $\dfrac{\tan x \cdot \sec^2 x}{1 + \tan^2 x}$ (c) $\dfrac{\cot x \cdot \operatorname{cosec}^2 x}{1 - \cot^2 x}$ (d) $\cos^5 x$

(e) $\sec^4 x \tan x$ (f) $\tan^3 2x \cdot \sec 2x$ (g) $\dfrac{\sqrt{\tan x}}{\sin x \cdot \cos x}$ (h) $\dfrac{1 - \tan x}{1 + \tan x}$

(i) $\dfrac{1}{x^2} \sin \dfrac{1}{x}$ (j) $\dfrac{1}{x^2} \cos \dfrac{1}{x}$ (k) $\dfrac{1}{x^2} \tan \dfrac{1}{x}$.

VI. Integrate the following w.r.t x

(a) $\dfrac{\sin 2x}{4 + 9 \sin^2 x}$ (b) $\dfrac{\sin x \cos x}{4 \cos^2 x + 9 \sin^2 x}$ (c) $\dfrac{\sin 2x}{1 + \sin^4 x}$ (d) $\dfrac{x^3}{\left(1 + x^2\right)^2}$

(e) $\sec^4 x$ (f) $\cot^4 x$ (g) $\dfrac{\tan x}{\sec x + \tan x}$ (h) $\dfrac{\sec^2 x \cdot \tan x}{\sec^2 x + \tan^2 x}$

(i) $\dfrac{\cos x - \sin x}{1 + \sin 2x}$ (j) $\dfrac{\cos x + \sin x}{1 - \sin 2x}$ (k) $\dfrac{1}{1 + \tan x}$ (l) $\dfrac{1}{1 + \cot x}$

(m) $\dfrac{\sin x}{\sin x - \cos x}$.

VII. Integrate the following w.r.t x

(a) $\dfrac{1}{\sqrt{x} + x}$ (b) $\dfrac{1}{x - \sqrt{x}}$ (c) $\dfrac{x}{\sqrt{1 + x}}$ (d) $\dfrac{2x - 1}{(x + 1)^2}$.

VIII. Integrate the following w.r.t x

(a) $\dfrac{\cot(\log x)}{x}$ (b) $\dfrac{1}{\sin x \cos^2 x}$ (c) $\tan^3 x$ (d) $\cot^3 x$

(e) $\sqrt{1 + 2 \tan x (\tan x + \sec x)}$.

Answers

Constant of integration to be included.

I. (a) $\dfrac{1}{2} \log (x^2 + 2x + 4)$ (b) $\dfrac{1}{2} \log (x^2 + 6x + 7)$ (c) $4\sqrt{x^2 + x + 1}$ (d) $\dfrac{2}{3}\left(x^2 + x + 1\right)^{3/2}$

(e) $\dfrac{2}{3}\left(x^2 - 3x + 2\right)^{3/2}$ (f) $-\sqrt{4 - x^2}$ (g) $\dfrac{1}{30}\left(3 + 5x^4\right)^{3/2}$ (h) $\dfrac{1}{2}(6 \cos x + 4 \sin x)$

(i) $\dfrac{1}{5}(\tan^{-1} x)^5$

II. (a) $\dfrac{2}{3}(4 + \log x)^{3/2}$ (b) $2\sqrt{1 + \log x}$ (c) $\dfrac{2}{3}(2 + \log x)^{3/2}$

(d) $\dfrac{1}{2}\left(\log x - \dfrac{\sin (2 \log x)}{2}\right)$ (e) $\tan (\log x)$ (f) $-\cot (1 + \log x)$

III. **(a)** $-\frac{1}{2}\cos(x^2+1)$ **(b)** $\frac{1}{3}\log(\sec x^3+\tan x^3)$ **(c)** $-\cot(e^x)$ **(d)** $\frac{1}{3}\sin\left(e^{x^3}\right)$

(e) $2\sin\sqrt{x}$ **(f)** $-2\cos\sqrt{x}$ **(g)** $-\frac{1}{2}\cos(2\tan^{-1}x)$ **(h)** $-\frac{1}{4}\cos(\tan^{-1}x^4)$

(i) $\log(x+\log\sec x)$ **(j)** $\frac{1}{2}[\log(\sec x+\tan x)]^2$ **(k)** $\frac{2}{9}(2+\sin 3x)^{3/2}$ **(l)** $-\frac{1}{3}\cos^3 x+\frac{1}{5}\cos^5 x$.

IV. **(a)** $-\log(e^{-x}+1)$ **(b)** $\log(1-e^{-x})$ **(c)** $-\frac{4}{3}\log(3e^{-x}+7)$ **(d)** $\tan^{-1}e^x$

(e) $\frac{-1}{2\left(e^{2x}+1\right)}$ **(f)** $e^x-\log(1+e^x)$ **(g)** $\log(e^x-e^{-x})$ **(h)** $2\sec^{-1}\left[\sqrt{2e^x}\right]$

(i) $x+\log(1-e^{-2x})$ **(j)** $2\sqrt{e^x-1}-2\tan^{-1}\sqrt{e^x-1}$.

V. **(a)** $\log(10^x+x^{10})$ **(b)** $\frac{1}{2}\log(1+\tan^2 x)$ **(c)** $\frac{1}{2}\log(1-\cot^2 x)$

(d) $\sin x-\frac{2}{3}\sin^3 x+\frac{1}{5}\sin^5 x$ **(e)** $\frac{\sec^4 x}{4}$ **(f)** $\frac{\sec^3 2x}{6}-\frac{1}{2}\sec 2x$

(g) $2\sqrt{\tan x}$ **(h)** $-\log\left[\sec\left(\frac{\pi}{4}-x\right)\right]$ **(i)** $\cos\frac{1}{x}$ **(j)** $-\sin\frac{1}{x}$

(k) $\log\left(\cos\frac{1}{x}\right)$.

VI. **(a)** $\frac{1}{9}\log(4+9\sin^2 x)$ **(b)** $\frac{1}{10}\log(4\cos^2 x+9\sin^2 x)$ **(c)** $\tan^{-1}(\sin^2 x)$

(d) $\log(1+x^2)+\frac{1}{1+x^2}$ **(e)** $\tan x+\frac{\tan^3 x}{3}$ **(f)** $-\frac{\cot^3 x}{3}+\cot x+x$

(g) $-\tan^{-1}(\cos x)$ **(h)** $\frac{1}{2}\log(1+2\tan^2 x)$ **(i)** $\frac{-1}{\cos x+\sin x}$

(j) $\frac{-1}{\sin x-\cos x}$ **(k)** $\frac{1}{2}x+\log(\sin x+\cos x)$ **(l)** $\frac{1}{2}x-\log(\sin x+\cos x)$

(m) $\frac{1}{2}x+\frac{1}{2}\log(\sin x-\cos x)$

VII. **(a)** $2\log\left(1+\sqrt{x}\right)$ **(b)** $2\log\left(\sqrt{x}-1\right)$ **(c)** $\frac{2}{3}(1+x)^{3/2}-2\sqrt{1+x}$

(d) $2\log x+\frac{3}{x+1}$

VIII. **(a)** $\log[\sin(\log x)]$ **(b)** $\sec x+\log(\operatorname{cosec} x-\cot x)$ **(c)** $\frac{\tan^2 x}{2}-\log(\sec x)$

(d) $-\frac{\cot^2 x}{2}-\log(\sin x)$ **(e)** $\log(\sec x+\tan x)+\log(\sec x)$.

1.6 Integration by Parts

In this section we shall see an another technique of integration called **integration by parts**.

Let u and v be two functions of x such that $v=\frac{dw}{dx}$.

We have from differential calculus $\frac{d}{dx}(u \cdot w) = u\frac{dw}{dx} + w\frac{du}{dx}$

Integrating both sides w.r.t x, we get,

$$\int \frac{d}{dx}(u \cdot w)\, dx = \int \left[u\frac{dw}{dx} + w\frac{du}{dx}\right] dx$$

$$\Rightarrow \qquad u\,w = \int u \cdot \frac{dw}{dx}\, dx + \int w \cdot \frac{du}{dx}\, dx$$

$$\Rightarrow \qquad \int u \cdot \frac{dw}{dx}\, dx = u \cdot w - \int w\frac{du}{dx}\, dx \qquad \text{.... (1)}$$

But we have, $\quad v = \frac{dw}{dx} \quad \Rightarrow \quad \int v\, dx = w$

Thus, (1) becomes

$$\int \boldsymbol{u \cdot v\, dx} = \boldsymbol{u \cdot \int v\, dx - \int\left[\int v\, dx\right] \cdot \frac{du}{dx}\, dx}$$

This result is known as formula for **integration by parts.**

In words,

Integral of the product of two functions = (1st function × integral of the 2nd)

– Integral of [integral of 2nd × diff. coeff. of 1st]

While applying the technique of integration by parts, care must be taken to choose the first function and the second function. By and large the following guidelines are use full.

1. If the product under the integral sign is of the form $x^n \cdot f(x)$, where n is a positive integer and the integral of $f(x)$ is directly known, then choose x^n as first function.

2. If the product under the integral sign is the product of x^n with a logarithmic function or with an inverse trigonometric function, then choose x^n as the second function.

3. Supposing we have a single function $f(x)$, like logarithmic function or inverse trignomic function, under the integral sign we write it as $f(x) \cdot 1$ and integrate it by parts by taking $f(x)$ as 1st function and 1 as 2nd function. $\left(\because \int 1\, dx = x\right)$.

Different examples worked out below will illustrate the guide lines.

Example 1. Evaluate

(a) $\int x \sin x\, dx$ (b) $\int x \sec^2 2x\, dx$ (c) $\int x^2 \cos^2 x\, dx$ (d) $\int x^2 \cdot e^{3x}\, dx$

(e) $\int x \tan^2 x\, dx$ (f) $\int x \sin^3 x\, dx$

Solution : Consider,

(a) $I = \int x \sin x\, dx$, Integrating by parts, we get,

$$= x \cdot \left(\int \sin x\, dx\right) - \int\left[\int \sin x\, dx\right] \cdot \frac{d}{dx}(x)\, dx$$

$$= x(-\cos x) - \int -\cos x \cdot 1\, dx = \boldsymbol{-x\cos x + \sin x + c}$$

$$\int u \cdot v\, dx = u \cdot \left(\int v\, dx\right) - \int \left(\qquad\right) \cdot \frac{du}{dx}\, dx$$

For example,

$$\int x \cdot \sin x\, dx = x \cdot (-\cos x) - \int (-\cos x) \cdot 1\, dx = \mathbf{-x \cos x + \sin x + c}$$

(b) $\int x \cdot \sec^2 2x\, dx = x \cdot \int \left(\sec^2 2x\right) dx - \int \left(\int \sec^2 2x\, dx\right) \cdot 1\, dx$

$$= x \cdot \frac{\tan 2x}{2} - \int \frac{\tan 2x}{2}\, dx = \frac{1}{2} x \tan 2x - \frac{1}{2} \cdot \frac{1}{2} \log(\sec 2x) + c$$

$$= \mathbf{\frac{1}{4}\left(2x \tan 2x - \log(\sec 2x)\right) + c}$$

(c) $\int x^2 \cdot \cos^2 x\, dx = \int x^2 \left(\frac{1 + \cos 2x}{2}\right) dx$

$$= \frac{1}{2}\int x^2\, dx + \frac{1}{2}\int x^2 \cos 2x\, dx$$

$$= \frac{x^3}{6} + \frac{1}{2}\left[x^2\left(\frac{\sin 2x}{2}\right) - \int \frac{\sin 2x}{2} \cdot 2x\, dx\right]$$

$$= \frac{x^3}{6} + \frac{1}{4} x^2 \sin 2x - \frac{1}{2}\int x \sin 2x\, dx$$

$$= \frac{x^3}{6} + \frac{1}{4} x^2 \sin 2x - \frac{1}{2}\left[x\left(-\frac{\cos 2x}{2}\right) - \int -\frac{\cos 2x}{2} \cdot 1\, dx\right]$$

$$= \mathbf{\frac{x^3}{6} + \frac{1}{4} x^2 \sin 2x + \frac{1}{4} x \cos 2x - \frac{1}{4}\frac{\sin 2x}{2} + c}$$

(d) $\int x^2 \cdot e^{3x}\, dx = x^2 \cdot \frac{e^{3x}}{3} - \int \frac{e^{3x}}{3} \cdot 2x\, dx = \frac{x^2 \cdot e^{3x}}{3} - \frac{2}{3}\int x \cdot e^{3x}\, dx$

$$= \frac{x^2 \cdot e^{3x}}{3} - \frac{2}{3}\left[x \cdot \frac{e^{3x}}{3} + \int \frac{e^{3x}}{3} \cdot 1\, dx\right]$$

$$= \mathbf{\frac{x^2 \cdot e^{3x}}{3} - \frac{2}{9} x e^{3x} - \frac{2}{27} e^{3x} + c} \qquad \left(\because \int e^{3x}\, dx = \frac{e^{3x}}{3}\right)$$

(e) $\int x \tan^2 x\, dx = \int x\left(\sec^2 x - 1\right) dx = \int x \sec^2 x\, dx - \int x\, dx,$

$$= x \cdot \tan x - \int \tan x \cdot 1\, dx - \frac{x^2}{2} \qquad \text{(integrating by parts)}$$

$$= \mathbf{x \cdot \tan x - \log(\sec x) - \frac{x^2}{2} + c}$$

(f) $\int x \sin^3 x\, dx = \frac{1}{4}\int x \cdot (3 \sin x - \sin 3x)\, dx \qquad (\because 4 \sin^3 x = 3 \sin x - \sin 3x)$

Now integrating by parts taking $(3 \sin x - \sin 3x)$ as 2nd function, we get

$$= \frac{1}{4}\left[x\left(-3\cos x + \frac{\cos 3x}{3}\right) - \int\left(-3\cos x + \frac{\cos 3x}{3}\right)\cdot 1\, dx\right]$$

$$= \frac{1}{4}\left[-3x\cos x + x\frac{\cos 3x}{3} + 3\sin x - \frac{\sin 3x}{9}\right] + c$$

Note : Consider $\int u \cdot v\, dx$. Integrating by parts, we get

$$\int u \cdot v\, dx = u \cdot \int v\, dx - \int\left[\int v\, dx\right]\cdot \frac{du}{dx}\, dx$$

Denoting, $\int v\, dx$ by v_1 and $\frac{du}{dx}$ by u', we have, $\int u \cdot v\, dx = u \cdot v_1 - \int u' \cdot v_1\, dx$

Again integrating the second term of r.h.s by parts, we get,

$$\int u \cdot v\, dx = u \cdot v_1 - \left[u' \cdot \int v_1\, dx - \int\left(\int v_1\, dx\right)\cdot \frac{du'}{dx}\, dx\right]$$

Denoting, $\int v_1\, dx$ by v_2 and $\frac{du'}{dx}$ by u'' we have,

$$\int u \cdot v\, dx = u \cdot v_1 - u' \cdot v_2 + \int u'' \cdot v_2\, dx$$

Continuing this way, we get a formula known as **generalised integration by parts:**

$$\int u \cdot v\, dx = u \cdot v_1 - u' \cdot v_2 + u''\, v_3 - u'''\, v_4 + \ldots\ldots$$

where u', u'', u''', ... denotes the successive differentiation of u and v_1, v_2, v_3, ... denotes the "successive integrations" of v.

The above formula has its own limitations. We can not apply it to any product $u \cdot v$ under the integral sign - for the reason the process of repeated application may not terminate at any stage. However, it is useful when the function under the integral sign is any one of the form, $x^n \cdot \sin ax$, $x^n \cdot \cos ax$ or $x^n \cdot e^{ax}$.

In what follows, we use the generalised integration parts where ever it suits.

Example 2. Evaluate

(a) $\int x^2 \cdot \sin 3x\, dx$ **(b)** $\int x^3 \cdot e^{2x}\, dx$ **(c)** $\int x \sec^2 x\, dx$ **(d)** $\int x \cos^2 x\, dx$

Solution : (a) $\int x^2 \cdot \sin 3x\, dx$, Here, $u = x^2$, $v = \sin 3x$

By generalised integration parts, we have,

$$= x^2 \cdot \left(-\frac{\cos 3x}{3}\right) - 2x \cdot \left(-\frac{\sin 3x}{9}\right) + 2\left(\frac{\cos 3x}{27}\right) + 0 + c$$

$$= -\frac{1}{3}x^2\cos 3x + \frac{2}{9}x\sin 3x + \frac{2}{27}\cos 3x + c$$

(b) $\int x^3 \cdot e^{2x}\, dx$, Here, $u = x^3$, $v = e^{2x}$

$$= x^3 \cdot \frac{e^{2x}}{2} - 3x^2 \cdot \frac{e^{2x}}{4} + 6x \cdot \frac{e^{2x}}{8} - 6 \cdot \frac{e^{2x}}{16} + 0 + c$$

$$= \left(\frac{1}{2}x^3 - \frac{3}{4}x^2 + \frac{3}{4}x - \frac{3}{8}\right)e^{2x} + c = \frac{1}{8}\left(4x^3 - 6x^2 + 6x - 3\right)e^{2x} + c$$

(c) $\int x \sec^2 x\, dx = x \cdot (\tan x) - 1 \cdot \log(\sec x) + 0 \cdot (\,) = \mathbf{x \tan x - \log(\sec x) + c}$

(d) $\int x \cos^2 x\, dx = \frac{1}{2}\int x(1 + \cos 2x)\, dx$; Here, $u = x$, $v = 1 + \cos 2x$

$$= \frac{1}{2}\left[x\left(x + \frac{\sin 2x}{2}\right) - 1 \cdot \left(\frac{x^2}{2} - \frac{\cos 2x}{4}\right)\right]$$

$$= \frac{1}{2}\left[x^2 + \frac{1}{2}x \sin 2x - \frac{x^2}{2} + \frac{\cos 2x}{4}\right] = \mathbf{\frac{1}{2}\left[\frac{x^2}{2} + \frac{x}{2}\sin 2x + \frac{\cos 2x}{2}\right]}$$

Example 3. Evaluate (a) $\int \log x\, dx$ **(b)** $\int \sin^{-1} x\, dx$ **(c)** $\int \tan^{-1} x\, dx$

Solution : (a) Consider, $I = \int \log x\, dx = \int \log x \cdot 1\, dx$

Integrating by parts taking 1 as second function we get,

$$I = \log x \cdot x - \int x \cdot \frac{1}{x}\, dx \Rightarrow I = x \log x - \int 1\, dx \Rightarrow I = \log x - x + c \Rightarrow \mathbf{I = x(\log x - 1) + c}$$

(b) Consider, $I = \int \sin^{-1} x\, dx = \int \sin^{-1} x \cdot 1\, dx$

Integrating by parts taking 1 as second functions we get,

$$I = \sin^{-1} x \cdot x - \int x \cdot \frac{1}{\sqrt{1 - x^2}}\, dx$$

$$\Rightarrow \quad I = x \sin^{-1} x + \frac{1}{2}\int \frac{(-2x)\, dx}{\sqrt{1 - x^2}} \qquad \text{(note this step)}$$

$$\Rightarrow \quad I = x \cdot \sin^{-1} x + \frac{1}{2} \cdot 2\sqrt{1 - x^2} \qquad \left(\because \int \frac{f'(x)}{\sqrt{f(x)}}\, dx = 2\sqrt{f(x)}\right)$$

$$\Rightarrow \quad \mathbf{I = x \sin^{-1} x + \sqrt{1 - x^2} + c}$$

(c) Consider, $I = \int \tan^{-1} x\, dx = \int \tan^{-1} x \cdot 1\, dx$

Integrating by parts taking 1 as second functions we get

$$I = \tan^{-1} x \cdot x - \int x \cdot \frac{1}{1 + x^2}\, dx$$

$$I = x \cdot \tan^{-1} x - \frac{1}{2}\int \frac{2x}{1 + x^2}\, dx \qquad \text{(note this step)}$$

$$\Rightarrow \quad \mathbf{I = x \tan^{-1} x - \frac{1}{2}\log\left(1 + x^2\right) + c} \qquad \left(\because \int \frac{f'(x)}{f(x)}\, dx = \log f(x)\right)$$

Exercise

I. Integrate the following w.r.t x

(a) $x \sin x$ **(b)** $x \cos 3x$ **(c)** $x \sin 3x$ **(d)** $x^2 \sin^2 x$ **(e)** $x \cos^3 x$

(f) $(1 + x) \cos 2x$ **(g)** $x^2 e^{2x}$ **(h)** $x^2 \cdot e^{3x}$ **(i)** $x \cot^2 x$

(j) $x^2 \log x$ **(k)** $x^2 (\log x)^2$ **(l)** $\frac{\log x}{x^2}$ **(m)** $(\log x)^2$ **(n)** $\log (1 + x^2)$

(o) $x^3 \cdot e^{x^2}$

II. Integrate the following w.r.t x

(a) $e^x \left(\frac{x-1}{x^2}\right)$ **(b)** $e^x \left(\frac{\sin x \cos x - 1}{\sin^2 x}\right)$ **(c)** $e^x \sec x (1 + \tan x)$

(d) $e^{-x} \left(\frac{\cos x - \sin x}{\cos^2 x}\right)$ **(e)** $e^{3x} \left(\frac{3 + \tan x}{\cos x}\right)$

Answers

I. **(a)** $-x \cos x - \sin x$ **(b)** $\frac{1}{3} x \sin 3x + \frac{1}{9} \cos 3x$ **(c)** $-\frac{1}{3} x \cos 3x + \frac{1}{9} \sin 3x$

(d) $\frac{1}{2}\left[\frac{x^3}{3} - \frac{x^2}{2} \sin 2x - \frac{x}{2} \cos 2x + \frac{1}{4} \sin 2x\right]$

(e) $\frac{1}{4}\left[3x \sin x + \frac{1}{3} x \sin 3x + 3 \cos x + \frac{\cos 3x}{9}\right]$ **(f)** $\frac{1}{2}(1 + x) \sin 2x - \frac{1}{4} \cos 2x$

(g) $\frac{1}{2} x^2 e^{2x} - \frac{1}{2} x e^{2x} + \frac{1}{4} e^{2x}$ **(h)** $\frac{1}{27}\left(9x^2 - 6x - 2\right) e^{3x}$

(i) $-x \cot x + \log \sin x - \frac{x^2}{2}$ **(j)** $\frac{x^3}{3} \log x - \frac{x^3}{9}$

(k) $\frac{1}{27}\left[9 (\log x)^2 - 6 \log x + 2\right] x^3$ **(l)** $-\frac{\log x}{x} - \frac{1}{x}$

(m) $(\log x - 1)^2 x$ **(n)** $x \log (1 + x^2) - 2x + 2 \tan^{-1} x$ **(o)** $\frac{1}{2}\left(x^2 - 1\right) e^{x^2} + c$

II. **(a)** $\frac{1}{x} e^x$ **(b)** $e^x \cot x$ **(c)** $e^x \sec x$ **(d)** $-e^{-x} \sec x$ **(e)** $e^{3x} \sec x$

Chapter 2

Definite Integral and its Applications

2.1 Definite Integral

Definition : Let $f(x)$ be a continuous function for $a \le x \le b$ and such that $\int f(x)\, dx = F(x)$, then the definite integral of $f(x)$ over $[a, b]$ is denoted and defined as

$$\int_a^b f(x)\, dx = F(b) - F(a)$$

Here a and b are respectively called lower limit and upper limit of the integral.

Working Rule to evaluate $\int_a^b f(x)\, dx$

Step 1. Evaluate $\int f(x)\, dx$, let it be $F(x)$ and write, $\int_a^b f(x)\, dx = F(x)\Big]_a^b$

Step 2. Find the value of $F(x)$ at $x = b$, then the value of $F(x)$ at $x = a$. That is find $F(b)$ and $F(a)$

Step 3. Write $\int_a^b f(x)\, dx = F(b) - F(a)$, which is the required value.

Example 1. Evaluate

(a) $\int_1^2 x^2\, dx$ **(b)** $\int_{\pi/6}^{\pi/3} \frac{\sin x}{\cos^2 x} dx$ **(c)** $\int_0^{\pi/3} \frac{2 + 3 \sin x}{\cos^2 x} dx$

(d) $\int_0^{\pi/2} \sin^2 x\, dx$ **(e)** $\int_0^{\pi/4} \tan^2 x\, dx$ **(f)** $\int_0^1 \frac{1 - x}{1 + x} dx$

Solution : Consider,

(a) $\int_1^2 x^2\, dx = \frac{x^3}{3}\Big]_1^2 = \frac{2^3}{3} - \frac{1^3}{3} = \frac{8}{3} - \frac{1}{3} = \mathbf{\frac{7}{3}}$

(b) $\int_{\pi/6}^{\pi/3} \frac{\sin x}{\cos^2 x} dx = \int_{\pi/6}^{\pi/3} \sec x \cdot \tan x\, dx = \sec x\Big]_{\pi/6}^{\pi/3} = \sec \frac{\pi}{3} - \sec \frac{\pi}{6} = \mathbf{2 - \frac{2}{\sqrt{3}}}$

(c) $\int_0^{\pi/3} \frac{2 + 3 \sin x}{\cos^2 x} dx = \int_0^{\pi/3} \left(2 \sec^2 x + 3 \sec x \cdot \tan x\right) dx$

$$= 2\tan x + 3\sec x\Big]_0^{\pi/3} = \left(2\tan\frac{\pi}{3} + 3\sec\frac{\pi}{3}\right) - (2\tan 0 + 3\sec 0)$$

$$= 2\sqrt{3} + 6 - 3 = \mathbf{2\sqrt{3} + 3} \qquad (\because \sec 0 = 1)$$

(d) $$\int_0^{\pi/2} \sin^2 x\,dx = \frac{1}{2}\int_0^{\pi/2} (1 - \cos 2x)\,dx$$

$$= \frac{1}{2}\left[x - \frac{\sin 2x}{2}\right]_0^{\pi/2} = \frac{1}{2}\left[\left(\frac{\pi}{2} - \frac{\sin\pi}{2}\right) - (0-0)\right] = \mathbf{\frac{\pi}{4}} \qquad (\because \sin\pi = 0)$$

(e) $$\int_0^{\pi/4} \tan^2 x\,dx = \int_0^{\pi/4}\left(\sec^2 x - 1\right)dx = [\tan x - x]_0^{\pi/4} = \left(1 - \frac{\pi}{4}\right) - 0 = \mathbf{1 - \frac{\pi}{4}}$$

(f) $$\int_0^1\left(\frac{1-x}{1+x}\right)dx = \int_0^1 \frac{2-(1+x)}{1+x}dx = 2\int_0^1 \frac{1}{1+x}dx - \int_0^1 1\,dx$$

$$= 2\left[\log(1+x)\right]_0^1 - [x]_0^1$$

$$= 2[\log 2 - \log 1] - (1-0) = \mathbf{2\log 2 - 1} \quad (\because \log 1 = 0)$$

Example 2. Evaluate

(a) $\int_0^{\pi} \sin 3x \cos 2x\,dx$ **(b)** $\int_0^4 \frac{dx}{\sqrt{16 - x^2}}$ **(c)** $\int_0^3 \frac{dx}{x^2+9}$ **(d)** $\int_0^{\pi/2}\sqrt{1 - \cos 2x}\,dx$

Solution : Consider,

(a) $$\int_0^{\pi} \sin 3x\cos 2x\,dx = \frac{1}{2}\int_0^{\pi}(\sin 5x + \sin x)\,dx$$

$$= \frac{1}{2}\left[-\frac{\cos 5x}{5} - \cos x\right]_0^{\pi}$$

$$= \frac{1}{2}\left[\left(-\frac{\cos 5\pi}{5} - \cos\pi\right) - \left(-\frac{1}{5} - 1\right)\right] \qquad (\because \cos 0 = 1)$$

$$= \frac{1}{2}\left[\left(\frac{1}{5} + 1\right) + \left(\frac{1}{5} + 1\right)\right] = \mathbf{\frac{6}{5}} \qquad (\because \cos 5\pi = -1)$$

(b) $$I = \int_0^4 \frac{dx}{\sqrt{16 - x^2}} = \int_0^4 \frac{dx}{\sqrt{4^2 - x^2}} = \sin^{-1}\frac{x}{4}\Big]_0^4 = \sin^{-1} 1 - \sin^{-1} 0 = \mathbf{\frac{\pi}{2}}$$

(c) $$I = \int_0^3 \frac{dx}{x^2 + 9} = \int_0^3 \frac{dx}{3^2 + x^2} = \frac{1}{3}\tan^{-1}\frac{x}{3}\Big]_0^3 = \frac{1}{3}(\tan^{-1} 1 - \tan^{-1} 0) = \mathbf{\frac{\pi}{12}}$$

(d) $$I = \int_0^{\pi/2}\sqrt{1 - \cos 2x}\,dx = \int_0^{\pi/2}\sqrt{2}\sin x\,dx \qquad (\because 1 - \cos 2x = 2\sin^2 x)$$

$$= \sqrt{2}(-\cos x)_0^{\pi/2} = \sqrt{2}\left[-\left(\cos\frac{\pi}{2} - \cos 0\right)\right] = \mathbf{\sqrt{2}}$$

$$\left(\because \cos\frac{\pi}{2} = 0,\ \cos 0 = 1\right)$$

Example 3. Evaluate

(a) $\int_0^{\pi} x \sin x\, dx$ (b) $\int_0^{\pi/2} x^2 \cos x\, dx$ (c) $\int_0^1 x^2 e^x\, dx$ (d) $\int_1^2 \log x\, dx$

Solution : In evaluating the integral we shall use the generalised integration by parts.

(a) $$I = \int_0^{\pi} x \sin x\, dx = \left[x(-\cos x) - 1\cdot(-\sin x)\right]_0^{\pi}$$

$$= \left[-x\cos x + \sin x\right]_0^{\pi}$$

$$= (-\pi\cos\pi + \sin\pi) - (0+0) = \pi \qquad (\because \cos\pi = -1,\ \sin\pi = 0)$$

(b) $$I = \int_0^{\pi/2} x^2 \cos x\, dx = x^2(\sin x) - 2x(-\cos x) + 2(-\sin x)\Big]_0^{\pi/2}$$

$$= \left(\frac{\pi^2}{4}\cdot 1 - 0 - 2\cdot 1\right) - (0+0-0) \qquad \left(\because \cos\frac{\pi}{2} = 0,\ \sin 0 = 0\right)$$

$$= \frac{\pi^2}{4} - 2$$

(c) $$I = \int_0^1 x^2 e^x\, dx = x^2 e^x - 2x e^x + 2e^x\Big]_0^1 = (1\cdot e^1 - 2\cdot e^1 + 2\cdot e^1) - (0 - 0 + 2) = e - 2$$

(d) $$I = \int_1^2 \log x\, dx = \int_1^2 (\log x)\cdot 1\, dx = \log x\cdot x\Big]_1^2 - \int_1^2 x\frac{1}{x}\, dx$$

$$= (2\log 2 - 0) - \int_1^2 1\, dx \qquad (\because \log 1 = 0)$$

$$= 2\log 2 - [x]_1^2 = \log 4 - (2-1) = \log 4 - 1$$

Alternative method.

Consider, $\int \log x\, dx = \int \log x\cdot 1\, dx = \log x\cdot x - \int x\frac{1}{x}\, dx = \log x\cdot x - x$

Now, $\int_1^2 \log x\, dx = [x\log x - x]_1^2 = (2\log 2 - 2) - (0 - 1) = \mathbf{\log 4 - 1}$

Note : When the method of substitution is used to evaluate the definite integral, the limits of integration also changes correspondingly.

Example 4. Evaluate

(a) $\int_0^{\pi/2} \frac{\cos\theta}{\sqrt{4-\sin^2\theta}}\, d\theta$ (b) $\int_0^{\pi/2} \frac{dx}{9 + 16\cos^2 x}$ (c) $\int_0^{\pi/2} \sqrt{\cos x}\cdot\sin^3 x\, dx$

Solution : Consider

(a) $$I = \int_0^{\pi/2} \frac{\cos\theta}{\sqrt{4-\sin^2\theta}}\, d\theta$$ Put, $\sin\theta = t \Rightarrow \cos\theta\, d\theta = dt$;

Also, $\theta = 0 \Rightarrow t = \sin 0 \Rightarrow t = 0;\ \theta = \frac{\pi}{2} \Rightarrow t = \sin\frac{\pi}{2} \Rightarrow t = 1.$

$\therefore \quad I = \int_0^1 \frac{1}{\sqrt{2^2 - t^2}}\, dt$ (note the changes in the limits of integration)

$= \sin^{-1}\frac{t}{2}\Big]_0^1 = \sin^{-1}\frac{1}{2} - \sin^{-1} 0 = \frac{\pi}{6}$ $(\because \sin^{-1} 0 = 0)$

(b) $I = \int_0^{\pi/2} \frac{dx}{9 + 16\cos^2 x} = \int_0^{\pi/2} \frac{dx}{\cos^2 x\left(9\sec^2 x + 16\right)} = \int_0^{\pi/2} \frac{\sec^2 x\, dx}{9 + 25\tan^2 x}$ $(\because \sec^2 x = 1 + \tan^2 x)$

Put $\tan x = t \Rightarrow \sec^2 dx = dt$. Also $\mathbf{x = 0} \Rightarrow t = \tan 0 \Rightarrow \mathbf{t = 0}$; $\mathbf{x = \frac{\pi}{2}} \Rightarrow t = \tan\frac{\pi}{2} \Rightarrow \mathbf{t = \infty}$.

$\therefore \quad I = \int_0^{\infty} \frac{dt}{3^2 + (5t)^2} = \frac{1}{3}\cdot\frac{1}{5}\tan^{-1}\frac{5t}{3}\Big]_0^{\infty} = \frac{1}{15}\left(\frac{\pi}{2} - 0\right) = \frac{\pi}{30}$

(c) $I = \int_0^{\pi/2} \sqrt{\cos x}\cdot\sin^3 x\, dx = \int_0^{\pi/2} \sqrt{\cos x}\cdot\sin^2 x \sin x\, dx = \int_0^{\pi/2} \sqrt{\cos x}\cdot\left(1 - \cos^2 x\right)\sin x\, dx$

Put, $\cos x = t \Rightarrow -\sin dx = dt$

Also, $\mathbf{x = 0} \Rightarrow t = \cos 0 \Rightarrow \mathbf{t = 1}$; $\mathbf{x = \frac{\pi}{2}} \Rightarrow t = \cos\frac{\pi}{2} \Rightarrow \mathbf{t = 0}$

$\therefore \quad I = \int_1^0 -\sqrt{t}\left(1 - t^2\right) dt = -\int_1^0 \left(\sqrt{t} - t^{5/2}\right) dt$

$= -\left(\frac{2}{3}t^{3/2} - \frac{2}{7}t^{7/2}\right)_1^0 = -\left[(0) - \left(\frac{2}{3} - \frac{2}{7}\right)\right] = \frac{8}{21}$

Exercise

I. Evaluate the following.

(a) $\int_0^1 \left(x^2 + 1\right) dx$ **(b)** $\int_1^2 \left(x + \frac{1}{x}\right)^2 dx$ **(c)** $\int_0^2 \left(4x^3 - 5x^2 + 6x\right) dx$ **(d)** $\int_0^1 \frac{1}{1 + x^2}\, dx$

(e) $\int_1^2 \frac{x^3 + 3x - 2}{x}\, dx$ **(f)** $\int_1^{\infty} \frac{1}{x^2}\, dx$ **(g)** $\int_0^4 \frac{x + 1}{\sqrt{x}}\, dx$ **(h)** $\int_0^1 \frac{x}{x + 1}\, dx$

(i) $\int_0^{\pi/4} \tan^2 x\, dx$ **(j)** $\int_0^{\pi} \cos x\, dx$ **(k)** $\int_0^{\pi/4} \sin 2x\, dx$ **(l)** $\int_0^{\pi/2} \cos^2 x\, dx$

(m) $\int_0^{\pi/2} \sin^3 x\, dx$ **(n)** $\int_0^{\pi/3} \tan x\, dx$ **(o)** $\int_0^{\pi/4} \sec x\, dx$ **(p)** $\int_0^3 \frac{dx}{\sqrt{x^2 + 9}}$

(q) $\int_0^5 \frac{dx}{25 + x^2}$ **(r)** $\int_0^{\pi/4} \sqrt{1 - \sin x}\, dx$

II. Evaluate the following.

(a) $\int_0^{\pi} x\cos x\, dx$ **(b)** $\int_0^1 x\, e^x\, dx$ **(c)** $\int_0^{\pi/2} x^2 \sin x\, dx$

(d) $\int_0^{\pi/2} x \sin\frac{x}{2}\cos\frac{x}{2}\, dx$ **(e)** $\int_0^{\pi} x\sin^2 x\, dx$ **(f)** $\int_0^{\pi/3} x\tan^2 x\, dx$ **(g)** $\int_1^2 x\log x\, dx$

III. Evaluate the following.

(a) $\int_0^{\pi/2} \cos 5x \cos 2x \, dx$ **(b)** $\int_0^{\pi/2} \sin 2x \sin x \, dx$ **(c)** $\int_0^{\pi/2} \sin 4x \cos 2x \, dx$

(d) $\int_0^{\pi/2} \cos^3 x \sin x \, dx$ **(e)** $\int_0^{\pi/4} \tan^4 x \sec^2 x \, dx$ **(f)** $\int_0^1 \sqrt{\frac{\cos^{-1} x}{1 - x^2}} \, dx$

(g) $\int_{1/e}^{e} \frac{(\log x)^2}{x} \, dx$ **(h)** $\int_0^{\pi/4} \frac{\left(\tan^{-1} x\right)^2}{1 + x^2} \, dx$ **(i)** $\int_0^{\pi/2} \frac{\sin \theta}{\sqrt{1 + \cos \theta}} \, d\theta$

(j) $\int_0^{\pi/2} (3 \sin x - 4)^2 \cos x \, dx$ **(k)** $\int_0^{\pi/2} \frac{\sin x}{1 + \cos^2 x} \, dx$ **(l)** $\int_0^1 x \, e^{x^2} \, dx$

(m) $\int_0^{\pi/2} \frac{\cos x}{\sqrt{4 - \sin^2 x}} \, dx$ **(n)** $\int_1^2 \frac{dx}{(x + 1)(x + 2)}$ **(o)** $\int_0^{\pi/2} \frac{\cos x \, dx}{(1 + \sin x)(2 + \sin x)}$

IV. Evaluate the following.

(a) $\int_0^1 x^2 \sqrt{1 - x^2} \, dx$ **(b)** $\int_0^{\pi/2} \frac{\sin x \cos x}{1 + \sin^4 x} \, dx$ **(c)** $\int_0^1 \frac{\cos\left(\tan^{-1} x\right)}{1 + x^2} \, dx$

(d) $\int_0^1 \frac{dx}{e^x + e^{-x}}$ **(e)** $\int_4^9 \frac{1 - \sqrt{x}}{1 + \sqrt{x}} \, dx$ **(f)** $\int_0^{\pi/4} \tan^4 x \, dx$

(g) $\int_0^{\pi/4} \sec^4 x \, dx$ **(h)** $\int_0^{\pi/2} \frac{\sin 2x}{\sin^4 x + \cos^4 x} \, dx$ **(i)** $\int_0^{\pi/2} \frac{4 \sin x + 3 \cos x}{\cos x + \sin x} \, dx$

(j) $\int_0^{\pi/2} \frac{dx}{a^2 \cos^2 x + b^2 \sin^2 x}$ **(k)** $\int_{-\pi/4}^{\pi/4} \frac{1}{1 + \sin x} \, dx$ **(l)** $\int_0^a \frac{dx}{x + \sqrt{a^2 - x^2}}$

(m) $\int_0^{\pi} \frac{dx}{3 + 2 \cos x}$ **(n)** $\int_0^{\pi/2} \frac{\sin x \cos x}{\cos^2 x + 3 \cos x + 2} \, dx$ **(o)** $\int_0^{\infty} \frac{dx}{1 + e^{3x}}$

Answers

I. **(a)** $\frac{4}{3}$ **(b)** $\frac{29}{6}$ **(c)** $\frac{44}{3}$ **(d)** $\frac{\pi}{4}$ **(e)** $\frac{16}{3} - \log 4$ **(f)** 1 **(g)** $\frac{28}{3}$

(h) $1 - \log 2$ **(i)** $1 - \frac{\pi}{4}$ **(j)** 0 **(k)** $\frac{1}{2}$ **(l)** $\frac{\pi}{4}$ **(m)** $\frac{2}{3}$ **(n)** $\log 2$

(o) $\log\left(\sqrt{2} + 1\right)$ **(p)** $\frac{1}{2}\left(e - \frac{1}{e}\right)$ **(q)** $\frac{1}{5}$ **(r)** $\sqrt{2} - 1$

II. **(a)** -2 **(b)** 1 **(c)** $\pi - 2$ **(d)** $\frac{1}{2}$ **(e)** $\frac{\pi^2}{4}$

(f) $\frac{\pi}{\sqrt{3}} - \log 2 - \frac{\pi^2}{18}$ **(g)** $\log 4 - \frac{3}{4}$

III. (a) $-\frac{5}{21}$ (b) $\frac{2}{3}$ (c) $\frac{2}{3}$ (d) $\frac{1}{4}$ (e) $\frac{1}{5}$ (f) $\frac{\pi^{3/2}}{3\sqrt{2}}$

(g) $\frac{2}{3}$ (h) $\frac{1}{3}$ (i) $2\sqrt{2}-2$ (j) $\frac{21}{3}$ (k) $\frac{\pi}{4}$ (l) $\frac{1}{2}(e-1)$

(m) $\frac{\pi}{6}$ (n) $\log\frac{9}{8}$ (o) $\log\frac{4}{3}$

IV. (a) $\frac{\pi}{16}$ (b) $\frac{\pi}{8}$ (c) $\frac{1}{\sqrt{2}}$ (d) $\tan^{-1} e-\frac{\pi}{4}$ (e) $4\log\frac{3}{4}-1$

(f) $\frac{\pi}{4}-\frac{2}{3}$ (g) $\frac{4}{3}$ (h) $\frac{\pi}{2}$ (i) $\frac{7\pi}{4}$ (j) $\frac{1}{ab}\cdot\frac{\pi}{2}$

(k) 2 (l) $\frac{\pi}{4}$ (m) $\frac{\pi}{\sqrt{5}}$ (n) $\log\frac{9}{8}$ (o) $\frac{\log 2}{3}$

2.2 Application of Definite Integral

One of the applications of definite integrals is to find the area bounded by the curve $y = f(x)$, x - axis and the ordinates drawn at $x = a$ and $x = b$. The process of finding the area under a given curve is called **quadrature**.

We shall establish the following result.

The area bounded by the curve $y = f(x)$, the x - axis, the ordinates at $x = a$ and $x = b$, is given by, $\int_a^b y\,dx = \int_a^b f(x)\,dx$, where $y = f(x)$ is finite continuous function of in $a \le x \le b$.

Proof : Consider the arc of the given curve $y = f(x)$ in $a \le x \le b$.

Let AC and BD be the ordinates drawn at $x = a$ and $x = b$.

We are supposed to find the area $ABDCA$.

Let $P(x, y)$ be any point on the curve and $Q(x + \delta x, y + \delta y)$ be a neighbouring point on the curve. PE and QF be perpendicular to x - axis.

Let the area $AEPCA$ be denoted by A_x and that of $AFQCA$ by $A_x + \delta A_x$. Thus the area $EFQPE = (A_x + \delta A_x) - A_x = \delta A_x$.

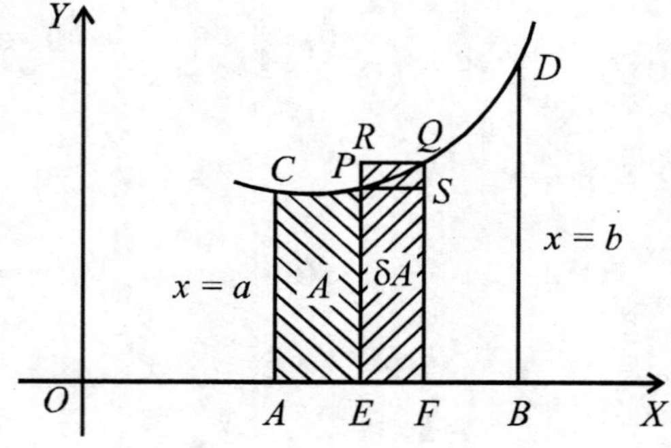

From the figure we have

Area *EFSPE* < Area *EFQPE* < Area *EFQRE*

$\Rightarrow \qquad y \cdot \delta x < \delta A_x < (y + \delta y)\, \delta x$ (because *EFSPE* and *EFQR* are rectangles)

$$\Rightarrow \qquad y < \frac{\delta A_x}{\delta x} < y + \delta y$$

Taking the limit as $\delta x \to 0$, we have,

$$\lim_{\delta x \to 0} y < \lim_{\delta x \to 0} \frac{\delta A_x}{\delta x} < \lim_{\delta x \to 0} (y + \delta y) \;\Rightarrow\; y \le \frac{dA_x}{dx} \le y \qquad (\because \delta y \to \text{ as } \delta x \to 0)$$

$$\Rightarrow \quad \frac{dA_x}{dx} = y = f(x)$$

$$\Rightarrow \quad dA_x = y\, dx$$

$$\Rightarrow \quad A_x = \int y\, dx = F(x) + c \text{ say.} \qquad \text{.... (1)}$$

Now, when $x = a$, then *PE* coincides with *CA* and the area *A* becomes zero. Thus, we have

$$0 = F(a) + c \qquad \text{.... (2)}$$

Again when $x = b$, then *PE* coincides with *DE*, then the area becomes the required area say *A*. Thus we have,

$$A = F(b) + c \qquad \text{.... (3)}$$

From (2) and (3) we have

$$F(b) - F(a) = A. \quad \text{But,} \quad F(b) - F(a) = \int_a^b f(x)\, dx$$

Thus, the area *A* bounded by the curve $y = f(x)$, *x* - axis and the ordinates at $x = a$ and $x = b$ is given by, $$A = \int_a^b y\, dx = \int_a^b f(x)\, dx$$

Similarly, we can prove the following three results.

1. **The area bounded by the curve $x = g(y)$, the *y* - axis and two abscissae $y = c$ and $y = d$ is given by,** $\int_c^d x\, dy = \int_c^d g(y)\, dy$ **(Fig. 2)**
2. **The area bounded by two intersecting curves $y = f(x)$ and $y = g(x)$ (such that $f(x) > g(x)$) and the ordinates $x = a$ and $x = b$ is given by**

$$\int_a^b \{f(x) - g(x)\}\, dx \qquad \textbf{(Fig. 3)}$$

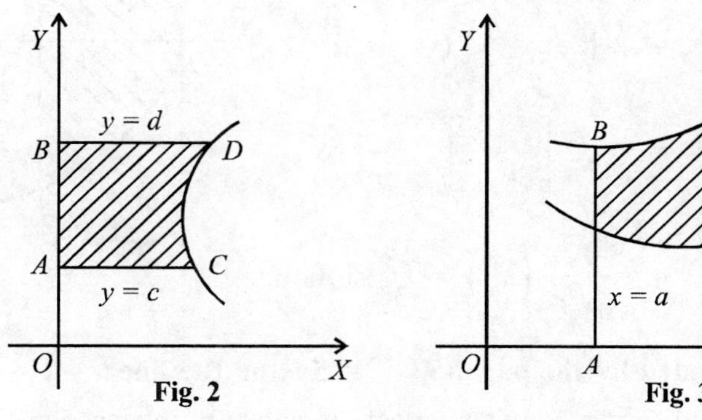

Fig. 2 **Fig. 3**

3. The area bounded by two intersecting curves $x = f(y)$ and $x = g(y)$ and the abscissas $y = c$ and $y = d$ is given by, $\int_c^d \{f(y) - g(y)\}\, dy$

Example 1. Find the area bounded by the curve $y = x^3$, x - axis and the ordinates $x = 1$ and $x = 4$.

Solution : The required area A is given by

$$A = \int_1^4 y\, dx = \int_1^4 x^3\, dx \qquad (\because y = x^3)$$

$$= \left.\frac{x^4}{4}\right]_1^4 = 64 - \frac{1}{4} = \frac{\mathbf{255}}{\mathbf{4}} \textbf{ sq. units}$$

Example 2. Find the area bounded by the curve $y = x^2 - 7x + 10$ and x - axis.

Solution : The equation of the curve is $y = x^2 - 7x + 10$.

To find the x - coordinates of the points where the curve cuts the x - axis we shall put $y = 0$.

$\Rightarrow \qquad x^2 - 7x + 10 = 0 \quad \Rightarrow \quad (x - 5)(x - 2) = 0 \quad \Rightarrow \quad x = 2, x = 5$

Thus, the required area is the area bounded by the curve, x - axis and the ordinates $x = 2$ and $x = 5$.

i.e., $$A = \int_2^5 y\, dx = \int_2^5 \left(x^2 - 7x + 10\right) dx = \left.\frac{x^3}{3} - \frac{7x^2}{2} + 10x\right]_2^5$$

$$= \left(\frac{125}{3} - \frac{175}{2} + 50\right) - \left(\frac{8}{3} - \frac{28}{2} + 20\right)$$

$$= \frac{25}{6} - \frac{52}{6} = -\frac{27}{6} = -\frac{9}{2} \Rightarrow A = \frac{9}{2} \text{ sq. units}$$

(numerically)

Note : The negative sign in the final answer indicates the curve lies below the x - axis. Since the area is a positive quantity, we take the absolute value of the integral.

Example 3. Find the area of the region bounded by parabola $y^2 = 8x$ and its latus rectum.

Solution : We have $y^2 = 8x$, Here $4a = 8 \Rightarrow a = 2$.

Thus the ends of latus rectum are $L = (2, 4)$, $L' = (2, -4)$

Required area = Area $OL'SLO$ (shaded region)

$= 2$ (area $OSLO$)

$$= 2 \cdot \int_0^2 y\, dx = 2\int_0^2 \sqrt{8x}\, dx \quad (\because y^2 = 8x)$$

$$= 4\sqrt{2}\left[\frac{2}{3}x^{3/2}\right]_0^2$$

$$= \frac{8\sqrt{2}}{3} \cdot \left(2^{3/2}\right) = \frac{\mathbf{32}}{\mathbf{3}} \textbf{ sq.units.}$$

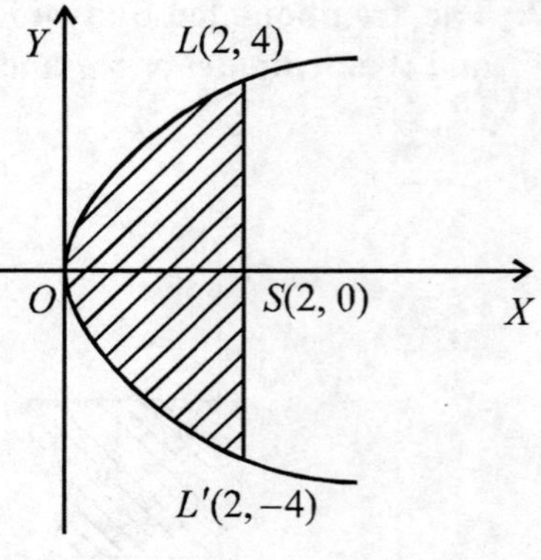

Example 4. Find the area bounded by the parabola $y^2 = 5x$ and the line $x - y = 0$.

Solution : We have, $y^2 = 5x$ and $y = x$. Solving we get the points of intersection.

Put, $y = x$ in $y^2 = 5x$, we get, $x^2 = 5x \Rightarrow x(x-5) = 0 \Rightarrow x = 0,\ x = 5$

Now, $y^2 = 5x \Rightarrow \boldsymbol{y = \sqrt{5x} = f(x)}$, $\boldsymbol{y = x = g(x)}$

Thus, the required area is

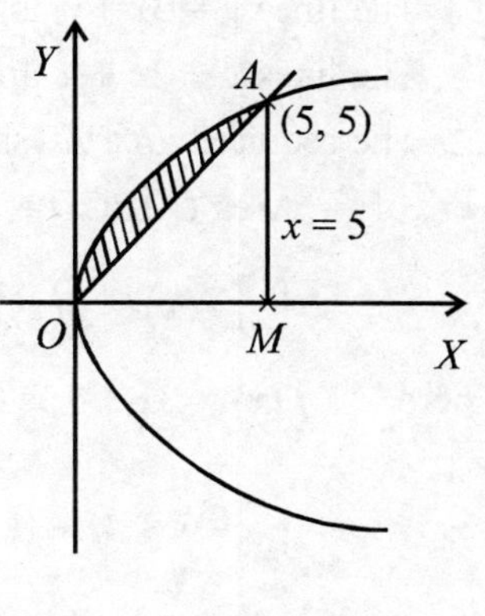

$$A = \int_0^5 [f(x) - g(x)]\,dx$$

$$= \int_0^5 \left[\sqrt{5}\,(x)^{1/2} - x\right] dx = \left[\sqrt{5}\cdot\frac{2}{3}x^{3/2} - \frac{x^2}{2}\right]_0^5$$

$$= \frac{2\sqrt{5}}{3}\,5^{3/2} - \frac{25}{2}$$

$$= \frac{50}{3} - \frac{25}{2} = \frac{\mathbf{25}}{\mathbf{6}}\ \textbf{Sq.units.}$$

Example 5. Find the area enclosed between the parabolas $x^2 = y$ and $y^2 = 8x$.

Solution : We have, $y = x^2$ and $y^2 = 8x$.

Put $y = x^2$ in $y^2 = 8x$, we get, $x^4 = 8x \Rightarrow x(x^3 - 8) = 0 \Rightarrow \boldsymbol{x = 0,\ x = 2}$

Now, $y^2 = 8x \Rightarrow \boldsymbol{y = 2\sqrt{2}\,\sqrt{x} = f(x)}$; $\boldsymbol{y = x^2 = g(x)}$

The required Area A is given by

$$A = \int_0^2 [f(x) - g(x)]\,dx = \int_0^2 \left[2\sqrt{2}\,\sqrt{x} - x^2\right] dx$$

$$= \left[2\sqrt{2}\cdot\frac{2}{3}x^{3/2} - \frac{x^3}{3}\right]_0^2 = \frac{4\sqrt{2}}{3}\cdot 2^{3/2} - \frac{8}{3} = \frac{16}{3} - \frac{8}{3} = \frac{\mathbf{8}}{\mathbf{3}}\ \textbf{sq. units.}$$

Example 6. Find the area bounded by the curve $y = \sin x$, the x - axis and the ordinates $x = 0$ and $x = 2\pi$.

Solution : The curve $y = \sin x$, cuts the x - axis at $(\pi, 0)$.

Thus, the required area

$$A = \text{Area } OAB + \text{Area } BCD$$

Since curve OAB and curve BCD lies on opposite sides of x - axis these areas must be evaluated separately.

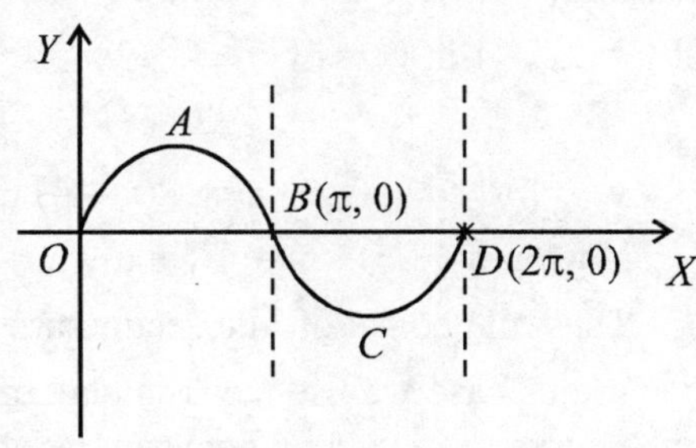

$$\text{Area } OAB = \int_0^{\pi} y\,dx = \int_0^{\pi} \sin x\,dx = (-\cos x)_0^{\pi}$$

$$= -(-1-1) = \mathbf{2}$$

$$\text{Area } BCD = \int_{\pi}^{2\pi} y\,dx = \int_{\pi}^{2\pi} \sin x\,dx = (-\cos x)_{\pi}^{2\pi}$$

$$= -(1+1) = \mathbf{-2}$$

$\therefore$ Area $BCD = |-2| = 2$. (reason for the –ve sign is the curve lies below x - axis)

$\therefore$ required Area $= 2 + 2 =$ **4 sq. units.**

Example 7. Find the area of the triangular region whose sides have the equation $y = 2x + 1$, $y = 3x + 1$ and $x = 4$.

Solution : The lines $y = 2x + 1$ and $y = 3x + 1$ intersect at A (0, 1) (by solving)

The lines $y = 3x + 1$ and $x = 4$ intersect at B (4, 13).

The lines $y + 2x + 1$ and $x = 4$ intersect at C (4, 9)

The required area A (shaded in the fig) is

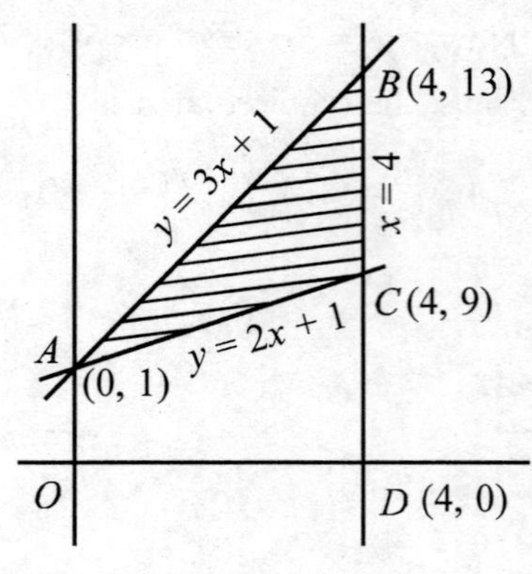

A = Area $OABCD$ – Area $OACD$

$$= \int_0^4 [f(x) - g(x)]\, dx,$$

Where $\quad$ $\boldsymbol{f(x) = 3x + 1}, \quad \boldsymbol{g(x) = 2x + 1}$

$$= \int_0^4 [(3x + 1) - (2x + 1)]\, dx = \int_0^4 x\, dx = \left.\frac{x^2}{2}\right]_0^4$$

$$= \textbf{8 sq. units.}$$

Example 8. Find the area enclosed between the curves $y = 8 - x^2$ and $y = x^2$.

Solution : We have, $\quad y = 8 - x^2 \quad$ and $\quad y = x^2$. $\quad$ Put, $\quad y = x^2$ in $y = 8 - x^2$

We get, $\qquad x^2 = 8 - x^2 \Rightarrow 2x^2 = 8 \Rightarrow x^2 = 4 \Rightarrow x = \pm 2$

Now, $\qquad \boldsymbol{y = 8 - x^2 = f(x)} \quad$ and $\quad \boldsymbol{y = x^2 = g(x)}$

The required area is given by

$$A = \int_{-2}^{2} [f(x) - g(x)]\, dx = \int_{-2}^{2} (8 - x^2 - x^2)\, dx$$

$$= \int_{-2}^{2} (8 - 2x^2)\, dx = \left. 8x - \frac{2x^3}{3}\right]_{-2}^{2} = \left(16 - \frac{16}{3}\right) - \left(-16 + \frac{16}{3}\right)$$

$$= 32 - \frac{32}{3} = \frac{\mathbf{64}}{\mathbf{3}} \textbf{ sq.units.}$$

Example 9. Find the area bounded by the curve $y^2 = 4x$ and the line $y = 2x - 4$.

Solution : We have $y^2 = 4x$ and $y = 2x - 4$.

Put $y = 2x - 4$ in $y^2 = 4x$, $\Rightarrow (2x - 4)^2 = 4x \Rightarrow 4x^2 - 16x + 16 = 4x$

$\Rightarrow x^2 - 5x + 4 = 0$

$\Rightarrow (x - 4)(x - 1) = 0 \Rightarrow x = 4, x = 1$

Now, $x = 4 \Rightarrow y = 4$ and $x = 1 \Rightarrow y = -2$

Thus, the points of intersection are (1, –2) and (4, 4) (see fig)

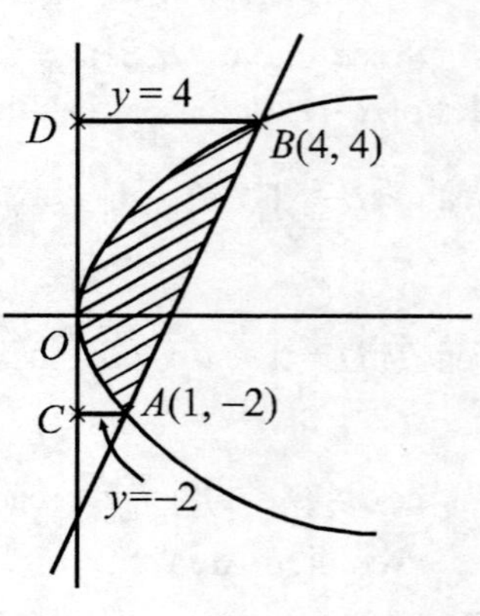

In this case we use the formula corresponding to area bounded by the curve, y - axis and between the abscisic $y = c$ and $y = d$.

i.e., $\qquad A = \int_c^d [f(y) - g(y)]\, dy$

where $x = f(y)$ and $x = g(y)$ are the equations to the curves.

Now, $\quad y^2 = 4x \Rightarrow x = \dfrac{\mathbf{y^2}}{\mathbf{4}} = \boldsymbol{g(y)}$

and $y = 2x - 4 \Rightarrow x = \frac{1}{2}(y + 4) = f(y)$

The required area is given by

$$A = \int_{-2}^{4} [f(y) - g(y)]\, dy \Rightarrow A = \int_{-2}^{4} \left[\frac{1}{2}(y+4) - \frac{y^2}{4}\right] dy$$

$$= \left[\frac{y^2}{4} + 2y - \frac{y^3}{12}\right]_{-2}^{4} = \left(4 + 8 - \frac{16}{3}\right) - \left(1 - 4 + \frac{2}{3}\right)$$

$$= \frac{20}{3} + \frac{7}{3} = \mathbf{9\ sq.\ units.}$$

Note : While finding the area between the curves, if the y - coordinates of the points of intersection are of same sign use the formula,

$$A = \int_b^a [f(x) - g(x)]\, dx$$

where $x = a$ and $x = b$ are the ordinates drawn at the points of intersections;

If the y - coordinates of the points of intersection are of opposite signs then use the formula.

$$A = \int_c^d [f(y) - g(y)]\, dy$$

where $y = c$ and $y = d$ are the abscissa drawn at the points of intersections.

Example 10. Find the area of the circle $x^2 + y^2 = a^2$ by the method of integration.

Solution : $x^2 + y^2 = a^2$ is a circle with centre at the origin and radius equal to a units.

The required area is 4 times the area OAB.

We have, $O = (0, 0)$ and $A = (a, 0)$

Also, $x^2 + y^2 = a^2 \Rightarrow y = \sqrt{a^2 - x^2}$

Thus, Area of the circle $= 4\int_0^a y\, dx$

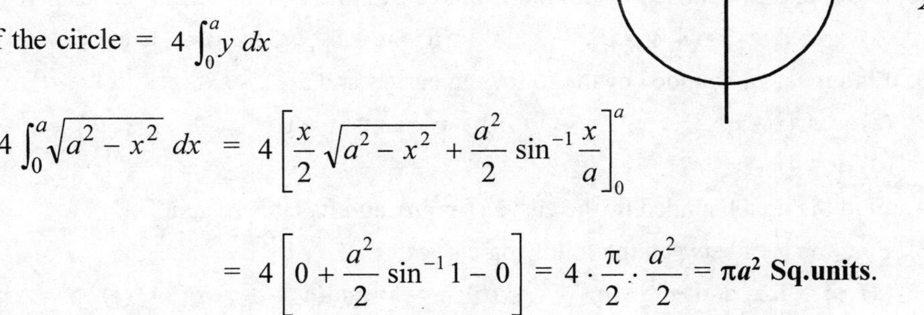

$$= 4\int_0^a \sqrt{a^2 - x^2}\, dx = 4\left[\frac{x}{2}\sqrt{a^2 - x^2} + \frac{a^2}{2}\sin^{-1}\frac{x}{a}\right]_0^a$$

$$= 4\left[0 + \frac{a^2}{2}\sin^{-1} 1 - 0\right] = 4 \cdot \frac{\pi}{2} \cdot \frac{a^2}{2} = \pi a^2 \text{ Sq.units.}$$

Example 12. Find the area of the ellipse $\frac{x^2}{a^2} + \frac{y^2}{b^2} = 1$ by the method of integration.

Solution : We have, $\frac{x^2}{a^2} + \frac{y^2}{b^2} = 1.$

$A = (a, 0), \quad A' = (-a, 0).$

Clearly the required area is 4 times the area OAB.

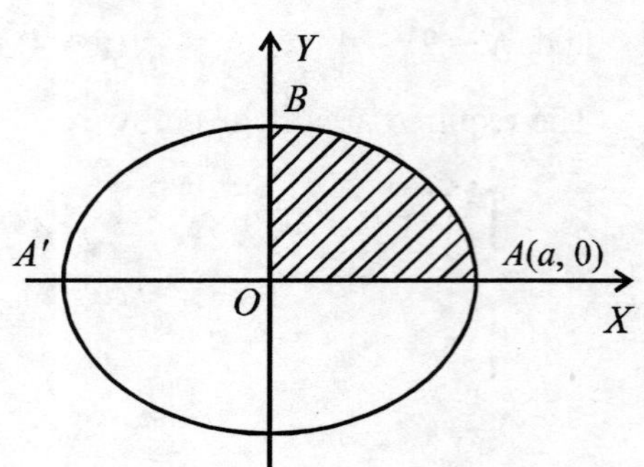

Now, $\dfrac{x^2}{a^2} + \dfrac{y^2}{b^2} = 1 \Rightarrow \dfrac{y^2}{b^2} = 1 - \dfrac{x^2}{a^2}$

$\Rightarrow y^2 = \dfrac{b^2}{a^2}(a^2 - x^2)$

$\Rightarrow y = \dfrac{b}{a}\sqrt{a^2 - x^2}$

Thus, the required area

$$A = 4\int_0^a y\,dx = 4\int \frac{b}{a}\sqrt{a^2 - x^2}\,dx = \frac{4b}{a}\left[\frac{x}{2}\sqrt{a^2 - x^2} + \frac{a^2}{2}\sin^{-1}\frac{x}{a}\right]_0^a$$

$$= \frac{4b}{a}\left[0 + \frac{a^2}{2}\frac{\pi}{2} + 0\right] = \frac{4b}{a}\cdot\frac{\pi a^2}{4} = \pi ab \text{ Sq.units.}$$

Exercise

1. Find the area bounded by the following curves, the x - axis and the ordinates given.

(a) $y = x^2 + 1;\ x = 0, x = 2$ **(b)** $y = 9 - x^2;\ x = 0, x = 3$ **(c)** $y = 4\sqrt{x - 1};\ x = 1, x = 3$

(d) $xy = 4;\ x = 1, x = 3$ **(e)** $y = \dfrac{x^2}{1 + x^2};\ x = 0, x = 1$ **(f)** $y = x^2 - 1;\ x = 0, x = 2$

(g) $y = \cos 3x;\ x = 0, x = \dfrac{\pi}{6}$ **(h)** $y = \sin x;\ x = 0, x = \pi$ **(i)** $y^2 = 4x;\ x = 0, x = 1$

(j) $xy = 16;\ x = 4, x = 8$

2. Find the area bounded by the following curves, the y - axis and abscissas given

(a) $x^2 = 8y;\ y = 4, y = 9$ **(b)** $3x = 4y + 12;\ y = -1, y = 6$

3. Find the area bounded by the following curves and the x - axis

(a) $y = 4x - x^2 - 3$ **(b)** $y = x^2 - 5x + 6$ **(c)** $y = x^2 - 3x + 2$

(d) $y = 4x - x^2$ **(e)** $y = 5x - x^2 - 4$.

4. Find the area bounded by the curve $y^2 = 4ax$ and its latus rectum.

5. Find the area between the following curves

(a) $y^2 = 4ax$ and $x^2 = 4ay$ **(b)** $x^2 = y$ and $y^2 = 8x$ **(c)** $y^2 = 4ax$ and $y = mx$

(d) $y^2 = 4x$ and $x = 2y$ **(e)** $y^2 = 4x$ and $y = 2x$ **(f)** $y = 11x - 24 - x^2$ and $y = x$

(g) $y^2 = 6x$ and $x^2 = 6y$ **(h)** $x^2 = 4y$ and $x = 4y - 2$

6. Find the area enclosed by the curve $y^2 = 2x - 2$ and the line $y = x - 5$

7. Find the are enclosed between the parabola $y^2 = x$ and the line $x + y = 2$

8. Find the area of the circle $x^2 + y^2 = 16$ by the method of integration

9. Find the area bounded by the ellipse $\frac{x^2}{9}+\frac{y^2}{16}=1$ by the method of integration.

10. Find the area between the circle $x^2+y^2=1$ and the line $x+y=1$

11. Find the area between the ellipse $\frac{x^2}{25}+\frac{y^2}{16}=1$ and the line $\frac{x}{5}+\frac{y}{4}=1$.

Answers

1. **(a)** $\frac{14}{3}$ **(b)** 18 **(c)** $\frac{16\sqrt{2}}{3}$ **(d)** 4 log 3 **(e)** $1-\frac{\pi}{4}$

(f) $\frac{2}{3}$ **(g)** $\frac{1}{3}$ **(h)** 2 **(i)** $\frac{4}{3}$ **(j)** 16 log 2

2 **(a)** $\frac{76}{3}\sqrt{2}$ **(b)** $\frac{154}{3}$.

3. **(a)** $\frac{4}{3}$ **(b)** $\frac{1}{6}$ **(c)** $\frac{1}{6}$ **(d)** $\frac{32}{3}$ **(e)** $\frac{9}{2}$ **4.** $\frac{8a^2}{3}$

5. **(a)** $\frac{16a^2}{3}$ **(b)** $\frac{8}{3}$ **(c)** $\frac{8a^2}{3m^3}$ **(d)** $\frac{64}{3}$ **(e)** $\frac{1}{3}$ **(f)** $\frac{4}{3}$

(g) 12 **(h)** $\frac{9}{8}$

6. 18 **7.** $\frac{9}{2}$ **8.** 16π **9.** 12π **10.** $\frac{\pi}{4}-\frac{1}{2}$ **11.** $5(\pi-2)$

UNIT VII

Differential Equations

Chapter 1

Differential Equations

1.1 Introduction

In this chapter we shall see the concept of differential equation, the formation of differential equations and a method of solving the differential equations.

1.2 Differential Equations – Order and Degree

Definition (Differential equation) : An equation in which the independent variable, dependent variable and the derivatives of dependent variable or the differentials of the variables are involved is called a differential equation.

Following are few examples of differential equations.

1. $(2y - 3x)\dfrac{dy}{dx} + (7x^2 + 2y) = 0$

2. $\dfrac{dy}{dx} = \sin x + \cos x$

3. $y\left(\dfrac{dy}{dx}\right)^2 = x\left(\dfrac{dy}{dx}\right) + 1$

4. $\dfrac{d^2y}{dx^2} - 4\left(\dfrac{dy}{dx}\right)^2 + 3y = 4x$

5. $\left[1 + \left(\dfrac{dy}{dx}\right)^2\right]^{3/2} = \dfrac{d^2y}{dx^2}$

6. $x^2\, dx + 2y\, dy = 0$

7. $x\dfrac{dy}{dx} + \dfrac{3}{(dy/dx)} = y^2$

Definition (Order of differential equation) : The order of a differential equation is the highest order derivative appearing in the equation.

Definition (Degree of a differential equation) : The degree of a differential equation is the power of the derivative of the highest order occurring in the equation, after it has been expressed in the form of a polynomial in the derivatives.

Note : From the definition of the degree of a differential equation, it follows that to obtain the degree of a differential equation, first the equation must be expressed in a form free from the radicals and fractional powers as far as the derivatives are concerned.

In the above examples,

(1) is of **first order** and **first degree.**

(2) is of **first order** and **first degree.**

(3) is of **first order** and **second degree.**

(4) is of **second order** and **first degree.**

(6) is of **first order** and **first degree.**

To obtain the degree of the equation (5), we square both sides of the equation

i.e.,
$$\left[1+\left(\frac{dy}{dx}\right)^2\right]^3 = \left(\frac{d^2y}{dx^2}\right)^2$$

Now, we can see that the powers of the derivatives are positive integers. Thus, the **order is 2** and **degree 2** – because the power of the highest order derivative is 2.

Again consider the equation (7)

$$x\frac{dy}{dx} + \frac{3}{(dy/dx)} = y^2 \Rightarrow x\left(\frac{dy}{dx}\right)^2 + 3 = y^2\frac{dy}{dx}$$

Thus, it is of **first order** and **second degree**.

Definition (Solution of a differential equation) : Any relation between the dependent variable and the independent variable, which satisfies the given differential equation is called a solution or an integral of the differential equation.

The process of finding a solution of the given differential equation is called **"solving the equation"**.

Definition : A solution of a differential equation containing as many independent constants as the order of the equation is called the general solution of the equation.

The solution obtained from the general solution by giving particular values to the arbitrary constants that are involved is called a **particular solution** of the equation.

Example 1. Find the order and the degree of the following differential equations.

(a) $a^2\frac{d^2y}{dx^2} = 1+\left(\frac{dy}{dx}\right)^2$ **(b)** $1+\left(\frac{dy}{dx}\right)^2 = 7\left(\frac{d^2y}{dx^2}\right)^3$ **(c)** $\left(\frac{d^2y}{dx^2}\right)^{3/2} + y = x$

(d) $\left(x+\frac{dy}{dx}\right) = \sqrt{1+\frac{dy}{dx}}$ **(e)** $(1-y^2)\,dx + y\,(1-x^2)\,dy = 0$ **(f)** $\frac{d^2y}{dx^2} + \frac{dy}{dx} = \log\left(\frac{d^2y}{dx^2}\right)$

Solution : (a) Clearly, order = **2**, degree = **1**.

Degree is the power of the highest ordered derivative involved – here highest order derivative is $\frac{d^2y}{dx^2}$ and its power is 1.

(b) Order = **2** and degree = power of $\frac{d^2y}{dx^2}$ = **3.**

(c) We have $\left(\frac{d^2y}{dx^2}\right)^{3/2} + y = x$. This should be expressed as a polynomial in the derivatives.

The equation is $\left(\frac{d^2y}{dx^2}\right)^{3/2} = x - y$, Squaring both sides, we get,

$$\left(\frac{d^2y}{dx^2}\right)^3 = (x-y)^2 \Rightarrow \text{Order} = \mathbf{2},\ \text{degree} = \mathbf{3}.$$

(d) Consider, $x + \frac{dy}{dx} = \sqrt{1 + \frac{dy}{dx}}$

Squaring both sides, we get, $\left(x + \frac{dy}{dx}\right)^2 = 1 + \frac{dy}{dx}$ $\Rightarrow$ Order = **1,** degree = **2.**

(e) Consider, $(1 - y^2)\, dx + y\,(1 - x^2)\, dy = 0$; $(1 - y^2) + y\,(1 - x^2)\frac{dy}{dx} = 0$ (dividing by dx)

Thus, Order = **1,** degree = **1.**

(f) Consider, $\frac{d^2y}{dx^2} + \frac{dy}{dx} = \log\left(\frac{d^2y}{dx^2}\right)$. Clearly the order of the equation is **2.**

The degree of the differential equation can not be determined as we can not write the equation as a polynomial in all the differentials. In such cases we say the **degree is not defined.**

Example 2. Verify, **(a)** $y = \frac{1}{x}$ **is a solution of the equation** $\frac{dy}{dx} + y^2 = 0$.

(b) $y = ae^{3x}$ **is a solution of the equation** $\frac{dy}{dx} = 3y$.

(c) $y = 4ax$ **satisfies the equation** $y = x\frac{dy}{dx} + a\frac{dy}{dx}$.

Solution : (a) We have $y = \frac{1}{x}$ $\Rightarrow$ $\frac{dy}{dx} = -\frac{1}{x^2}$

Putting this in the equation $\frac{dy}{dx} + y^2 = 0$, we get, $-\frac{1}{x^2} + \frac{1}{x^2} = 0$ $\left(\because y = \frac{1}{x}\right)$

$\Rightarrow$ 0 = 0 which is true.

Thus, $y = \frac{1}{x}$ satisfies the equation, **hence a solution.**

(b) Consider, $y = ae^{3x}$ $\Rightarrow$ $\frac{dy}{dx} = 3ae^{3x}$ $\Rightarrow$ $\frac{dy}{dx} = 3y$ $(\because y = ae^{3x})$

which is the given differential equation. Hence $y = ae^{3x}$ satisfies the given equation, **hence a solution.**

(c) Consider, $y^2 = 4ax$ $\Rightarrow$ $2y\frac{dy}{dx} = 4a$ $\Rightarrow$ $\frac{dy}{dx} = \frac{2a}{y}$ and $\frac{dx}{dy} = \frac{y}{2a}$

Consider, $x\frac{dy}{dx} + a\frac{dx}{dy} = x \cdot \frac{2a}{y} + a \cdot \frac{y}{2a} = \frac{2ax}{y} + \frac{y}{2}$

$= \frac{y^2}{2y} + \frac{y}{2}$ $\left(\because 2ax = \frac{y^2}{2}\right)$

$= \frac{y}{y} + \frac{y}{2} = y$ = r.h.s of the equation.

$\therefore$ $x\frac{dy}{dx} + a\frac{dx}{dy} = y$ Thus, $y^2 = 4ax$ is a solution of the equation.

Example 3. Show that

(a) $y = A \sin 4x + B \cos 4x$ **is a solution of the differential equation** $\dfrac{d^2y}{dx^2} + 16y = 0.$

(b) $y \sec x = \tan x + c$ **is a solution of** $\dfrac{dy}{dx} + y \tan x = \sec x.$

(c) $x^2 + y^2 = r^2$ **is a solution of the differential equation** $y = x\dfrac{dy}{dx} + r\sqrt{1 + \left(\dfrac{dy}{dx}\right)^2}.$

Solution : (a) Consider

$$y = A \sin 4x + B \cos 4x \quad \Rightarrow \quad \frac{dy}{dx} = 4A \cos 4x - 4B \sin 4x$$

$$\Rightarrow \quad \frac{d^2y}{dx^2} = -16A \sin 4x - 16B \cos 4x$$

$$\Rightarrow \quad \frac{d^2y}{dx^2} = -16\,(A \sin 4x + B \cos 4x)$$

$$\Rightarrow \quad \frac{d^2y}{dx^2} = -16y \qquad (\because y = A \sin 4x + B \cos 4x)$$

$$\Rightarrow \quad \frac{d^2y}{dx^2} + 16y = 0$$

which is the given differential equation. Hence the problem.

(b) Consider, $\quad y \sec x = \tan x + c$

$$\Rightarrow \quad y \cdot \sec x \cdot \tan x + \sec x \cdot \frac{dy}{dx} = \sec^2 x$$

$$\Rightarrow \quad \sec x \left[y \tan x + \frac{dy}{dx} \right] = \sec^2 x$$

$$\Rightarrow \quad \frac{dy}{dx} + y \tan x = \sec x \qquad (\because \sec x \neq 0)$$

which is the given differential equation. Hence the problem.

(c) Consider, $\quad x^2 + y^2 = r^2$

$$\Rightarrow \quad 2x + 2y\frac{dy}{dx} = 0$$

$$\Rightarrow \quad \frac{dy}{dx} = -\frac{x}{y}$$

Now, consider

$$x\frac{dy}{dx} + r\sqrt{1 + \left(\frac{dy}{dx}\right)^2} = x\left(-\frac{x}{y}\right) + r\sqrt{1 + \frac{x^2}{y^2}}$$

$$= -\frac{x^2}{y} + r\sqrt{\frac{x^2 + y^2}{y^2}}$$

$$= -\frac{x^2}{y} + r\left(\frac{r}{y}\right) \qquad (\because x^2 + y^2 = r^2)$$

$$= \frac{1}{y}\left(-x^2 + r^2\right)$$

$$= \frac{1}{y}(y^2) = y \qquad (\because r^2 = x^2 + y^2)$$

$$\therefore \quad x\frac{dy}{dx} + r\sqrt{1 + \left(\frac{dy}{dx}\right)^2} = y \Rightarrow$$ **$x^2 + y^2 = r^2$ is a solution of a given equation.**

Exercise

I. 1. Determine the order and the degree of the following differential equations.

(a) $\left(\frac{dy}{dx}\right)^3 + 5y = 0$ **(b)** $x + \left(\frac{dy}{dx}\right)^2 = \sqrt{1 + \left(\frac{dy}{dx}\right)^2}$ **(c)** $\frac{d^2y}{dx^2} + \left(\frac{dy}{dx}\right)^2 + xy = 0$

(d) $x\frac{dy}{dx} + \frac{3}{(dy/dx)} = y^2$ **(e)** $2\frac{d^2y}{dx^2} + \sqrt[3]{1 - \left(\frac{dy}{dx}\right)^2} = 0$ **(f)** $\left(\frac{d^2z}{dx^2}\right)^{5/2} = 2z - 5$

(g) $\sqrt{1 - x^2}\, dy + \sqrt{1 - y^2}\, dx = 0$ **(h)** $(xy^2 - x)\, dx + (y - x^2y)\, dy = 0$

2. Verify that

(a) $y = a \cos x$ is a solution of the differential equation $\frac{dy}{dx} + y \tan x = 0$.

(b) $y = Ae^{Bx}$ is a solution of the equation $\frac{d^2y}{dx^2} - \frac{1}{y}\left(\frac{dy}{dx}\right)^2 = 0$.

(c) $y = ce^{-x}$ is a solution of the differential equation $\frac{dy}{dx} + y = 0$.

(d) $y = A \cos 2x + B \sin 2x$ is a solution of $\frac{d^2y}{dx^2} + 4y = 0$.

(e) $y = ae^x + be^{2x}$ is a solution of $\frac{d^2y}{dx^2} - 3\frac{dy}{dx} + 2y = 0$.

(f) $y = cx + \frac{a}{e}$ is a solution of $x\frac{dy}{dx} + a\frac{dy}{dx} = y$.

Answers

1. **(a)** 1, 3 **(b)** 1, 4 **(c)** 2, 1 **(d)** 1, 2 **(e)** 2, 3 **(f)** 2, 5 **(g)** 1, 1 **(h)** 1, 1

1.3 Formation of Differential Equations

In this section we shall see the method of formation of differential equation for the family curves represented by a relation between the variables x and y and containing some arbitrary constants.

A relation between x and y with two arbitrary constants is given by $f(x, y, c_1, c_2) = 0$. In general it represents a family of plane curves. For example the equation $x^2 + y^2 = a^2$ represents a family of concentric circles with centre at the origin. This represents a relation between x and y with one arbitrary constant. Similarly, the equation $y = mx$ represents a family of lines passing through the origin. The equation $\frac{x^2}{a^2} + \frac{y^2}{b^2} = 1$, $(a > b)$ is an expression having two arbitrary constants. This represents a family of ellipses, whose foci lie along the x-axis. Similarly the equation $y = Ae^{x} + Be^{-x}$ is an another example of family of curves with two arbitrary constants.

The constants that are involved in the above equations are called **parameters** of the equation. By giving different values to the parameters, involved in the equation, we get the equations of different members of the family (i.e., particular curves that are in the family of curves).

For example consider the equation $y^2 = 4ax$. This represents a family of parabolas with the vertex at the origin and focus on the x-axis. In this equation 'a' is the parameter. For different values of 'a' the equation gives the equations of different parabolas, with verter at the origin and focus on the x-axis.

Similarly, in the equation $y = mx$, 'm' is the parameter. For different values of m, we get the equation of different lines through the origin.

Now, given a relation between the variables x and y, with some arbitrary constants (parameters), we are supposed to form a differential equation, whose solution is the given relation. Further this differential equation should be independent of the arbitrary constants involved in the given equation. This process is what we call "the formation of differential equation by eliminating the arbitrary constants" involved.

Below we shall give a working procedure to form a differential equation.

Working Rule :

Let the given family of curves is represented by the equation $f(x, y, c_1, c_2, ... c_n) = 0$, where $c_1, c_2, ... c_n$ are n arbitrary constants (parameters).

Step 1: Differentiate the given equation

$$f(x, y, c_1, c_2, ... c_n) = 0$$

n times successively, to get n equations.

Step 2: Eliminate the n constants $c_1, c_2, ..., c_n$, from the n relations obtained in step 1 and the given relation. The resultant equation is the required differential equation.

Note : 1. If the given equation has n arbitrary constants, then, we need at least n independent equations to eliminate these n constants. That is why we differentiate the given equation as many times as the number of constants involved.

2. The order of the differential equation thus obtained from the given equation, is equal to the number of independent arbitrary constants that are involved in the equation. For example, if the given equation has one arbitrary constant, the order of the differential equation thus formed will be one. Similarly the order of the differential equation thus formed will be two, if the given equation has two arbitrary constants and so on.

The following examples illustrate the procedure.

Example 1. Form the differential equation for the equation $x^2 + y^2 = a^2$.

Solution : Consider, $x^2 + y^2 = a^2$. Differentiating w.r.t. 'x', we get

$$2x + 2y\frac{dy}{dx} = 0 \quad \Rightarrow \quad y\frac{dy}{dx} + x = 0$$

The equation obtained does not contain the arbitrary constant 'a'.

Thus the required differential equation is $\boldsymbol{y\frac{dy}{dx} + x = 0}$.

Example 2. Form the differential equation by eliminating the constants in the equation $(x-a)^2 + y^2 = a^2$.

Solution : The given equation is

$$(x-a)^2 + y^2 = a^2 \qquad \text{.... (1)}$$

Differentiating w.r.t. x, we get

$$2(x-a) + 2y\frac{dy}{dx} = 0 \qquad \text{.... (2)}$$

We have to eliminate 'a' between (1) and (2).

Now, (2) $\Rightarrow$ $(x-a) = -y\dfrac{dy}{dx}$, putting this in (1), we get

$$\left(-y\frac{dy}{dx}\right)^2 + y^2 = \left(x + y\frac{dy}{dx}\right)^2 \qquad \left(\because a = x + y\frac{dy}{dx}\right)$$

$$\Rightarrow \quad y^2\left(\frac{dy}{dx}\right)^2 + y^2 = x^2 + 2xy\frac{dy}{dx} + y^2\left(\frac{dy}{dx}\right)^2$$

$$\Rightarrow \quad 2xy\frac{dy}{dx} = y^2 - x^2 \Rightarrow \boldsymbol{\frac{dy}{dx} = \frac{y^2 - x^2}{2xy}}$$

Alternative method.

The given equation is

$$(x-a)^2 + y^2 = a^2 \Rightarrow x^2 + y^2 - 2ax = 0 \qquad \text{.... (1)}$$

Differentiating w.r.t. x, we get

$$2x + 2y\frac{dy}{dx} - 2a = 0 \Rightarrow 2a = 2x + 2y\frac{dy}{dx}$$

$$\therefore \quad (1) \Rightarrow x^2 + y^2 - x\left(2x + 2y\frac{dy}{dx}\right) = 0$$

$$\Rightarrow \quad y^2 - x^2 - 2xy\frac{dy}{dx} = 0 \Rightarrow \boldsymbol{\frac{dy}{dx} = \frac{y^2 - x^2}{2xy}}$$

which is required differential equation.

Example 3. Form the differential equation by eliminating the arbitrary constants from the equation $y = ae^{3x} + be^{-3x}$.

Solution : The given equation is, $\quad y = ae^{3x} + be^{-3x}$ .... (1)

The given equation as two constants. Thus we have to differentiate twice to find the required differential equation.

Now, $\quad y = ae^{3x} + be^{-3x} \Rightarrow \dfrac{dy}{dx} = 3ae^{3x} - 3be^{-3x}$

$$\Rightarrow \quad \frac{d^2y}{dx^2} = 9ae^{3x} + 9be^{-3x}$$

$$\Rightarrow \quad \frac{d^2y}{dx^2} = 9\,(ae^{3x} + be^{-3x})$$

$$\Rightarrow \quad \frac{d^2y}{dx^2} = 9y \qquad \text{(from (1))}$$

This is required differential equation.

Example 4. Form the differential equation, whose general solution is $y = c^2 + \dfrac{c}{x}$.

Solution : We have

$$y = c^2 + \frac{c}{x} \quad \Rightarrow \quad \frac{dy}{dx} = -\frac{c}{x^2} \quad \Rightarrow \quad c = -x^2\frac{dy}{dx}$$

Now,
$$y = c^2 + \frac{c}{x} \quad \Rightarrow \quad y = x^4\left(\frac{dy}{dx}\right)^2 + \frac{1}{x}\left(-x^2\frac{dy}{dx}\right)$$

$$\Rightarrow \quad y = x^4\left(\frac{dy}{dx}\right)^2 - x\frac{dy}{dx}$$

This is the required differential equation.

Example 5. Form the differential equation of the family of curves, $y = ae^{-x} + b$.

Solution : We have

$$y = ae^{-x} + b \quad \Rightarrow \quad \frac{dy}{dx} = -ae^{-x} \quad \Rightarrow \quad \frac{d^2y}{dx^2} = ae^{-x}.$$

Now,
$$\frac{d^2y}{dx^2} = ae^{-x} \quad \Rightarrow \quad \frac{d^2y}{dx^2} = -\left(\frac{dy}{dx}\right) \qquad \left(\because\ ae^{-x} = -\frac{dy}{dx}\right)$$

$$\Rightarrow \quad \frac{d^2y}{dx^2} + \frac{dy}{dx} = 0$$

This is the required differential equation.

Exercise

Form the differential equation by eliminating the arbitrary constants from the following equations:

1. $y = x^2 + ax$ **2.** $y^2 = 4ax$ **3.** $x^2 + 4ay = 0$ **4.** $x^2 + 3y^2 = ay$

5. $xy = c^2$ **6.** $y = ae^{-x}$ **7.** $y = c \tan x$ **8.** $x^2 + cy^2 = 4$

9. $y = k \sin^{-1} x$ **10.** $y = k \cdot e^{\sin^{-1} x}$ **11.** $y = x + 2ax^2$ **12.** $x^2 = 4ay$

13. $x^3 + y^3 = 4ax$ **14.** $y = cx + \dfrac{a}{c}$ (a is a fixed constant)

Answers

I. 1. $y = x\dfrac{dy}{dx} - x^2$ **2.** $y = 2x\dfrac{dy}{dx}$ **3.** $\dfrac{dy}{dx} = \dfrac{2y}{x}$ **4.** $\dfrac{dy}{dx} = \dfrac{2xy}{x^2 - 3y^2}$

5. $x\dfrac{dy}{dx} + y = 0$ 6. $\dfrac{dy}{dx} + y = 0$ 7. $\cot x \dfrac{dy}{dx} - y \operatorname{cosec}^2 x = 0$

8. $(4 - x^2)\dfrac{dy}{dx} + xy = 0$ 9. $\sin^{-1}x \dfrac{dy}{dx} = \dfrac{y}{\sqrt{1 - y^2}}$ 10. $\sqrt{1 - x^2} \cdot \dfrac{dy}{dx} = y$ 11. $x\dfrac{dy}{dx} = 2y - x$

12. $x\dfrac{dy}{dx} = 2y$ 13. $3xy^2 \dfrac{dy}{dx} = y^3 - 2x^3$ 14. $x\left(\dfrac{dy}{dx}\right)^2 - y\dfrac{dy}{dx} + a = 0$

1.4 Solution of First Order Differential Equation, by the Method of Separation of Variables

A first order differential equation of first degree in general can be written in the following two forms.

(i) $\dfrac{dy}{dx} = f(x, y)$ **(ii)** $M\,dx + N\,dy = 0$ **where both** M **and** N **are functions of** x **and** y.

An equation in one form may be written in the other form.

For example, consider, $\dfrac{dy}{dx} = \dfrac{x^2 + 2yx}{x^2 + y^2}$, which is the form (i).

This can be written as, $(x^2 + 2xy)\,dx - (x^2 + y^2)\,dy = 0$ which is of the form (ii).

There are several methods of solving a first order, first degree equations. We shall see a method of solving a first order differential equation when it is in the form called "**Variable separable**" form.

A differential equation is said to be of the type variable separable, if it can be expressed in such a way, so that the coefficient of dx **is a function of** x **alone and the coefficient of** dy **is a function of** y **alone.**

The general form of such differential equation can be written as

$$f(x)\,dx = g(y)\,dy \qquad \text{.... (1)}$$

Integrating both sides and introducing an arbitrary constant (to any one side), we get the solution of the equation.

Thus the general solution (1) is given by, $\int f(x)\,dx = \int g(y)\,dy + c$

A *particular solution* of the equation can be obtained by giving a particular value for constant c.

Note : The process of expressing the given first order and first degree equation in the form

$$f(x)\,dx + g(y)\,dy = 0$$

is called **separation of variables.**

Working rule of solving first order differential equation by the method of separation of variables

Step 1: Write the given differential equation in the form, $f(x)\,dx = g(y)\,dy$

Step 2: Integrate both sides and add constant to any one side and write the general solution.

Step 3: If a particular solution is required, use the condition given in the problem and evaluate the value of c **and write the required particular solution.**

Example 1. Solve $(x^2 + 1)\dfrac{dy}{dx} = 1.$

Solution : The given equation can be written as, $dy = \dfrac{dx}{1+x^2}$; On integration, we get,

$$\int dy = \int \frac{dx}{1+x^2} + c \Rightarrow \mathbf{y = \tan^{-1} x + c,} \quad \text{which is the required solution.}$$

Example 2. Solve, $\sec^2 x \tan y\, dx + \sec^2 y \tan x\, dy = 0.$

Solution : Dividing by $\tan x \cdot \tan y$, we get, $\dfrac{\sec^2 x}{\tan x}\, dx + \dfrac{\sec^2 y}{\tan y} = 0$

On integration we get, $\displaystyle\int \frac{\sec^2 x}{\tan x}\, dx + \int \frac{\sec^2 y}{\tan y}\, dy = \log c$

$\Rightarrow \quad \log(\tan x) + \log(\tan y) = \log c \Rightarrow \log(\tan x \cdot \tan y) = \log c$

Thus the required solution $\mathbf{\tan x \cdot \tan y = c.}$

Note : The constant of integration can be written in any form, suitable for final form of the solution. In fact, in the above problem we have taken it as log c - reason is clear from the working.

Example 3. Solve by the method of separation of variables $(x^2 - yx^2)\, dy + (y^2 + x^2y^2)\, dx = 0$

Solution : The given equation is, $x^2(1 - y)\, dy + y^2(1 + x^2)\, dx = 0$

To separate the variables, divide throughout by x^2y^2, we get,

$$\left(\frac{1-y}{y^2}\right) dy + \left(\frac{1+x^2}{x^2}\right) dx = 0$$

On integration we get, $\displaystyle\int \left(\frac{1}{y^2} - \frac{1}{y}\right) dy + \int \left(\frac{1}{x^2} + 1\right) dx = c$

$$\Rightarrow \quad -\frac{1}{y} - \log y - \frac{1}{x} + x = c \Rightarrow \left(\frac{1}{x} + \frac{1}{y} - x\right) + \log y = k \qquad (k = -c)$$

which is the required solution.

Example 4. Solve, $(e^x + 1)y\, dy = (y + 1)\, e^x\, dx.$

Solution : The given equations is, $(e^x + 1)\, y\, dy = (y + 1)\, e^x\, dx$

Dividing throughout by $(y + 1)(e^x + 1)$ we get, $\dfrac{y}{y+1}\, dy = \dfrac{e^x}{e^x+1}\, dx$

On integration, we get,

$$\int \frac{y}{y+1}\, dy = \int \frac{e^x}{e^x+1}\, dx + \log c \Rightarrow \int \frac{(y+1)-1}{y+1}\, dy = \log(e^x+1) + \log c$$

$$\Rightarrow \int \left(1 - \frac{1}{1+y}\right) dy = \log(e^x + 1) + \log c$$

$$\Rightarrow y - \log(1 + y) = \log(e^x + 1) + \log c$$

$$\Rightarrow \mathbf{y = \log[c(e^x + 1)(1 + y)]}$$

which is the required solution.

Example 5. Solve, $\dfrac{dy}{dx} = e^{x-y} + x^2 e^{-y}$ **given that** $y = 1$ **when** $x = 0$.

Solution : The given equation can be written as, $dy = e^{-y}(e^x + x^2)\,dx$

Separating the variables and integrating we get,

$$\int e^y\,dy = \int \left(e^x + x^2\right) dx + c \;\Rightarrow\; e^y = e^x + \frac{x^3}{3} + c$$

This is the general solution of the equation.

To find a particular solution, we shall use the condition that $y = 1$ when $x = 0$.

$$\Rightarrow \qquad e^1 = e^0 + \frac{0}{3} + c \;\Rightarrow\; c = (e - 1)$$

Thus, the required particular solution is $\boldsymbol{e^y = e^x + \dfrac{x^3}{3} + (e - 1)}$

Example 6. Solve, $\boldsymbol{x^{-1}\cos^2 y\,dy + y^{-1}\cos^2 x\,dx = 0.}$

Solution : The given equation is $\dfrac{\cos^2 y}{x}\,dx + \dfrac{\cos^2 x}{y}\,dx = 0$

Multiplying throughout by xy we get, $\quad y\cos^2 y\,dy + x\cos^2 x\,dx = 0$

On integration we get,

$$\int y\cos^2 y\,dy + \int x\cos^2 x\,dx = c \;\Rightarrow\; \frac{1}{2}\int y\,(1 + \cos 2y)\,dy + \frac{1}{2}\int x\,(1 + \cos 2x)\,dx = c$$

Integrating by parts using generalised integration by parts, we get,

$$\frac{1}{2}\left[y\left(y + \frac{\sin 2y}{2}\right) - 1\left(\frac{y^2}{2} - \frac{\cos 2y}{4}\right)\right] + \frac{1}{2}\left[x\left(x + \frac{\sin 2x}{2}\right) - 1\left(\frac{x^2}{2} - \frac{\cos 2x}{4}\right)\right] = c$$

$$\Rightarrow \qquad \frac{1}{2}\left[\frac{y^2}{2} + \frac{y}{2}\sin 2y + \frac{\cos 2y}{4} + \frac{x^2}{2} + \frac{x}{2}\sin 2x + \frac{\cos 2x}{4}\right] = \text{c}$$

$$\Rightarrow \qquad \frac{1}{8}\,(2y^2 + 2x^2 + 2y\sin 2y + 2x\sin 2x + \cos 2y + \cos 2x) = c$$

$$\Rightarrow \qquad \boldsymbol{2x^2 + 2y^2 + 2y\sin 2y + 2x\sin 2x + \cos 2y + \cos 2x = k} \qquad (k = 8c)$$

Example 7. Solve, $\dfrac{dy}{dx} + \sqrt{\dfrac{1 - y^2}{1 - x^2}} = 0$ **given that** $y = 1$ **when** $x = 0$.

Solution : The given equation is, $\dfrac{dy}{dx} = -\dfrac{\sqrt{1 - y^2}}{\sqrt{1 - x^2}}$

Separating the variable and integrating we get, $\displaystyle\int \frac{dy}{\sqrt{1 - y^2}} = \int \frac{-dx}{\sqrt{1 - x^2}} + c \;\Rightarrow\; \sin^{-1} y + \sin^{-1} x = c$

This is the general solution. To find particular solution, we have $y = 1$ when $x = 0$.

$$\Rightarrow \qquad \sin^{-1} 1 + \sin^{-1} 0 = c \;\Rightarrow\; c = \frac{\pi}{2}$$

Thus the required solution is $\boldsymbol{\sin^{-1} y + \sin^{-1} x = \dfrac{\pi}{2}}$.

Example 8. Solve, $a^x (y^2 + 1)\, dx = y\, dy$.

Solution : The given differential equation is

$$a^x (y^2 + 1)\, dx = y\, dy$$

Dividing throughout by $(y^2 + 1)$ we get

$$a^x\, dx = \frac{y}{1+y^2}\, dy$$

On integration, we get

$$\int a^x\, dx = \frac{1}{2}\int \frac{2y}{1+y^2}\, dy + c \qquad \text{(Note this step)}$$

$$\Rightarrow \qquad \frac{a^x}{\log a} = \frac{1}{2}\log\left(1+y^2\right) + c \;\Rightarrow\; \mathbf{2a^x = \log a \cdot \log (1 + y^2) + k}$$

(where $k = c \log a$)

This is the required solution.

Example 9. Solve, $y (1 + \log x)\, dx - x \log x\, dy = 0$.

Solution : The given equation is

$$y (1 + \log x)\, dx - x \log x\, dy = 0$$

Separating the variables, we have

$$\frac{1+\log x}{x \log x}\, dx - \frac{1}{y}\, dy = 0$$

On integration, we have

$$\int \frac{1+\log x}{x \log x}\, dx - \int \frac{1}{y}\, dy = \log k$$

Observe that, $\qquad \dfrac{d}{dx}(x \cdot \log x) = x \cdot \dfrac{1}{x} + 1 \cdot \log x = (1 + \log x)$

$\therefore$ We have, $\qquad \log (x \log x) - \log y = \log k \qquad \left(\because \int \dfrac{f'(x)}{f(x)}\, dx = \log f(x)\right)$

$$\Rightarrow \qquad \log\left(\frac{x \log x}{y}\right) = \log k \;\Rightarrow\; \frac{x \log x}{y} = k \;\text{ or }\; \mathbf{x \log x = ky}$$

This is the required solution.

Example 10. Solve, $(x + 1)\dfrac{dy}{dx} + 1 = 2e^{-y}$.

Solution : The given differential equation is

$$(x + 1)\frac{dy}{dx} + 1 = 2e^{-y} \qquad \Rightarrow \qquad (x+1)\frac{dy}{dx} = 2e^{-y} - 1$$

Separating the variables, we have

$$\frac{dy}{2e^{-y}-1} = \frac{dx}{x+1} \Rightarrow \int \frac{dy}{2e^{-y}-1} = \int \frac{dx}{x+1} + \log k$$

$$\Rightarrow \int \frac{e^y}{2-e^y}\, dy = \log (x+1) + \log k \qquad \text{(Note)}$$

$$\Rightarrow -\log (2-e^y) = \log k\,(x+1)$$

$$\Rightarrow \log \frac{1}{2-e^y} = \log k\,(x+1)$$

$$\Rightarrow \mathbf{\frac{1}{2-e^y} = k\,(x+1)}$$

which is the solution of the equation.

Example 11. Find the particular solution of the equation, $dy = x\,(100 - x^2)^{1/2}\, dx$ given that $y = 0$ and $x = 0$.

Solution : The given equation is

$$dy = x\,(100-x^2)^{1/2}\, dx \Rightarrow \int dy = -\frac{1}{2}\int (100-x^2)^{1/2} \cdot (-2x) + k \qquad \text{(Note)}$$

$$\Rightarrow y = -\frac{1}{2}\,\frac{(100-x^2)^{3/2}}{(3/2)} + k$$

$$\Rightarrow y = -\frac{1}{3}(100-x^2)^{3/2} + k$$

By data, $y = 0$ when $x = 0$ $\Rightarrow$ $0 = -\frac{1}{3}(100)^{3/2} + k$

$$\Rightarrow k = \frac{1000}{3}$$

Thus the required particular solution is

$$y = -\frac{1}{3}(100-x^2)^{3/2} + \frac{1000}{3} \Rightarrow \mathbf{3y + (100-x^2)^{3/2} = 1000}$$

Example 12. Find the general solution of $y\,(2 \log y + 1)\, dy = (\sin x + x \cos x)\, dx$.

Solution : We have, $y\,(2\log y + 1)\, dy = (\sin x + x\cos x)\, dx$

$$\Rightarrow \int y\,(2\log y + 1)\,dy = \int (\sin x + x\cos x)\, dx$$

$$\Rightarrow \int (2\log y)\cdot y\, dy + \int y\, dy = \int \sin x\, dx + \int \cos x\, dx$$

$$\Rightarrow (2\log y)\frac{y^2}{2} - \int \frac{y^2}{2}\cdot\left(\frac{2}{y}\right) dy + \frac{y^2}{2} = -\cos x + \left[x\sin x - \int \sin x \cdot 1\, dx\right]$$

$$\Rightarrow y^2 \log y - \frac{y^2}{2} + \frac{y^2}{2} = -\cos x + x\sin x + \cos x + k$$

$$\Rightarrow \mathbf{y^2 \log y = x\sin x + k,} \text{ which is the general solution.}$$

Example 13. Find the particular solution of the equation $(x + 1)(y - 1)\,dx + (x - 1)(y + 1)\,dy = 0$, given $y = 2$ when $x = 2$.

Solution : The given equation is

$$(x - 1)(y - 1)\,dx + (x - 1)(y + 1)\,dy = 0$$

Dividing, throughout by $(y - 1)(x - 1)$ we get

$$\frac{x+1}{x-1}\,dx + \frac{y+1}{y-1}\,dy = 0$$

On integration, we get

$$\int \frac{x+1}{x-1}\,dx + \int \frac{y+1}{y-1}\,dy = k \quad \Rightarrow \quad \int \frac{(x-1)+2}{(x-1)}\,dx + \int \frac{(y-1)+2}{(y-1)}\,dy = k \qquad \text{(Note)}$$

$$\Rightarrow \quad \int \left(1 + \frac{2}{x-1}\right) dx + \int \left(1 + \frac{2}{y-1}\right) dy = k$$

$$\Rightarrow \quad x + 2\log(x - 1) + y + 2\log(y - 1) = k$$

$$\Rightarrow \quad (x + y) + 2\log(x - 1)(y - 1) = k$$

By data, $y = 2$ when $x = 2$. Thus we have

$$(2 + 2) + 2\log(2 - 1)(2 - 1) = k \quad \Rightarrow \quad k = 4 \qquad (\because \log 1 = 0)$$

Thus, the required particular solution is

$$\mathbf{(x + y) + 2\log(x - 1)(y - 1) = 4}$$

Example 14. Solve, $\dfrac{dy}{dx} + xy = xy^2$, when $y = 4$ and $x = 1$.

Solution : We have, $\dfrac{dy}{dx} + xy = xy^2 \quad \Rightarrow \quad \dfrac{dy}{dx} = xy(y - 1)$

Separating the variables, we get

$$\frac{dy}{y(y-1)} = x\,dx \quad \Rightarrow \quad \int \left(\frac{1}{y-1} - \frac{1}{y}\right) dy = \int x\,dx + k \quad \text{(Resolving into partial fraction)}$$

$$\Rightarrow \quad \log(y - 1) - \log y = \frac{x^2}{2} + k$$

$$\Rightarrow \quad \log\left(\frac{y-1}{y}\right) = \frac{x^2}{2} + k$$

By data, $y = 4$ and $x = 1$. Thus we have,

$$\log\left(\frac{4-1}{4}\right) = \frac{1}{2} + k \quad \Rightarrow \quad \log\frac{3}{4} = \frac{1}{2} + k \Rightarrow k = \log\left(\frac{3}{4}\right) - \frac{1}{2}$$

$\therefore$ the required particular solution is $\log\left(\dfrac{y-1}{y}\right) = \dfrac{x^2}{2} + \log\left(\dfrac{3}{4}\right) - \dfrac{1}{2}$

Example 15. Solve, $y\cos^2x\,dy = x\cos^2y\,dx$.

Solution : The given equation is

$$y\cos^2x\,dy = x\cos^2y\,dx$$

$$\Rightarrow \quad \frac{y\,dy}{\cos^2 y} = \frac{x}{\cos^2 x}\,dx$$

$$\Rightarrow \quad \int y\cdot\sec^2 y\,dy = \int x\cdot\sec^2 x\,dx + k$$

Integrating by parts we get

$$y\cdot\tan y - \int \tan y\cdot 1\,dy = x\cdot\tan x - \int \tan x\cdot 1\,dx + k$$

$$\Rightarrow \quad \mathbf{y\cdot\tan y - \log(\sec y) = x\tan x - \log(\sec x) + k}$$

which is required solution.

Example 16. Solve, $\quad y - x\dfrac{dy}{dx} = 3\left(1 + x^2\dfrac{dy}{dx}\right)$.

Solution : The given equation is

$$y - x\frac{dy}{dx} = 3\left(1 + x^2\frac{dy}{dx}\right)$$

$$\Rightarrow \quad \frac{dy}{dx}(3x^2 + x) = y - 3$$

$$\Rightarrow \quad \frac{dy}{y-3} = \frac{dx}{x(3x+1)} \quad \Rightarrow \quad \frac{dy}{y-3} = \left(\frac{1}{x} - \frac{3}{3x+1}\right)dx$$

$$\Rightarrow \quad \int\frac{dy}{y-3} = \int\left(\frac{1}{x} - \frac{3}{3x+1}\right)dx + \log k$$

$$\Rightarrow \quad \log(y-3) = \log x - \log(3x+1) + \log k$$

$$\Rightarrow \quad \log(y-3) = \log\left(\frac{kx}{3x+1}\right) \quad \Rightarrow \quad \mathbf{y - 3 = \frac{kx}{3x+1}}$$

This is the general solution.

Example 17. Solve, $\cot y\,\dfrac{dy}{dx} = \cos(x+y) + \cos(x-y)$.

Solution : We have $\quad \cot y\,\dfrac{dy}{dx} = \cos(x+y) + \cos(x-y)$

$$\Rightarrow \quad \cot y\,\frac{dy}{dx} = 2\cos\left(\frac{x+y+x-y}{2}\right)\cos\left(\frac{x+y-x+y}{2}\right) \quad \text{(applying transformation formula)}$$

$$\Rightarrow \quad \cot y\,dy = 2\cos x\cdot\cos y \quad \Rightarrow \quad \frac{1}{\sin y}\,dy = 2\cos x\,dx$$

$$\Rightarrow \quad \int \operatorname{cosec} y\,dy = 2\int\cos x\,dx + k$$

$$\Rightarrow \quad \mathbf{\log(\operatorname{cosec} y - \cot y) = 2\sin x + k}$$

which is the required solution.

Exercise

I.

Solve the following differential equations by separating the variables.

1. $\frac{dy}{dx} = 3x^2 + 2$ **2.** $\frac{dy}{dx} = 4x + 6$ **3.** $\frac{dy}{dx}\ (e^x + 1)\, y = 0$

4. $x^2 \frac{dy}{dx} = 4$ **5.** $\frac{dy}{dx} + 2x = e^{3x}$ **6.** $x^2 \frac{dy}{dx} = y^2 + 1$ **7.** $(e^x + e^{-x}) \frac{dy}{dx} = e^x - e^{-x}$.

II.

Solve the following differential equations by separating the variables.

1. $(x^2 + 4x + 9) \frac{dy}{dx} = x + 2$

2. $(1 - x^2)\, dy + xy\, dx = xy^2\, dx$

3. $e^x \tan y\, dx + (1 - e^x) \sec^2 y\, dy = 0$

4. $(e^y + 1) \cos x\, dx + e^y \sin x\, dy = 0$

5. $e^y \frac{dy}{dx} + x^2 = x^2 e^y$

6. $x \sqrt{1 + y^2}\, dx + y \sqrt{1 + x^2}\, dy = 0$

7. $(y^2 + y)\, dx + (x^2 + x)\, dy = 0$

8. $(1 - x^2) \frac{dy}{dx} - xy = y$

9. $\frac{dy}{dx} = \frac{x\,(2 \log x + 1)}{\sin y + y \cos y}$

10. $\cos y \cdot \log (\sec x + \tan x)\, dx = \cos x \cdot \log (\sec y + \tan y)\, dy$.

11. $\frac{dy}{dx} = 3x + 4y + 6xy + 2$

12. $(1 - \cos 2x) dy + (1 + \cos 2y)\, dx = 0$

13. $3e^x \tan y\, dx + (1 - e^x) \sec^2 y\, dy = 0$

14. $(\tan^2 x + 2 \tan x + 5) \frac{dy}{dx} = 2\,(1 + \tan x) \sec^2 x$

15. $(x^2y - x^2)\, dx + (xy^2 - y^2)\, dy = 0$

16. $a \left(x \frac{dy}{dx} + 2y \right) + 2xy \frac{dy}{dx} = 0$

17. $\tan y \frac{dy}{dx} = \sin (x + y) + \sin (x - y)$

18. $(1 - x^2)\,(1 - y)\, dx = xy\,(1 + y)\, dy$.

19. $\tan x \cdot \sec y\, dx + dy = 0$

20. $\cos^2 y\, dx - \operatorname{cosec} x\, dy = 0$

21. $xy\, dx + \sqrt{1 + x^2}\, dy = 0$

22. $\log \left(\frac{dy}{dx} \right) = ax + by$

23. $(x - 2)\, dy - (6x + 5)\, dx = 0$

24. $y^2\, dx - dy = x^2\, dy$

25. $\left(1 + y^2\right) \sin^{-1} x\, dx + \sqrt{1 - x^2}\, dy = 0$

26. $(1 + y^2) \tan^{-1} x\, dx + 2y\,(1 + x^2)\, dy = 0$

27. $\sec^2 y \cdot \tan x\, dy + \sec^2 x \tan y\, dx = 0$

28. $3e^x \tan y\, dx + (1 + e^x) \sec^2 y\, dy = 0$

29. $y^2 - \frac{dy}{dx} = x^2 \frac{dy}{dx}$

30. $\sec x\, dx + \operatorname{cosec} y\, dy = 0$

31. $e^{(dy/dx)} = x$

32. $x\,(1 + y^2)\, dx + y\,(1 + x^2)\, dy = 0$

33. $y - x \frac{dy}{dx} = a \left(y^2 + \frac{dy}{dx} \right)$

34. $\frac{dy}{dx} = 1 + x + y + xy$

III.

1. Solve $\frac{dy}{dx} = (1 + x)(1 + y^2)$, given that $y = 1$ and $x = 0$.

2. Solve $(1 + y)\,x\,\frac{dy}{dx} + (1 + x)\,y = 0$, given that $y = 1$ when $x = 1$.

3. Solve $(1 + x^2)\frac{dy}{dx} + (1 + y^2) = 0$, given that $y = 1$ when $x = 0$.

4. Solve $xy\frac{dy}{dx} = y + 2$, given that $y = 0$ when $x = 2$.

5. Solve $\frac{dy}{dx} = e^{x+y}$, given that $y = 1$ when $x = 1$.

6. Solve $\cos y\, dy + \cos x \cdot \sin y\, dx = 0$, given that $y = \frac{\pi}{2}$ when $x = \frac{\pi}{2}$.

7. Solve $\log\left(\frac{dy}{dx}\right) = 3x + 4y$, given that $y = 0$ when $x = 0$.

8. Solve $(x - 1)\frac{dy}{dx} = 2xy$, given $y = 1$ when $x = 2$.

9. Solve $(1 + e^{2x})\, dy + (1 + y^2)\, e^x\, dx = 0$, given that $y = 1$ when $x = 0$.

Answers

I.

1. $y = x^3 + 2x + c$
2. $y = 2x^2 + 6x + c$
3. $\log y = e^x + x + c$
4. $y = -\frac{4}{x} + c$
5. $y = \frac{1}{3}e^{3x} - x^2 + c$
6. $\tan^{-1} y = -\frac{1}{x} + c$
7. $y = \log(e^x + e^{-x}) + c$.

II.

1. $2y = \log(x^2 + 4x + 9) + k$
2. $(y - 1)\sqrt{1 - x^2} = cy$
3. $\tan y = c(1 - e^y)$
4. $(e^y + 1)\sin x = c$
5. $\log(e^y - 1) = \frac{x^3}{3} + c$
6. $\sqrt{1 + x^2} + \sqrt{1 + y^2} = c$
7. $xy = c(x + 1)(y + 1)$
8. $y(1 - x) = c$
9. $y \sin y = x^2 \log x + c$
10. $[\log(\sec x + \tan x)]^2 = [\log(\sec y + \tan y]^2 + c$
11. $\log(1 + 2y) = 3x^2 + 4x + k$
12. $\tan y - \cot x = k$
13. $\tan y = c(1 - e^x)^3$
14. $y = \log(\tan^2 x + 2\tan x + 5) + c$
15. $x^2 + y^2 + 2(x + y) + \log(x - 1)^2 (y - 1)^2 = k$
16. $\log(xy^2) + \frac{2y}{a} = c$
17. $y = x + \frac{1}{2}\sin 2x + c$
18. $2\log x - x^2 = -4\log(1 - y) - y^2 - 4y + k$
19. $\sin y - \log(\cos x) = c$
20. $\cos x + \tan y + c = 0$
21. $\sqrt{1 + x^2} + \log y = c$
22. $\frac{e^{ax}}{a} + \frac{e^{-by}}{b} = c$
23. $y = 6x + 17\log(x - 2) + c$
24. $\tan^{-1} x + \frac{1}{y} = c$
25. $(\sin^{-1} x)^2 = 2\tan^{-1} y + c$

26. $\left(\tan^{-1} x\right)^2 + 2 \log\left(1 + y^2\right) = c$ **27.** $(\tan x) \cdot (\tan y) = c$

28. $(1 + e^x)^3 \cdot \tan y = k$ **29.** $y \tan^{-1} x + 1 = ky$

30. $(\sec x + \tan x)(\operatorname{cosec} y - \cot y) = c$ **31.** $y = x \log x - x + k$

32. $(1 + x^2)(1 + y^2) = k$ **33.** $(y + a)(1 - ax) = Ax$ **34.** $\log(1 + y) = \frac{x^2}{2} + x + c$

III.

1. $\tan^{-1} y = x + \frac{x^2}{2} + \frac{\pi}{4}$ **2.** $x + y + \log xy = 2$ **3.** $\tan^{-1} x + \tan^{-1} y = \frac{\pi}{4}$

4. $y - 2 \log(y + 2) = \log x - 3 \log 2$ **5.** $e^x + e^{-y} - \frac{1 + e^2}{e} = 0$ **6.** $\log(\sin y) + \sin x = 1$

7. $4e^{3x} + 3e^{-4y} + 7 = 0$ **8.** $\log y = 2x + 2 \log(x - 1) - 4$ **9.** $\tan^{-1} y + \tan^{-1} e^x = \frac{\pi}{2}$.

1.5 Equations Reducible to Variable Separable Form

Some of the differential equations can be transformed to a variable separable form, by suitable substitution.

For example, equation of the form, $\frac{dy}{dx} = f(ax + by + c)$, can be reduced to an equation in which the variables can be separated, by putting $ax + by + c = v$.

Example 1. Solve, $\frac{dy}{dx} = (3x + 2y + 4)^2$.

Solution : Put, $3x + 2y + 4 = v \Rightarrow 3 + 2\frac{dy}{dx} = \frac{dv}{dx} \Rightarrow \frac{dy}{dx} = \frac{1}{2}\left(\frac{dv}{dx} - 3\right)$

Thus, the given equation becomes

$$\frac{1}{2}\left(\frac{dv}{dx} - 3\right) = v^2 \Rightarrow \frac{dv}{dx} = 2v^2 + 3 \Rightarrow \int \frac{dv}{2v^2 + 3} = \int dx + c \quad \text{(separating the variable)}$$

$$\Rightarrow \int \frac{dv}{\left(\sqrt{3}\right)^2 + \left(\sqrt{2}\,v\right)^2} = x + c$$

$$\Rightarrow \frac{1}{\sqrt{3}} \frac{1}{\sqrt{2}} \tan^{-1} \frac{\sqrt{2}\,v}{\sqrt{3}} = x + c$$

Thus, the solution is, $\frac{1}{\sqrt{6}} \tan^{-1}\left[\sqrt{\frac{2}{3}}\,(3x + 2y + 4)\right] = x + c.$

Example 2. Solve, $(x + y + 1)\frac{dy}{dx} = 1$.

Solution : Put, $x + y + 1 = v \Rightarrow 1 + \frac{dy}{dx} = \frac{dv}{dx} \Rightarrow \frac{dy}{dx} = \frac{dv}{dx} - 1$

Thus, the given equation becomes, $v\left(\frac{dv}{dx} - 1\right) = 1 \Rightarrow \frac{dv}{dx} = \frac{1}{v} + 1 \Rightarrow \frac{dv}{dx} = \frac{1 + v}{v}$

Separating the variables and integrating we get,

$$\int \frac{v}{1+v}\,dv = \int dx + c \quad \Rightarrow \quad \int \frac{(1+v)-1}{1+v}\,dv = \int dx + c$$

$$\Rightarrow \quad \int \left(1 - \frac{1}{1+v}\right) dv = x + c \quad \Rightarrow \quad v - \log(1+v) = x + c$$

$$\Rightarrow \quad \mathbf{(x+y+1) - \log(x+y+2) = x + c.}$$

Example 3. Solve, $\dfrac{dy}{dx} + 1 = e^{x+y}$.

Solution : Put, $x + y = v \;\Rightarrow\; 1 + \dfrac{dy}{dx} = \dfrac{dv}{dx}$. Thus the equation becomes,

$$\frac{dv}{dx} = e^v \quad \Rightarrow \quad \int e^{-v}\,dv = \int dx + c \quad \Rightarrow \quad -e^{-v} = x + c \quad \Rightarrow \quad \mathbf{x + e^{-(x+y)} + c = 0}$$

Example 4. Solve, $\dfrac{dy}{dx} = \cos(x+y)$.

Solution : Put, $x + y = v \;\Rightarrow\; 1 + \dfrac{dy}{dx} = \dfrac{dv}{dx} \;\Rightarrow\; \dfrac{dy}{dx} = \dfrac{dv}{dx} - 1$

Thus the equation becomes, $\dfrac{dv}{dx} - 1 = \cos v \;\Rightarrow\; \dfrac{dv}{dx} = 1 + \cos v$

Separating the variables and integrating we get,

$$\int \frac{dv}{1+\cos v} = \int dx + c \quad \Rightarrow \quad \int \frac{1}{2\cos^2 \frac{v}{2}}\,dv = x + c$$

$$\Rightarrow \quad \frac{1}{2}\int \sec^2 \frac{v}{2}\,dv = x + c \quad \Rightarrow \quad \tan\frac{v}{2} = x + c \quad \Rightarrow \quad \mathbf{\tan\left(\frac{x+y}{2}\right) = x + c.}$$

Example 5. Solve, $\dfrac{dy}{dx} = \dfrac{x+y+1}{2x+2y+3}$.

Solution : The given equation is, $\dfrac{dy}{dx} = \dfrac{(x+y)+1}{2(x+y)+3}$

Put, $x + y = v \;\Rightarrow\; 1 + \dfrac{dy}{dx} = \dfrac{dv}{dx} \;\Rightarrow\; \dfrac{dy}{dx} = \dfrac{dv}{dx} - 1$

Thus, the equation becomes, $\dfrac{dv}{dx} - 1 = \dfrac{v+1}{2v+3} \;\Rightarrow\; \dfrac{dv}{dx} = \dfrac{v+1}{2v+3} + 1 \;\Rightarrow\; \dfrac{dv}{dx} = \dfrac{3v+4}{2v+3}$

Separating the variable and integrating we get,

$$\int \frac{2v+3}{3v+4}\,dv = \int dx + c \quad \Rightarrow \quad \frac{2}{3}\int dv + \frac{1}{3}\int \frac{1}{3v+4}\,dv = x + c \qquad \left(\because 2v+3 = \frac{2}{3}(3v+4) + \frac{1}{3}\right)$$

$$\Rightarrow \quad \frac{2}{3}v + \frac{1}{9}\log(3v+4) = x + c$$

i.e., $\quad \mathbf{\frac{2}{3}(x+y) + \frac{1}{9}\log(3x+3y+4) = x + c,}$ which is a solution.

Example 6. Solve, $\frac{dy}{dx} = \sin(x+y) + \cos(x+y)$.

Solution : Put, $x + y = v \Rightarrow 1 + \frac{dy}{dx} = \frac{dv}{dx} \Rightarrow \frac{dy}{dx} = \frac{dv}{dx} - 1$

Thus, the equation becomes,

$$\frac{dv}{dx} - 1 = \sin v + \cos v \quad \Rightarrow \quad \frac{dv}{dx} = 1 + \cos v + \sin v$$

$$\Rightarrow \quad \frac{dv}{dx} = 2\cos^2\frac{v}{2} + 2\sin\frac{v}{2}\cos\frac{v}{2} \quad \Rightarrow \quad \frac{dv}{dx} = 2\cos^2\frac{v}{2}\left(1 + \tan\frac{v}{2}\right)$$

Separating the variables and integrating we get,

$$\int \frac{dv}{2\cos^2\frac{v}{2}\left(1+\tan\frac{v}{2}\right)} = \int dx + c \quad \Rightarrow \quad \int \frac{\frac{1}{2}\sec^2\frac{v}{2}}{\left(1+\tan\frac{v}{2}\right)}dv = x + c$$

$$\Rightarrow \quad \log\left(1 + \tan\frac{v}{2}\right) = x + c \qquad \left(\because \int \frac{f'(x)}{f(x)}dx = \log f(x)\right)$$

$$\Rightarrow \quad \mathbf{\log\left[1 + \tan\left(\frac{x+y}{2}\right)\right] = x + c.}$$

Example 7. Solve, $\left(x\frac{dy}{dx} - y\right)e^{y/x} = x^2\sec^2 x$, by putting $\frac{y}{x} = v$.

Solution : We have, $\frac{y}{x} = v \quad \Rightarrow \quad y = vx \quad \Rightarrow \quad \frac{dy}{dx} = v + x\frac{dv}{dx}.$

Thus the equation becomes

$$\left[x\left(v + x\frac{dv}{dx}\right) - xv\right]e^v = x^2\sec^2 x$$

$$\Rightarrow \quad xe^v\left[v + x\frac{dv}{dx} - v\right] = x^2\sec^2 x$$

$$\Rightarrow \quad xe^v\frac{dv}{dx} = x\sec^2 x \quad \Rightarrow \quad e^v\frac{dv}{dx} = \sec^2 x$$

$$\Rightarrow \quad e^v\,dv = \sec^2 x\,dx$$

$$\Rightarrow \quad \int e^v\,dv = \int \sec^2 x\,dx + c$$

$$\Rightarrow \quad e^v = \tan x + c$$

$$\Rightarrow \quad \mathbf{e^{(y/x)} = \tan x + c} \qquad \left(\because v = \frac{y}{x}\right)$$

Example 8. Solve, $\left(x \tan \frac{y}{x} + y \sec^2 \frac{y}{x}\right) dx = x \sec^2 \frac{y}{x}\, dy$**, using** $y = vx$**.**

Solution : The given equation can be written as

$$\frac{dy}{dx} = \frac{1}{x \sec^2 (y/x)}\left[x \tan\left(\frac{y}{x}\right) + y \sec^2\left(\frac{y}{x}\right)\right]$$

$$\frac{dy}{dx} = \frac{\tan (y/x)}{\sec^2 (y/x)} + \frac{y}{x}$$

Now, $\quad y = vx \quad \Rightarrow \quad \frac{dy}{dx} = v + x\frac{dv}{dx}$. Thus, we have

$$v + x\frac{dv}{dx} = \frac{\tan v}{\sec^2 v} + v \quad \Rightarrow \quad x\frac{dv}{dx} = \frac{\tan v}{\sec^2 v}$$

$$\Rightarrow \quad \frac{\sec^2 v}{\tan v}\, dv = \frac{dx}{x}$$

$$\Rightarrow \quad \int \frac{\sec^2 v}{\tan v}\, dv = \int \frac{dx}{x} + \log c$$

$$\Rightarrow \quad \log (\tan v) = \log x + \log c$$

$$\Rightarrow \quad \log (\tan v) = \log xc$$

$$\Rightarrow \quad \mathbf{\tan (y/x) = xc} \qquad \left(\because v = \frac{y}{x}\right)$$

Exercise

Solve the following differential equation

1. $(x - y)^2 \frac{dy}{dx} = 1$
2. $\frac{dy}{dx} = (4x + y + 1)^2$
3. $\frac{dy}{dx} = (2x + 3y - 4)^2$
4. $(x + y)^2 \frac{dy}{dx} = 4$
5. $\frac{dy}{dx} + 1 = e^{x-y}$
6. $\sin^{-1}\left(\frac{dy}{dx}\right) = x + y$
7. $\frac{dy}{dx} = \sec (x + y)$
8. $\frac{dy}{dx} = \frac{4x + 6y + 5}{2x + 3y + 4}$
9. $\frac{dy}{dx} = \frac{x + y + 1}{x + y - 1}$
10. $\frac{dy}{dx} = (x + y)^2$
11. $\frac{dy}{dx} = \tan^2 (x + y)$
12. $(4x + y)^2 = \frac{dy}{dx}$ by putting $4x + y = v$
13. $\sin^{-1}\left(\frac{dy}{dx}\right) = x + y$ by putting $x + y = v$
14. $\cos^2 (x - 2y) = 1 - 2\frac{dy}{dx}$ by putting $x - 2y = u$
15. $(x - y)\left(1 - \frac{dy}{dx}\right) = e^x$ by putting $x - y = v$
16. $(x + y)\frac{dy}{dx} + y = 0$ by putting $x + y = v$
17. $x + y\frac{dy}{dx} = x^2 + y^2$ by putting $x^2 + y^2 = v$
18. $x + y\frac{dy}{dx} = \sec (x^2 + y^2)$ by putting $x^2 + y^2 = v$
19. $(3x + y)^2 \frac{dy}{dx} = 1$ putting $3x + y = v$

20. $1 - \frac{dy}{dx} = \sec(x - y)$ by putting $x - y = v$

21. $\left(y + x\frac{dy}{dx}\right)\cos xy = \sin^2 x$ by putting $xy = v$

22. $\frac{dy}{dx} = \frac{y}{x} + \tan\frac{y}{x}$ by putting $y = vx$

23. $\left(1 + e^{x/y}\right)dx + e^{x/y}\left(1 - \frac{x}{y}\right)dy = 0$ by putting $x = vy$

24. $\left(x\frac{dy}{dx} - y\right)\sin\left(\frac{y}{x}\right) = x^2 e^x$ by putting $y = vx$

25. $(2x - y)e^{y/x}\,dx + (y + xe^{y/x})dy = 0$ by putting $y = vx$

Answers

1. $x - y + \frac{1}{2}\log\left(\frac{x - y - 1}{x + y + 1}\right) = x + c$

2. $\frac{1}{2}\tan^{-1}\left(\frac{4x + y + 1}{2}\right) = x + c$

3. $\frac{1}{\sqrt{6}}\tan^{-1}\left[\frac{\sqrt{3}\,(2x + 3y - 4)}{\sqrt{2}}\right] = x + c$

4. $(x + y) - 2\tan^{-1}\left(\frac{x + y}{2}\right) = x + c$

5. $2x + \log(2e^{y-x} - 1) + k = 0$

6. $\tan(x + y) - \sec(x + y) = x + c$

7. $y - \tan\left(\frac{x + y}{2}\right) = c$

8. $3y + 9\log(16x + 24y + 23) = 6x + k$

9. $y - x - \log(x + y) = k$

10. $\tan^{-1}(x + y) = x + c$

11. $y - x + \frac{1}{2}\sin 2(x + y) = c.$

12. $\frac{1}{2}\tan^{-1}\left(\frac{4x + y}{2}\right) = x + c$

13. $\tan(x + y) - \sec(x + y) = x + c$

14. $\tan(x - 2y) = x + c$

15. $\frac{(x - y)^2}{2} = e^x + c$

16. $y^2 + 2xy = c$

17. $\log(x^2 + y^2) = 2x + c$

18. $\sin(x^2 + y^2) = 2x + c$

19. $\frac{3x + y}{3} - \frac{\sqrt{3}}{9}\tan^{-1}\left[\sqrt{3}\,(3x + y)\right] = x + c$

20. $\sin(x - y) = x + c$

21. $4\sin xy = 2x - \sin 2x + c$

22. $\sin\frac{y}{x} = kx$

23. $x + ye^{x/y} = c$

24. $e^x + \cos\frac{y}{x} = c$

25. $2x^2 e^{y/x} + y^2 = c$

1.6 Homogeneous Differential Equations

In this section we shall see an another method of solving a first order first degree equation, when it is in a particular form called homogeneous differential equation. Even in this case we look for a substitution (which is uniform for this type of equation) to bring the equation to variable separable form.

Definition : A function $f(x, y)$ in the variables x and y is called a homogeneous function, of degree n, if the degree of each term is n.

Example : $f(x, y) = 3x^2 - xy + y^2$, $g(x, y) = 4x^3 - 3x^2y + 7y^3$

are homogeneous function of degree 2 and 3 respectively. Observe that the degree of each term in $f(x, y)$ is 2, where as the degree of each term in $g(x, y)$ is 3.

The function $h(x, y) = 4x^3 - 3x^2y + 4xy^2 - 3xy + 4x^2$, is not a homogeneous function - for the reason, the first, second and third terms are of degree 3, where as the next two terms is of degree two.

A homogeneous function $f(x, y)$ of degree n, can always be expressed as,

$$f(x, y) = x^n g\left(\frac{y}{x}\right)$$

Where $g\left(\frac{y}{x}\right)$ is an expression in $\frac{y}{x}$.

For example,

$$3x^2 - xy - y^2 = x^2\left[3 - \frac{y}{x} - \left(\frac{y}{x}\right)^2\right]$$

Here, $f(x, y) = 3x^2 - xy - y^2$ and $g\left(\frac{y}{x}\right) = 3 - \frac{y}{x} - \left(\frac{y}{x}\right)^2$.

Note : Any function $f(x, y)$ which can be expressed is the form $x^n\ g\left(\frac{y}{x}\right)$ is a homogeneous function of degree n.

Example : $x^2 \sin\left(\frac{y}{x}\right)$ or $x \tan\left(\frac{y}{x}\right)$ are homogeneous function of degree 2 and 1 respectively.

Defintion : The differential equation of the form $\frac{dy}{dx} = \frac{f(x, y)}{g(x, y)}$, where $f(x, y)$ and $g(x, y)$ are homogeneous function of same degree, is called a homogeneous first order equation.

For example :

(i) $\frac{dy}{dx} = \frac{3x^2 + y^2}{2xy}$ **(ii)** $\frac{dy}{dx} = x \sin\left(\frac{y}{x}\right)$ **(iii)** $\frac{dy}{dx} = \frac{x^2 - y^2}{x^2 + y^2}$

To solve a homogeneous equation of first order, we put $y = vx$ and procede. This substitution, make the differential equation to take variable separable form, so that it can be solved.

When $y = vx$ then $\frac{dy}{dx} = v + x\frac{dv}{dx}$

Working rule to solve homogeneous differential equation.

Step 1: Put $y = vx$ and replace $\frac{dy}{dx}$ by $v + x\frac{dv}{dx}$.

Step 2: Separate the variables v and x.

Step 3: Integrate and introduce the constant of integration.

Step 4: Replace v by $\frac{y}{x}$.

Example 1. Solve, $\dfrac{dy}{dx} = \dfrac{x+y}{x}$.

Solution : Clearly the given equation is a homogeneous differential equation.

Put, $y = vx \Rightarrow \dfrac{dy}{dx} = v + x \cdot \dfrac{dv}{dx}$. Thus the equation becomes

$$v + x \cdot \frac{dv}{dx} = \frac{x + vx}{x} \quad \Rightarrow \quad v + x\frac{dv}{dx} = 1 + v$$

$$\Rightarrow \quad x\frac{dv}{dx} = 1$$

$$\Rightarrow \quad dv = \frac{dx}{x}$$

$$\Rightarrow \quad \int dv = \int \frac{dx}{x} + c$$

$$\Rightarrow \quad v = \log x + c \Rightarrow \frac{y}{x} = \log x + c \qquad \left(\because\ v = \frac{y}{x}\right)$$

which is the required solution.

Example 2. Solve, $\dfrac{dy}{dx} = \dfrac{x^2 + y^2}{xy}$.

Solution : Clearly, the given equation is a homogeneous equation.

Put, $y = vx \quad \Rightarrow \quad \dfrac{dy}{dx} = v + x\dfrac{dv}{dx}$

$$\therefore \quad v + x\frac{dv}{dx} = \frac{x^2 + v^2x^2}{x \cdot vx}$$

$$\Rightarrow \quad v + x\frac{dv}{dx} = \frac{x^2\left(1 + v^2\right)}{x^2 \cdot v}$$

$$\Rightarrow \quad v + x\frac{dv}{dx} = \frac{1+v^2}{v} \quad \Rightarrow \quad x\frac{dv}{dx} = \frac{1+v^2}{v} - v$$

$$\Rightarrow \quad x\frac{dv}{dx} = \frac{1 + v^2 - v^2}{v}$$

$$\Rightarrow \quad x\frac{dv}{dx} = \frac{1}{v}$$

$$\Rightarrow \quad v\,dv = \frac{dx}{x}$$

$$\Rightarrow \quad \int v\,dv = \int \frac{dx}{x} + c$$

$$\Rightarrow \qquad \frac{v^2}{2} = \log x + c$$

$$\Rightarrow \qquad \mathbf{\frac{y^2}{2x^2} = \log x + c} \qquad \left(\because v = \frac{y}{x}\right)$$

This is the required solution.

Example 3. Solve, $(x^2 + 3xy + y^2)\,dx - x^2\,dy = 0$.

Solution : The given equation can be written as

$$\frac{dy}{dx} = \frac{x^2 + 3xy + y^2}{x^2}$$

This is a homogeneous differential equation.

Put, $\qquad y = vx \qquad \Rightarrow \qquad \frac{dy}{dx} = v + x\frac{dv}{dx}$

$$\therefore \qquad v + x\frac{dv}{dx} = \frac{x^2 + 3x \cdot vx + v^2x^2}{x^2}$$

$$\Rightarrow \qquad v + x\frac{dv}{dx} = \frac{\cancel{x^2}\left(1 + 3v + v^2\right)}{\cancel{x^2}}$$

$$\Rightarrow \qquad v + x\frac{dv}{dx} = 1 + 3v + v^2 \qquad \Rightarrow \qquad x\frac{dv}{dx} = v^2 + 2v + 1$$

$$\Rightarrow \qquad x\frac{dv}{dx} = (v+1)^2$$

$$\Rightarrow \qquad \int \frac{dv}{(v+1)^2} = \int \frac{dx}{x} + c$$

$$\Rightarrow \qquad -\frac{1}{v+1} = \log x + c$$

$$\Rightarrow \qquad -\frac{1}{\left(\frac{y}{x} + 1\right)} = \log x + c \qquad \left(\because v = \frac{y}{x}\right)$$

$$\Rightarrow \qquad \mathbf{\left(\frac{x}{x+y}\right) + \log x + c = 0}$$

This is the required solution.

Example 4. Solve, $x^2\frac{dy}{dx} = x^2 + xy + y^2$.

Solution : The given equation can be written as

$$\frac{dy}{dx} = \frac{x^2 + xy + y^2}{x^2}$$

This is a homogeneous differential equation.

Put, $y = vx \Rightarrow \dfrac{dy}{dx} = v + x\dfrac{dv}{dx}$

$$\therefore \quad v + x\frac{dv}{dx} = \frac{x^2 + x \cdot vx + v^2x^2}{x^2} \Rightarrow v + x\frac{dv}{dx} = \frac{\cancel{x^2}\left(1 + v + v^2\right)}{\cancel{x^2}}$$

$$\Rightarrow \quad v + x\frac{dv}{dx} = 1 + v + v^2$$

$$\Rightarrow \quad x\frac{dv}{dx} = 1 + v^2$$

$$\Rightarrow \quad \int\frac{dv}{1+v^2} = \int\frac{dx}{x} + c$$

$$\Rightarrow \quad \tan^{-1} v = \log x + c$$

$$\Rightarrow \quad \mathbf{\tan^{-1}\left(\frac{y}{x}\right) = \log x + c} \qquad \left(\because v = \frac{y}{x}\right)$$

This is the required solution.

Example 5. Solve, $\dfrac{dy}{dx} = \dfrac{y + \sqrt{x^2 - y^2}}{x}$.

Solution : The given equation can be written as

$$\frac{dy}{dx} = \frac{y}{x} + \frac{x}{x}\sqrt{1 - \frac{y^2}{x^2}} \Rightarrow \frac{dy}{dx} = \frac{y}{x} + \sqrt{1 - \frac{y^2}{x^2}}$$

R.H.S. is a function of $\dfrac{y}{x}$. Thus it is a homogeneous function.

Put $y = vx \Rightarrow \dfrac{dy}{dx} = v + x\dfrac{dv}{dx}$

$$\therefore \quad v + x\frac{dv}{dx} = v + \sqrt{1 - v^2} \qquad \left(\because \frac{y}{x} = v\right)$$

$$\Rightarrow \quad x\frac{dv}{dx} = \sqrt{1 - v^2} \Rightarrow \int\frac{dv}{\sqrt{1 - v^2}} = \int\frac{dx}{x} + c$$

$$\Rightarrow \quad \sin^{-1} v = \log x + c$$

$$\Rightarrow \quad \mathbf{\sin^{-1}\left(\frac{y}{x}\right) = \log x + c}$$

This is the required solution.

Exercise

Solve the following differential equation

1. $\frac{dy}{dx} = \frac{y^2 - x^2}{2xy}$ 2. $\frac{dy}{dx} = \frac{y}{x + y}$ 3. $x\frac{dy}{dx} = 3x + 2y$ 4. $\frac{dy}{dx} = \frac{xy - y^2}{x^2}$

5. $\frac{dy}{dx} = \frac{x + y}{x - y}$ 6. $\frac{dy}{dx} = \frac{x^2 y}{x^3 + y^3}$ 7. $\frac{dy}{dx} = \frac{x^2 - xy + y^2}{xy}$ 8. $x^2 dy + y(x + y)\, dx = 0$

Answers

1. $x^2 + y^2 = cx$ 2. $y = ke^{x/y}$ 3. $y = kx^2 - 3x$ 4. $\frac{x}{y} + \log x + c = 0$

5. $\tan^{-1}\frac{y}{x} - \frac{1}{2}\log\left(x^2 + y^2\right) = c$ 6. $-\frac{x^2}{3y^3} + \log y = c$

7. $\frac{y}{x} + \log\left(\frac{x - y}{x}\right) + \log x = c$ 8. $x^2 y = k\,(y + 2x)$

1.7 Applications of Differential Equations

In this section we shall consider few applications of first order differential equations to solve some practical physical problems which involves rate of growth or rate of decay.

The mathematical formulation of certain physical problems often leads to differential equations of first order. Solving these equations we get the solution of the physical problems under given boundary conditions. Most of the times we have to derive the concerned differential equation.

It is known that the radio active element disintegrates (decays) at a rate proportional to the amount present. If m is the amount of material at any time t, then

$$\frac{dm}{dt} \propto m \quad \Rightarrow \quad \frac{dm}{dt} = -km$$

where k is the constant of proportionality. The negative sign is to show that the element is disintegrating (i.e., not a growth).

Example 1. It is given that the radium decays at a rate proportional to the amount present. If the initial mass of radium is N_0, find an expression for the mass of radium at time 't'.

Solution : Let N be the mass of the radium at time t. Then by data, $\frac{dN}{dt} = -kN$, where k is a constant, known for the given material. The negative sign is taken as the material is disintegrating.

$$\text{Now,} \quad \frac{dN}{dt} = -kN \quad \Rightarrow \quad \frac{dN}{N} = -kdt$$

$$\Rightarrow \quad \int\frac{dN}{N} = -k\int dt + \log c$$

$$\Rightarrow \quad \log N = -kt + \log c$$

$$\Rightarrow \quad \log N - \log c = -kt$$

$$\Rightarrow \qquad \log \frac{N}{c} = -kt$$

$$\Rightarrow \qquad N = c \cdot e^{-kt}$$

By data, $N = N_0$ when $t = 0$. Thus $N_0 = ce^0 \Rightarrow N_0 = c$.

Thus we have, $\boldsymbol{N = N_0 e^{-kt}}$, which is the expression for the mass of radium at time t.

Example 2. A radio active substance disintegrates at a rate proportional to its mass. When mass is 10 mgm the rate of disintegration is 0.051 mgm per day. How long will it take for the mass to be reduced from 10 to 5 mgm?

Solution : We have by data $\frac{dN}{dt} = -kN$, where N is the mass of the material at time 't'.

Again by data, when $N = 10$ mgm, $\frac{dN}{dt} = 0.051$.

$$\Rightarrow \quad 0.051 = -k \cdot 10 \quad \Rightarrow \quad k = -\frac{0.051}{10} = -0.0051$$

$$\text{Now,} \quad \frac{dN}{dt} = -kN \quad \Rightarrow \quad \int \frac{dN}{N} = -k \int dt + \log c$$

$$\Rightarrow \quad \log N = -kt + \log c$$

$$\Rightarrow \quad \log N - \log c = -kt \quad \Rightarrow \quad N = ce^{-kt}$$

Now, $N = 10$, when $t = 0 \quad \Rightarrow \quad 10 = c$

Thus, $\quad N = 10e^{-kt}$

Let $t = T$ when $N = 5$. Thus $5 = 10e^{-kT} \quad \Rightarrow \quad e^{kT} = 2$

$$\text{Now,} \quad e^{kT} = 2 \quad \Rightarrow \quad kT = \log_e^2 = 0.69$$

$$\Rightarrow \quad T = \frac{0.69}{0.0051} \qquad (\because k = 0.0051)$$

$$\Rightarrow \quad T = 135 \text{ days approximately.}$$

Example 3. The rate of increase of bacteria in a culture proportional to the number of bacteria present and it is found that the number doubles in 5 hours. Calculate how many times the bacteria may be expected to grow at the end of 15 hours.

Solution : Let the original number of bacteria present be x_0. That is the number of bacteria at $t = 0$ is x_0.

Let x be the number of bacteria at the end of t hours.

Then, by data, $\frac{dx}{dt} \propto x \quad \Rightarrow \quad \frac{dx}{dt} = kx$, where k is the constant of proportion.

$$\text{Now,} \quad \frac{dx}{dt} = kx \quad \Rightarrow \quad \int \frac{dx}{x} = \int k\,dt + c$$

$$\Rightarrow \quad \log x = kt + c$$

We have, $x = x_0$, when $t = 0 \quad \Rightarrow \quad \log x_0 = c \qquad \therefore \quad \log x = kt + \log x_0$

Again, by data, $x = 2x_0$ when $t = 5$.

$$\Rightarrow \quad \log 2x_0 = 5k + \log x_0 \Rightarrow \log \frac{2x_0}{x_0} = 5k \Rightarrow k = \frac{1}{5}\log 2$$

Thus we have, $\quad \log x = \left(\frac{1}{5}\log 2\right) t + \log x_0.$

Now we have to find x, when $t = 15$

$$\therefore \quad \log x = \left(\frac{1}{5}\log 2\right) 15 + \log x_0 \Rightarrow \log x = \log 8 + \log x_0$$

$$\Rightarrow \log x = \log 8x_0 \Rightarrow x = 8 \cdot x_0$$

Hence the number of bacteria will be 8 times the original number at the end of 15 hours.

Example 4. A population grows at the rate of 5% per year. How long does it take for the population to double?

Solution : Let the initial population be x_0 and the population at the end of t years be x.

By data, $\quad \frac{dx}{dt} = \frac{5}{100}x \Rightarrow \frac{dx}{x} = \frac{1}{20}dt$

$$\Rightarrow \int \frac{dx}{x} = \frac{1}{20}\int dt + c$$

$$\Rightarrow \log x = \frac{1}{20}t + c$$

When $t = 0$, we have $x = x_0$.

$$\therefore \quad \log x_0 = \frac{1}{20}\cdot 0 + c \Rightarrow \boldsymbol{c = \log x_0}$$

Now, we have to find t when $x = 2x_0 \Rightarrow \frac{x}{x_0} = 2.$

Now, $\quad \log x = \frac{1}{20}t + \log x_0 \Rightarrow \log x - \log x_0 = \frac{1}{20}t$

$$\Rightarrow \log \frac{x}{x_0} = \frac{1}{20}t$$

Now, when $\frac{x}{x_0} = 2$, we have, $\log 2 = \frac{1}{20}t \Rightarrow$ $\boldsymbol{t = 20 \cdot \log 2.}$

Thus the population will be double in **20 · log 2 years.**

Example 5. A radio active substance has half-life of *h* days. Find a formula for its mass m interms of *t*, if the initial mass is m_0. What is its initial decay rates?

Solution : By data, the half of the mass decays in h days.

Let m be the mass of substance after t days.

We have, $\quad \frac{dm}{dt} = -km,\quad$ where k is a constant. .... (1)

$$\Rightarrow \quad \frac{dm}{m} = -k\,dt \quad \Rightarrow \quad \int \frac{dm}{m} = -k \int dt$$

$$\Rightarrow \quad \log m = -kt + c$$

By data, $m = m_0$ when $t = 0$. $\therefore$ $\mathbf{\log m_0 = c}$.

$\therefore$ We have, $\log m = -kt + \log m_0$

Again, we have $m = \frac{1}{2} m_0$ when $t = h$. Thus we have

$$\log\left(\frac{m_0}{2}\right) = -kh + \log m_0 \quad \Rightarrow \quad kh = \log\left(\frac{m_0}{2m_0}\right) \quad \Rightarrow \quad \mathbf{k = \frac{1}{h}\log 2}$$

Thus, $\log m = -kt + \log m_0$ takes the form

$$\log m = -\frac{t}{h}\log 2 + \log m_0$$

$\Rightarrow$ $\log m = \log (2^{-t/h}) + \log m_0 \Rightarrow \mathbf{m = 2^{(-t/h)}\, m_0}$, which is the required formula.

Initial decay rate can be obtained by putting $k = \frac{1}{h}(\log 2)$ in (1)

i.e., $$\mathbf{\frac{dm}{dt} = -\frac{1}{h}(\log 2)\, m_0}.$$

Example 6. A wet porous substance in the open air loses its moisture at rate proportional to the moisture content. If a sheet hung in the wind loses half of its moisture during the first hour, when will it have lost 90% moisture, weather conditions remaining the same.

Solution : Let the moisture content be M at time t.

Thus we have, $$\frac{dM}{dt} = -kM \quad \Rightarrow \quad \int \frac{dM}{M} = -\int k\,dt + c$$

$$\Rightarrow \quad \log M = -kt + c \qquad \text{.... (1)}$$

Let, $M = M_0$ at $t = 0$, then $\mathbf{c = \log M_0}$.

Also, $M = \frac{1}{2} M_0$ at $t = 1$, then $\log \frac{M_0}{2} = -k + \log M_0 \Rightarrow \mathbf{k = \log 2}$.

Thus (1) becomes

$$\log M = (-\log 2)\, t + \log M_0 \qquad \text{.... (2)}$$

We have to find the value of t when the substance has lost 90% of moisture. That is we have to find the value of t, when the substance has only 10% of moisture.

Let, $M = 10\%$ of M_0 at $t = t_1$ i.e., let $M = \frac{1}{10} M_0$ at $t = t_1$

$$\therefore (2) \Rightarrow \quad \log\left(\frac{M_0}{10}\right) = (-\log 2)\, t_1 + \log M_0$$

$$\Rightarrow \quad (\log 2)\, t_1 = \log\left(\frac{10\, M_0}{M_0}\right) \quad \Rightarrow \quad \mathbf{t_1 = \frac{\log 10}{\log 2}\ \text{hours}}$$

Exercise

1. A population grows at the rate of 8% per year. How long does it take for the population to double?
2. Experiments show that radium disintegrates at a rate proportional to the amount of radium present at the moment. If half life is 1590 years, what percentage will disappear in one year?
3. A certain radio active material has a half life of 2 hours. Find the time interval required for a given amount of this material to decay to one tenth of its original mass.
4. A population of a country doubles in 40 years. Assuming that the rate of increase is proportional to the number of inhabitants, find the number of years in which it will treble itself.
5. The rate of increase of bacteria in a culture is proportional to the number of bacteria present and it is found that the number doubles in 6 hours. Calculate how many times the bacteria may be expected to grow at the end of 18 hours.
6. The rate of increase of bacteria in a culture is proportional to the number of bacteria present and it is found that the number doubles in 5 hours. Express this mathematically, using rate of increase of bacteria with respect to time. Hence calculate how many times the bacteria may be expected to grow at the end of 15 hours.
7. A wet porous substance in the open air loses its moisture at rate proportional to the moisture content. If a sheet hung in the wind loses half of its moisture during the first hour, when will it have lost 95% moisture, weather conditions remaining the same.
8. The rate of growth of a population is proportional to the number present. If the population of a city has doubled in the past 25 years and the present population is 1,00,000, when will the city have a population of 5,00,000.
9. In a culture, the bacteria count is 1,00,000. The number is increased by 10% in 2 hours. In how many hours will be count reach 2,00,000, if the rate of growth of bacteria is proportional to the number present.
10. Radium decomposes at rate proportional to the quantity of radium present. It is found that in 25 years, approximately 1.1 present of a certain quantity of radium has decomposed. Determine approximately how long it will take one-half of the original amount of radium to decompose ($\log_e(.989) = 0.01106$, $\log_e 2 = 0.6931$).
11. The rate of increase of bacteria in a certain culture is proportional to the number present. Given that the number triples in 5 hours, find how many bacteria will be present after 10 hours. Also find the time necessary for the number of bacteria to be 10 times the number initially present.
12. It is given that the rate at which some bacteria multiply is proportional to the instantaneous number present. If the original number of bacteria doubles in two hours, in how many hours will it be five times?
13. It is given that radium decomposes at a rate proportional to the amount present. If p percent of the original amount of radium disappears is l years, what percentage of it will remain after $2l$ years?
14. A radio active substance disintegrates at a rate proportional to the amount of the substance present. 50% of the given amount disintegrates in 1600 years. What percentage of the substance disintegrates in 10 years?

Answers

1. $\frac{25}{2}\log 2$ years **2.** 0.04% **3.** 30 min. **4.** $\frac{40\,(\log 3)}{\log 2}$ **5.** 8 times

6. 8 times **7.** $\frac{\log 50}{\log 2}$ hours **8.** 58 years **9.** $\frac{2\log 2}{\log\,(11/10)}$ hours

10. 1567 years **11.** 9 times, $\frac{5\log 10}{\log 3}$ hours **12.** $2\left(\frac{\log 5}{\log 2}\right)$ hours

13. $\left(10 - \frac{p}{10}\right)^2$% **14.** 0.43%

UNIT VIII

Laplace Transforms

Chapter 1

Laplace Transforms

1.1 Introduction

In this chapter we shall introduce basic concepts of Laplace Transform and develope its fundamental properties.

1.2 Laplace Transform

Definition : Let $f(t)$ be a function of real variable t, defined for $t \geq 0$. Laplace transform of the function $f(t)$ is denoted and defined by

$$L[f(t)] = \int_0^\infty e^{-st} f(t)\, dt$$

provided the r.h.s integral exists; here s is a parameter real or complex.

The operator L is called the **Laplace transform operator**.

Clearly, right hand side integral, when it exists, is a function of the parameter 's'. Thus $L[f(t)]$ is a function of 's'. This function is denoted by $\tilde{f}(s)$ or $F(s)$. In what follows we shall use the notation $F(s)$.

Thus, $$L\{f(t)\} = \int_0^\infty e^{-st} \cdot f(t)\, dt = F(s)$$

If $F(s)$ is the Laplace transform of the function $f(t)$, then the function $f(t)$ itself is called the **inverse Laplace transform of $F(s)$**. This is denoted by $L^{-1}\{F(s)\}$.

Thus, $$L\{f(t)\} = F(s) \quad \Leftrightarrow \quad f(t) = L^{-1}\{F(s)\}$$

1.3 Properties of Laplace Transform

We shall see few basic properties of Laplace transforms of the functions, in the form of theorems.

Theorem 1 (Linearity Property) : If $f(t)$ and $g(t)$ are two functions whose Laplace transforms exist and if a and b are any two constants, then

$$L\{a\,f(t) + b\,g(t)\} = a \cdot L\{f(t)\} + b \cdot L\{g(t)\}$$

Proof : By the definition of Laplace transform, we have,

$$L\{a\,f(t) + b\,g(t)\} = \int_0^\infty e^{-st}\left[a\,f(t) + b\,g(t)\right] dt$$

$$= \int_0^\infty e^{-st}\left[a \cdot f(t)\right] dt + \int_0^\infty e^{-st}\left[b \cdot g(t)\right] dt$$

$$= a\int_0^\infty e^{-st} \cdot f(t)\, dt + b\int_0^\infty e^{-st} \cdot g(t)\, dt$$

$$= a \cdot L\{f(t)\} + b \cdot L\{g(t)\}$$

Hence, $\quad \boldsymbol{L\{a \cdot f(t) + b \cdot g(t)\} = a \cdot L\{f(t)\} + b \cdot L\{g(t)\}}$

Note : In particular, we have,

(i) $\quad L\{f(t) + g(t)\} = L\{f(t)\} + L\{g(t)\}$

(ii) $\quad L\{c \cdot f(t)\} = c \cdot L\{f(t)\}$, where c is any constant

Theorem 2 (First shifting property) : If $L\{f(t)\} = F(s)$ then $L\{e^{at} \cdot f(t)\} = F(s - a)$.

Proof : By data, $\quad L\{f(t)\} = F(s)$

Now by the definition of Laplace transform, we have,

$$L\{e^{at} \cdot f(t)\} = \int_0^\infty e^{-st} \cdot \left[e^{at} \cdot f(t)\right] dt = \int_0^\infty e^{-(s-a)t} \cdot f(t)\, dt$$

If we put $s - a = r$ then we have,

$$L\{e^{at} \cdot f(t)\} = \int_0^\infty e^{-rt} \cdot f(t)\, dt = F(r) \quad \text{(by the definition Laplace transform)}$$

$$\therefore \quad \boldsymbol{L\{e^{at} \cdot f(t)\} = F(s-a)} \qquad (\because\ r = s - a)$$

The above result implies that, if we know the Laplace transform of a function $f(t)$ as $F(s)$, then we can write Laplace transform of the function $e^{at} \cdot f(t)$ by replacing s by $s - a$ in $F(s)$.

Theorem 3 (Change of scale property) : If $L\{f(t)\} = F(s)$ then $L\{f(at)\} = \dfrac{1}{a} F\left(\dfrac{s}{a}\right)$.

Proof : By data, $\quad L\{f(t)\} = F(s)$

Now by the definition of Laplace transform form, we have, $\quad L\{f(at)\} = \int_0^\infty e^{-st} \cdot f(at)\, dt$

Put, $at = u \Rightarrow dt = \dfrac{1}{a} du$. Now, $t = 0 \Rightarrow u = 0$ and $u \to \infty$ as $t \to \infty$.

$$\therefore \quad L\{f(at)\} = \int_0^\infty e^{-\frac{su}{a}} \cdot f(u) \cdot \frac{1}{a}\, du = \frac{1}{a}\int_0^\infty e^{\left(-\frac{s}{a}\right)u} \cdot f(u)\, du$$

$$= \frac{1}{a} F\left(\frac{s}{a}\right) \qquad \text{(by the definition)}$$

Thus, $\quad \boldsymbol{L\{f(at)\} = \dfrac{1}{a} F\left(\dfrac{s}{a}\right)}$

1.4 Laplace Transforms of Elementary Functions

We shall consider Laplace transform of certain elementary functions. During the process of finding transforms of elementary functions under consideration, we assume the existence of integral involved.

1. To show that, $L\{1\} = \frac{1}{s}$, $(s > 0)$.

Consider, $$L\{1\} = \int_0^\infty e^{-st} \cdot 1\, dt = \left(-\frac{1}{s}\right) e^{-st}\Big]_0^\infty = \frac{1}{s} \qquad (\because\ s > 0,\ e^{-st} \to 0 \text{ as } t \to \infty)$$

$$\therefore \quad \mathbf{L\{1\} = \frac{1}{s}}$$

2. To show that, $L\{e^{at}\} = \frac{1}{s-a}$, $(s > a)$.

$$L\{e^{at}\} = \int_0^\infty e^{-st} \cdot e^{at}\, dt$$

$$= \int_0^\infty e^{-(s-a)t}\, dt$$

$$= \left(-\frac{1}{s-a}\right) e^{-(s-a)t}\Big]_0^\infty$$

$$= 0 + \frac{1}{s-a} \qquad (\because\ s > a \Rightarrow s - a > 0 \Rightarrow e^{-(s-a)t} \to 0 \text{ as } t \to \infty)$$

$$\therefore \quad \mathbf{L\{e^{at}\} = \frac{1}{s-a}}$$

Note : 1. $L\{e^{-at}\} = \frac{1}{s+a}$, provided $s + a > 0$.

2. If $a = 0$, we have, $L\{1\} = \frac{1}{s}$, $s > 0$.

3. To show that, $L\{\sin at\} = \frac{a}{s^2 + a^2}$, $s > a$.

$$L\{\sin at\} = \int_0^\infty e^{-st} \cdot \sin at\, dt$$

$$= \frac{e^{-st}}{s^2 + a^2}(-s \sin at - a \cos at)\Big]_0^\infty$$

$$\left[\because \int_0^\infty e^{ax} \cdot \sin(bx + c)\, dx = \frac{e^{ax}}{a^2 + b^2}\big(a \sin(bx + c) - b \cos(bx + c)\big)\right]$$

$$\Rightarrow \quad L\{\sin at\} = 0 - \frac{1}{s^2 + a^2}(0 - a) \qquad (\because\ e^{-st} \to 0 \text{ as } t \to \infty,\ s > 0)$$

$$\Rightarrow \quad \mathbf{L\{\sin at\} = \frac{a}{s^2 + a^2}}$$

4. To show that, $L\{\cos at\} = \dfrac{s}{s^2 + a^2}$.

$$L\{\cos at\} = \int_0^{\infty} e^{-st} \cdot \cos at\, dt = \frac{e^{-st}}{s^2 + a^2}\{-s\cos at + a \sin at\}\Bigg]_0^{\infty}$$

$$\left[\because \int e^{ax} \cdot \cos(bx + c)\, dx = \frac{e^{ax}}{a^2 + b^2}\{a \cos(bx + c) + b \sin(bx + c)\}\right]$$

$$\Rightarrow \quad L\{\cos at\} = 0 - \frac{1}{s^2 + a^2}(-s + 0)$$

$$\Rightarrow \quad L\{\cos at\} = \frac{s}{s^2 + a^2}$$

5. To show that $L\{t^n\} = \dfrac{n!}{s^{n+1}}$, when n is a positive integer.

By the definition of Laplace transform, we have

$$L\{t^n\} = \int_0^{\infty} e^{-st} \cdot t^n\, dt$$

$$= t^n\left(-\frac{1}{s}e^{-st}\right)\Bigg]_0^{\infty} - \int_0^{\infty} n \cdot t^{n-1}\left(-\frac{1}{s}e^{-st}\right)dt = \frac{n}{s}\int e^{-st} \cdot t^{n-1}\, dt$$

$$L\{t^n\} = \frac{n}{s} L\{t^{n-1}\}$$

$$L\{t^n\} = \frac{n-1}{s} L\{t^{n-2}\}$$

....

$$L\{t^n\} = \frac{n}{s} \cdot \frac{n-1}{s} \cdot \frac{n-2}{s} \cdots\cdots \frac{2}{s} \cdot \frac{1}{s} L\{t^0\} = \frac{n!}{s^n} \cdot \frac{1}{s} = \frac{n!}{s^{n+1}}$$

Now by applying the first shifting property, we obtain the following results.

1. $L\{e^{at} \cdot t^n\} = \dfrac{n!}{(s-a)^{n+1}} \qquad \because \qquad L\{t^n\} = \dfrac{n!}{s^{n+1}}$

2. $L\{e^{at} \sin bt\} = \dfrac{b}{(s-a)^2 + b^2} \qquad \because \qquad L\{\sin bt\} = \dfrac{b}{s^2 + b^2}$

3. $L\{e^{at} \cos bt\} = \dfrac{s-a}{(s-a)^2 + b^2} \qquad \because \qquad L\{\cos bt\} = \dfrac{s}{s^2 + b^2}$

1.5 Existence of Laplace Transform

In this section we shall state a condition for the existence of Laplace transform for a given function.

If $f(t)$ is continuous and lim $[e^{-at} \cdot f(t)]$ is finite, the Laplace transform of $f(t)$ exists for $s > a$.

i.e., $$\int_0^{\infty} e^{-st} f(t)\, dt, \quad \text{exists for } s > a$$

The above condition is only sufficient condition but not necessary.

Example 1. Find the Laplace transform of

(a) $t^2 - 3t + 5$ (b) $e^{-4t} + 3e^{-2t}$ (c) $\cos 3t \cdot \sin 2t$ (d) $\cos^2 2t$

(e) $\sin^2 4t$ (f) $\sin^3 2t$ (g) $\cos^3 t$

Solution : (a) Let, $f(t) = t^2 - 3t + 5.$

Now, $L\{f(t)\} = L\{t^2\} - 3L\{t\} + 5L\{1\}$

$$= \frac{2}{s^3} - 3\frac{1}{s^2} + 5\frac{1}{s}, \ (s > 0) \qquad \left(\because L\left\{\frac{1}{t^n}\right\} = \frac{n!}{s^{n+1}}\right)$$

$$= \frac{1}{s^3}\left(2 - 3s + 5s^2\right)$$

(b) Let, $f(t) = e^{-4t} + 3e^{-2t}$

$$L\{f(t)\} = L\{e^{-4t}\} + 3\{e^{-2t}\}$$

$$= \frac{1}{s+4} + 3 \cdot \frac{1}{s+2} \qquad \left(\because L\{e^{at}\} = \frac{1}{s-a}\right)$$

$$= \frac{2(2s+7)}{(s+4)(s+2)}$$

(c) Let, $f(t) = \cos 3t \cdot \sin 2t$

$\Rightarrow$ $$f(t) = \frac{1}{2}[\sin 5t - \sin t] \qquad (\because 2\cos A \sin B = \sin(A+B) - \sin(A-B))$$

$$L\{f(t)\} = \frac{1}{2}[L\{\sin 5t\} - L\{\sin t\}]$$

$$= \frac{1}{2}\left[\frac{5}{s^2+25} - \frac{1}{s^2+1}\right] \qquad \left(\because L\{\sin at\} = \frac{a}{s^2+a^2}\right)$$

$$= \frac{2\left(s^2-5\right)}{\left(s^2+25\right)\left(s^2+1\right)}$$

(d) Now, $$\cos^2 2t = \frac{1}{2}(1 + \cos 4t) \qquad (\because 2\cos^2\theta = (1 + \cos 2\theta))$$

Let, $$f(t) = \cos^2 2t = \frac{1}{2}(1 + \cos 4t)$$

$$L\{f(t)\} = \frac{1}{2}[L\{1\} + L\{\cos 4t\}]$$

$$= \frac{1}{2}\left[\frac{1}{s} + \frac{s}{s^2+16}\right] \qquad \left(\because L\{\cos at\} = \frac{s}{s^2+a^2}\right)$$

$$= \frac{1}{2}\cdot\frac{2s^2+16}{s(s^2+16)} = \frac{\mathbf{s^2+8}}{\mathbf{s(s^2+16)}}$$

(e) Now, $\sin^2 4t = \frac{1}{2}(1 - \cos 8t)$ $\quad (\because 2\sin^2\theta = 1 - \cos 2\theta)$

Let, $f(t) = \sin^2 4t = \frac{1}{2}(1 - \cos 8t)$

$$L\{f(t)\} = \frac{1}{2}[L\{1\} - L\{\cos 8t\}] = \frac{1}{2}\left[\frac{1}{s} - \frac{s}{s^2+64}\right] = \frac{\mathbf{32}}{\mathbf{s(s^2+64)}}$$

(f) Consider, $\sin^3 2t = \frac{1}{4}(3\sin 2t + \sin 6t)$ $\quad \left[\because \sin^3\theta = \frac{1}{4}(3\sin\theta - \sin 3\theta), \text{ here } \theta = 2t\right]$

Let, $f(t) = \sin^3 2t$

$\therefore$ $L\{f(t)\} = \frac{1}{4}[3L\{\sin 2t\} + L\{\sin 6t\}]$

$$= \frac{1}{4}\left[3\cdot\frac{2}{s^2+4} + \frac{6}{s^2+36}\right]$$

$$= \frac{3}{2}\left[\frac{2s^2+40}{(s^2+4)(s^2+36)}\right] = \frac{\mathbf{3(s^2+20)}}{\mathbf{(s^2+4)(s^2+36)}}$$

(g) Consider, $\cos^3 t = \frac{1}{4}(3\cos t + \cos 3t)$

Let, $f(t) = \cos^3 t$

$\therefore$ $L\{f(t)\} = \frac{1}{4}[3L\{\cos t\} + L\{\cos 3t\}]$

$$= \frac{1}{4}\left[3\cdot\frac{s}{s^2+1} + \frac{s}{s^2+9}\right] = \frac{\mathbf{s(s^2+7)}}{\mathbf{(s^2+1)(s^2+9)}}$$

Example 2. Find the Laplace transforms of

(a) $e^{2t}(2t^2 - 3t + 4)$ **(b)** $e^{-3t}(2\cos 5t - 3\sin 5t)$ **(c)** $e^{3t}\cos^2 t$ **(d)** $e^{2t}\sin 3t\cos 2t$

Solution : (a) Let, $f(t) = 2t^2 - 3t + 4$

$$L\{f(t)\} = 2L\{t^2\} - 3L\{t\} + 4L\{1\}$$

$$= 2\frac{2}{s^3} - 3\frac{1}{s^2} + 4\cdot\frac{1}{s} = \frac{4 - 3s + 4s^2}{s^3} = F(s)$$

Now applying first shifting property weget,

$$L\{e^{2t}\cdot f(t)\} = F(s-2) = \frac{4-3(s-2)+4(s-2)^2}{(s-2)^3}$$

$$\therefore \quad L\{e^{2t}(2t^2-3t+4)\} = \frac{\mathbf{4s^2-19s+26}}{\mathbf{(s-2)^3}}$$

(b) Let,

$$f(t) = 2\cos 5t - 3\sin 5t$$

$$L\{f(t)\} = 2L\{\cos 5t\} - 3L\{\sin 5t\}$$

$$= 2\cdot\frac{s}{s^2+25} - 3\cdot\frac{5}{s^2+25} = \frac{2s-15}{s^2+25} = F(s)$$

By applying first shifting property we have,

$$L\{e^{-3t}\cdot f(t)\} = F(s+3) = \frac{2(s+3)-15}{(s+3)^2+25} = \frac{\mathbf{2s-9}}{\mathbf{s^2+6s+34}}$$

(c) Consider,

$$\cos^2 t = \frac{1}{2}(1+\cos 2t)$$

Let,

$$f(t) = \cos^2 t$$

$$L\{f(t)\} = \frac{1}{2}[L\{1\} + L\{\cos 2t\}] = \frac{1}{2}\left[\frac{1}{s} + \frac{s}{s^2+4}\right] = \frac{s^2+2}{s\cdot(s^2+4)} = F(s)$$

By applying first shifting property, we have,

$$L\{e^{3t}\cos^2 t\} = F(s-3) = \frac{(s-3)^2+2}{(s-3)\left[(s-3)^2+4\right]} = \frac{\mathbf{s^2-6s+11}}{\mathbf{(s-3)(s^2-6s+13)}}$$

(d) Consider,

$$2\sin 3t\cos 2t = \sin 5t + \sin t \quad (\because 2\sin A\cos B = \sin(A+B) + \sin(A-B))$$

Now

$$L\{\sin 3t\cos 2t\} = \frac{1}{2}\left[L\{\sin 5t\} + L\{\sin t\}\right]$$

$$= \frac{1}{2}\left[\frac{5}{s^2+25} + \frac{1}{s^2+1}\right]$$

$$= \frac{1}{2}\cdot\frac{6s^2+30}{(s^2+1)(s^2+25)} = \frac{3s^2+15}{(s^2+1)(s^2+25)} = F(s)$$

By applying first shifting property, we get,

$$L\{e^{2t}\sin 3t\cos 2t\} = \frac{3(s-2)^2+15}{\left[(s-2)^2+1\right]\left[(s-2)^2+25\right]}$$

$$= \frac{\mathbf{3s^2-12s+27}}{\mathbf{(s^2-4s+5)(s^2-4s+29)}}$$

Example 3. Find the Laplace transform of (a) a^t (b) 6^t.

Solution : (a) We know that $A^B = e^{B \log A}$. Thus $a^t = e^t \log a$, $a^t = e^{t \log a}$.

Now,
$$L\{a^t\} = L\{e^{(\log a)\cdot t}\} = \frac{1}{s - \log a}$$

(b)
$$L\{6^t\} = L\{e^{(\log 6)\cdot t}\} = \frac{1}{s - \log 6}$$

Exercise

Find the Laplace transform of the following functions.

I. **1.** $t^2 - 1$ **2.** $t^2 + 2t + 3$ **3.** $(1 + t)^3$ **4.** $t^3 + 3t^2 - 6t + 8$ **5.** $(2t - 1)^2$

II. **1.** e^{-4t} **2.** e^{-3t} **3.** $(1 + 3t)^2$ **4.** 4^t **5.** 10^t **6.** $5t - \frac{1}{e^{2t}}$

III. **1.** $\cos 2t$ **2.** $\sin 2t$ **3.** $\sin^2 4t$ **4.** $\cos^2 3t$ **5.** $2 \sin t \cdot \sin 3t$
6. $\sin 3t \cos 4t$ **7.** $\sin^3 2t$ **8.** $\cos^3 3t$

Answers

I. **1.** $\frac{2 - s^2}{s^3}$ **2.** $\frac{3s^2 + 2s + 2}{s^3}$ **3.** $\frac{s^3 + 3s^2 + 6s + 6}{s^4}$

4. $\frac{8s^3 - 6s^2 + 6s + 6}{s^4}$ **5.** $\frac{1}{s^3}\left(s^2 - 4s + 8\right)$

II. **1.** $\frac{1}{s + 4}$ **2.** $\frac{1}{s + 3}$ **3.** $\frac{4s^3 - 8s + 2}{s(s - 1)(s - 2)}$

4. $\frac{1}{s - \log 4}$ **5.** $\frac{1}{s - \log 10}$ **6.** $\frac{1}{s - \log 5} - \frac{1}{s + 2}$

III. **1.** $\frac{s}{s^2 + 4}$ **2.** $\frac{s}{s^2 + 9}$ **3.** $\frac{64}{s\left(s^2 + 64\right)}$ **4.** $\frac{s^2 + 18}{s\left(s^2 + 36\right)}$

5. $\frac{12s}{\left(s^2 + 16\right)\left(s^2 + 4\right)}$ **6.** $\frac{1}{2}\left[\frac{7}{s^2 + 49} - \frac{1}{s^2 + 1}\right]$ **7.** $\frac{48}{\left(s^2 + 4\right)\left(s^2 + 36\right)}$ **8.** $\frac{s\left(s^2 + 63\right)}{\left(s^2 + 9\right)\left(s^2 + 81\right)}$